Weather Matters for Energy

Alberto Troccoli · Laurent Dubus
Sue Ellen Haupt
Editors

Weather Matters for Energy

Springer

Editors
Alberto Troccoli
Pye Laboratory
CSIRO Marine and Atmospheric Research
Canberra, ACT
Australia

Sue Ellen Haupt
Research Applications Laboratory
Haupt Associates
Boulder, CO
USA

Laurent Dubus
EDF-R&D
Chatou
France

ISBN 978-1-4614-9220-7   ISBN 978-1-4614-9221-4 (eBook)
DOI 10.1007/978-1-4614-9221-4
Springer New York Heidelberg Dordrecht London

Library of Congress Control Number: 2013958148

© Springer Science+Business Media New York 2014

This work is subject to copyright. All rights are reserved by the Publisher, whether the whole or part of the material is concerned, specifically the rights of translation, reprinting, reuse of illustrations, recitation, broadcasting, reproduction on microfilms or in any other physical way, and transmission or information storage and retrieval, electronic adaptation, computer software, or by similar or dissimilar methodology now known or hereafter developed. Exempted from this legal reservation are brief excerpts in connection with reviews or scholarly analysis or material supplied specifically for the purpose of being entered and executed on a computer system, for exclusive use by the purchaser of the work. Duplication of this publication or parts thereof is permitted only under the provisions of the Copyright Law of the Publisher's location, in its current version, and permission for use must always be obtained from Springer. Permissions for use may be obtained through RightsLink at the Copyright Clearance Center. Violations are liable to prosecution under the respective Copyright Law.
The use of general descriptive names, registered names, trademarks, service marks, etc. in this publication does not imply, even in the absence of a specific statement, that such names are exempt from the relevant protective laws and regulations and therefore free for general use.
While the advice and information in this book are believed to be true and accurate at the date of publication, neither the authors nor the editors nor the publisher can accept any legal responsibility for any errors or omissions that may be made. The publisher makes no warranty, express or implied, with respect to the material contained herein.

Printed on acid-free paper

Springer is part of Springer Science+Business Media (www.springer.com)

Foreword 1

In France, the energy sector has been for a long time among the main users of meteorological, hydrological, and other climate information. Météo-France enjoys the benefit of a long-term strategic cooperation with this sector. This is why I am happy to congratulate the team of Editors of this International Conference Energy & Meteorology (ICEM) 2011 book and to contribute to it with this short introduction.

The public and the industry expect more secure and cleaner energy, which means that the vulnerability of energy systems has to be minimized in accordance with the possible hazards impacting them, in compliance with sustainable development. Meteorological, hydrological, and other climate data are indeed essential both in day-to-day energy management and for the definition of production and distribution infrastructures. For instance, provision of electricity to users can be endangered by extreme meteorological and hydrological events such as thunderstorms with unusually strong winds, flash floods, severe icing, severe colds, heat waves, sea-level elevation associated to storm surges, floods, or other hazards.

To be protected against such events, it is not sufficient to act after their impacts have happened. It is also necessary to identify precisely the potential impacts in advance and to assess their probabilities and manage the risk through regularly updated and seamless Weather-Water and Climate Services such as the ones identified within the World Meteorological Organization (WMO)-led World Climate Conference 3 (WCC-3, 2009) and now being implemented through the Global Framework for Climate Services (GFCS-2012).

The recent advances in atmospheric sciences and in the field of energy production and management offer a huge range of new possibilities. Considering that further developments will benefit essentially from the dialog between experts from both fields, Météo-France as a member of the world meteorological community is willing to take the lead and foster this dialog within the European energy and climate agenda and even more widely. The present book will be most useful both for meteorologists and for energy practitioners to get acquainted with the recent progresses in that direction and to learn for instance from the successful cooperation between EDF and Météo-France, and many others around the globe.

The weather, water and climate community coordinated under WMO is strongly committed to improve continuously its services to the energy sector. More generally, building on the WCC-3 recommendations, WMO and many partners have agreed to establish a GFCS and are since October 2012 working on its implementation. The GFCS has the objective to ensure that climate information and predictions are made available to decision-makers enduring the increasing impacts of climate variability and change. A historic extraordinary session of the World Meteorological Congress took place on 29–31 October 2012, which approved the governance structure and implementation plan for the GFCS. This is a sweeping initiative to capitalize on scientific advances and roll out user-driven services starting from improved weather, water short-term forecasts to longer term climate predictions such as seasonal climate outlooks and El Niño watches, flood prediction and drought monitoring tools, and extreme events occurrence probabilities.

The GFCS is a necessary step in view to define and implement scientifically sound measures of adaptation to climate variability and climate change. It will contribute to improve the field observations in the world, to make them easily accessible under a readily utilizable form. It will also help to develop numerical modeling of the atmosphere and the Earth in general, and to produce information tailored to the users needs.

I take this opportunity to commend the work done by all the experts involved, then all the partners who contributed with resources and NATO for being the initiator and supporting the first workshop in 2008 and the first publication in 2010 which in 2011 provided the framework for the first international conference on Energy and Meteorology, conference which has assembled the material for the present publication. In view of the determination, commitment, and professionalism of the organizers of this Energy and Meteorology partnership process, I have no doubt that this endeavor, which started very simply in 2008 as an advanced research workshop will continue on a long term as a capacity building program. Météo-France was happy to host the ICEM 2013 conference, which took place in the Conference Centre of the "Meteopole" in South–West France (Toulouse, 25–28 June 2013, < https://www.icem2013.org >).

François Jacq

François Jacq studied at the École Polytechnique (Paris) and at the École Nationale Supérieure des Mines de Paris. He holds a Ph.D. in History and Sociology of Science. Before joining Météo-France in April 2009 as CEO, he was adviser to the Prime Minister on Sustainable Development, Research and Industry. He was previously a researcher at École des Mines de Paris, Director of the Department on Energy, Transportation, Environment and Natural resources in the Ministry of Research, Chief executive of the National Agency for the Management of Radioactive Waste, and Director for Energy demand and markets in the Ministry of Industry of France. Dr. Jacq is also a member of the Executive Council of the World Meteorological Organization.

Foreword 2

When Laurent Dubus asked me to write a foreword for the book he is editing, and contributing to, with Alberto Troccoli and Sue Ellen Haupt, I accepted at once. Indeed, this is a topic which is key in my mind! I do believe that the relationships between weather, weather forecast, climate, and our job, which is to deliver safe, reliable, and affordable electricity, will increase in the future.

The interactions make sense on all aspects of our business: production, transport and distribution, and consumption. My sense is that the effect that meteorology has on energy systems can be represented by four important categories, roughly related to different meteorological time scales. I believe these four categories are as follows.

Crisis management—By improving our knowledge of severe meteorological events such as floods, strong winds, heavy snowfalls, heat and cold waves, we can better manage both demand side (with DSM—Demand Side Management) and supply side and thus we can manage crises more effectively. This means being able to access short-term forecasts but with very high resolution especially around areas of critical interest like power plants, rivers, towns, and so on. It is particularly important to try to anticipate these severe events and provide appropriate alerts, when the risk of extreme events is above a tolerable threshold, even if this means accepting some false alarms.

Day-to-day operations—My understanding about the strong relationship between electricity production, electricity use, and weather initially came through my on the ground experience. I have gained a good appreciation about the limits of forecasts so as to be able to use them properly; I worked toward improving the forecast quality and reliability to reduce the gaps between predictions and observed weather, particularly for air temperature as linked to consumption, wind speed or rainfall as linked to production. In the future, the development of electricity from renewable sources will further increase the dependence of power systems on weather variability, and hence the importance of climate uncertainty.

Planning of production facilities—Power production requires having an idea of the weather in the middle term (warm or cold weather, drought or excess in river discharge). This means not only the use of forecasts of mean weather conditions but also their probabilities. More research is required to be able to produce reliable midterm forecasts. Exactly like farmers who would like to know what cereals they can grow under certain rain patterns, we need to know if the coming summer will be sunny or rainy as this knowledge will shape the consumption (more or less cooling) and the balance between the different ways to produce electricity (e.g., do we need to "save" water in the reservoirs or can we release some?), also taking into account the necessary maintenance of power plants.

Facility construction and maintenance—We rely on climate knowledge to build our facilities: floods, temperature of rivers, winds etc…These data are critical in the dimensioning of plants. We use data from the past to assess the events and to define the characteristics of our plants. This point is certainly the most important for me as it requires a complete change in our relationship with stakeholders including people working on climate. The changing world, in which we currently live, including climate variations and changes, will require a serious rethinking of our "business as usual approach". We will need to expand our knowledge and practice with an eye not only on the past but also critically, on the future. We built our infrastructures to resist to the extreme events we knew through past climate data; today we need to redefine the limits between resistance and resilience of our facilities and this will require understanding of interactions with meteorological variables under climate change.

The cooperation EDF develops with the scientific community, including meteorologists, to fulfill its needs on weather and climate forecasts is more alive than ever. We need meteorologist and climatologists to increase the knowledge, provide new and more accurate data, and develop new models, forecasts, and projection at all time scales, from the real time to the end of the century.

Fulfilling our needs is only possible with a close collaboration between the weather and climate sector on the one side, and the energy sector on the other side. I do believe that this volume, and the International Conference Energy and Meteorology (ICEM) conference cycle, is an important component of such an efficient partnership.

Claude Nahon

Claude Nahon was named Group's Executive Vice President for Sustainable Development at Electricité de France (EDF) in January 2003. Prior to her appointment, she was Head of Hydro and Renewable Energies. A graduate of France's prestigious École Polytechnique, Claude Nahon has been with EDF since 1978, holding managerial positions in both generation and distribution. In addition to her responsibilities at EDF, Claude Nahon represents EDF Group in different occasions and institutions. She is Liaison Delegate of Henri Proglio in the World Business Council for Sustainable Development (WBCSD). She sits on Institut du Développement Durable et des Relations Internationales (IDDRI) and Entreprises pour l'Environnement (EPE).

Preface

A storm is mounting. You were alerted about the approaching severe weather conditions by the weather forecaster. You're now sitting in front of your monitors trying to decide how many power units to commit. The precise power mix, whether it is more economical to run more hydro instead of wind for instance, is also key but you don't have much time to take a decision. With a combination of modeling tools and expert knowledge, you're trying to figure out whether the thick deck of clouds will make the 100 MW solar power plant ramp down within minutes or hours. And can some of this shortfall be partly compensated by the newly installed 150 MW wind farm? Meanwhile, gas generators will most likely be safe, but will there be enough time to start them up and avoid that horrific black out?

It has been a long day, but luckily the effects of this weather event have been well handled, as many tools were available to assess this situation and act on it efficiently. The seasonal forecast had predicted that such storms would be likely this month; thus the maintenance department had performed preventive maintenance in advance. However, it makes you wonder whether storms like this will become more frequent and impactful in the future. Already we are seeing an increasing number of meteorologically derived disasters. Warmer temperatures, as projected decades from now by climate models, will very likely bring even more severe events. It is almost a no-brainer to factor these climate effects into energy sector risk management decision-making. But exactly how to do this may not be straightforward.

It is the purpose of this book to provide the meteorological knowledge and tools to improve the risk management of energy industry decisions, ranging from the long-term finance and engineering planning assessments to the short-term operational measures for scheduling and maintenance. Most of the chapters in this book are based on presentations given at the inaugural International Conference Energy and Meteorology (ICEM), held in the Gold Coast, Australia, 8–11 November 2011 (see http://www.icem2011.org). The main aim of the conference was to strengthen the link between Energy and Meteorology, so as to make meteorological information more relevant to the planning and operations of the energy sector. The ultimate goal would be to make the best use of weather and climate data in order to achieve a more efficient use of energy sources. This book seeks to realize the same objective.

It is worth highlighting the close connection, in terms of temporal and spatial scales, between decisions in the energy industry on the one hand and natural meteorological events on the other. Indeed, decisions in the energy industry extend from the tiny temporal scales related to electrical instabilities, to the short-term typical of supply and demand balancing, to the longer terms typical of maintenance through to planning. Similarly, meteorological phenomena occur over essentially all time and space scales from the tiny scales of turbulence, to the short-term and small-scale events such as tornadoes, to the longer/larger hurricanes, through to climate change via El Niño. Often there is a strong link between a particular meteorological phenomenon and its implications for the energy sector. Thus, for instance, a hurricane, which typically lasts several days and affects an area of a few thousand kilometres, will be factored in energy operational and maintenance decisions over the span of a number of days, and with an advance notice of several days (hurricanes can be predicted several days in advance).

The book is structured to emphasize the role of the energy industry in terms of meteorological requirements. Indeed, unlike the more standard approach, which begins by presenting meteorological information, as if the latter was in search of a purpose, here we have genuinely attempted to put energy in the driver's seat. Such order is also reflected in the titles of the book parts, as with *Why Should the Energy Industry be Concerned About Weather Patterns?* of Part One.

There is no doubt that Energy and Meteorology is a burgeoning inter-sectoral discipline. It is also clear that the catalyst for the stronger interaction between these two sectors is the renewed and fervent interest in renewable energies, especially wind and solar power. This connection is also apparent from the content of the book. However, it must be realised that weather and climate information is also critical to managing the energy supply from other energy sectors (e.g., off-shore oil operations) as well as understanding and estimating energy demand. We have tried to stress this broader dependency in various parts of the book.

The book could not have come together without the ICEM 2011 meeting. Hence, we are indebted to the superb support of Elena Bertocco, the extraordinary organizational role of Aurélie Favennec, the efficient work of the steering and scientific organising committee, and the keen contributions of the conference delegates during an intense and fascinating week along the beautiful ocean shore of the Gold Coast of Australia.

It is a great pleasure to acknowledge the tremendous assistance of Danielle Stevens who has indefatigably and very diligently been assembling the book, even when prodding was needed to obtain responses from chapter authors. We are also indebted to Pierre Audinet for his strong support in shaping this book and for suggesting what we believe is an appropriate and attractive book title. We also wish to thank all the authors of the book for chipping in and helping to carry out thorough reviews of all chapters of this book.

We do hope you will enjoy this book. Happy reading!

Alberto Troccoli
Laurent Dubus
Sue Ellen Haupt

Contents

Part I Why Should the Energy Industry be Concerned About Weather Patterns?

A New Era for Energy and Meteorology 3
Beverley F. Ronalds, Alex Wonhas and Alberto Troccoli
1 The Energy Picture Today 3
2 Vulnerability of Energy Systems to the Current Climate 6
3 Transforming Our Energy System: Challenges and Opportunities 8
4 A Step-Change in Meteorology and Energy Linkages 10
5 Conclusions 14
References 14

Climate Risk Management Approaches in the Electricity Sector: Lessons from Early Adapters. 17
Pierre Audinet, Jean-Christophe Amado and Ben Rabb
1 The Need to Strengthen the Resilience of Electricity
   Systems in Developing Countries 20
2 Existing Research on Climate Data and Information,
   Associated Risks, and Adaptation Solutions 23
3 Early Adaptation Efforts in the Electricity Sector 33
4 Current Focus of Adaptation Efforts and Gaps 49
5 Lessons Learnt and the Way Forward 56
6 Appendix 59
References 62

Climatic Changes: Looking Back, Looking Forward. 65
Alberto Troccoli
1 Introduction 65
2 Methodology... 68
3 Historical Climate Relevant to Energy 73
4 Projected Climate Changes Relevant to Energy. 79
5 Extreme Weather Events Relevant to Energy 85
6 Summary 86
References 87

**Renewable Energy and Climate Change Mitigation:
An Overview of the IPCC Special Report** 91
Ralph E. H. Sims
1 Introduction 92
2 Trends and Future Scenarios.............................. 93
3 Climate Change Impacts on Renewable Resources 96
4 Costs 99
5 Integration 103
6 Sustainable Development 106
7 Co-Benefits, Barriers and Policies 108
8 Conclusions 109
References.. 110

Part II   How is the Energy Industry Meteorology-Proofing Itself?

**Improving Resilience Challenges and Linkages of the Energy
Industry in Changing Climate** 113
Shanti Majithia
1 Introduction 113
2 Energy 115
3 Infrastructure Components 117
4 Potential Risk to the Infrastructure..................... 119
5 Business Preparedness: Infrastructure and Corporate Resilience 121
6 Information Gap Analysis in Climate Science.............. 127
7 Market Mitigation 2020/2050 127
8 Linking Meteorology, Climate and Energy 128
9 Conclusion 129
References.. 130

**Combining Meteorological and Electrical Engineering Expertise
to Solve Energy Management Problems** 133
Giovanni Pirovano, Paola Faggian, Paolo Bonelli, Matteo Lacavalla,
Pietro Marcacci and Dario Ronzio
1 Introduction 133
2 The Meteorological Demand from Energy Community 134
3 Climate Change Impact on the Electric System............. 135
4 Forecasting "Weather Energy" 139
5 Weather Risks for the Power System....................... 142
6 Conclusion 152
References.. 153

**Weather and Climate Impacts on Australia's National
Electricity Market (NEM)** 155
Tim George and Magnus Hindsberger
1 Introduction 155
2 Setting the Scene 157
3 Characteristics of Weather-Dependent Generation 158
4 Operational Considerations 164
5 Planning for the Longer Term 168
6 Conclusions 174
References 175

**Bioenergy, Weather and Climate Change in Africa:
Leading Issues and Policy Options** 177
Mersie Ejigu
1 Introduction 177
2 Africa's Energy Profile and Vulnerability to Climate Change .. 178
3 Bioenergy Vis-à-Vis Other Renewables:
   The Significance of Weather 183
4 Bioenergy: Climate Opportunities and Risks 184
5 The Direct and Indirect Impacts of Weather/Climate Change .. 187
6 Main Policy Issues and Options 191
7 Conclusion 196
References 196

Part III   What can Meteorology Offer to the Energy Industry?

**Weather and Climate Information Delivery within
National and International Frameworks** 201
John W. Zillman
1 Introduction 201
2 The Nature of Weather and Climate 202
3 Meteorological and Related Services 204
4 The National Meteorological Service System 208
5 The Global Framework for Meteorological Service Provision .. 211
6 Future Directions 214
7 Conclusion 218
References 218

Meteorology and the Energy Sector 221
Geoff Love, Neil Plummer, Ian Muirhead, Ian Grant
and Clinton Rakich
1 Introduction 221
2 Factors Affecting Electricity Production 222

3	The Energy Sector's Requirement for Meteorological Services	224
4	Overview: Uses of Meteorological Information in the Energy Sector	227
5	An Energy Sector Case Study: Solar Resource Assessment	229
6	A New Global Information Service	232
7	Components of the Global Framework for Climate Services	233
8	Some Concluding Remarks	234
References		234

Earth Observation in Support of the Energy Sector 237
Pierre-Philippe Mathieu

1	The View from Space: A Unique Perspective to Help the Energy Sector	237
2	EO Demonstration Pilot Projects in Support of the Energy Sector	242
3	Conclusions	253
References		254

Emerging Meteorological Requirements to Support High Penetrations of Variable Renewable Energy Sources: Solar Energy ... 257
David S. Renné

1	Introduction	258
2	Global Trends in PV Development	258
3	Perspectives from the Utility Industry	260
4	Challenges for the Solar Resource Community	262
5	Summary and Conclusions	272
References		273

Current Status and Challenges in Wind Energy Assessment 275
Sven-Erik Gryning, Jake Badger, Andrea N. Hahmann and Ekaterina Batchvarova

1	Introduction	276
2	Global Wind Resources	276
3	Dynamical Downscaling for Wind Applications Using Meteorological Models	283
4	Summary and Outlook	291
References		292

Wind Power Forecasting 295
Sue Ellen Haupt, William P. Mahoney and Keith Parks

1	The Need for Renewable Energy Power Forecasts	295
2	A System's Approach to Forecasting	297
3	Numerical Weather Prediction	300
4	Statistical Postprocessing	302

5	Short-Term Forecasting	307
6	Utility Grid Integration	310
7	Concluding Remarks	315
References		316

Regional Climate Modelling for the Energy Sector 319
Jack Katzfey

1	Introduction	319
2	Methods/Techniques	323
3	Application of Regional Climate Models for the Energy Industry	329
4	Summary and Discussion	330
References		331

**In Search of the Best Possible Weather Forecast
for the Energy Industry** 335
Pascal Mailier, Brian Peters, Devin Kilminster and Meghan Stephens

1	Introduction	335
2	What Makes a 'Good' Forecast?	336
3	The Fallacy of Accuracy	338
4	The Value of a Probability Forecast	341
5	Best Versus Most Useful	347
6	Conclusions	348
References		349

Part IV How is the Energy Industry Applying State-of-the-Science Meteorology?

**A Probabilistic View of Weather, Climate,
and the Energy Industry** 353
John A. Dutton, Richard P. James and Jeremy D. Ross

1	Introduction	353
2	Probability Methods	355
3	Probability Forecasts of Atmospheric Events	357
4	Modeling Probabilities of Business Results	368
5	Atmospheric Informatics	373
6	Conclusion	377
References		377

**Weather and Climate and the Power Sector: Needs, Recent
Developments and Challenges** 379
Laurent Dubus

1	Introduction: The Power Sector is Increasingly Weather Dependent	380

2	Probabilistic Temperature Forecasts of a Few Days to One Month	383
3	Improvement in Monthly River Flow Forecasts	387
4	Some Challenging Problems	390
5	Conclusion: Importance of Collaboration Between Users and Providers	396
References		396

Unlocking the Potential of Renewable Energy with Storage 399
Peter Coppin, John Wood, Chris Price, Andreas Ernst and Lan Lam

1	Introduction	399
2	The Technologies	405
3	A Case Study: Wind Farm Smoothing	407
4	A More Advanced Algorithm	410
5	Advanced Algorithm Results	410
6	Conclusions	411
References		412

Improving NWP Forecasts for the Wind Energy Sector 413
Merlinde Kay and Iain MacGill

1	Introduction	414
2	Wind Farm Data and the Forecast Model Used in This Study	415
3	Bias Correction Methodology	418
4	Results	420
5	Conclusion	427
References		427

Overview of Irradiance and Photovoltaic Power Prediction 429
Elke Lorenz, Jan Kühnert and Detlev Heinemann

1	Introduction	429
2	Typical Outline of PV Power Prediction Systems	431
3	Irradiance Forecasting	433
4	Evaluation of Irradiance Forecasts	440
5	PV Power Forecasting	447
6	Evaluation of PV Power Forecasts	450
7	Summary and Outlook	452
References		453

Spatial and Temporal Variability in the UK Wind Resource: Scales, Controlling Factors and Implications for Wind Power Output 455
Steve Dorling, Nick Earl and Chris Steele

1	Introduction	455
2	Tools	456

3	Analysis of 10 m Station Measurements...................	457
4	Sea Breeze modelling	458
5	Conclusions	464
References...		464

Reducing the Energy Consumption of Existing Residential Buildings, for Climate Change and Scarce Resource Scenarios in 2050 467
John J. Shiel, Behdad Moghtaderi, Richard Aynsley and Adrian Page

1	Introduction	467
2	Method...	469
3	Results...	478
4	Discussion	482
5	Conclusion	489
References...		490

Part V Concluding Chapter

Energy and Meteorology: Partnership for the Future............. 497
Don Gunasekera, Alberto Troccoli and Mohammed S. Boulahya

1	Introduction	497
2	Current Forms of Meteorological Service Provision............	500
3	Challenges Faced by the National Meteorological and Hydrological Services	503
4	Energy Services: Current and Future Trends..................	505
5	Energy and Meteorology Interaction	506
6	Way Forward	507
References...		509

Index 513

Part I
Why Should the Energy Industry be Concerned About Weather Patterns?

A New Era for Energy and Meteorology

Beverley F. Ronalds, Alex Wonhas and Alberto Troccoli

Abstract In this chapter it is argued that the successful transformation of the world's energy systems depends on enhanced interplay between the meteorological and energy sciences. Key drivers of the energy transformation are described and the likely attributes and challenges of our future energy system are outlined. We identify a framework and give examples of ways in which a new cross-disciplinary science that truly combines energy and meteorological expertise will significantly reduce the risks and costs inherent in energy infrastructure. The need for this cross-disciplinary science is urgent given the scale and complexity of current and future energy infrastructure and its increasing vulnerability to the vagaries of the weather. Short-term opportunities to foster a much closer collaboration between energy and meteorology are also discussed.

1 The Energy Picture Today

The world runs on energy. The developed and, increasingly, the developing world relies on oil for personal mobility and for global trade. The equipment in offices and homes is powered by electricity. Industry often uses natural gas for process heat in manufacturing. As a result, some of the largest global companies are in the energy business. Nonetheless, around 20 % of the world's population does not yet have adequate access to energy (IEA 2011).

Global energy demand in 2010 reached 12,000 Mtoe, an increase of about 70 % compared to 30 years earlier (Fig. 1, IEA 2011). Australia, for instance, is an important exporter of coal, gas and uranium. The value of Australia's coal and liquefied natural gas (LNG) exports has increased four-fold over the last decade

---

B. F. Ronalds · A. Wonhas · A. Troccoli (✉)
Commonwealth Scientific and Industrial Research Organisation (CSIRO),
Highett, VIC, Australia
e-mail: Alberto.Troccoli@csiro.au

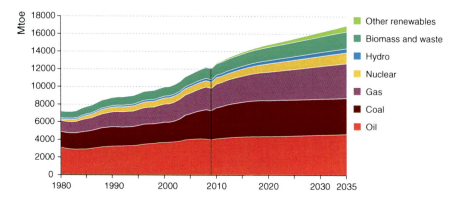

Fig. 1 Historical and projected changes in World primary energy demand by fuel. Fossil fuels are projected to maintain a central role in the primary energy mix but their share is projected to decline. The projections are based on the so-called New Policies Scenario, a set of broad policy commitments and plans which address energy security, climate change and local pollution, and other pressing energy-related challenges (IEA 2011)

(Australian Government 2011), and LNG projects currently under development will again more than triple Australia's export capacity in the next 5 years (Australian Government 2012). This growth in export mainly feeds energy demand from Asia.

With a number of countries including Australia having experienced 'peak oil' in their domestic production, global oil trade particularly from the Middle East and North Africa will also remain vital, and will cause an increased balance of trade burden for importers if the price of oil rises over time as forecast.

The 2011 World Energy Outlook (WEO) (IEA 2011) predicts global demand for energy will continue to grow, mainly because of the expected robust growth in population in Asia and Africa (Fig. 2). Such increase in demand is projected to be around 35 % over the quarter-century to 2035, corresponding to an estimated total energy demand of about 16,500 Mtoe (Fig. 1). However, factors like energy efficiency and carbon penalty policies introduce uncertainties to this projection.

What sources of energy will be used to meet such a substantial increase in demand? This is an area where uncertainties are even larger, especially because we live at a time when technological development is occurring at a fast pace. Consider for example the recent dramatic increase of shale or coal seam gas production or the strong decline in the cost of photovoltaic (PV) modules. However, to counterbalance these rapid technological advances, the lifetime of some power plants can extend to 40–50 years, as in the case of coal power stations, and this longevity slows the overall transition of the energy system.

In spite of these uncertainties, projections of how various technologies could contribute to the energy mix to meet the expected demand a few decades out, such as those produced by the International Energy Agency (IEA), help guide planning and investment in the energy sector. For instance, the IEA's projection shows that demands for all types of energy technologies are expected to increase in

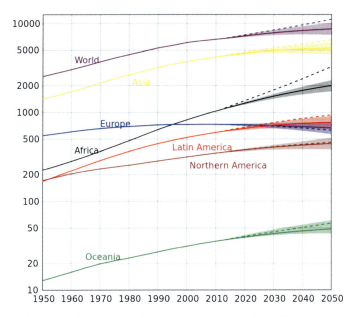

Fig. 2 Historical and projected population evolution globally and in different continents (World Population Prospects, UN Development Office, from http://en.wikipedia.org/wiki/World_population). The *vertical axis* is logarithmic and is in millions of people. Note the wider x-axis range compared to that in Fig. 1

non-OECD (Organisation for Economic Cooperation and Development) countries, while demand for coal and oil should decline in the OECD (Fig. 3). In particular, gas, nuclear and renewable energies are projected to partially replace coal and oil in OECD countries. Non-OECD countries, however, will require energy from all these sources to meet their increasing demand. It is important to note that over the same period renewable energy capacity is expected to triple globally. This is especially driven by the power sector where their share should rise from 19 % in 2008 to 32 % in 2035 (IEA, WEO 2010).

Energy systems interact with the climate in both directions. Mainly through their emissions, energy systems may modify the global temperature and other meteorological variables. Meteorology, hence both weather and climate, in turn affect planning and operations of energy systems. These interrelationships will only become more complex as both energy systems and the climate continue to evolve, and will impact energy demand (load characteristics) as well as supply (mitigation measures, water supplies, etc.). Adapting to climate change will also have consequences for energy supply.

Fossil fuels are the major source of greenhouse gas emissions. According to the Global Carbon Project (GCP 2011), emissions in 2010 were the highest in human history and the concentration of CO_2 in the atmosphere is about 40 % above that at the start of the Industrial Revolution, with a corresponding increase in global temperatures of about 0.8 °C (IPCC 2007).

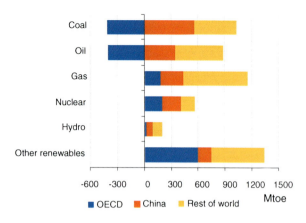

Fig. 3 Projected changes in incremental global primary energy demand. Demand for all types of energy increases in non-OECD countries, while demand for coal and oil declines in the OECD. The projections are based on the so-called New Policies Scenario, a set of broad policy commitments and plans which address energy security, climate change and local pollution, and other pressing energy-related challenges (IEA, World Energy Outlook 2010)

With the expansion of energy systems there is also an increased vulnerability of these systems to severe weather and climate events. This chapter explores ways in which the energy industry can benefit from a deeper interaction with meteorology—which includes weather and broader climate science. Indeed, the intersection between energy and meteorology is the leitmotif of this book.

2 Vulnerability of Energy Systems to the Current Climate

Leaving aside projected changes in climate means and extremes, energy systems are already exposed to the vagaries of the weather. Naturally, most renewable energies are affected by weather and climate, but oil, coal, gas and even nuclear energy systems are, in different ways, also vulnerable to meteorological conditions (e.g. hurricane Katrina's impact on the Gulf of Mexico's oil production). Indeed, intense weather and climate conditions are expected to increasingly and critically influence the efficiency and economics of all energy systems.

According to Munich Re statistics, meteorological, hydrological and climatological catastrophic events have seen a marked increase since the 1980s (Fig. 4). While such events may not directly impact energy systems, they can be so large that they create shocks to the energy supply–demand balance. As a simple example, an extremely high temperature will induce exceptional air-conditioning use. However, to date focus has been on increasing energy supplies to satisfy industrial and societal demand for energy, by managing the risks perceived to be of immediate concern but without systematically accounting for risks posed by current and future climate changes (Ebinger and Vergara 2011; Schaeffer et al.

A New Era for Energy and Meteorology 7

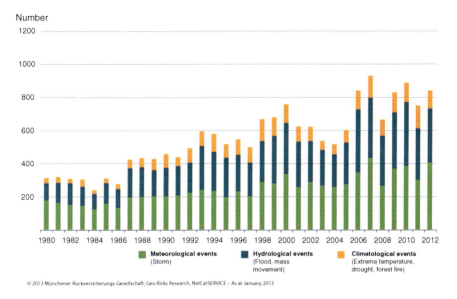

Fig. 4 Weather catastrophes worldwide 1980–2012: number of events (NatCatService, Munich Re, 2012, available from http://www.munichre.com/touch)

2012). It is important, therefore, to highlight some specific examples of the many ways in which weather and climate affect energy systems: these and others will also be expanded in the rest of the book.

2.1 Coal Production and Flooding

Coal mines in Queensland, Australia, experienced widespread disruptions in late 2010 to early 2011 because of heavy rains and floods caused by an unusually strong La Niña event. As a consequence of this event, and the projection of similar ones to come, one large mine built a new bridge and a levee designed for a 1 in 1,000 year flood event to prepare for the eventuality that these conditions become more typical (Johnston et al. 2012, "Meteorology and the Energy Sector" by Plummer et al.).

2.2 Oil Industry and Storms

Hurricanes, and storms in general, can cause widespread disruption to the oil and gas industry. For instance hurricane Katrina in 2005, the costliest natural disaster in the history of the United States, caused, amongst the many damages, oil spills from 44 facilities throughout southeastern Louisiana (http://en.wikipedia.org/wiki/Hurricane_Katrina). In 1982, the North Atlantic Ocean Ranger storm caused an oil rig to capsize and resulted in the death of the entire 84 people crew (Froude and Gurney 2010).

2.3 Nuclear Energy and Heatwaves

Nuclear power stations rely on water flows for their cooling. Warm and hot weather may cause cooling water to reach temperatures too high for the water to be effective. In France in 2003, the very low river flows and increased water temperature led to reductions in power production and exceptional exemptions from legal limits on the temperature at which water may be returned to rivers (Dubus 2010). In addition, such higher temperatures returned to rivers can result in damage to flora and fauna.

2.4 Gas Demand and Market Response

Sudden changes in meteorological variables such as temperature can have a large impact on energy price, induced by factors such as supply shortages or speculation. For instance, day-ahead gas price can vary by 80 % or more within days as a response to severe cold weather events which lead to a sudden increase in demand for heating (Troccoli 2010).

2.5 Wind and Solar Energy Grid Integration

Wind and solar power are entirely reliant on meteorological conditions to function. Highly variable weather conditions on timescales of minutes or shorter can cause disruptions to grid operations. For instance, excessive changes, or ramp rates, in power supply from wind or solar energy installations may lead to their curtailment to avoid the grid reaching its capacity limits (e.g. Sayeef et al. 2012).

3 Transforming Our Energy System: Challenges and Opportunities

In shaping our energy future we need to address risks of, and barriers to, the transformation of the energy sector. We can specify three very simple goals for energy: to be 'clean' (low emissions), 'secure' (available and affordable) and to

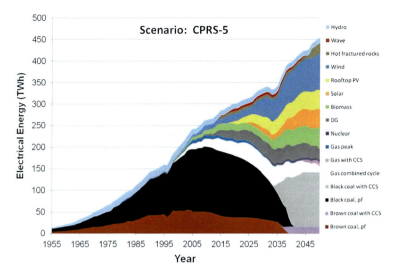

Fig. 5 One scenario for achieving 60 % emissions cuts in Australia by 2050

underpin prosperity. Achieving these goals in the coming decades will require a transformation of the energy system. IEA (2011) estimates the investment required worldwide in energy supply infrastructure over the period to 2035 to be $38 trillion (2010 dollars), that is, around 60 % of the current global annual gross domestic product (GDP), or an average of more than 2 % of GDP annually. Exactly how this transformation will develop depends on many factors including the physical environment, community opinion, government policies, market forces and technological innovation. Indeed, we have already been witnessing part of this transformation with, for instance, the burgeoning of the renewable power sector.

There is no silver bullet to achieve these energy goals. Modelling globally, and for particular economies, gives the common message that a range of new technologies will be required. CSIRO's modelling in Fig. 5, for example (Hayward et al. 2011) shows the evolution of the electricity generation mix in Australia under one particular scenario. The specific scenario presented here is based on the 'carbon pollution reduction scheme' (CPRS), a cap and trade scheme designed to achieve a 5 % GHG emissions reduction from 2000 levels by 2020 and a 60 % reduction by 2050. The scheme was proposed in Australia in 2008 and has since been replaced by a different carbon pricing scheme. The other key scenario assumptions about future demand and technology costs were the mid-range assumptions at the time. For a detailed overview of the scenario assumptions, see (Hayward et al. 2011).

Historically, Australia has relied on its abundant coal resources supplemented by hydroelectricity. Over the past decade, wind and biomass have entered the mix, as a result of targets for renewable energy generation set by the Government. The use of natural gas for electricity generation has also grown significantly due to both its lower emissions and its flexibility in supporting uptake of variable generation.

In future, we may still see significant fossil fuel use through the development of carbon capture and storage (CCS) technologies, but we will also see dramatic increases in renewable energy generation, including wind, biomass, large and small-scale PV, concentrated solar thermal, geothermal (e.g. hot-fractured rock technologies) and ocean renewable energy (e.g. wave energy).

The rainbow of future supply options in Fig. 5 is very exciting for energy engineers but it also poses large challenges. The first challenge is that variability in the energy generation will play a greatly increased role in our future energy mix. In the scenario depicted in Fig. 5, 38 % of electricity generated in 2050 will be from variable sources (Hayward et al. 2011). Energy will also come from more diverse and geographically distributed sources than in the past. This diversity may also lead to markedly different grid infrastructure. Furthermore, it will depend on less mature and, at least initially, more expensive technologies: a key goal therefore is to accelerate the reduction of costs through 'learning by doing'. A rapidly evolving technology landscape creates high technical and commercial risks, which are particularly difficult to deal with given that investments in energy infrastructure are necessarily large (of the order of billions of dollars) and long term (of the order of a few decades).

In Australia, the daily power demand profile is becoming more 'peaky', largely due to increasing use of electrical goods such as air-conditioners. In turn, these larger peaks have led to a grid with comparatively poor utilisation, where large investments accommodate just a few large peaks every year to 'keep the lights on' during those rare occasions. Emerging demand management techniques—at an individual home or facility or across the grid—are needed to improve network asset utilisation and energy end-use efficiency. CSIRO, for example, has developed building energy management software that 'learns' a building's energy requirements and then balances thermal comfort, running costs and greenhouse gas emissions associated with commercial air-conditioning systems. Energy savings of up to 30 % have been achieved in buildings in Australia and the USA.

Solving the energy puzzle involves more than new energy supply and demand technologies. It will also require public acceptance of new technologies and behaviour change to fully utilise the benefits these technologies offer. Indeed, energy investors and research organisations regularly undertake various outreach activities including publications (e.g. Wright et al. 2009) and community-based discussion groups designed to encourage individuals to be better informed and empowered about energy options.

4 A Step-Change in Meteorology and Energy Linkages

There have always been important links between energy and meteorology but these interactions need to be stepped up to a new level of inter-connectedness to facilitate an efficient transformation of the energy sector and underpin the associated capital investment.

A New Era for Energy and Meteorology   11

Tables 1 and 2 outline the energy value chain and give examples of key energy–meteorology linkages. Table 1 is presented from the perspective of the life cycle of a single energy facility, whereas Table 2 focuses mostly on the electricity system, which is where much of the enhanced interdependency between energy and meteorology lies. In each element of the energy value chain, better understanding of energy–meteorological linkages will drive multi-billion dollar investment decisions: several examples of these are given in the tables and below, including referring to later chapters in this book.

Table 1 Energy facility life cycle: some interactions with meteorology

Phase	PLANNING	DEVELOPMENT		OPERATION	
		Design	Construction	Production	Export
Energy Supply	Resource characterisation	Design loads: extreme cyclic	Schedule optimisation	Operability / Disruptions / Efficiency	
Balance				Supply / Demand forecasting	
Energy Demand	Usage patterns Systems response			Grid integration Price optimisation	

4.1 Energy Facility, Policy and Planning

Resource characterisation is a key element of the planning process for an energy production plant and, for many energy sources, the resource parameters are closely related to climate factors. This is more evident for weather-based resources such as wind and solar energy. In these cases, atmospheric observations and modelling are combined and used in conjunction with prospecting tools in order to identify high-power-yield wind and solar farm sites, as undertaken by companies such as WindLab Systems, a CSIRO spinoff. Energy supply needs to be balanced with demand, and climate considerations are also important in understanding the ongoing role of an energy facility over its lifetime in satisfying energy demand; examples include the evolution of usage patterns and attributes of the overall energy system to which the facility is contributing. Climate variability across the full range of temporal and spatial scales is also a critical consideration for governments in setting advantageous policy parameters to ensure an optimum future energy system.

4.2 Energy Facility Construction

In the facility development phase, offshore oil and gas platforms (Fig. 6) are a good example of where meteorological and oceanographic (met-ocean) factors dictate design and construction. Critical met-ocean actions include extreme loads (e.g. the 100-year return storm), and cyclic loads that can cause material fatigue

Table 2 Energy–meteorology linkages in the electricity system. Links to other chapters in this volume have been highlighted

Electricity value chain	Examples of energy–meteorology linkages
Generation	• Understand effect of meteorological events on non-renewable energy resources and their production (oil, gas, coal, nuclear) (Audinet et al. 2013, Love et al. 2013, Katzfey 2013)
	• Understand current and future (variable) renewable energy resources (wind, solar, wave, hydro, biomass) and their implications for energy markets to underpin long-term investment decisions (Troccoli 2013, Sims 2013, Renné 2013, Gryning et al. 2013, Dorling et al. 2013)
	• Maximise returns of variable generators through better minute to inter-annual resource forecasting (Lorenz et al. 2013, Kay et al. 2013, Dubus 2013, Dutton et al. 2013, Haupt et al. 2013)
	• Understand and where possible maximise benefit of renewable energy facilities on local meteorological and oceanographic conditions, e.g. impact of wave farms on coastal wave resource and erosion (Coppin 2013, Ejigu 2013)
Transmission and distribution	• Understand spatial and temporal correlation between different variable renewable resources as well as with non-renewable resources in order to optimise grid infrastructure investment and maximise asset utilisation for a set reliability level (Pirovano et al. 2013)
	• Enable operation of the transmission network closer to its operational limit through an improved understanding of weather variables such as local temperature and wind profiles (Majithia 2013)
Use	• Understand spatial and temporal (daily, seasonal and decadal) demand patterns especially as they relate to, for instance, increased penetration of air-conditioning systems (Mailier et al. 2013, Shiel et al. 2013)
	• Understand spatial and temporal "negative demand" implications of embedded variable generators (e.g. roof-top PV) (George 2013)

failure. Being able to accurately forecast 'weather windows' is essential in scheduling offshore installation activities, especially when the day-rate of a large crane vessel might be millions of dollars. These challenges are becoming more acute as petroleum exploration and production moves to harsher environments to sustain global demand.

4.3 Energy Facility Operation and Maintenance

Weather patterns again have many influences on an energy facility's operation phase. Weather forecasts are important in tuning production for demand, in scheduling maintenance and in maximising plant efficiency. Weather can also cause sudden disruptions. Shifting our focus again from the specific facility to the overall energy system, weather plays a key role in forecasting and hence balancing energy supply and demand across a diverse set of energy sources.

Fig. 6 Floating offshore petroleum production platform

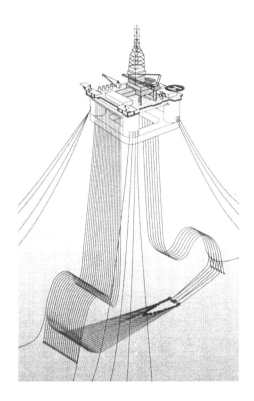

4.4 Energy Transport

Transmission and transfer of energy can extend over very long distances (of the order of thousands of kilometres) and are therefore likely to be exposed to a variety of meteorological events. Exceedingly strong winds, icing, avalanches, landslides, flooding as well as high temperatures are all examples of weather events which can cause transmission power line malfunctioning or even failures (Schaeffer et al. 2012; "Combining Meteorological and Electrical Engineering Expertise to Solve Energy Management Problems" by Pirovano et al.).

Moreover, injection of growing quantities of variable energy sources into the grid has an impact on the electrical load of transmission lines. Variability in generation occurs across a range of timescales, from sub-second flicker to climate change-induced decadal trends. Such variability has to be understood, predicted and managed at an individual facility to maximise financial returns to the operator as well as system-wide to ensure regional security and reliability of supply. Energy storage (e.g. hydro or batteries) or demand-side management techniques can also be used to improve dispatch of variable energy sources to the grid. Again, weather forecasting tools provide a way to ensure optimum incorporation of storage in the grid (see "Unlocking the potential of renewable energy with storage" by Coppin).

4.5 Energy Use

Finally, the demand side of energy use, already affected by meteorological variables such as temperature and humidity, will become increasingly dependent on weather and climate impacts, in particular due to the large-scale adoption of air-conditioning and roof-top PV systems. In Brisbane Australia, for example, air-conditioning penetration has increased from 23 to 72 % over a period of 12 years (Simshauser et al. 2011). Such increased reliance on air conditioning creates a significant exposure to future climate change.

Together, these various factors will significantly extend energy–meteorology linkages across the whole value chain. Tailored weather and downscaled climate information will be required to fully understand and manage our future energy system. Investors, facility designers and system operators alike will request reliable data to inform the large future investments into our energy system and to ensure security and reliability of the system at affordable prices.

5 Conclusions

Energy and meteorology are central to understanding and solving our global sustainability needs. Indeed, the interdependencies between energy and meteorology are amplifying as we develop, build and operate new technologies and philosophies for energy production and use to deliver the triple goals of clean, secure and wealth-creating energy. Bringing energy and meteorological experts together in new ways to create a new depth of understanding—perhaps a new interdisciplinary science—is therefore critical to improved risk management of the energy value chain.

References

Audinet P (2013) Climate risk management approaches in the electricity sector: Lessons from early adapters. Weather Matters for Energy, Springer, New York, USA

Australian Government (2011) Draft energy white paper 2011: Strengthening the Foundation for Australia's Energy Future

Australian Government (2012) Resources and energy quarterly, March quarter 2012, Bureau of Resources and Energy Economics

Coppin P, Wood J, Price W, Ernst A, Lam L (2013) Unlocking the potential of renewable energy with storage. Weather matters for energy, Springer, New York, USA

Dorling S, Earl N, Steele C (2013) Spatial and temporal variability in the uk wind resource: scales, controlling factors and implications for wind power output. Weather matters for energy, Springer, New York, USA

Dubus (2010) Practices, needs and impediments in the use of weather/climate information in the electricity energy sector. In: Troccoli A (ed) Management of weather and climate risk in the energy industry, NATO Science Series, Springer Academic Publisher, pp 175–188

Dubus L (2013) Weather and climate and the power sector: needs, recent developments and challenges. Weather matters for energy. Springer, New York, USA

Dutton JA, James RP, Ross JD (2013) A probabilistic view of weather, climate and the energy industry. Weather matters for energy. Springer, New York, USA

Ebinger J and Vergara W (eds) (2011) Climate impacts on energy systems: key issues for energy sector adaptation, World Bank publication

Ejigu M (2013) Bioenergy, weather and climate change in africa: leading issues and policy options. Weather matters for energy, Springer, New York, USA

Froude and Gurney (2010) Storm prediction: research and its application to the oil/gas industry. In: Troccoli A (ed) Management of Weather and Climate Risk in the Energy Industry, NATO Science Series, Springer Academic Publisher, pp 241–252

George T (2013) Weather and climate impacts on australia's national electricity. Weather matters for energy. Springer, New York, USA

Global Carbon Project (2011) Carbon Budget 2010

Gryning SE, Badger J, Hahmann AN, Batchvarova E (2013) Current status and challenges in wind energy assessment. Weather matters for energy, Springer, New York, USA.

Haupt SE, Mahoney WP, Parks K (2013) Wind power forecasting. Weather matters for energy. Springer, New York, USA

Hayward J, Graham P, Campbell P (2011) Projections of the future costs of electricity generation technologies, CSIRO Report

International Energy Agency (2010) World Energy Outlook 2010

International Energy Agency (2011) World Energy Outlook 2011

IPCC (2007) Climate Change 2007: Synthesis Report, Summary for Policymakers

Johnston PC, Gomez JF, Laplante B (2012) Climate risk and adaptation in the electric power sector. Asian Development Bank publication. Available at: http://www.iadb.org/intal/intalcdi/PE/2012/12152.pdf

Katzfey J (2013) Regional climate modelling for the energy sector. Weather matters for energy, Springer, New York, USA.

Kay M and MacGill I (2013) Improving NWP forecasts for the wind energy sector. Weather matters for energy. Springer, New York, USA

Lorenz E, Kühnert J, Heinemann D (2013) Overview of irradiance and photovoltaic power prediction. Weather matters for energy, Springer, New York, USA

Love G, Plummer N, Muirhead I, Grant I and Rakich C (2013) Meteorology and the energy sector. Weather matters for energy, Springer, New York, USA.

Mailier P, Peters B, Kilminster D, Stephens M (2013) In search of the best possible weather forecast for the energy industry. Weather matters for energy, Springer, New York, USA

Majithia S (2013) Improving resilience challenges and linkages of the energy industry in a changing. Weather matters for energy, Springer, New York, USA

Pirovano G, Faggian P, Bonelli P, Lacavalla M, Marcacci P, Ronzio D (2013) Combining meteorological and electrical engineering expertise to solve energy management problems. Weather matters for energy, Springer, New York, USA

Renné DS (2013) Emerging meteorological requirements to support high enetrations of variable renewable energy sources: solar energy. Weather matters for energy. Springer, New York, USA.

Sayeef S, Heslop S, Cornforth D, Moore T, Percy S, Ward JK, Berry A, Rowe D (2012) Solar intermittency: Australia's clean energy challenge. Characterising the effect of high penetration solar intermittency on Australian electricity networks. CSIRO Technical report. Available at: http://www.csiro.au/en/Organisation-Structure/Flagships/Energy-Transformed-Flagship/Solar-Intermittency-Report.aspx

Schaeffer R, Szklo A, Frossard Pereira de Lucena A, Soares Moreira Cesar Borba B, Pinheiro Pupo Nogueira L, Pereira Fleming F, Troccoli A, Harrison A, Boulahya MS (2012) "Energy Sector Vulnerability to Climate Change: a review", The Int Energy J, 38, 1–12. doi:10.1016/j.energy.2011.11.056

Shiel JJ, Moghtaderi B, Aynsley R, Page A (2013) Reducing the energy consumption of existing, residential buildings, for climate change and scarce resource scenarios in 2050. Weather matters for energy. Springer, New York, USA

Sims REH (2013) Renewable energy and climate change mitigation: an overview of the IPCC special report. Weather matters for energy. Springer, New York, USA.

Simshauser P, Nelson T, Doan T (2011) The Boomerang Paradox, Parts I and II, The Electricity Journal, Vol 24, Issues 1 and 2

Troccoli (2010) Weather and climate predictions for the energy sector. In: Troccoli A (ed) Management of weather and climate risk in the energy industry, NATO Science Series, Springer Academic Publisher, pp 25–37

Troccoli A (2013) Climatic changes: looking back, looking forward. Weather matters for energy. Springer, New York, USA.

Wright J, Osman P, Ashworth P (2009) The CSIRO Home Energy Saving Handbook. Pan Macmillan, Sydney

Climate Risk Management Approaches in the Electricity Sector: Lessons from Early Adapters

Pierre Audinet, Jean-Christophe Amado and Ben Rabb

Abstract Climate change adds a new source of unknowns for the electricity sector. Despite considerable risks and opportunities, energy sector actions to manage climate change risks and take advantage of future opportunities remain limited and patchy. An estimate of the sums spent since 2000 and planned out to the 2020s by five utilities on climate risk management totals US$1.5 billion. Considering that these investments are to address climate change risks or opportunities of a considerable magnitude, they are relatively modest. The sector has focused on climate data analysis and research on impacts rather than on concrete capital, technological and/or behavioral adaptation responses. Further, most of this research is concentrated for the most part in the developed world and on a handful of climate change impacts. Analysis of early adapters in the electricity sector offers a number of useful lessons for power utilities, regulators and stakeholders in thedeveloping world, for instance: (i) joint efforts between the electricity sector and hydrometeorological offices to develop high quality and tailored climate data and information are needed to avoid 'wait-and-see' strategies among power

---

The authors are grateful to Dr. Jeanne Ng and Dorothy Chan (CLP Holdings), Lwandle Mqadi (Eskom), René Roy (Hydro-Québec), Steve Wallace (National Grid), Jonathan Rhodes and John Rixham (E.ON), and Channa Perera (Canadian Electricity Association) for their willingness to share their insights and experiences for this paper; Istvan Dobozi, Silvia Martinez, Vanessa Lopes, Rohit Khanna (ESMAP), and Peter Fraser (Ontario Energy Board) for earlier reviews; Pedzi Makumbe (ESMAP), Richenda Connell, Bastien Fournier-Peyresblanques and Michelle Colley (Acclimatise) for their assistance.

P. Audinet (✉)
ESMAP–World Bank Group, Washington DC, USA
e-mail: paudinet@worldbank.org

J.-C. Amado
Deloitte (formerly Acclimatise), 100 Queen street, Ottawa, Ontario, K1P 5T8, Canada
e-mail: jamado@deloitte.ca

B. Rabb
Acclimatise, Pascoe House, 54 Bute Street, Cardiff Bay, Cardiff CF10 5AF, UK
e-mail: b.rabb@acclimatise.uk.com

utilities; (ii) energy sector adaptation requires going beyond high level research on impacts and adaptation to produce information that can be applied operationally; (iii) without a business environment favorable to climate change adaptation, power utilities have little incentive to go beyond 'business-as-usual' weather risk management; and (iv) it is by building the economic case for adaptation that utilitiescan be incentivized to take action.

Climate change adds a new source of unknowns for electricity sector decision makers. Currently, use of present-day or historical weather and seasonal climate data and information is part of everyday risk management for many utilities and regulators across the world. However, the integration of forward-looking information on climate change in decision-making known as 'climate change adaptation' remains limited. Yet, long-term changes in climate and short-term increases in climate variability are increasingly impacting generation, transmission and distribution of electricity, forcing the industry to consider new ways to manage the associated risks. In some cases, climate change creates opportunities for the electricity sector, for example increased hydropower generation potential in the near- to medium-term in glacial-fed river basins.

This chapter takes stock of initiatives to assess those risks and manage future impacts based on interviews with five utilities, one power regulator and electricity one industry association, and a review of published energy sector information. The purpose of this chapter is to extract lessons with a particular focus on what can be learnt for developing countries where progress to date on managing climate change risks and opportunities is lagging behind. This chapter builds upon the work of the World Bank Energy Sector Management Assistance Program (ESMAP) on energy sector vulnerability to climate change. In 2010, ESMAP and the World Bank's Global Expert Team for Adaptation published a compendium of what is known about weather variability and projected long-term changes, and their impacts on energy systems.

Overall, the electricity sector is at a very early stage of its "climate change adaptation journey." Research on climate data and information, risks and adaptation solutions is concentrated, for the most part, in the developed world and on a handful of climate change impacts. Considerable uncertainties remain about the likelihood, severity, and timing of these impacts for the electricity sector industry. Adaptation responses in the electricity sector remain patchy. Utilities have started investing in adaptation to reduce climate change vulnerability, avoid some future impacts, or seize future opportunities. A rough estimate of the capital expenditures for climate change adaptation since 2000—and presently planned up to the 2020s—by six large electricity utilities surveyed[1] amounts to a cumulated US$ 1.5

---

[1] Authors interviewed China Light and Power (China), Electricité de France (France), E-ON (Germany), ESKOM (South Africa), Hydro-Québec (Canada), National Grid (United Kingdom) by email and phone.

billion. Considering that these investments address future climate impacts or opportunities potentially of a considerable magnitude, those investments are relatively modest. For instance, the repair costs for hurricane Isaac which hit southeastern United States are estimated to have reached around $400 million for Entergy alone in four states (Arkansas, Louisiana, Mississippi and New Orleans) for Entergy alone. This suggests that there could be a strong business case to scale up climate change adaptation investments in the electricity sector.

Four lessons are drawn from this stock-taking exercise.

1. *Joint operations between the electricity sector and hydrometeorological offices to develop high-quality and tailored climate data and information are needed.*
Electric utilities require data and information on observed and future climate conditions on a range of timescales, spatial resolutions, and statistical variables other than climatic averages. Currently such data and information are often not immediately available, which often leads to a "wait-and-see" attitude in the industry. Evidences point to a limited engagement of the electricity industry with the producers of climate scenarios and meteorological institutions.

2. *Electric utilities, regulators, industry associations, governments, and the academic world need to coordinate the expansion of the knowledge base on climate change risks and adaptation solutions in the electricity sector.*
A lot of the information available on climate change impacts and adaptation is too aggregated to be of direct use in the electricity sector's operations. Too few utilities are doing work to identify cost-effective adaptation measures based upon assessments of material risks for generation, transmission and distribution infrastructure.

3. *Governments, international institutions, and professional bodies can incentivize measures that build resilience against to future climate change in the electricity sector by developing standards, regulations and technical guidance.*
The majority of today's climate change adaptation responses in the electricity sector are usual risk management measures that are strengthened by considering how the climate is changing. There are very few examples of "new" technological, behavioral, or institutional measures implemented solely to manage future climate change impacts.

4. *Industry and government need to support research to raise awareness on the nature and possible range of future industry costs with and without adaptation, as well as on the methods and tools to take into account uncertainties in climate change adaptation planning and cost-benefit analysis.*
The ability to build a strong economic case for climate change adaptation remains constrained by a number of factors: lack of reliable climate data and information, low confidence in the return on investment of adaptation expenditure due to multiple uncertainties, and short-termism.

1 The Need to Strengthen the Climate Change Resilience of Electricity Systems in Developing Countries

Natural resource endowment for electricity production (such as river runoff, wind, and solar radiation), transmission and distribution, and electricity demand are all sensitive to weather conditions, climate variability, and long-term changes in climate (see Table 1) (Ebinger and Vergara 2011; European Observation Network, Territorial Development, and Cohesion 2010; Troccoli 2009; Williamson et al. 2009).

However, little is known about what potential impacts will constitute material risks for electricity regulators or utilities, regulators and stakeholders (such as customers). Some impacts have the potential to affect electricity system reliability, security of supply, affordability, and environmental performance (including greenhouse gas emission reductions), while others will have very little effect on electricity assets or systems (Ebinger and Vergara 2011; European Observation Network, Territorial Development, and Cohesion 2010; Troccoli 2009; Williamson 2009). Existing literature shows that integration of climate change information in energy sector decision-making remains limited, for the most part, to large-scale electricity infrastructure in developed countries (Urban and Mitchell 2011).

Due to a combination of several factors, electricity systems in developing countries are particularly vulnerable to a changing climate. First, in many countries, the electricity sector is highly sensitive to weather variability and extreme events. For example, in 2005 alone, extremely hot and cold weather explained 13 % of the variability in energy productivity[2] in developing countries, though much of this variability remains unexplained (MacKinsey 2009). Countries that have repeatedly experienced difficulties maintaining a reliable electricity supply in the past due to weather conditions or that routinely suffer considerable financial loss due to extreme weather events are at risk from climate change (Mechler 2009). Second, high reliance on hydropower, design and construction of power assets on the basis of poor hydrometeorological data, presence of aging assets, and insufficient infrastructure compared to existing needs all constitute factors of climate change vulnerability (Ebinger et al. 2011). This means that many electricity systems are not even adapted to present-day climate, let alone future climate. Finally, developing countries have low capacity to improve their resilience. For example, only a small percentage of costs due to natural disasters are absorbed by insurance in developing countries (International Institute for Applied Systems Analysis (IIASA) 2010). Utilities often have limited ability to finance capital investments, electricity systems often operate with no or few interconnections with other systems, and access to high-quality hydrometeorological and climate data is often poor (Veit 2009).

---

[2] Energy output divided by energy consumed (aka "supply-side efficiency").

Table 1 Electricity sector vulnerability to climate change

Electricity sector value chain	Relevant climate impacts			Impacts on the energy sector
	General	Specific	Additional	
Resource endowment				
Hydropower	Runoff	Quantity (+/-), seasonal high flows and low flows, extreme event	Erosion, siltation	Reduced firm energy, increased variability, increased uncertainty
Wind power	Wind field characteristics, changes in wind resource	Changes in density, wind speed, increased wind variability	Changes in vegetation (might change roughness and available wind)	Increased uncertainty
Solar power	Atmospheric transmissivity	Water content, cloudiness, cloud characteristics	Pollution/dust and humidity absorb part of the solar spectrum	Positive or negative impacts
Wave and tidal energy	Ocean climate	Wind field characteristics, no effect on tides	Strong linearity between wind speed and wave power	Increased uncertainty, increased frequency of extreme events
Supply				
Thermal power plants	Generation cycle efficiency, cooling water availability	Reduced efficiency, increased water needs (e.g., during heat waves)	Extreme events	Reduced energy generated, increased uncertainty
Hydropower	Water availability and seasonality	Water resource variability, increased uncertainty of expected energy output	Impact on the grid, wasting excessive generation, extreme events	Increased uncertainty, revision of system reliability, revision of transmission needs
Wind power	Alteration in wind speed frequency distribution	Increased uncertainty of energy output	Short life span reduces risk associated with climate change, extreme events	Increased uncertainty on energy output
Solar power	Reduced solar cell efficiency	Solar cell efficiency reduced by higher temperatures	Extreme events	Reduced energy generated, increased uncertainty

(continued)

Table 1 (continued)

Electricity sector value chain	Relevant climate impacts			Impacts on the energy sector
	General	Specific	Additional	
Transmission and distribution	Increased frequency of extreme events, sea level rise	Wind and ice, landslides and flooding, coastal erosion, sea level rise	Erosion and siltation, weather conditions that prevent transport	Increased vulnerability of existing assets
Demand	Increased demand for indoor cooling	Reduced growth in demand for heating, increased energy use for indoor cooling	Associated efficiency reduction with increased temperature	Increased demand and peak demand, taxing transmission and distribution systems
Support or connected infrastructure, and local communities	Competition for water resources, competition for adequate siting locations, transport disruptions	Conflicts in water allocation during stressed weather conditions, competition for good siting locations	Potential competition between energy and non-energy crops for land and water resources	Increased vulnerability and uncertainty, increased costs, increased disruptions

Source adapted from Ebinger et al. 2011

2 Existing Research on Climate Change Data and Information, Associated Risks, and Adaptation Solutions

2.1 Research on Climate Change Data and Information

Utilities, regulators and professional bodies in the electricity sector have initiated research efforts to improve hydrometeorological data and climate normals[3] used by electricity sector decision makers. The variables that have received attention are temperature, wind, rainfall, ice, snow, and runoff, because electricity regulators and utilities use information about these climatic factors to make planning, risk management, asset design, and operational decisions.

A lot of work is ongoing to improve observed data and future simulations for rainfall, ice, snow, and runoff in order to optimize hydropower plant asset design and operations. For example, Hydro-Québec has improved the capacity of its hydrological model to provide future runoff simulations that are driven by climate model projections. Such work has started to inform new asset design and the operation of existing hydropower plants (see Box 1). Hydro-Tasmania has also done work to assess future water inflows in a changing climate (Brewster et al. 2009). Canadian electric utility BC Hydro is working with university researchers to develop specialized weather prediction services for icing, precipitation, and lightning. Research is also underway to improve ice-loading forecasting systems (Musilek et al. 2009).

Box 1. Hydro-Québec

> Hydro-Québec is one of the largest electric utilities in North America, with a total installed generation capacity of 36,671 MW and the longest transmission system in the US and Canada. More than 90 % of its installed capacity comes from hydropower installations. The company owns and manages 59 hydroelectric generating stations, 26 large reservoirs, 571 dams, four thermal and one nuclear power plants, 33 453 kilometers of electric lines and 514 electric substations. The company supplies electricity in Québec, Canada, and trades with other Canadian suppliers (in Ontario and New Brunswick) and US Northeast states.
> *Research to understand future changes in climate and impacts on water inflows*
> In response to a series of adverse weather conditions (the 1996 Saguenay-Lac-Saint-Jean flooding, the 1998 ice storm and drought conditions) that caught the attention of Hydro-Québec's executives and government,

---

[3] *Climate normals* are decadal or multidecadal datasets used to summarize or describe the average climatic conditions of a particular location.

Hydro-Québec developed a research program in 2002 to improve knowledge of future climate change, business impacts, and adaptation solutions in the mid- to long-term, so that risks can be managed and opportunities exploited. Joining efforts with the Québec government, Hydro-Québec set up a scientific consortium with the mandate to study regional climate, impacts and adaptation solutions: the Consortium on Regional Climatology and Adaptation to Climate Change (Ouranos). Hydro-Québec partly finances Ouranos by contributing annually CDN$1 million and 5 full-time equivalent researchers and engineers. As part of Ouranos, Hydro-Québec cooperates with other electricity producers (Rio Tinto Alcan, Ontario Power Generation, and Manitoba Hydro) on climate change risk and adaptation issues and has access to the data and information created by Ouranos.

Using its in-house hydrological model and climate change scenarios from Ouranos, Hydro-Québec produced an extensive set of future runoff projections for each of the watersheds where it has hydropower facilities. To increase the level of confidence in its projections, Hydro-Québec analyzed the sensitivity of its hydroclimatic simulations to the choice of greenhouse gas (GHG) emission scenario, climate model, hydrological model, and climate impact methodologies. This "multi-method" approach has shown that, for the 2050 time horizon, the choice of climate models influences hydroclimatic simulations much more than the choice of GHG emission scenarios, hydrological models, or methodology. Hydro-Québec has consequently revised its set of future watershed runoff projections, drawing upon a large number of climate change scenarios. Results are presented in Fig. 1.

Despite some uncertainty between climate models, these projections point at an overall increase in runoff by the 2050, with a higher increase in the north of the Québec province, where most of the company's production facilities are located, compared with the southeast. Projections also suggest more sustained winter inflows from November to April and reduced summer inflows. Furthermore, high river flows will occur earlier in the Spring due to increased temperatures and peak river flows will be lower on average due to reduced winter snow mass.

Hydro-Québec has planned further refinements to its hydroclimatic modeling. For instance, the company wants to:

- Analyze the role of Canadian bogs and fens in Northern Québec in water flow and balance.
- Improve the resolution of its future runoff simulations using direct outputs from regional climate model projections using dynamical downscaling techniques.

Hydro-Québec has started using these future runoff projections at an operational level. For example, the Equipment division assessed the impacts of climate change on hydrological conditions for the "Eastmain 1A—Dérivation Rupert" project. The results were presented at a public hearing to

Climate Risk Management Approaches in the Electricity Sector

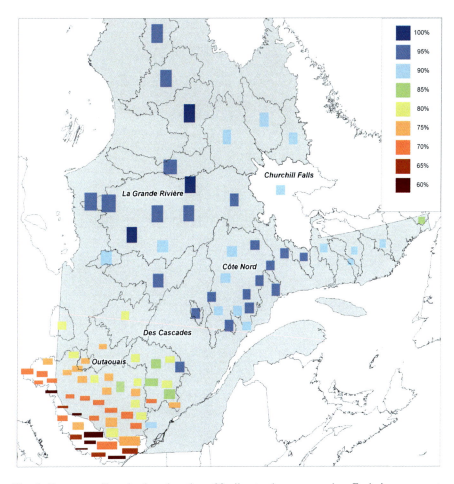

Fig. 1 Future runoff projections based on 90 climate change scenarios. Each *bar* represents results for one watershed; *height* indicates the amplitude of the future increase between 1970–2000 and 2040–2071; *width* indicates the difference in results between different climate change scenarios; and *color* indicates the degree of convergence between the different scenarios, in percentage of scenarios projecting an increase (*blue* for strong, *brown/red* for weak). *Source* Roy et al. 2008 (updated since to 94 watersheds and 90 climate scenarios)

answer questions from the public on the cumulative impact of climate change for fisheries (Comité Provincial d'Examen (COMEX) 2006).

The company is planning to use these runoff projections to assess the benefits associated with adapting the design of different hydropower infrastructure assets for both new equipment and the refurbishment of existing facilities.

Recognizing that the 1990–2000 decade was much warmer in comparison to the 1961–1990 time period, Hydro-Québec's Distribution division

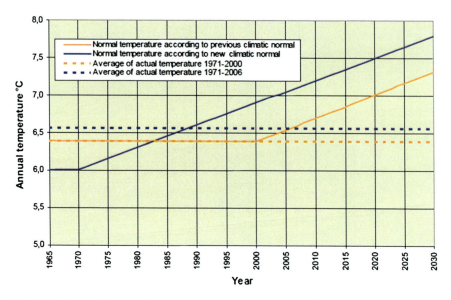

Fig. 2 Evolution of the annual temperature normal for Montréal after updating climate normals to include recent observations and the effect of climate change. *Source* Nadeau 2008

initiated work to update the climate normals it uses to forecast long-term electricity demand.

In 2007, working with the Consortium on Regional Climatology and Adaptation to Climate Change, Hydro-Québec updated the reference period of its climate normals to 1971–2006 and introduced a warming trend of +0.3 °C per decade into the calculation of its climate normals. That resulted in an increase in the average annual temperature normals for the city of Montréal of 0.51 °C for the period 2001–2030 (see Fig. 2). This increase reduced load energy requirements by almost 1 TWh (−0.5 %) per year as a result of reduced heating needs. Furthermore, Hydro-Québec had a 350 MW (−1 %) peak load reduction under the new climate normals. These reduced electricity demand forecasts impacted the required Hydro-Québec electricity production rate and the 10-year procurement plan, which were approved by the Québec regulator.

Incorporate risk management into Environmental Impact Assessments (EIA)
Hydro-Québec considered the effect of future climate change on hydrological regimes as part of the EIA for a new hydroelectric development in Québec, Canada ('Complexe La Romaine'). This helped Hydro-Québec understand future runoff changes and the implications for hydropower production optimization and operational safety.

Using future climate projections, the EIA found that annual average runoff could increase by up to 19 % by 2050 compared to 1961–1990. Considering that current design standards already take into account a high

> interannual and monthly variability of water inflows, Hydro-Québec concluded that the projected impacts of climate change could easily be managed thanks to adaptive management measures, for example adapting operating rules.[4] The Canadian Environmental Assessment Agency approved this approach in a report of April 2008.[5]

Electric utilities and regulators usually rely on normalized average temperature to forecast customers' future electricity needs, which influences long-term supply/demand balance and investment plans up to 20 years in the future. Though different methods are used, the North American power sector usually uses 30-year average temperature datasets from government organizations, such as the US National Oceanic and Atmospheric Administration or Environment Canada. However, these datasets seldom take into account future temperature rises, which could affect the reliability of demand forecasts. Some climate data providers, electricity regulators, and utilities are developing alternative methods or climate datasets, to better take into account future climate variability: (Electric Power Research Institute and (EPRI), North American Electric Reliability Corporation (NERC), and Power System Engineering Research Center (PSERC) 2008). Some organizations have started using more recent and shorter observed time periods or rolling decadal averages, which already reflect some degree of climate change (for example, Hydro-Québec, see Box 1). BC Hydro is also developing high resolution wind projections for British Columbia, to improve its wind forecast capability and management practices (Toth 2011).

2.2 Research on Risks Associated with Changes in Climate

Few organizations have taken steps to assess how future changes in climate or climate-related variables will likely affect their systems, operations, and assets.

2.2.1 Hydropower Generation

BC Hydro and Hydro-Québec have put considerable effort into simulating future water inflows (see Box 1). The companies plan to use this data in the coming years to estimate future electricity output, and inform asset design, potential changes in operations and long-term planning.

---

[4] See the environmental impact assessment for the La Romaine hydrolectric complex (Vol 7, Chap. 49, pp. 49–6 to 49–19), available at http://www.hydroquebec.com/romaine/documents/etude.html (accessed 12/09/2011).

[5] See http://www.ceaa.gc.ca/050/documents/26480/26480E.pdf (accessed 12/09/2011).

Hydro-Tasmania has modeled future water inflows into its reservoirs based on rainfall and evaporation projections. Results indicate lower annual inflows, changed runoff seasonality and an overall reduction in output, especially in run-of-the-river plants and in central Tasmania (Brewster et al. 2009).

2.2.2 Transmission and Distribution

Recognizing the risk of stronger winds for transmission and distribution, BC Hydro has begun to study the relationship between tree fall, transmission and distribution damage, and storm intensity, to better understand the risk of infrastructure damage (Toth 2011).

To comply with safety obligations, utilities must operate electric equipment and hardware below maximum design temperatures. Thermal ratings of electrical lines are calculated on the basis of ambient conditions (usually based on a long-term average), and maximum current that can be safely carried, in order to ensure that maximum design temperature is never exceeded. Higher average temperatures and changes in natural cooling (rainfall, wind, or cloud cover) could reduce current ratings and constrain transmission or distribution capacity (Pytlak et al. 2011; Toth 2011; National Grid Electricity Transmission plc. 2010; South West Climate Change Impacts partnership 2010). Several electricity sector groups have engaged in research showing future thermal de-ratings due to changes in climate conditions (Pytlak et al. 2011; Toth 2011; National Grid Electricity Transmission plc. 2010; South West Climate Change Impacts partnership 2010) (see the example of National Grid, Box 4). While this may not represent a risk for well-designed systems, it could be an issue in countries with aging assets operating close to their design ranges. Utilities wanting to keep electricity supply efficiency levels above a certain level have to consider the trade-off between the cost of upgrading line thermal rating and the price of electricity sold.

2.2.3 Demand Loads

The Electric Power Research Institute (EPRI), the North American Electric Reliability Corporation (NERC), and the Power System Engineering Research Center (PSERC) organized a technical summit on climate change and demand load forecasts in 2008, which called for strengthening climate change impact data collection and impact assessment methodologies, including methods to measure the benefits of increasing the adaptive flexibility of electric systems and compare those to the associated costs. (EPRI, NERC, and PSERC 2008). A number of electricity regulators and utilities, such as the Canadian Independent Electricity Operator, Hydro-Québec, and Electricité de France, have started revising their own demand load forecasting methods using climate model projections, in response to future higher temperatures and associated long-term electricity demand changes (see Sect. 2) (Mirza 2011; Dubus 2009; Minville 2009).

For example, the number of cooling degree days[6] in London, UK, increased by an additional 30–34 days over the period 1961–2006 (Jenkins et al. 2007). The UK project GENESIS (Generic Process for Assessing Climate Change Impacts on the Electricity Supply Industry and Utilities) found that by the 2080s summer electricity demand for cooling will exceed winter demand for heating during the daytime (Watson 2006). This will aggravate system overloading in certain parts of London, and will require additional supply responses.

2.2.4 Asset-Level Risks

Drawing on existing information on specific climate change impacts and adaptation measures, CLP Holdings and Eskom have assessed climate risks and identified adaptation measures for a few pilot assets, to build the business case internally for adaptation-related decisions, and improve operational risk management (see Box 2 and 3).

Box 2. CLP Holdings

> For over 100 years, CLP Holdings, formerly known as China Light and Power, has played an active role in powering Asia's growth. From its home base of Hong Kong, CLP's operations have expanded to include mainland China, Australia, India, Southeast Asia, and Taiwan. As of 31 December 2010, CLP has invested in 13,635 MW of electricity generation (from a range of energy sources including coal, gas, nuclear, hydro, wind, solar, and biomass), 6,599 MW of generation capacity across the Asia–Pacific region, 22PJ of gas storage capacity in Australia, 13,767 km of electricity transmission and distribution lines, 13,421 substations in Hong Kong, and a number of electricity and gas retail businesses serving over 3 million customers in Hong Kong and Australia.
>
> As stated in CLP's report *Climate Vision 2050*, the organization is committed to addressing climate change: "[CLP will] ensure that our business develops only commercially viable, environmentally responsible energy generation assets to meet rising market demand" (CLP's Climate Vision 2050). CLP has engaged on a "climate change adaptation journey" to address risks within the company's "fenceline." As shown in Table 2, CLP has identified a number of potential risks at a corporate-level.
>
> *Adaptation pilot study*
> CLP began to work with a specialist consultancy firm to understand the possible future impacts of climate change on its assets, with an initial pilot study focusing on two existing operations that have already experienced

---

[6] Cooling degree day (CDD) is the number of days when average temperature is above 65 degrees Fahrenheit/18 degrees Celsius and people start to use air conditioning to cool buildings.

Table 2 Corporate-level climate change risks identified by CLP in its response to the 2011 Carbon Disclosure Project investor survey

Risk driver	Description	Potential impact	Timeframe	Likelihood	Magnitude
Change in mean (average) temperature	Affects operations as demand for electricity for, say, air conditioning, will not follow established patterns	Increased operational cost	Current	Very likely	Medium
Change in temperature extremes	Affects operations as demand for electricity for cooling or heating could fluctuate dramatically	Increased operational cost	Current	Very likely	Medium
Change in precipitation extremes and droughts	Extreme precipitation could lead to flooding and drought could affect availability of process water	Increased operational cost	Current	More likely than not	Medium
Sea level rise	Our facilities that are located near the coast could be affected	Increased capital cost	Current	More likely than not	Medium–high
Tropical cyclones	Our facilities may not be built to sustain cyclones that have become stronger than historical events	Increased capital cost	Current	More likely than not	Medium–high

Source Carbon Disclosure Project website

> physical impacts from extreme weather conditions. The main objective was to help the company to develop a methodology that can be applied across CLP for assessing what climate change adaptation measures can and should be taken from a business case perspective for existing operations, and perhaps provide some insight for possible future acquisitions or greenfield developments. The two pilot sites chosen comprised a gas-fired power station in India (GPEC) and a coal-fired power station in Taiwan (Ho-Ping), both located near coastal areas.
>
> The potential loss arising from climate change was quantified for each site and adaptation options identified where possible. This was followed by a cost-benefit analysis of the various adaptation options and the testing of a decision-making process that involves not only taking mandatory regulatory requirements as "must dos", but also CLP's company values which dictate what it perceives to be the "right things".

CLP analyzed the following sources of climate data and information to determine future climate scenarios: historical trends, local minima and maxima, Global Circulation Model (GCM) simulations, and regional climate projections.

Climate projections are useful for determining the direction of future changes, however they contain considerable uncertainties. Large-scale projections of future climate utilize Global Circulation Models (GCMs) to describe the physical circulation of the coupled atmosphere/ocean/land system. Furthermore, greenhouse gas emission scenarios are used to model a range of futures considering different global and regional driving forces. GCMs are highly sophisticated, however inherent uncertainties mean that the projections have to be applied with care. Regional climate maxima were used wherever possible to reduce uncertainty. However, as the Ho-Ping coal-fired power plant showed, history is not always a good predictor of the future, as future climate poses challenges outside historical experience. The study identified and assessed adaptation options to determine the most effective and efficient options in response to the identified risks.

Risk and adaptation findings

The coal storage domes and coal conveyor at the Ho-Ping coal-fired power station were designed to withstand wind speeds during typhoons of up to 60 m/s for a 3 s duration (CLP 2011). The strongest gust speed ever recorded at the time of design was 56 m/s. However, since asset construction, wind speeds have exceeded this threshold on a number of occasions, during which all three coal domes have experienced damage and caused coal supply disruptions. High winds associated with typhoons have also caused power outages on four occasions between 2005 and 2008. On each occasion, electricity output from the plant was lost, causing extended power cuts for customers. During typhoon Jangmi in September 2009, wind damage to transmission lines caused 17 days of outages.

Recent research suggested that on a global scale tropical cyclones will see an increase in intensity of between 2 and 11 % by 2100. Given a maximum historical wind speed of 89.8 m/s during Typhoon Alex in 1984, this could give rise to gust speeds of 100 m/s (CLP 2011). For the West Pacific, this figure could be even higher, given possible average projected increases of up to 20 % (Knutson et al. 2010). CLP also investigated the risk of landslides, erosion and high wind speeds to transmission lines infrastructure at Ho-Ping.

Similarly, climate risks were assessed for the gas-fired power station in India (GPEC), which is also vulnerable to tropical cyclones (see Table 3). In particular, GPEC was found to be at risk of saline intrusion and flooding as a result of storms and sea level rise. In the case of GPEC, a number of adaptation options have already been implemented, for example raising the floor level of buildings housing critical infrastructure, building flood levees around low-lying parts of the site, increasing drainage capacity, and

Table 3 Climate risks identified at Ho-Ping coal-fired power station in Taiwan, and GPEC gas-fired power station in India

Impact	Ho-Ping	GPEC
Wind	Coal dome and conveyor damage Transmission line outages	Damage to third party gas terminals
Erosion/landslides	Transmission tower damage Fresh water supply cut-off	n/a
Coastal flooding	Coal conveyor damage Coal dome inundation	Damage to third party gas terminals reducing supply
Pluvial flooding	Coal dome inundation	n/a
Sea level rise	n/a	Fresh water salinity increase
Temperature increase	Cooling water temperature : increased heat rate and fuel cost	Ambient temperature : decreased output

Source CLP Holdings

diverting cooling water pipes to access fresh water in the event of saline intrusion.

In light of the risk assessment, the pilot study identified the following climate change adaptation options: (CLP 2011):

- Inspect tower on or close to erosion/landslide risk slopes;
- Commission a wave action study to estimate maximum wave height during typhoons;
- Reinforce the base of towers on landslide risk slopes;
- Reinforce the base of five towers close to landslide risk slope;
- Strengthen five towers and transmission line sections to withstand strong gusts;
- Investigate emergency coal delivery by rail;
- Reinforce coal conveyor cladding;
- Protect domes from water ingress;
- Reinforce fresh water pipeline/secure alternate sources; and
- Increase drainage capacity on-site.

Conclusions

Climate change poses site-specific risks to current and future assets which can be managed through a range of adaptation measures. Through the pilot study, CLP Holdings identified a number of situations, which are site specific, but may possibly be applied generally to a wide range of facilities. For example, the company found that the availability of historical data and climate projections varies across Asia. The incomplete information sometimes makes decision-making based on quality data difficult and the robustness of the assessments could be compromised. In light of this, scenario analysis is helpful for determining the "what-if" impacts given the inherent uncertainties. It also helps encompass a wider spectrum of factors (engineering, managerial, legal, cost, company's standards).

UK-based electricity transmission and distribution companies, Western Power Distribution and National Grid, have both assessed climate change impacts on weather-related faults and supply interruptions (South West Climate Change Impacts partnership 2010). In the case of National Grid, this has included assessing flood risk using climate change projections for all its substations (see Box 4).

2.3 Research on Adaptation Solutions

Some work has been undertaken by the utilities interviewed to improve understanding of the technological or behavioral solutions that could reduce climate change vulnerability and attached costs and benefits.

For example, a researcher from the Hydro-Québec research institute assessed the avoided hydropower loss due to unproductive reservoir water spills if operating rules are adapted. By incorporating future runoff projections into an optimization model, which calculates weekly operating rules according to simulated runoff, this work found that for one hydroelectric plant in Québec (Chute-des-passes) (Silver and Roy 2010):

- Annual mean hydropower generation could decrease by up to 14 % if operating rules are not adapted between the control period and 2050; and
- Adapting operating rules could increase hydropower generation between 1 and 15 % in the same period.

Utility interest in distribution loss reduction technologies has increased in recent years. BC Hydro (ex BC Transmission Corporation) is testing the benefits of Dynamic Thermal Rating Systems to operate electricity distribution equipment closer to their design limits, taking into consideration *observed* weather conditions rather than *static* (or normalized) climate data (Janos and Gurney 2008).

3 Early Adaptation Efforts in the Electricity Sector

Adaptation to future climate risks in the electricity sector can take three principal forms:

- Behavioral, whereby utilities relocate their assets, or modify their emergency, maintenance and operating plans;
- Institutional, whereby utilities and regulators adopt climate change adaptation strategies, assign staff responsibility, incorporate climate risk management into existing systems and standards, or disclose information on climate change impacts and adaptation; and
- Technological, whereby utilities invest in new or adapted technologies, or improve the design of assets.

Table 4 presents some generic examples of each of these three types of adaptation actions.

Climate change adaptation being at its very early stages in the electricity sector, there are only few examples of electricity utilities, regulator or industry associations that are taking technological, behavioral, or institutional actions to manage future climate risks.

3.1 Behavioral and Institutional Responses

Behavioral and institutional responses to a changing climate in the electricity sector are limited, though the following real-life examples show leading efforts from utilities.

The integrated UK gas company, BG Group, which operates thermal power plants across the world, has a climate change adaptation strategy in place which requires that staff assess the risks to operations from foreseeable environmental changes arising from climate change.[7] BG Group has designed a Climate Risk Management Framework to support assets and projects in delivering against this requirement. E.ON also has a comprehensive climate change adaptation strategy in place, as explained in Box 5. Other utilities are in the process of developing strategies, for example Eskom and BC Hydro.

To comply with a directive issued by the Department for Environment, Food, and Rural Affairs under the 2008 Climate Change Act, the UK's National Grid published its first Climate Change Adaptation report in 2011, disclosing how the organization has embedded its climate change policy in its risk management procedure, assessed future climate risks, and identified adaptation measures (see Box 4).

To increase resilience against a possible increase in extreme weather events, such as storms and flooding, several utilities, and regulators have invested in climate change adaptation. For example, Entergy in the southeast USA has relocated some of its data centers away from flood risk areas (U.S. Department of Energy 2010). National Grid is investing in flood mitigation work to raise the standard of protection of its substations from a 1:100-year to a 1:1,000-year fluvial or tidal flood event (see Box 4). Recognizing that a changing climate affects weather risk, the Independent Electricity System Operator (IESO) is auditing and assessing the adequacy of existing processes and standards at the local, regional and North American level, to evaluate and manage high-impact, low-frequency weather-related events (Ontario 2011).

Some utilities have also started to recognize that changing the way they operate their assets can reduce vulnerability to future climate change, and is often more

---

[7] See http://www.bg-group.com/sustainability09/climate_change/Pages/climate_change_our_strategy.aspx (accessed 16/10/2011).

Table 4 Generic examples of adaptation responses in the electricity sector

Electricity sector value chain	Technological "Hard" (structural)	"Soft" (technology and design)	Behavioral Re(location)	Anticipation	Operation and maintenance	Institutional
Supply						
Thermal power plants	Improve robustness of installations to withstand storms (offshore), and flooding/drought (inland)	Replace water cooling systems with air cooling, dry cooling, or recirculating systems Improve design of gas turbines (inlet guide vanes, inlet air fogging, inlet air filters, compressor blade washing techniques, etc.)	(Re)locate in areas with lower risk of flooding or drought (Re)locate to safer areas, build dikes to contain flooding, reinforce walls and roofs	Emergency planning	Manage on-site drainage and runoff Changes in coal handling due to increased moisture content Adapt regulations so that a higher discharge temperature is allowed Consider water reuse and integration technologies at refineries	Adopt a corporate- or business-level climate change adaptation strategy Review internal codes of practice and manuals
Hydropower	Build de-silting gates and buffers Increase dam height Construct small dams in upper basins Adapt capacity to flow regime	Changes in water reserves and reservoir management Regional integration through transmission connections	(Re)locate based on changes in flow regime		Adapt plant operations to changes in river flow patterns Operational complementarities with other sources (for example natural gas)	
Wind power		Improve design of turbines to withstand higher wind speeds	(Re)locate based on expected changes in wind speeds (Re)locate based on anticipated sea level rise and changes in river flooding			

(continued)

Table 4 (continued)

Electricity sector value chain	Technological		Behavioral		Institutional
	"Hard" (structural)	"Soft" (technology and design)	Re(location)	Anticipation	Operation and maintenance
Solar power		Improve design of panels to withstand storms	(Re)locate based on expected changes in cloud cover	Repair plans to ensure functioning of distributed solar systems after extreme events	
Transmission and distribution	Improve robustness of infrastructure to withstand more extreme weather events Burying or cable re-rating of the power grid		Emergency planning	Regular inspection of vulnerable infrastructure such as wooden utility poles	
Demand	Invest in high-efficiency infrastructures and equipment Invest in decentralized power generation, such as rooftop PV generators or household geothermal units			Efficient use of energy through good operating practice	
Support or connected infrastructure, and local communities		Consider underground fossil fuel transfers and transport structures			Engage in community forums

Source adapted from Ebinger et al. 2011

cost-effective than making physical changes to existing assets. For instance, Hydro-Tasmania has changed its seasonal operating rules and turbine outage management methods to cope with reduced inflows and changing seasonality (Brewster et al. 2009).

There are a few examples of utilities that have started mainstreaming climate risk management into everyday business. Hydro-Québec considered climate change impacts as part of the Environmental Impact Assessment (EIA) for a new hydropower complex on the La Romaine river in Québec, Canada (see Box 1). Australian utility, ActewAGL, appraised how climate change may affect flood risk as part of its EIA, and elevated electrical equipment in new substations to ensure the integrity of the network during peak flood events (AECOM/Purdon Associates 2009). There is one precedent in Ontario, Canada, whereby the review panel of an Environmental Assessment Report for a new nuclear power plant requested that the project promoter does more in-depth and localized modeling of climate change impacts, based on high resolution data, to ensure adequate consideration of climate change risks before a construction license is issued (Darlington New Nuclear Power Plant Project 2011).

In relation to long-term resource planning, utilities surveyed have started to discuss possible exemptions or differential prices with government and regulators to improve their capacity to cope with climate change or to cover adaptation costs. For instance, during the summer 2003 European heat wave, Electricité de France negotiated exemptions on maximum water discharge temperature obligations, to avoid shutting down too many nuclear plants. An exceptional exemption from these legal requirements was granted to four conventional thermal power plants and sixteen nuclear reactors, permitting them to exceed the maximum discharge water temperature (Letard 2004). Hydro-Québec has integrated the effects of warming into its future demand load forecasts presented to the Québec regulator (see Box 1).

Finally, a handful of professional bodies (e.g., the Canadian Electricity Association) have also recognized that climate change poses a risk to the electricity industry, and are supporting initiatives that help to better understanding climate risks and how to reduce vulnerability.

Box 3. Eskom

Eskom was established in South Africa in 1923 as the Electricity Supply Commission. In July 2002, it was converted into a public limited liability company, wholly owned by the government. Eskom is one of the top 20 utilities in the world by generation capacity, with a net maximum self-generated capacity of 41,194 MW.

Eskom is the largest electricity generation, transmission and distribution company in Africa. It generates approximately 95 and 45 % of the electricity used in South Africa and the whole African continent, respectively. It buys and sells electricity in the countries of the Southern African Development

Community (SADC). Eskom operates coal- and gas-fired power stations, nuclear plants, hydropower facilities, and wind turbine sites. In 2011, the company looked after 28,000 and 46,000 km of high and low-voltage electric lines respectively.

Eskom's Climate Change Response
Eskom supports South Africa's government approach of contributing to global efforts to combat climate change whilst ensuring the sustainability of its economy and society. South Africa is a signatory to the United Nations Framework Convention on Climate Change (UNFCCC) and its Kyoto Protocol, and has been leading a number of global and regional initiatives which promote sustainable responses to climate change adaptation. The National Climate Change Response Policy Development process has been initiated by the Department of Environmental Affairs (DEA). This has included the development of the National Climate Change Response (NCCR) White Paper for South Africa approved by the South African Cabinet in October 2011.

The NCCR has highlighted the following objectives for South Africa on climate change:

- Make a fair contribution to the global effort to stabilize greenhouse gas concentrations (i.e., mitigation of climate change); and
- Effectively manage unavoidable climate change impacts through interventions that build and sustain South Africa's social, economic, and environmental resilience and emergency response capacity (i.e., climate change adaptation).

Eskom supports this approach in the form of its corporate Climate Change Response Strategy which highlights Eskom's commitment to deal with climate change in the following six key areas:

- Diversification of the generation mix to lower carbon emitting technologies;
- Energy efficiency measures to reduce demand and greenhouse gas and other emissions;
- Innovation through research, demonstration, and development;
- Adaptation to the negative and positive impacts of climate change;
- Investment through carbon market mechanisms; and
- Progress through advocacy, partnerships, and collaboration.

In the financial year 2010 to 2011, Eskom reviewed its Climate Change Response Strategy, with a view to better address the company's resilience to climate variability and long-term changes in climate.

Eskom's research to understand changes in climate, future impacts, and adaptation
The Climate Change and Sustainability Department in partnership with various business units within Eskom has been hosting a number of applied

research, workshops, and conducting surveys to feed into the process of developing Eskom's adaptation strategy. This work includes the following:

- Definition of climate change impacts for South Africa and its specific impacts on Eskom;
- Investment in applied climate change research;
- Redefinition, benchmarking and continuous review of business planning assumptions and continuously review;
- Cost-benefit analysis of adaptation interventions (adaptation costs curves);
- Integrated risk and resilience management (including the identification of climate-related thresholds in business systems, unacceptable levels of climate-related risk, and required levels of adaptive capacity and business resilience); and
- Integration of climate change adaptation imperatives into business operations.

By 2011, Eskom had done research in two separate areas, as explained below:

(a) Case studies on weather impacts

In order to understand vulnerability to changes in climate, the company analyzed historical and current weather conditions, extreme events and climate variability, and their impacts on a number of business areas, including two coal-fired power stations (Hendrina and Kendal), the NorthEast transmission grid and the Eastern region distribution.

(b) Weather surveys

Eskom has also undertaken weather surveys asking the different Eskom businesses the following key questions:

- What weather data is Eskom already monitoring across all operating units and strategic functions?
- Which business processes and operations are affected by weather phenomena and will benefit from appropriate weather data integration?
- What aspects of weather do we need to monitor in real-time to warn about extreme weather events for situational awareness and response purposes in the control rooms and customer nerve centers?
- What aspects of weather data do we need to have in a long-term climate data warehouse for research and analysis purposes?

Results from these surveys have been assessed and have also informed Eskom's support in long-term research activities and strategies relevant to climate change resilience.

(c) Thresholds, adaptive capacity and vulnerability assessments of Eskom systems

Eskom is currently investing in research to further define climate-related thresholds, as well as the vulnerability and adaptive capacity of its systems to future climate change impacts. These studies are undertaken in collaboration with the University of Cape Town (UCT), the University of Kwazulu-Natal (UKZN), and the Council for Scientific and Industrial Research (CSIR). They include the following activities:

- Identifying climate-sensitive thresholds of vulnerable systems within Eskom, possible adaptation measures, and associated costs and benefits;
- Developing climate change projections for rainfall, temperature, lightning and storms;
- Assessing the impacts of climate change on water resources in the Waterberg area;
- Modeling the hydrology of four catchments in the Waterberg area and around the Hartebeespoort pipeline; and
- Modeling of summer convection (thunderstorms, lightning and rainfall intensity and frequency) over Southern Africa (Table 5).

Promote dry cooling systems to reduce reliance on freshwater for thermoelectric or nuclear power plant cooling

Although the company has started investing in dry cooling in the mid-1980s, Eskom identifies dry cooling systems as a short-term climate change adaptation measure for new power stations in its "Climate Change Commitment" (Eskom 2009). Eskom recognizes that dry cooling involves higher costs at the construction and operation stages, reduced overall plant efficiency, and lower plant output. Eskom accepts that sustainability and adaptation will override economic considerations in certain cases when choosing between wet or dry cooling.

In 2000, Eskom operated the largest power plants with a dry cooling system in the world (Matimba, Kendal, and Majuba). Eskom's investment in dry cooling has resulted in an estimated combined saving in excess of 90 million cubic meters of water per annum (Pather 2004).

3.2 Technological Responses

Only a few utilities have begun to implement technological responses to climate change adaptation, by making structural changes to existing assets (e.g., increasing energy efficiency of electrical equipment), building new assets (e.g., building a

Table 5 Results from Eskom case studies on weather impacts for four assets

Assets	Vulnerability and levels of adaptive capacity	Existing and potential adaptation to climate change measures
Hendrina Power Station	A: *Increased heavy rainfall and wet coal* Heavy rain causes wet coal which has the following impacts: – Blockages in bunkers – Clogging of milling plant – Reduction in the amount of electricity generated if load losses result B: *Increased heavy rainfall and dams overflow (normal and ash)* – Increased water levels may result in bursting of the dams. Overflows yes, but how possible is actual "bursting" of dam walls?	– Coal blending – Reclaim the coal – Use of sensors and redirecting wet coal back to the stockpile. However, this causes uncertainty when it stays wet for a prolonged period – Increase the dam capacity – Build additional dams – Change dam management strategy – Drainage pipes – Reuse more water (treating water)—also impacts on other sustainability issues and should be a priority
	Floods – Ash dams overflow	– Build new dams – Replace the dam lines – Increase the frequencies of dam level inspections
	Low temperature – Delayed combustion process resulting in load losses due to boiler losses	– Use of more coal to get the same MW output, but at high cost
	High temperature – Increased condenser and vacuum temperatures	– Use of more coal to get the same MW output—at high cost
	Lightning – Stack pollution monitors	– Use lightning arrestors

(continued)

Table 5 (continued)

Assets	Vulnerability and levels of adaptive capacity	Existing and potential adaptation to climate change measures
Kendal Power Station	*Heavy rainfall* Wet coal results in the following impacts: – Blockages in bunkers – Clogging of milling plant – Reduction in the amount of electricity generated	– Coal blending – Reclaim the coal – Use of sensors and redirecting wet coal back to the stockpile – Cover the coal stockpile
	Drought Station runs out of recycled (potable) water from the three dams	– Station uses water from reservoirs pumped from Vaal River which is dirty and expensive to clean
	High temperature – Air heater packs affected on the units, resulting in high particulate emissions (especially during summer) – Affects the performance of the cooling tower (if dirty) – Heat exchanges on the Auxiliary cooling plant are affected resulting in potential multiple-unit trips – Increased condenser and vacuum temperatures – Load losses above certain temperature because of drop in efficiency	– Replace the design layers that were removed – Apply for exemptions to obtain waiver under license to operate (for the high emissions) – Use more coal to get the same output—at high cost
	Low temperature – Delayed combustion process, load losses due to boiler losses	– Use of more coal to get the same output—at a high cost
North East Transmission Grid	*Heavy rainfall* – Excessive vegetation growth on the transmission line servitudes, interfering with the lines' clearance and may cause fires resulting in line faults – Submerged tower footing and resultant rusting of structures	– Increase the frequency of cutting the grass, i.e., cutting the grass three times a year instead of twice – Use stainless steel material to reduce corrosion – Build foundations as per the design specifications
	Ice/mist and fog – Ice/mist or fog covers insulator sheds in the presence of pollution, causing flashovers on the substation transformers and lines resulting in tripping	– Coat the insulators with silicone – Use water-repellent composite polymer insulators—expensive – Increase spacing of insulator sheds – Install shed extenders – Pollution deposits on the insulators
	High temperature – Conductor sagging which may cause fires	– Contact national control to switch off the affected line, until the fires are extinguished – Send fire-fighters to the site to extinguish the fire – Carry out planned and controlled burning

(continued)

Table 5 (continued)

Assets	Vulnerability and levels of adaptive capacity	Existing and potential adaptation to climate change measures
Eastern Region Distribution	*Heavy rainfall* – Damage to insulators, transformers, surge arresters and circuit breakers *Ice/mist* – Damage to lines and substations – Conductor sagging which may cause fires *Floods* – Substations and lines inland are affected by floods – Foundations of towers are damaged resulting in their collapse *Sea swells* – Traction substations that supply railway lines are impacted	– Use special type of insulator – Coat insulators with a special silicone grease – Use silicon rubber coating – Use steel instead of wooden poles – Use shells of different diameters to avoid accumulation of ice to reduce the flashovers – Increase the distance between insulators – Use cranes, helicopters and vehicles to displace the snow/ice – Increase the height of tower—increasing the clearance distance between the rower and vegetation – Reduce the spans if the highest tower is used already – Increase the tension of the conductors – Increase the elevation of the substations – Increase the distance between the coast and the substations

dam in upper basins to regulate future runoff increase), or modifying asset design (e.g., improving the design of transmission pylons to withstand higher ice or snow loads).

Hydro-Tasmania's *Climate Change Response Strategy* includes a number of technological adaptation measures aimed at compensating part of the projected future loss of inflows, such as:

- Increasing water storage capacity to capture higher inflows during winter and release water during drier periods;
- Replacing the coating of existing water canals (also known as "relining") and increasing the capacity of a few water canals to reduce evaporation and increase inflow efficiency;
- Confirming that existing weirs are operating efficiently to ensure that water flows are well-regulated, and considering new water diversion schemes that could pass more water through existing hydropower plants; and
- Developing new projects to make up for the expected loss of output, such as mini-hydro schemes or refurbishing old power stations (Brewster et al. 2009).

These adaptation measures could increase production capacity by over 1,000 GWh, at a total cost of approximately AU$420 million. Building new small hydropower plants, increasing storage capacity, and building new dams are expected to have a significant cost (AU$320 million), with a generation potential of 700 GWh (Brewster et al. 2009). Conversely, Hydro-Tasmania has identified measures that are relatively low-cost, but have a considerable generation potential: for example, raising dam height and upgrading water canals would cost AU$48 and AU$10 million respectively, with a corresponding total generation potential of 300 GWh (Brewster et al. 2009).

Since the 1980s, South African utility Eskom has made considerable investments in dry cooling systems: the company has increased its air-cooled thermoelectric generation capacity by 10,000 MW between the mid-1980s and the early 2000s (Lennon 2010) Fig. 3. Eskom has identified this as an adaptation measure to cope with future reduced cooling water availability in South Africa (see Box 3).

Finally, there are a few tangible technological investments in transmission and distribution. Following the 1998 Eastern Canada ice storm, which cost Hydro-Québec CDN$ 725 million in damages,[8] the utility reviewed its design standards for high voltage lines. An internal technical committee recommended an increase in maximum ice and hourly wind loads, and cumulative ice/wind loads, on transmission hardware, and the installation of special pylons at standard intervals to avoid cascades of falling pylons during high ice load events.[9] The Québec utility

---

[8] Extreme icing damaged 116 transmission lines and 3,110 support structures (including 1,000 steel pylons), as well as 350 low-voltage lines and 16,000 wood posts. To restore service rapidly to its customers following the disaster, Hydro-Québec spent CDN$725 million repairing the lines and support structures with the least damage and building temporary transmission and distribution equipment. See Turcotte et al., 2008, ibid.

[9] René Roy, Hydro-Québec, personal communication, 25/10/2011.

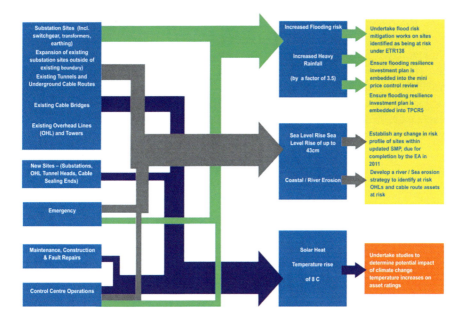

Fig. 3 Actions (on the *right*) in response to climate change risks identified (on the *left*). Source National Grid

spent close to CDN$ 1 billion between 1998 and 2008 replacing old pylons and structures with stronger ones (Turcotte 2008). The utility was planning to spend CDN$200 million by 2012 on the electricity system interconnection between the Canadian provinces of Québec and Ontario and on anti-icing equipment, which are expected to reduce vulnerability to future extreme winter weather events (Turcotte 2008). BC Hydro is testing high performance corrosion-resistant materials. For example, it has installed cross arms made of high performance composite material, which have much longer lifetimes and higher resistance to extreme weather conditions (Toth 2011). The utility has also started replacing wooden electricity poles, damaged by mountain pine beetle infestations[10] with metal ones, and is testing Dynamic Thermal Rating Systems to improve the capacity of its transmission system. In Australia, the Victorian Bushfires Royal Commission recommended burying electric lines to reduce the risk of network failures due to damage during extreme weather events and wildfires (Linnenluecke et al. 2011).

---

[10] Climate change has exacerbated recent mountain pine beetle outbreaks in Western North America. Unusually hot and dry summers (favorable for beetle reproduction), and mild winters (which allow beetle larvae to survive), have contributed to infestations destroying more than 700 million m3 of pine in British Columbia, Canada, which represents more than 50 % of the province's pine. See Carroll et al. 2003.

3.3 Adaptation Investment Estimates

Though uncertainty remains about the exact cost of future climate change impacts, the weather-related costs suffered by some electric utilities in the recent past provides a glimpse of the potential magnitude of future climate change challenges for the electricity sector. For example, Hydro-Québec spent US$679 million in emergency measures and repairs after a destructive ice storm in 1998 to restore service to customers (Turcotte 2008).

Interviews with electricity utilities operating in different geographical areas and markets show a wide range of capital expenditures being allocated to manage climate change impacts. This brief survey of five utilities accounted a cumulative total of approximately US$ 1.5 billion worth of allocated and planned expenditures for climate change adaptation. This is a modest figure compared to the magnitude of the expected risks and opportunities they aim to address, such as repair costs following a single extreme weather event, loss of hydropower output due to reduced inflows or reduced cooling water availability. Climate change damage cost estimates for the power sector could reach US$ 1.9 billion per year globally for the 2010–2050 time period, provided the world remains on a 2 °C global average warming trajectory by 2050 (World Bank 2010). As such, it is likely that the benefit to cost ratio of climate change adaptation investments in the electricity sector is attractive. This suggests that there could be a strong business case to scale up such investments in the electricity sector.

Box 4. National Grid

> National Grid owns and operates the high voltage electricity transmission system in England, Wales and operates the power transmission system in Scotland. It also owns the UK's gas transmission network and four of the eight gas distribution networks. In addition National Grid owns and operates electricity assets in New England and New York.
>
> National Grid is at a "very advanced" stage of embedding climate change adaptation into its systems and everyday operations. Responsibility for climate change risk management is distributed throughout the organization.
>
> Following the introduction of the UK Climate Change Act 2008 and a directive to report in March 2010, National Grid was required by law to present a climate change adaptation report to the Department for Energy, Flood, and Rural Affairs. The National Grid Electricity Transmission Climate Change Adaptation Report is the culmination of the first phase of adaptation reporting (National Grid Electricity Transmission plc 2010). National Grid also produced a separate report for its gas businesses in the UK.
>
> In its Climate Change Adaptation Report, National Grid presents the result of its enterprise-wide climate risk assessment and adaptation identification work (Fig. 3). The extreme scenarios of the official UK Climate

Projections of 2009 (UKCIP09) were used to allow for a "worst case scenario," in association with specific electricity transmission characteristics. The report suggested that National Grid's assets and procedures are generally "resilient to climate change that is projected to occur" up to 2080. Where they exist, risks are localized and do not threaten loss of supply on a large scale. See Table 6 for a summary of climate impact and risk studies carried out by National Grid.

National Grid's risk process has found that there is currently little justification to support adjusting network or asset design standards except for the areas of flooding, and potentially the thermal ratings of equipment and apparatus (see below for more details on flood risk and thermal ratings), although factors other than climate change dominate the latter.

In order to address climate change, National Grid's risk register has been updated in response to specific risks. Actions with associated timelines have been formulated to address the identified risks. It acknowledges that climate risk management must be flexible to accommodate new information when it comes to light. An example may be the development of more accurate climate change projections for the UK. As such National Grid's risk process is "constantly reviewed and updated with appropriate actions and targets."

Flood resilience
National Grid considers that flooding is an important climate change issue that could cause considerable risks to its businesses. Following severe flood events in the UK during the summer of 2007, an industry Engineering Technical Report (ETR 138) was developed setting out a common approach to the assessment of flood risk. The task group that produced ETR 138 was made up of representatives from networks companies (including National Grid), the UK Department of Energy and Climate Change (DECC), the UK Office of Gas and Electricity Markets (OFGEM), the UK Environment Agency (EA), the Scottish Environment Protection Agency (SEPA), the UK Meteorological Office (Met Office), and the Pitt Review Team. Since then, power transmission companies have agreed to protect the grid and primary substations against flooding by 2022. As part of this process, flood risk assessments explicitly integrate the impact of climate change (i.e., changes in precipitation regimes and sea level rise) on the delineation of flood zones.

National Grid has begun flood mitigation work at all its substations at risk of a 1:100 year fluvial or tidal flood event taking into consideration projected climate change. Until work to defend sites is complete, National Grid has emergency plans to utilize a 1.7 km mobile flood defense system, which can be deployed at short notice. In addition, new substation designs take into account projected flood risks and include design features such as placing critical plant and equipment in elevated positions, for example on "stilts." The costs of flood management schemes will be highlighted as part of

Table 6 Ongoing and past National Grid climate change impact studies

Project title	External body	Energy participants
Vegetation management	ADAS	National grid, EDF, SP, ENW, CN
Pluvial flood risk modeling	ADAS	CN
Future network resilience	Met office	ENA
Dynamic ratings project	Met office	CN
EP1/2 impact of climate change on the UK energy industry	Met office	ENA
Urban heat Island study	Birmingham university	CN
Earthing information systems	BGS and NSA	EDF, CN
Flooding risk reduction	Mott McDonald	National grid
Investigation to network resilience to weather events	EA, Met office	ENA
Reappraisal of seasons and temperature thresholds for the power rating of electrical plant—a pilot study considering transformers only 2006	Met office/Southampton dielectric consultants ltd	National grid
Reappraisal of seasons and temperature thresholds for the power rating of electrical plant—a pilot study considering transformers only 2007	Met Office/Southampton dielectric consultants ltd	National grid
Reappraisal of seasons and temperature thresholds for the power rating of electrical plant—a pilot study additional work 2008	Met office/Southampton dielectric consultants ltd	National grid
Flooding risk and severe weather mitigation demountable flood barrier facilitating work	N/A	National grid
Flood risk assessment	N/A	National grid
Flood risk mitigation works 1:100 risk sites	N/A	National grid
Flood risk mitigation studies 1:200 risk sites	N/A	National grid
Flood risk mitigation studies 1:1000 risk sites	N/A	National grid
Flood risk mitigation, towers and erosion studies	N/A	National grid

Source National grid

upcoming customer tariff negotiations with Ofgem (the UK energy sector regulator).

Transmission and distribution equipment ratings
The current thermal rating of electrical equipment is dependent on operating ambient temperature. It defines the maximum electrical current which can be passed safely without overheating (potentially leading to sagging of lines and breaching of clearance limits). If ambient temperature increases, the maximum current rating of overhead lines, cables, transformers, and switchgear is reduced. This restricts the transmission capacity of an electricity system.

Table 7 shows the range of percentage de-ratings across the UK for typical transmission and distribution line types based on UKCP09 projections (note that National Grid only has responsibility for the transmission system). The maximum percentage de-rating for transmission overhead lines in the UK is

Climate Risk Management Approaches in the Electricity Sector 49

Table 7 Reduction in asset capacity as a result of projected changes in temperature

EP2—Typical Reduction in asset capacity for high emissions at 90 % Probability Level

Equipment	UKCP09 Period 2070–2099 (%)
Overhead lines	3
Underground cables	5
Transformers	5

Source National grid

only 3 %, which is not expected to considerably affect operating costs and tariffs. However, in places where equipment is already operating close to its design range, for example in areas that currently experience extremely hot temperatures, this could cause non-negligible reductions in transmission capacity. Further, in developing countries where supply is already insufficient, a 3 % reduction in current rating could be significant.

In theory, reduced current rating could justify to re-conductoring some overhead lines, but, in practice, other concerns have priority (e.g., satisfying customer demand). For instance, growth of electricity demand in the UK is anticipated to be 0.2 % yearly until 2016–2017 (1.4 % in the high growth scenario). The associated required transmission network upgrade exceeds by far the improvements required to accommodate reduced current rating (Table 7).

National Grid suggests that "additional work is needed to study the potential impact of reduced ratings in order to ascertain potential effects on the system and associated costs." As de-rating is a function of demand and peak temperatures, National Grid is also investigating the use of real-time rating monitoring and management to increase the capacity of overhead lines and reduce the need for reconductoring.

4 Current Focus of Adaptation Efforts and Gaps

Electricity sector climate change research and adaptation efforts seem to have concentrated thus far on a handful of issues (see Tables 8 and 9). Further, there are a number of adaptation responses which could be promoted in the electricity sector at no or low additional cost. These are discussed below.

4.1 Hydropower Output

Generation of hydroelectricity is very vulnerable to climate change, because it relies directly on the climatically sensitive hydrological cycle.

Table 8 Electricity sector risks comprehensively assessed by the electricity industry

Electricity sector value chain	Climate risks	Industry examples of risk assessment
Resource endowment		
Hydropower	Runoff	Hydro-Québec, BC Hydro, Hydro-Tasmania
Wind power	Wind field characteristics, changes in wind resource	BC Hydro
Solar power	Atmospheric transmissivity	
Wave and tidal energy	Ocean climate	
Supply		
Thermal power plants	Generation cycle efficiency, cooling water availability, increased frequency of extreme events, sea level rise	E.ON, Eskom, CLP Holdings
Hydropower	Water inflows and seasonality	Hydro-Québec, BC Hydro, Hydro-Tasmania
Wind power	Alteration in wind speed frequency distribution	
Solar power	Reduced solar cell efficiency	
Transmission and distribution	Increased frequency of extreme events, sea level rise	National Grid, Hydro-Québec, ActewAGL, BC Hydro, Western Power Distribution
Demand	Increased demand for indoor cooling, reduced heating requirements	Hydro-Québec, Electricité de France, Canadian Independent Electricity Operator
Support or connected infrastructure, and local communities	Increased frequency of extreme events, sea level rise	Eskom, Entergy

Source authors, and adapted from Ebinger et al. 2011

Table 9 Examples of adaptation responses that have been implemented by surveyed electricity utilities

Electricity sector value chain	Adaptation responses implemented by electricity utilities or regulators
Supply	
Thermal power plants	Eskom, E.ON, CLP Holdings, Entergy, EDF, BG Group
Hydropower	Hydro-Québec, Hydro-Tasmania
Wind power	
Solar power	
Transmission and distribution	Hydro-Québec, UK National Grid, ActewAGL, BC Hydro, western power distribution
Demand	Hydro-Québec
Support or connected infrastructure, and local communities	Entergy

Source authors

As mentioned above, the hydropower sector is the focus of considerable research and adaptation investments. However, reliable methods and tools to appraise and plan for future changes in water inflows due to short- and long-term climate variations are still being developed. Efforts are needed to make sure such work is of value value to utilities in developing countries, where observed hydrometeorological data is often limited.

4.2 Transmission and Distribution Integrity

Extreme weather events—such as snow or ice storms, high winds, and flooding—are widely recognized as key risks for power transmission and distribution. For example, flooding can cause electricity supply interruptions, downtime, and serious infrastructure damage, by wetting electrical equipment in substations, and preventing staff from accessing equipment for maintenance or repair (Grynbaum 2011; National Grid Electricity Transmission plc. 2010; Williamson et al. 2009).

Work is underway in these areas to improve knowledge, and the industry is planning for more extreme weather by adopting stronger design standards, and improving maintenance, monitoring and emergency plans.

However, considerable uncertainties persist around the likelihood and severity of these future extreme hazards. This is partly due to the lack of reliable observed data at short timescales for these variables, and the limited capacity of climate and hydrological models to accurately simulate extreme weather events.

4.3 Efficiency Losses

Higher ambient temperatures and changes in climate variables that contribute to cool electrical equipment (wind, rainfall, and cloudiness) could reduce the current rating of electrical lines and other equipment, as explained in Sect. 3. Technical solutions exist to manage this risk: for example, replacing old equipment with equipment that has a higher thermal rating, or monitoring real-time data to estimate current rating more accurately.[11]

While this poses a challenge to operating conditions, well-designed systems will most likely cope with such efficiency losses. Yet, current business and regulatory environments for electricity utilities impose new demands and constraints on the development, operation, and maintenance of electricity systems. In this context, even small efficiency losses could affect overall power system efficiency objectives. Furthermore, in the case of systems operating close to their design

---

[11] Similarly, higher ambient temperatures will reduce the heat rate and power output of natural gas-based generating units. See Ebinger and Vergara 2011.

ranges, for example in areas that currently experience extremely hot temperatures, higher temperatures could jeopardize critical design thresholds and lead to unacceptable efficiency ratios.

4.4 Demand Load Forecasts

Short-term electricity demand depends on the time of day and weather conditions. Research indicates that climate change will also have an influence on long-term demand (Wilbanks et al. 2007; ESPON 2010). As such, rising temperatures should be a consideration in demand load forecasting and long-term investment plans.[12] The technical summit organized by the Electric Power Research Institute, the North American Electric Reliability Corporation, and the Power System Engineering Research Centre on this issue can be taken as evidence that this is seen as an industry-wide risk (EPRI et al. 2008).

A few utilities have started working on this, as shown in Sects. 2 and 3, but more collaboration between the industry and hydrometeorological institutes will be needed to agree on ways to revise long-term forecasting methods and take into account the warming trend and its effects on baseline and peak power demand loads.

4.5 Gaps in Research on Climate Risks

Table 8 shows that most electricity utilities have focused their attention on a handful of risk issues and that there are a number of risks which have not yet been comprehensively assessed by the industry.

For instance, efforts to understand how a changing climate will affect renewable electricity generation have largely concentrated on hydropower. This is not surprising considering that hydropower represented more than 80 % of the total installed capacity for renewable electricity worldwide in 2008. However, non-hydro renewable electricity has increased considerably in recent years, and it is expected that this trend will continue into the future (International Energy Agency (IEA) 2011).

Limited attention has been paid to indirect climate change risks, which arise through impacts on support or connected infrastructure (e.g., transport networks) or impacts on local communities. For example, disruptions suffered by customers during extreme weather events lead to reduced demand and sales for electricity

---

[12] Due to warming, less heating will be needed for industrial, commercial, and residential buildings and cooling demand will increase, though this will vary by region and season. However, overall net energy demand is influenced for the most part by the economy and the structure of the energy industry. See Wilbanks et al. 2007.

utilities. CLP Holdings and Entergy have recognized that indirect risks could, in some cases, be more important to utilities than direct climate change effects on assets or operations.

Box 5. E.ON

> E.ON, a global electricity supply company, has spent resources to understand and manage the future impacts of climate change on its operations for almost 10 years. Key activities taken to date by E.ON to manage climate risks in the United Kingdom are described in Table 10 below.
>
> *E.ON's assessment of climate change impacts on its UK generation business*
> E.ON designed a consequence/likelihood risk assessment process to determine the overall degree of risk from climate change at each of it sites. Over 150 individual risks were identified during the risk assessment process. These were used to generate a list of key climate change impacts. Figure 4 plots the key current and future climate change risk onto E.ON's consequence versus likelihood matrix.
>
> *Conclusions*
> Overall, climate change represents a low risk to E.ON's power generation business, for a number of key reasons:
>
> - A relatively small change in climate is projected during the lifetime of existing assets;
> - The diverse design and geographical locations of its power station fleet reduces the overall risk; and
> - The inherent flexibility of each station enables short-term responses to climatic pressures.
>
> The most significant risks identified relate to drought and high ambient temperature. E.ON's climate change adaptation plan contains a number of key actions, to:
>
> - Reduce uncertainty in future drought risk assessment, site flood risk assessments, and the interaction between the E.ON climate change plan and the plans of its stakeholders; and
> - Improve internal management systems, by (1) developing and regularly updating a climate change projection fact sheet; (2) assessing E.ON's consideration of climate change adaptation during the development of new power stations; and (3) enhancing E.ON's risk-based asset management framework to incorporate ongoing assessment and monitoring of climate change risks.

Table 10 Timeline of climate change adaptation research undertaken by E.ON

Study details	Utilization for E.ON	Date
Climate change impacts tracking activities	Informing E.ON UK of developments	2002–2007
Analysis of data in UK climate impacts programme (UKCIP)02	Informing E.ON UK of developments	2002
Participation in scoping study (known as EP1) on the impacts of climate change on the UK energy industry (joint venture with met office)	Highlights significance of climate change impacts for electricity generation sector	2006
Analysis of EP1 conclusions	Initial investigation of EP1 results for E.ON UK generation	2006
Assessment of external research programme building knowledge for a changing climate (BKCC): impacts of climate change on the built environment	Identifying relevance to E.ON UK	2007
The impact of climate change on the UK energy industry (joint venture) with Met Office—EP2 participation and analysis	Sector-wide identification of general impacts of climate change	2008
Climate change risks to generation plant—identification of main issues	Scoping study for E.ON UK generation	2008
The impacts of climate change on thermal power stations: a user requirement study. A study in collaboration with Met Office	Climate variable data source analysis	2008
Developments in CCA policy	Assessment of UK CCA policy and outline of E.ON UK CCA plan	2009
Analysis of UKCIP09 UK climate projections accessibility of data—review of usefulness to E.ON UK	Assessment of data pool; basic processing tools	2009
Project to produce methodology for climate change impact assessment for E.ON UK generation assets	To enable CCA report for E.ON UK generation to be produced to meet requirements placed on reporting authorities	2009–2011

Source E.ON

Climate Risk Management Approaches in the Electricity Sector 55

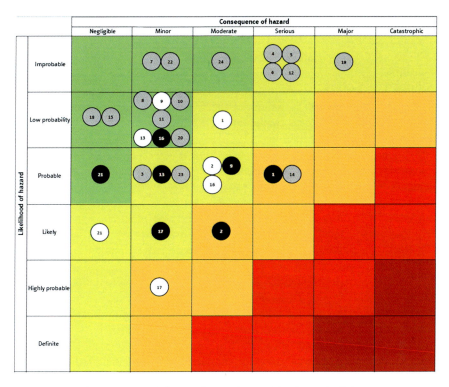

Fig. 4 Matrix of key current and future climate change risks. *Source* E.ON

Key
○ Current risk
● Future risk
◐ Current and future risk (where the future risk is not predicted to move into a different risk band)

1 Low river flow impact on station cooling/operation
2 Low river flow impact on compliance
3 Restricted supply of towns water
4 Extreme high river water levels
5 Coastal flooding
6 Flooding within site boundary
7 Impact on oil interceptors
8 Debris at water inlet
9 Impact on water quality
10 Impact on critical commodity access
11 Impact on staff access
12 High ambient temperature causing station trip
13 High air/water temperature impact on compliance
14 High temperature impact on performance
15 High temperature impact on occupational health
16 Freezing of water-containing equipment
17 Anti-icing impact on performance
18 Low temperature impact upon compliance
19 Impact on Operator safety
20 Impact of access of critical commodities
21 Impact on access of staff
22 Lightning
23 Meteorological conditions leading to cooling tower visible plume grounding
24 Subsidence/landslide

4.6 Gaps in Adaptation Action

As shown in Table 9, adaptation responses in the electricity sector remain patchy. More effort has been made in managing climate risks for transmission and distribution than for generation or demand. This is probably due to the fact that weather risk management is written in the DNA of the transmission and distribution industry, and climate change adaptation measures have often been taken following an extreme weather hazard.

Surprisingly, despite considerable industry research on climate change risks for hydropower output and demand loads, there is little evidence of technological adaptation responses in these areas, except for Hydro-Tasmania and Hydro-Québec. This is perhaps explained by the fact that water variability is an intrinsic component of hydropower equipment design and operation, and by the fact that there may be too much uncertainty to justify costly capital investments. This is why electricity producers, such as Hydro-Québec, prefer adapting their environmental impact management systems and operating rules.

Investments in technological adaptation measures remain limited to a few examples (for example, Dynamic Thermal Rating Systems, dry cooling and stronger design standards for transmission hardware), though in many cases they are primarily justified by considerations other than climate risks.

Electric utilities that are ahead of their peers on their "climate change adaptation journey" have started adopting climate change adaptation strategies and disclosing their actions to investors and stakeholders, as is the case for Eskom (see Box 3), E.ON (see Box 5), National Grid (see Box 4), and BG Group.

5 Lessons Learnt and Ways Forward

This stock-taking exercise helps point toward some research and adaptation gaps that would need to be filled to improve the climate change of electricity systems.

5.1 Quality and Tailored Climate Data and Information

Electricity organizations need accessible and high-quality data and information on observed and future climate conditions that are tailored to their needs, so they can plan ahead effectively. Data is required in a range of:

- Timescales, from short-term data for better management of existing supply/demand balance, through to data several years or decades ahead for planning and designing new energy assets;
- Spatial resolutions, from site specific to region- and country-wide; and

- Statistical variables other than averages (e.g., maximum consecutive days with no rain) and derived variables (e.g., Heating Degree Days), which are not directly given by climate models.

The current problem is that hydroclimate observations and future projections are not immediately available in a format that is easily adapted to electricity sector decisions (Troccoli 2009). For instance, utilities may require rainfall information at a high spatial resolution (e.g., a couple of square kilometers), and on short timescales (e.g., 12 h or daily), to assess future flood risk, but this is not immediately available from climate model projections. Furthermore, there is high model uncertainty about future changes in specific variables that are critical to electricity sector decision makers, such as rainfall, runoff, and wind. This leads to a "wait-and-see" attitude among most utilities which constrains climate risk management action.

It appears increasingly important for the electricity sector to work closely with hydrometeorological offices and research institutions to understand what data are available, and to identify gaps. The electricity sector can lobby government to fill these data gaps, in order to respond to industry needs. An example of successful collaboration on data development is provided by Hydro-Québec and Ouranos in Canada.

5.2 Operational Information on Impacts, Risks and Adaptation Strategies

A lot of the information available on climate change impacts and adaptation is often too *high-level* to be applied *operationally* within the electricity sector. For example, there are a number of resources explaining potential industry impacts and adaptation measures, without assessing the specific risk for the industry or explaining the methods, technological innovations, or cost-benefit ratios of different adaptation solutions. They are useful for establishing a holistic understanding of the challenges for the industry at a global, regional, or country level, and for building the case for climate change adaptation in the industry. However, examples of *applied* work influencing changes in planning, risk management, design, or operations, are limited.

A few utilities are doing work to assess a number of climate change risks and to identify cost-efficient adaptation measures. However, considerable uncertainties remain on the future likelihood and severity of different climate-related impacts. There is no clear view on *which* impacts will constitute material risks for power generation, transmission, and distribution, *when* these risks are likely to be felt, and *where* electricity assets or operations will most exposed.

Furthermore, findings from preliminary research efforts cannot easily be transferred to other utilities or locations. This is because the way a changing climate affects an electricity system will depend on a number of factors:

- The electricity system characteristics (e.g., asset design standards and operating rules);
- The exposure to climate and hydrometeorological variables and hazards, which depends on location (e.g., coastal, inland, by a river or a lake, etc.); and
- The level of adaptive capacity of the electricity sector concerned.

To enable the electricity sector to manage this complex issue, what is needed is significant collaboration across stakeholders to strengthen:

- Local climate data and information;
- Solid, but pragmatic, methods for assessing climate change impacts and risks and take advantage of opportunities in the face of uncertainty; and
- Technological, behavioral, and institutional good practice to manage risks and take advantage of opportunities despite uncertainty.

5.3 Favorable Environment for Adaptation Responses Beyond "Business-As-Usual"

Very few climate change adaptation measures are totally new. The majority of today's electricity sector adaptation responses are simply good practice risk management measures undertaken through a lens which considers how the climate is changing.

Most adaptation responses in the electricity sector are primarily motivated by factors other than climate change. In some cases, utilities recognize that these measures have benefits in terms of climate change resilience. This is the case for Eskom's investments in dry cooling technologies for thermoelectric plants, which the company began in the mid-1980s as a response to water scarcity issues. However, in many cases climate risk management measures are adopted as part of "normal" business and it is difficult to single these out as adaptation. For example, actions that improve electricity production, transmission, distribution, and end-use efficiency might also help to manage climate risks, such as reduced output, increased asset downtime, or higher supply disruptions.

With legislation and regulation on adaptation still in its infancy, climate change adaptation is considered optional at best in the electricity sector.[13] To incentivize electricity regulators and utilities in developing countries to go beyond "business-as-usual" and adopt climate change adaptation measures, there needs to be a favorable environment for adaptation, which includes the following elements:

---

[13] In a few examples, energy regulators have requested more in-depth analysis of climate change impacts as part of environmental assessment obligations. See for example recommendation 39 in Joint Review Panel Environmental Assessment Report—Darlington New Nuclear Power Plant Project. 2011. ISBN: 978-1-100-19116-4.

- Developing standards, regulations, and guidance—There are no obligations, standards, or guidelines in developing countries to manage climate change risks in the electricity sector, nor indeed in most developed countries. This is often a justification for inaction put forward by utilities. Governments, international institutions, and professional bodies (e.g., electricity associations) have roles to play in developing standards, regulations, and guidance which are favorable to adaptation, and can be applied in developing countries. For example, standards and guidance should be applicable in situations when there is a dearth of hydrometeorological data, and they should promote pragmatic approaches to climate risk management.
- Developing sources of finance—The lack of financial support from government and the impossibility of passing costs onto customers explain in many cases the lack of climate change adaptation action in the electricity sector. Developing sources of finance for research and development, or implementation of adaptation measures, by electric utilities in developing countries is a critical condition to improve climate change resilience. This can be done nationally, through the use of differentiated tariffs.[14] For example, the Ontario electricity regulator in Canada has approved a charge on customer bills for government-owned companies servicing remote companies and facing higher fuel costs due to reduced fuel transport on ice roads, and associated increase in air freight as a result of warming.[15]
- Integrating the electricity sector within national adaptation strategies—Finally, developing countries have an opportunity to include actions addressing electricity sector vulnerabilities within their national climate change adaptation strategies.[16] As international climate change adaptation financing for developing countries increases, the electricity sector should work with governments to develop adaptation measures that could be funded.

6 Appendix

Examples of capital expenditures for climate change adaptation investments by utilities (nominal US$)

---

[14] For further examples of possible funding arrangements see Troccoli, A. 2009. Weather and climate risk management for the energy sector: workshop recommendations. In: Troccoli, A. (ed.) 2009. *Management of Weather and Climate Risk in the Energy Industry. Proceedings of the NATO Advanced Research Workshop on Weather/Climate Risk Management for the Energy Sector Santa Maria di Leuca, Italy, 6-10 October 2008,* Springer.

[15] Peter Fraser, Ontario Energy Board, authors' communication, 22/10/2011.

[16] Presently, few developing countries have included the energy sector within their National Adaptation Plans of Action (NAPAs). A recent analysis found that only 3.7 % of 455 adaptation projects proposed by these NAPAs were related to the energy sector.

Adaptation measures	Climate risks addressed	Cost (million US$)	Timing	Assumptions and references
Budget support to research organization on climate change impacts and adaptation	Overall climate risks and opportunities	12	2002–today	– Financial contribution of CA$1 million and in-kind contribution of 5 full-time equivalent researchers and engineers[a] – Average annual salary of engineers working in universities in Québec: CA$ 66,655 (Institut de la Statistique de Québec 2009) – Average annual salary of technical staff, equivalent to researchers, working in universities in Québec: CA$ 45,653 (Institut de la Statistique de Québec 2009) – Exchange rate of 1 CA$ = 0.97 US$
System wide increase in maximum ice and wind loads, and install special pylons at predefined intervals to avoid cascades of falling pylons	Future increased extreme weather (e.g., wind or ice storms)	875	1999–2007	– Industry sources (Turcotte 2008) – Exchange rate of CA$ 1 = US$ 0.97
Build power grid interconnection with regional partners and invest in de-icing equipment	Future increased extreme weather (e.g., ice storms)	195	2008–2011	
Build new small hydropower plants	Future loss of inflows and hydropower output	125	Planned up to the 2020s	– Capital cost (CAPEX) figures from direct industry sources(Brewster et al. 2009) – Exchange rate of 1 AU$ = 1 US$
Increase storage capacity in existing plants	Future loss of inflows and hydropower output	100	Planned up to the 2020s	
Build new dams	Future loss of inflows and hydropower output	100	Planned up to the 2020s	
Raise height of existing dams	Future loss of inflows and hydropower output	50	Planned up to the 2020s	
Improve turbine runners	Future loss of inflows and hydropower output	17	Planned up to the 2020s	
Undertake major redevelopments on hydropower infrastructure	Future loss of inflows and hydropower output	15	Planned up to the 2020s	
Build new catchment water diversions	Future loss of inflows and hydropower output	11	Planned up to the 2020s	
Upgrade existing water canals	Future loss of inflows and hydropower output	10	Planned up to the 2020s	

(continued)

(continued)

Adaptation measures	Climate risks addressed	Cost (million US$)	Timing	Assumptions and references
Increase dry cooling capacity for thermoelectric generation assets	Reduced cooling water availability	500	1970–2000	– Overall CAPEX for dry cooling is 170 % higher than CAPEX for conventional wet cooling (Lennon 2011) – Average CAPEX to install an entire wet cooling system for a 550 MW coal plant: US$ 37.3 million – Average additional CAPEX to install an entire dry cooling system for a 550 MW coal plant: US$ 26.1 million – Eskom installed 10,500 MW of thermoelectric generation capacity between 1970 and 2000
Install cooling water diversion pipes in one existing power plant	Future increased risk of saline intrusion	0.035	Completed	– Equipment costs could amount to US$ 1.1 million for additional water pipes fitted to the water feed system – Retrofitting costs are estimated to be equivalent to standard industry maintenance costs, namely 3 % of original CAPEX
Raise floor level of buildings housing critical infrastructure in one existing power plant	Future increased flood risk	0.089	Completed	– Two scenarios are considered: one whereby only the floor level where generating motors are located is raised, and one whereby a larger surface area containing motors, pumps and other critical equipment is raised – No structural works are required to raise floor levels – CAPEX is estimated to be US$ 2 and 3.9 million in the two scenarios described above respectively – Retrofitting costs are estimated to be equivalent to standard industry maintenance costs, namely 3 % of original CAPEX
Inspect cooling tower at risk of erosion or landslide	Increased risk of land movement in coastal areas	1.33	N/A	– Average CAPEX for inspecting thermoelectric cooling towers, raising floor levels, and reinforcing the towers' bases as needed is US$ 4.4 million – The power plant concerned has 5 cooling towers – Retrofitting costs are estimated to be equivalent to standard industry maintenance costs, namely 3 % of original CAPEX
Mobile flood defense system	Future increased flood risk	1.25	Completed	– Average CAPEX for a 1-mile long mobile flood defense system is between US$ 1 and 1.5 million[b]

Source authors, based on utilities' interviews
[a] Authors' communication with René Roy, Hydro-Québec (24/10/2011).
[b] Authors' communications with five US vendors (20/12/2011).

References

AECOM/Purdon Associates (2009) East lake electrical infrastructure: draft environmental impact statement, vol 1. Canberra, Australia

Brewster M, Ling FLN, Connarty M (2009) Climate change response from a renewable electricity business. Int J HydroPower Dams 16:59–62

Carroll AL, Taylor SW, Regniere J, Safranyik L (2003) Effects of climate change on range expansion by the mountain pine beetle in British Columbia. In: Shore TL, Brooks JE, Stone JE (eds) Proceedings of mountain pine beetle symposium: challenges and solutions, pp 223–232. Information Report BC-X-399, Natural Resources Canada, Victoria, 2003

CLP (2011) CLP adaptation strategy

CLP's Climate Vision 2050: Our Manifesto on Climate Change (2007)

Comité Provincial d'Examen (COMEX) (2006) Project hydroélectrique Eastmain-1A et derivation Rupert. Rapport du comité provincial d'examen à l'administrateur du chapitre 22 de la Convention de la Baie-James et du Nord québécois. Ministère du développement durable, environnement et parcs, Québec

Darlington New Nuclear Power Plant Project (2011) See recommendation 39 in joint review panel environmental assessment report. ISBN: 978-1-100-19116-4

Dubus L (2009) Impacts of weather, climate and climate change on the electricity sector. Energy Week 2009, The World Bank Group

Ebinger J, Vergara W (2011) Climate impacts on energy systems: key issues for energy sector adaptation. World Bank, Washington, DC;European Observation Network, Territorial Development and Cohesion (ESPON) (2010) Discussion paper: impacts of climate change on regional energy systems

Ebinger et al (2011), ibid; and World Bank. 2010. Africa's development in a changing climate. Key policy advice from the World Development Report (2010) World Bank. Washington, DC

Electric Power Research Institute (EPRI), North American Electric Reliability Corporation (NERC) and Power System Engineering Research Center (PSERC) (2008) Proceedings of the joint technical summit on reliability impacts of extreme weather and climate change. EPRI, Palo Alto, CA, NERC, Princeton, NJ, and PSERC, Tempe, AZ: 2008. 1016095

Eskom (2009) ibid

Grynbaum MM (2011) Power substations in lower Manhattan are vulnerable to flooding. New York Times. 27 Aug 2011; and National Grid Electricity Transmission plc. 2010. Climate Change Adaptation Report; and Williamson, et al., 2010, ibid.

Institut de la Statistique de Québec (2009) Résultats de l'enquête sur la remuneration globale au Québec., Collecte

International Energy Agency (IEA) (2011) Climate and electricity annual 2011. IEA, Paris, p 90. ISBN: 978-92-64-11154-7

International Institute for Applied Systems Analysis (IIASA) (2010) Natural catastrophes and developing countries. IIASA, Vienna

Janos T, Gurney JH (2008) Impacts of climate change on the planning, operation and asset management of high voltage transmission system. Presentation to the CIGRE Canada (Canadian national committee of the international council on large electric systems) 2008 conference on power systems, Winnipeg, 19–21 Oct 2008

Jenkins GJ, Perry MC, Prior MJ (2007) The climate of the United Kingdom and recent trends. Met Office Hadley Centre, Exeter, UK

Knutson T, McBride JL, Chan J, Emanuel K, Holland G, Landsea C, Held I, Kossin JP, Srivastava AK, Sugi M (2010) Tropical cyclones and climate change. Nat Geosci 3:157–163

Lennon S (2010) Advances in dry cooling deployed at South Africa power stations. Presentation at the electric power research institute 2011 summer seminar, 1 Aug 2011

Lennon S (2011) Advances in dry cooling deployed at South Africa power stations. 2011 Electric Power Research Institute (EPRI) Summer Seminar, 1–2 Aug 2011

Létard V, Flandre H, Lepeltier S (2004) La France et les Français face à la canicule: les leçons d'une crise. Report No. 195 (2003–2004) to the Sénat, Government of France, pp 391

Linnenluecke, Stathakis, Griffiths (2011) Firm relocation as adaptive response to climate change and weather extremes. Glob Environ Change 21(1):123–133

MacKinsey (2009) Promoting Energy Efficiency in the Developing World. McKinsey Quarterly, February, New York

Mechler R, Hochrainer S, Pflug G, Williges K, Lotsch A (2009) Assessing financial vulnerability to climate-related natural hazards. Background paper for the WDR 2010

Minville M, Brissette F, Krau F, Leconte R (2009) Adaptation to climate change in the management of a Canadian water-resources system exploited for hydropower

Mirza M (2011) Climate change and the Canadian energy sector: A background paper

Musilek P, Arnold D, Lozowski EP (2009) An ice accretion forecasting system (IAFS) for power transmission lines using numerical weather prediction. SOLA 5:25–28

Nadeau Y (2008) Climate change and normals: planning issue initiative of Hydro-québec distribution. Presentation at the joint technical summit on reliability impacts of extreme weather and climate change. Electric Power Research Institute, North American Electric Reliability Corporation and Power System Engineering Research Center, Portland, 15 Oct 2008

National Grid Electricity Transmission plc. (2010) Climate change adaptation report

Ontario Government (2011) Climate ready: Ontario's Adaptation Strategy and Action Plan

Pather V (2004) Eskom and water. Proceedings of the 2004 water institute of Southern Africa Biennial conference, Cape Town, South Africa, 2–6 May 2004

Pytlak P, Musilek P, Lozowski E, Toth J (2011) Modelling precipitation cooling of overhead conductors. Electric Power Syst Res 81(12):2147–2154

Roy R, Pacher G, Roy L, Silver R (2008) Adaptive management for climate change in water resources planning and operation. Hydro Québec-IREQ, p 75

Silver R, Roy R (2010) Hydro-Québec's experience in adapting to climate change. Presentation for the international association for impact assessment special symposium on climate change and impact assessment, Washington, DC, 15–16 Nov 2010

South West Climate Change Impacts Partnership (2010) http://www.oursouthwest.com/climate/registry/100100-case-study-western-power-distribution.pdf

Toth J (2011) Risk/challenges for generation.Transmission and distribution infrastructure relative to climate change. BC Hydro, Vancouver, BC

Troccoli A (ed) (2009) Management of weather and climate risk in the energy industry. Proceedings of the NATO advanced research workshop on weather/climate risk management for the energy sector Santa Maria di Leuca, Springer, Italy, 6–10 Oct 2008

Turcotte C (2008) L'après-crise aura couté 2 milliards. *Le Devoir*. Available at http://www.ledevoir.com/environnement/actualites-sur-l-environnement/170771/l-apres-crise-aura-coute-deux-milliards. Accessed 25 Oct 2011

U.S. Department of Energy (2010) Hardening and Resiliency. U.S. Energy Industry Response to Recent Hurricane Seasons. U.S. Department of Energy, Washington, DC

Urban F, Mitchell T (2011) Climate change, disasters and electricity generation. Strengthening climate resilience discussion paper no 8, Institute of Development Studies, Brighton

Veit S (2009) Climate risk management for the energy sector in Africa: the role of the African Development Bank. In: Troccoli (ed.), 2009

Watson SJ (2006) Report on EPSRC project GR/18915/01: a generic process for assessing climate change impacts on the electricity supply industry and utilities (GENESIS)

Wilbanks TJ, Romero Lankao P, Bao M, Berkhout F, Cairncross S, Ceron J-P, Kapshe M, Muir-Wood R, Zapata-Marti R (2007) Industry, settlement and society. Climate change 2007: Impacts, adaptation and vulnerability. Contribution of working group II to the fourth assessment report of the intergovernmental panel on climate change, Parry ML, Canziani OF, Palutikof JP, van der Linden PJ, Hanson CE (eds), Cambridge University Press, Cambridge, pp 357–390

Williamson LE, Connor H, Moezzi M (2009) Climate proofing energy systems. Helio International
World Bank (2010) The costs to developing countries of adapting to climate change. New methods and estimates. The global report of the economics of adaptation to climate change study, consultation draft. The World Bank, Washington, DC

Climatic Changes: Looking Back, Looking Forward

Alberto Troccoli

Abstract Why do we need to be concerned about the role of meteorology in the energy system? How large a change in meteorological variables could have adverse effect on energy systems? These are the underlying questions explored in this chapter. In particular, we note that not only is the energy sector at risk from future climate changes, it is also at risk from current hydro-meteorological climate variability and change. It is important therefore to assess the climate observed in recent decades along with the changes we might expect in the future. By examining a selection of meteorological variables, this chapter exposes how climate has been, and will continue to be, variable on climatic timescales. The extent to which such variability (and extremes) could be modified under climate change, and therefore have an impact on the energy sector is discussed. Current understanding of pertinent changes in extreme weather events, and estimates of impacts on the energy sector given climate change are also summarised.

1 Introduction

In "A New Era for Energy and Meteorology" we have seen the growing influence meteorology is playing on energy systems. Such influence is displayed through a variety of weather events, some of which are so severe to have caused serious disruptions to energy systems, especially in the last few decades. So quite aside from risks due to expected potentially severe changes to the climate over the next decades, energy systems are already vulnerable to meteorological events, as also documented by the burgeoning literature on this issue (Troccoli et al. 2010; Ebinger and Vergara 2011; Johnston et al. 2012). Take for example the cases of

A. Troccoli (✉)
Commonwealth Scientific and Industrial Research Organisation (CSIRO), Highett, VIC, Australia
e-mail: Alberto.Troccoli@csiro.au

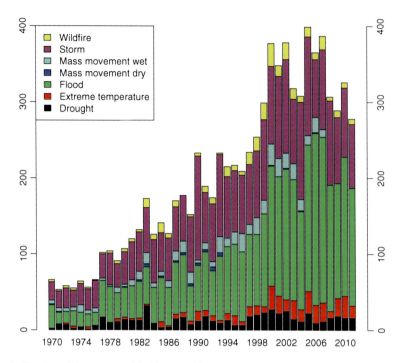

Fig. 1 Number of disasters worldwide caused by natural events for the period 1970–2011. The two main contributors are storms and floods. *Source* EM-DAT: The OFDA/CRED International Disaster Database—www.emdat.be, Université Catholique de Louvain, Brussels (Belgium)

the 2003 European heat wave or the 2010–2011 La Niña-related flooding in Australia which have caused widespread disruptions to the electrical supply in France in the first case and severe interruptions to coal mining operations in the second case.

Figure 1 shows a marked increase in weather-related severe events from the 1950s to the mid-2000s with a subsequent slight decline—due to storms, floods, extreme temperatures and alike—which have affected the population over the last 60 years. Although not all of these events are explicitly related to energy systems, it is fair to say that energy provision would have been affected in some way in most if not all of these events. But how are these increasing severe events related to actual changes in the climate? In other words, can we relate the main features of climatic changes observed in the recent past with such severe weather events? Also, looking ahead, how big are the projected changes in key climatic variables that could affect, and therefore increase the vulnerability of, energy systems?

This chapter addresses these questions by exploring the magnitude of changes both in retrospect (a recent 30-year period, 1981–2010) and in the future (projections over the 30-year period 2050–2079), for some of the meteorological variables more relevant to energy systems. A wide variety of meteorological factors have a strong influence on many aspects of the life cycle of energy systems

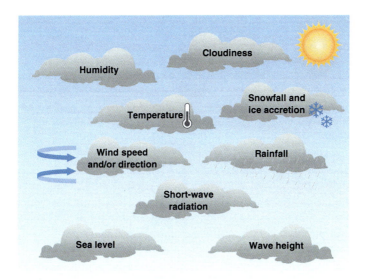

Fig. 2 Meteorological variables of relevance to the energy industry

such as exploration, extraction/production, transportation, refining, generation, transmission/delivery, disposal and demand. Thus, not only are the more obvious meteorological variables such as surface temperature and precipitation relevant to the energy sector, also other variables like humidity, mean sea level, wind and solar radiation should be considered, along with their statistical features like extremes.

In addition, derived quantities such as heating/cooling degree days, permafrost extent and sea ice cover play important roles. It should also be noted that while existing energy infrastructures (mostly fossil fuels in developing countries) are mainly affected by temperature and land/sea hydrology (e.g. changes in water temperature of rivers or in sea level), future infrastructures are likely to be impacted by a wider range of meteorological variables (Schaeffer et al. 2012). Figure 2 depicts a number of meteorological variables of relevance to energy systems. While an attempt is made here to touch on as many meteorological variables as possible, only a select number will be covered in greater detail.

It is apparent that by considering 30-year averages the focus of this chapter is mainly on climatic conditions, namely quantities that are especially relevant to the policy and planning of energy infrastructure (i.e. strategic decisions). As a consequence, specific severe weather events cannot be identified via the analysis presented here. However, by identifying significant meteorological changes on climatic timescales, they can provide the basis for more detailed analysis of specific types of events.

2 Methodology

An important baseline upon which to assess future climate changes is the observed climate over the recent past or the best available (global) representation of it. Direct observations, whenever available, certainly would provide the best assessment of the climate. However, on a global scale even surface temperature, the most observed meteorological variable, presents numerous gaps and even inconsistencies in the way it is measured.

A standard alternative to direct observations is the use of *reanalyses*. These provide meteorological variables on regular grids and are constructed by combining as many observations as possible with a dynamical model of the atmosphere (also referred to as Numerical Weather Prediction [NWP] model), which can also have an ocean component. In other words, reanalyses act as interpolators where observations are missing, subject to the strong constraints given by the laws of dynamics and thermodynamics of the atmospheric system. Thus, reanalyses, despite their limitations, offer the most effective instrument to analyse the recent climate on a global basis.

By contrast, climate models, being mostly data unconstrained, do not provide an adequate representation of the current climate. Instead they are invaluable tools to assess long-term changes in large-scale climate features.

Box 1—Predicting weather, simulating climate: the role of dynamical models

The most complete way to simulate weather and climate processes, be it for Numerical Weather Predictions (NWP), Seasonal Climate Forecasts, Climate Change Scenarios and Reanalyses, is by means of (coupled) general circulation models (GCMs). In such models the Earth system is subdivided into cells of sizes varying according to the purpose of the model, namely from O (10 km) in NWPs to O (100 km) for climate change scenarios. Dynamic and thermodynamic physical relationships are solved numerically, often on large supercomputers, for each cell as well as for the interactions amongst cells. GCMs calculate atmospheric, land and oceanic parameters, such as temperatures and winds, and their changes in time, at the surface and at various levels in the atmosphere, over the land and in the oceans.

Different components or attributes of the Earth system affect weather and climate in specific ways. By isolating certain aspects of the Earth system, predictions of weather and climate at different lead times become a more tractable problem. Generally speaking, physical processes can be divided into fast ones (e.g. atmospheric convection) and slow ones (e.g. circulation of the deep ocean), with an essentially continuous spectrum of processes in between. Given the presence of this wide spectrum, choices about which process is more relevant for a particular purpose have to be made. Thus, for instance, it would be of little use to run a complex sea ice component to produce forecasts for tomorrow's weather as the sea ice response is much longer than a day. Likewise, for a climate forecast several years hence, the precise details of today's weather are less relevant than for forecasts for the next few days/weeks. Thus, it is possible to distinguish between short to interannual range predictions for which initial conditions are essential (with diminishing importance with increasing lead time) and decadal projections for which boundary conditions (or radiative forcing effects) are dominant.

Although in principle a single system for all lead times would be desirable (and this is what some prediction centres are attempting to achieve), in practice predictions are made with built-for-purpose model configurations. However, as computers have become more powerful, enabling calculations to be run more quickly, increasing detail has been included and cross-fertilisation has been used often to improve models for these different targets.

In order to start a forecast from say a few hours ahead to about 1 year with any model, this needs to be given an initial condition (a.k.a. 'analysis'), i.e. fields that describe the current state of the relevant components of the Earth system. Creating an analysis is a complex and costly task on its own, involving collecting data from around the world (within a limited amount of time in the case of weather forecasting) and processing those data into the analysis through several stages, one of which is called 'assimilation'. Once the analysis is available the model can be run from it to produce a forecast. Through improvement in assimilation techniques and archiving of 'late' data, historical reanalyses are produced nowadays that can provide the most consistent and detailed pictures of past global climate available for recent decades, information of great value to the energy sector.

Despite marked advances in models and computer, the models are not, nor ever will be, perfect. Equally, analyses inevitably contain errors however carefully prepared. Both types of error feed into the predictions producing unavoidable inaccuracies. A way to alleviate these problems are: (i) to use historical statistics to provide forecast error estimates; (ii) to use an 'ensemble', i.e. several predictions each slightly different from all others; (iii) to combine several models using the so-called multi-model. Even with the latter two approaches, none of the, perhaps 50, ensemble members/multi-model realisations will contain the 'right' answer but the advantage of these multiple realisations is that they provide a probability distribution of forecasts. Taking an average across all ensemble members produces the optimal deterministic forecast, but use of this is never advisable without the additional information on the uncertainties revealed in the full ensemble.

The main features of the current climate which are relevant to the energy sector, including trends and extremes in variables such as wind and solar radiation, are explored here by using one of the most recent reanalyses, the so-called ERA-Interim (Dee et al. 2011), which is produced by the European Centre for Medium-

Range Weather Forecasts (ECMWF, http://data-portal.ecmwf.int/). It is standard practice to take 30-year averages to analyse climate characteristics. It could be argued that on climatic timescales a 30-year period is not particularly long; however, this is commensurable with the lifetime of most energy infrastructure. Here the 1981–2010 30-year period is considered. In addition, mean sea-level data from the TOPEX/Poseidon, Jason-1 and Jason-2 satellite missions, the first of which started in the early 1990s, have also been used. Due to their limited availability, the period considered for mean sea level is 1993–2011 (data obtained from http://www.cmar.csiro.au/sealevel/sl_data_cmar.html).

Box 2—What is a reanalysis?
Reanalysis is the result of the ingestion of as many meteorological observations as possible into a dynamical model aimed to provide a comprehensive record of how weather and climate are changing over time. The dynamical model basically acts as an interpolator to the meteorological observations, while respecting physical laws such as those of dynamics and thermodynamics. A reanalysis typically extends over several decades or longer, and covers the entire globe from the Earth's surface to well above the stratosphere. Reanalysis products are used extensively in climate research and services, including for monitoring and comparing current climate conditions with those of the past, identifying the causes of climate variations and change, and preparing climate predictions. Information derived from reanalyses is also being used increasingly in commercial and business applications in sectors such as energy, agriculture, water resources and insurance.

In principle, it would be preferable to use long-term direct observations to analyse climatic features for a specified location. However, and this is especially true for developing countries, these observations may not be available. Reanalyses thus become the principal tool available for assessing the statistics of climate in many parts of the globe. It should be noted that in regions where observations are scarce, different reanalyses are likely to produce large discrepancies.

The ERA-Interim reanalysis uses a December 2006 version of the ECMWF Integrated Forecast Model (IFS Cy31r2). It covers the period from 1 January 1979 to the present. The spectral resolution is T255 (about 80 km) and there are 60 vertical levels, with the model top at 0.1 hPa (about 64 km). The data assimilation is based on a 12-hourly four-dimensional variational analysis (4D-Var) with adaptive estimation of biases in satellite radiance data (VarBC). With some exceptions, ERA-Interim uses input observations prepared for the earlier ECMWF reanalysis, ERA-40, until 2002, and data from ECMWF's operational archive thereafter. The full description of the ERA-Interim system is given in Dee et al. (2011). More information on re-analyses can be found at: http://reanalyses.org/.

For future climate projections we use output from Coupled Model Intercomparison Project 3 (CMIP3) climate model runs, namely those used for the Intergovernmental Panel on Climate Change (IPCC) fourth assessment report (AR4, Meehl et al. 2007). Some 25 climate models form the CMIP3 dataset, but they are not all completely independent of each other (e.g. a model may be present with two similar versions). Thus, 15 models have been selected for this work (data obtained from https://esg.llnl.gov/), noting however that not all these models store all variable/period/scenario/etc. combinations. Meteorological fields from these 15 models have been used to compute the future climate in the 30-year period 2050–2079 as an anomaly with reference to the 1970–1999 period. Although the 2050–2079 period is beyond the current energy infrastructure planning, the advantage of analysing climate variables further ahead in time is that differences amongst scenario projections become more evident with longer lead times. It is worth pointing out that the accuracy of climate models is such that these should only be used for comparative studies (e.g. differences amongst periods), and not for absolute assessments, not even for past periods.

Box 3–Sources of uncertainty in IPCC/CMIP3 model results
In interpreting the climate model output it is important to be aware of their limitations. Despite climate models being the most complete approach available for making projections, models are particularly challenged when used to provide detailed regional and national projections. Continual improvements are however made to these models. Other factors, including the future concentrations of atmospheric greenhouse gases, provide further uncertainties in the projections.
Modelling of the Earth system—The complexity of the Earth system is such that uncertainty in climate change projections is unavoidable. Partly because of limited computational resources and partly because of our limited knowledge about the interaction amongst all the components of the Earth system (e.g. sea ice interaction with atmosphere and ocean), many approximations and short-cuts need to be made in order to be able to run climate change runs over long periods (O (10 yr) or more). One of the consequences is that regional details especially are not as accurate as global features and therefore it is important not to overemphasize small-scale signals. However, interpretation is also dependent on the variable considered: precipitation for instance is a more variable field than temperature and therefore statistics of the former are generally less significant (i.e. smaller signal-to-noise ratio) than the latter.
Greenhouse gas (GHG) emissions and concentrations—Changes in climate are dependent on future GHG concentrations. These are unknown, and will depend on human actions. As precise future emissions and consequent atmospheric gas concentrations are unknowable the IPCC uses a series of emissions scenarios in an attempt to bracket the likely range of reasonable possibilities, from non-curbing of the use of fossil fuels through to

progressive conversion to energy generation from non-carbon sources. Future climate projections from all models are critically dependent upon whichever emissions scenario is used, although as a rule of thumb all indicate larger changes in global and regional temperatures given higher emissions; similarly there is a tendency for projected total global rainfall to increase under higher emissions but, as mentioned, projected regional rainfall changes are more complex and cannot be so easily summarised.

The two emission scenarios used here have been selected to bracket the range used by the IPCC; these are 'typical' scenarios that provide a reasonable overview of all possibilities. To give an idea of the impact of these scenarios when combined with model uncertainties, the end-of-the-twenty-first century *globally-averaged* temperature ranges in the projections across about 5 °C. The two scenarios are named:

(a) SRESA2—a high future emissions scenario that results in a *best estimate* global temperature increase of over 3 °C by 2100.
(b) SRESB1—a relatively low future emissions scenario with a temperature increase of about 1.8 °C by 2100.

These are referred to sometimes as A2 and B1, respectively. Although runs for both scenarios are presented, the A2 results which depict the greatest simulated changes are the ones most closely fitting reality since these models were run.

In general terms, the frequency distribution of climate variables may be modified:

- Only in its mean value (first-order statistic, e.g. warmer temperatures).
- Only in its standard deviation (second-order statistic, an overall higher or lower variability).
- Only in its skewness (third-order statistic, e.g. longer upper tail due to increased heat waves).
- Only in any of the higher order statistics (e.g. kurtosis, fourth-order).
- As a combination of mean, standard deviation, skewness and higher order statistics (e.g. overall warmer but fewer extremes).

To assess the various variable/statistic combinations would require an extensive analysis. In this chapter, we therefore focus on some of the key indicators as to allow the reader to have an appreciation of the changes in select meteorological variables that have occurred in the recent past and those that are expected during the twenty-first century.

3 Historical Climate Relevant to Energy

In this section, we first analyse meteorological variables considered generally more relevant for the energy sector: near-surface temperature (taken at 2 m height), precipitation, the solar global irradiance (more commonly known as radiation) and wind speed (taken at 10 m height). Note that the 2 m and 10 m heights are determined by the meteorological data availability. We then assess another important climatic variable, mean sea-level height and touch on other more localised processes such as icing on cables and hail.

Global solar radiation, or simply solar radiation, is the sum of two components: direct beam or direct normal irradiance (DNI) and diffuse irradiance. This distinction is very important because Concentrating Solar Power (CSP) devices function solely when the DNI is non-zero. Photovoltaic (PV) devices, instead, respond to global radiation more generally, namely also when DNI is zero but the diffuse component is not. Normally the quality of DNI, when available, is inferior to that of global radiation. For this reason only the global radiation is considered here. In any case, PV devices are currently the dominant solar power technology.

Further, while 10 m (height) winds provide an indication of wind patterns and strength, current turbine hubs are considerably higher than 10 m (typically 80 m). Since wind increases exponentially from the surface upward, in a way that depends on the roughness of the surface and on the stability of the vertical thermal profile and is often not straightforward to quantify, the extrapolation from the 10 m wind to the hub height is not trivial. Hence higher level winds, at about 80 m, would really be needed but these are not readily available yet. In their absence, wind speeds on constant pressure surfaces, e.g. at 900 hPa, could be also considered as they would allow wind energy developers and operators to assess changes in wind patterns and speed that are less affected by surface processes, including anthropogenic ones such as land use change.

3.1 Historical Linear Trends of Select Meteorological Variables

The climate varies on all timescales and in different fashions, as discussed above. However, a simple and useful indicator of climatic changes—which responds to the question 'by how much is the mean distribution changing?'—is the linear trend. This statistic offers an immediate way to assess how stationary the climate is, even though it should also be noted that there is no guarantee that trends observed during the past will persist in the future. Indeed, trends are often sensitive to the period chosen. Further, linear trends are just approximations of changes that may actually be non-linear. Nonetheless, they often provide a valuable starting point for further analyses.

Table 1 Global linear trends for the 30-year period 1981–2010 for the ERA-Interim reanalysis

Linear trends	2 m temperature (°C 10 year^{-1})	Precipitation (mm day^{-1} 10 year^{-1})	Solar Radiation (W m^{-2} 10 year^{-1})	10 m wind (m s^{-1} 10 year^{-1})
Land + oceans	0.15 (1 %)	−0.030 (−1.0 %)	0.28 (0.1 %)	0.041 (0.7 %)
Land only	0.34 (3.8 %)	0.011 (0.5 %)	0.19 (0.1 %)	0.002 (0.1 %)
Ocean only	0.07 (0.4 %)	−0.047 (−1.5 %)	0.31 (0.2 %)	0.056 (0.8 %)

Trends are expressed in variable unit over a 10-year period (decade). In parenthesis the percentage changes (trends divided by their mean, see Table 2) are reported
Data source ECWMF

Table 2 Global averages for the 30-year period 1981–2010 for the ERA-Interim reanalysis

Averages	2 m temperature (°C)	Precipitation (mm day^{-1})	Solar radiation (W m^{-2})	10 m wind (m s^{-1})
Land + oceans	14.4	2.91	188	6.18
Land only	9.0	2.24	187	3.71
Ocean only	16.5	3.18	188	7.18

Data source ECWMF

In Table 1, the trend of four energy-relevant variables is shown, both in absolute terms and as a percentage of the average value of those variables (Table 2). The trend for near-surface (2 m) temperature (0.15 °C per decade increase globally) is just a confirmation that the ERA-Interim reproduces the well-documented global warming trend (e.g. Solomon et al. 2007 or the more recent Berkeley Earth Surface Temperature project, http://berkeleyearth.org/). Regionally, the positive trend is apparent over most of the globe as shown in Fig. 3, although statistical significance at the 95 % level (darker colours) over the 30-year period considered is displayed mostly over some of the Arctic and circumpolar current regions, and in several tropical areas. It can also be noted that trends over the land are mostly positive (also reflected by its global mean of 0.34 °C/decade, Table 1) and considerably higher than over the oceans (0.08 °C/decade).

The other variables shown in Fig. 3, precipitation, global solar radiation and 10 m wind speed, display some significant trends in the tropical regions. Although some of these trends may be linked to limitations in physical processes as simulated by reanalyses (as with the so-called tropical deep convection, with errors that would propagate to all these variables), it is unlikely that all these signals are spurious as they are present in other observational analyses (e.g. Trenberth et al. 2007).

On average, as shown in Table 1, all three variables have seen an increase in their values over land, although the percentage changes are relatively small (0.5 % or less over a decade). Changes are more pronounced over the oceans, at least twice as large as those over land, with precipitation showing a 1.5 % trend per decade but towards drier conditions.

Climatic Changes: Looking Back, Looking Forward

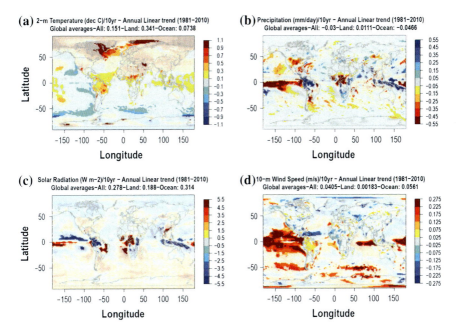

Fig. 3 Linear trends for the four select meteorological variables—2 m temperature (*top left*, in °C/10 yr), precipitation (*top right*, in [mm day^{-1}]/10 year), solar radiation (*bottom left*, in [W m^{-2}]/10 year) and 10 m wind speed (*bottom right*, in [m s^{-1}]/10 year)—over the period 1981–2010. *Darker colours* indicate significant values at the 95 % confidence level. Note reverse colour palette for precipitation. *Data source* ERA-Interim reanalysis (To be re-drawn)

More specifically, if we relate some of these changes to the estimated lifetime of wind and solar power plants (about 30 years), we can see from Table 1 that the increases in solar radiation and wind speed over land are 0.6 W m^{-2} and 6 10^{-3} m s^{-1}, respectively, and thus only account, on average, for a small relative change (0.3 % over 30 years in both cases).

Clearly these globally averaged values may hide regions with large relative trends, especially because, as seen in Fig. 3, marked trends of either sign are present. In particular, although wind speed trends are on average positive over both oceans and land, there are wide regions characterised by negative trends over North America, West Africa and Eastern Europe. Some of these trends have been highlighted by a flurry of the recent literature, summarised by McVicar et al. (2012). It is worth pointing out that most of the studies mentioned in McVicar et al. (2012) refer to measurements taken either at 10 m or even at lower levels. Troccoli et al. (2011a) showed that winds taken below 10 m can be affected by local obstacles and therefore are usually not good indicators of large-scale circulation changes. Although this latter study was carried out for the Australian continent, it is conceivable that this conclusion may apply globally. Indeed, a similar result was obtained by Vautard et al. (2010), who showed that observed trends at the 850 hPa level (about 1,300 m) can have a different character to those

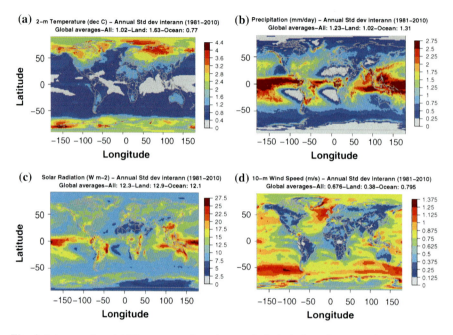

Fig. 4 Interannual variability computed as the standard deviation after removing the annual cycle for four select meteorological variables—2 m temperature (*top left*, in °C), precipitation (*top right*, in mm day^{-1}), solar radiation (*bottom left*, in W m^{-2}) and 10 m wind speed (*bottom right*, in m s^{-1})—over the period 1981–2010. *Data source* ERA-Interim reanalysis

for near-surface wind speed observations. Despite 850 hPa being normally much higher than wind turbine hubs, winds at a level like this should be considered as they provide a better understanding of variations in large-scale atmospheric flow.

3.2 Historical Interannual Variability of Select Meteorological Variables

Another important statistic for the planning of energy infrastructure, supply and trades is the interannual standard deviation (or variability) of the distribution. This measure is useful because it provides an indication of how a given meteorological variable varies from year to year. Indeed, decisions are often made on the basis of limited locally observed records (a year or two) and therefore climatic factors that vary on longer timescales such as El Niño may have an influence on energy infrastructure and its power yield.

The interannual variability statistic is computed by first removing the annual cycle, which, although often the principal factor in the standard deviation of the distribution, it is also relatively well predictable. The standard deviation of the distribution thus modified is then calculated. This statistic is shown in Fig. 4 for the same four select meteorological variables considered above.

In the case of 2 m temperature, large interannual variations of up to about 4 °C are observed at high latitudes, especially over the Siberian region. Since a large portion of this region is covered in permafrost (Schuur et al. 2008), the energy industry may be vulnerable to these large temperature variations. For instance, as discussed by Vlasova and Rakitina (2010), pipelines may be at risk of accidents when laid in permafrost as a result of permafrost thawing, particularly when the ice content is high, as this process can lead to uneven soil settling, displacement of piles, 'floating' condition of pipes and pipeline deformation.

Interannual variations in precipitation are especially marked in the tropical areas. Particularly relevant is the interannual variability over Brazil, West Africa and South Asia as these regions heavily rely on hydro power for their electricity production (e.g. "Bioenergy, Weather and Climate Change in Africa: Leading Issues and Policy Options"). Interannual changes of 2–3 mm/day are a large proportion of the annual mean precipitation, which peaks at about 12 mm/day in the wettest parts of the tropics like the western Pacific ocean, but over tropical land a more typical range is 6–9 mm/day. It should be borne in mind that, as mentioned before, some of this signal may be due to inaccuracies in the representation of rainfall processes in the reanalyses.

Even in the case of solar power the largest interannual variability is observed in the tropical areas, where the annual mean radiation is typically 220–280 W m^{-2}. Thus, variations of about 20–25 W m^{-2}, as seen over eastern Brazil, eastern Australia and South East Asia, represent around 10 % of the annual mean signal. Therefore, solar power yield estimates based on a short record may result in serious over/under-estimations. For instance, if a site observational campaign happened to coincide with a La Niña phase, solar power yield calculations that neglects longer term variations could be off by as much as 15 % over much of the Eastern part of Australia (Davy and Troccoli 2012).

The near-surface wind speed is an important indicator for several aspects of the energy industry activities, not only for the production of wind energy. Indeed, it is also critical for the planning and operations of oil rigs (cyclones, storm surges), delivery through power lines (heavy winds), transportation especially via sea (cyclones, storm surges), wind turbines themselves (wind gusts), etc. Although the largest interannual variations occur over the oceans, especially in the Arctic ocean and along the circumpolar current which also correspond to regions of high annual means, large variations are also observed in regions like the North Sea, the Mediterranean Sea, South-East Asia where the variations can be up to 20 % of their annual means. Shipping routes and offshore operations, such as for oil rigs maintenance, are therefore especially vulnerable to such variations.

3.3 Historical Linear Trend of Mean Sea Level

Mean sea level is a sensitive climate variable as it responds to changes in several components of the Earth system. For instance, as oceans warm up, mean sea level rises by thermal expansion; as glaciers or ice sheets melt due to increasing air

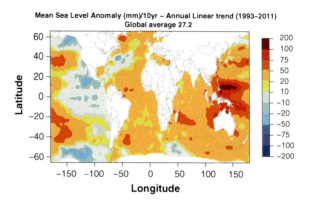

Fig. 5 Linear trend for the mean sea-level anomaly (in mm/10 year) over the period 1993–2011. *Darker colours* indicate significant values at the 95 % confidence level. *Data source* TOPEX/Poseidon, Jason-1 and Jason-2 satellite missions

temperature, mean sea level rises because of freshwater mass input to the oceans; modifications in the hydrological cycle lead to variations in river runoff, and ultimately to mean sea-level change (Pokhrel et al. 2012).

Changes in mean sea level may affect all coastal low-lying energy infrastructure (e.g. through inundation or saltwater intrusion) as well as offshore facilities (oil, gas, wind turbines). Mean sea level is therefore a critical variable to monitor and its changes should be accounted for in planning and operations. However, given the vastness of the oceans, systematic monitoring of mean sea-level variations only started in the early 1990s with the advent of satellite oceanography. In earlier decades the number of oceanographic observations available was orders of magnitude smaller than what satellites can now observe (Tribbia and Troccoli 2008).

Satellite altimetry has revealed considerable regional variability in the rates of mean sea-level change with some regions, such as the western Pacific and east of Madagascar, displaying mean sea-level rise rates of up to a few times larger than the global mean which is close to 3 mm yr^{-1} (Fig. 5). However, there are also places where the trend is negative, for instance off the western coast of Mexico. Such a non-uniform mean sea-level trend is caused principally by differential ocean thermal expansion and by wind-driven oceanic currents.

Sea level rise is a difficult climate parameter to determine using climate models because it involves interactions between all components of the climate system (oceans, ice sheets and glaciers, atmosphere, land water reservoirs) on a wide range of spatial and temporal scales. Even the solid Earth through its elastic response to changing crust and mantle parameters, as well as water mass redistribution, affects mean sea level. Systematic monitoring of oceans, cryosphere and land waters from in situ and space-observation systems are thus crucial to validate climate models, and hence improve future mean sea-level projections (Cazenave and Llovel 2010).

It is important to distinguish between 'sea level' and 'mean sea level' while the former is the instantaneous measure of sea level which is affected by both meteorological and oceanic processes, the latter filters out short-term meteorological processes (of the order of a few days) and focuses on longer term effects. As a result, phenomena such as storm surges, which are mainly atmospheric-driven (possibly with a contribution from sea surface temperature) but manifest themselves as anomalous sea levels, might not follow the same trends as that of mean sea level. In other words, higher mean sea levels might not be conducive to increased storm surge activity as shown in the case of Venice (Troccoli et al. 2011b).

3.4 Other Relevant Meteorological Factors

There are several other meteorological effects that impact the energy sector. For instance, Bonelli and Lacavalla (2010) investigated the effect of icing on power lines using proxy meteorological data for a few sites in Italy and highlighted some local climatic trends. Similar studies have provided evidence of the severity of icing on power lines (Makkonen 1998; Fikke et al. 2007; Leblond et al. 2006). Further discussion is provided in the "Combining Meteorological and Electrical Engineering Expertise to Solve Energy Management Problems."

Another example is the effect of hail on solar panels. Typical PV materials can withstand winds of up to 200 km h^{-1} and 2.5 cm hailstones at 80.5 km h^{-1}. A few studies have looked at possible changes in hailstones events under climate change. For instance, Botzen et al. (2010) found a 26–46 % increase in hail storm damage in the Netherlands associated with a 2 °C temperature increase. Instead, no significant change in hail storm risk for Australia was detected (Niall and Walsh 2005).

4 Projected Climate Changes Relevant to Energy

Climate projections as derived from two different greenhouse gas (GHG) emission scenarios are discussed in this section: the high-range A2 scenario and the low-range B1 scenario (see call out box 3 for further detail). Differences in terms of mean and standard deviation are discussed here between the projected 30-year period 2050–2079 and the baseline 30-year period 1970–1999 for the four main variables considered—2 m temperature, precipitation, solar radiation and 10 m wind speed. The projected variables are obtained from the 15 selected CMIP 3 models. Specific examples of selected impacts on the energy industry are also highlighted.

4.1 Projected Mean Changes to Select Meteorological Variables

As already remarked, temperature is a key variable in many, if not all, aspects of the energy sector. For instance, it can modify soil conditions and therefore crop growth for biofuels; it can affect energy demand, it can modify efficiency of

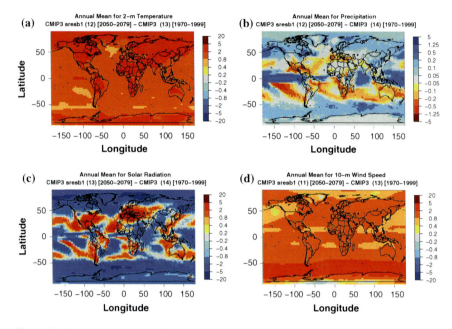

Fig. 6 Projected annual mean changes over the period 2050–2079 under the low-range B1 emission scenario for 2 m temperature (*top left*, in °C), precipitation (*top right*, in mm day^{-1}), solar radiation (*bottom left*, in W m^{-2}) and 10 m wind speed (*bottom right*, in m s^{-1}). Changes are computed as differences with respect to the 1970–1999 climate run results. Numbers in parenthesis in the title indicate the number of models used for the averaging of each of the two periods considered. *Data source* CMIP3 model output

photovoltaic cells, etc. Projected changes in temperature are amply documented in the AR4 and subsequent literature. For completeness the projected surface temperature changes under the B1 scenario are shown in Fig. 6 (top left panel). Under this low-range GHG emission scenario temperature is projected to increase globally by 2050–2079, with increases up to several degrees centigrade especially at high latitudes, but also over most continents.

Closely related to temperature are the derived quantities of heating/cooling degree days (HDDs/CDDs), which are quantitative indices designed to reflect the demand for energy needed to heat/cool a home or business. A few studies have investigated the potential impact of climate change on HDDs/CDDs so as to estimate energy requirements of residential houses. One such study included five regional climates varying from cold to hot humid in Australia and used projections from nine climate models (Wang et al. 2010). They found potentially significant climate change impacts on heating/cooling energy requirements within the lifespan of existing housing stock. Such requirements could vary in the range of −26 to 101 % by 2050 and −48 to 350 % by 2100 under mid-to-high range emission scenarios. "Reducing the Energy Consumption of Existing, Residential Buildings, for Climate Change and Scarce Resource Scenarios in 2050" also illustrates the challenge of meeting the comfort conditions in existing houses in a temperate climate in 2050. Thus, consideration will be needed in the planning of future energy requirements for buildings in warmer climates.

Precipitation, another meteorological variable amply discussed in the AR4, displays a mostly zonal response to climate change under the B1 scenario (Fig. 6, top right panel). Large increases of up to 5 mm day^{-1} are projected over most of the tropical regions especially over the oceans, whereas dryer conditions are expected for most of the subtropics with precipitation reductions of up to -5 mm day^{-1}, with Central America, Eastern Brazil, the Mediterranean and Southern Africa as the more notable regions potentially suffering drier conditions. High latitudes are projected to increase their mean precipitation by up to about 1 mm day^{-1}. It is also worth noting that, except for the central/eastern tropical Pacific ocean, these projected changes reflect the observed trends over the recent period (Fig. 3).

Large changes are projected also for solar radiation with values exceeding many W m^{-2}, both positive and negative, which can be as high as 10 % or more of the annual mean (Fig. 6, bottom left panel). The pattern of these changes appears highly correlated with that of precipitation. As it might be expected, wetter areas will have more cloud cover and therefore less solar radiation will reach the ground; but also the reverse seems to hold, with dryer areas receiving more solar radiation due to reduced cloud cover. As a consequence, the solar power industry might benefit by the projected increase in solar radiation in regions such as Central America, Eastern Brazil, the Mediterranean, Southern Africa and Eastern Australia.

Considerable positive changes are projected in 10 m wind speed, with increases of up to 2 m s^{-1} expected over most of the globe, with even larger changes of up to 5 m s^{-1} over the circumpolar current (Fig. 6, bottom left panel). With typical 10 m wind speed over land in the range of 1–5 m s^{-1}, the projected changes could be as large as 30 % of the current mean values. Unlike the trends in the observed period which can be either positive or negative (Fig. 3), these projected changes are all positive. For this 55-year period from now to 2065 (mid of 2050–2079), these marked changes are equivalent to a constant trend equal to about 0.2–0.3 m s^{-1} decade^{-1}, and are at the largest end of observed values in the 1981–2010 trends.

4.2 Projected Climatic Variability of Select Meteorological Variables

Figure 7 shows the expected change in climatic variability computed by means of the standard deviation for the B1 and A2 emission scenarios (left and right side respectively) and for the four select variables. In terms of near-surface temperature, the main changes under the low-range emission scenario, B1, are the reductions in variability at high latitudes. Under the high-range emission scenario, A2, however, wide and relatively large areas of positive changes up to 0.5 °C in variability emerge. Aside from the high latitudes where variability is normally large (in excess of 8 °C) these A2 scenario changes in temperature variability affect predominantly sub-Saharan Africa, the Middle East, Central and northern South America. Indeed, in these regions where the typical variability is up to 2.5 °C the variability could increase by about 20 %, therefore adding additional strain to energy systems and/or crop cultivations that may already be under stress. It would seem appropriate to put in place measures aimed at managing such expected increased variability, notwithstanding the fact that current interannual variability may already warrant attention.

Some projected changes in the precipitation variability under the B1 scenario are shown mainly over the equatorial Pacific and Indian oceans (Fig. 7). However, given that these are high variability regions, with current standard deviation ranging between 3 and 5 mm day^{-1}, the projected changes would be less than 10 %. Apart from these oceanic changes, the main land region affected by the higher projected variability would be parts of South East Asia. Under the higher emission A2 scenario, the pattern of change in precipitation variability remains essentially the same, but with magnitudes slightly strengthened.

For global solar radiation, the pattern of projected changes in variability appears noisy (Fig. 7). Nonetheless, some main features can still be identified such as the reduction in variability of up to −1 W m^{-2} in regions that correspond also to projected increases in annual mean (Fig. 6). Although these changes are relatively small (up to a few percent), a reduction in variability might be beneficial in the planning of solar power plants in these regions of abundant solar radiation. In fact, a reduced variability should lead to a reduced uncertainty in the solar radiation estimation and hence power yield. Similarly a reduced variability could lead to more accurate solar forecasts. As for precipitation, the pattern of solar radiation variability changes in the A2 scenario remains essentially the same, only with more pronounced magnitudes.

Variability in the 10 m wind speed is projected to generally increase under the B1 scenario especially over mid-high latitude regions such as North America and Europe, southern Africa, southern Australia and central south Asia with changes of up to 0.2 m s^{-1} (Fig. 7). Given the relatively low 10 m wind speed variability over land areas, such changes could be as large as 30 % of the current variability. Again, as for precipitation and solar radiation, the wind speed variability A2 scenario presents an analogous pattern to the B1 scenario, but with larger magnitudes. It is therefore important to account for possible large changes in energy system management such as in wind farm operations.

4.3 Projected Changes in Mean Sea Level and Sea Ice

The IPCC AR4 concluded that the projected mean sea-level rise is likely to be 0.26–0.59 m by the 2100 for their highest emissions scenario (Meehl et al. 2007). New research suggests that the possibility of sea level rise of up to 2 m by 2100 should be given serious consideration (Lowe and Gregory 2010), even if Pfeffer

Climatic Changes: Looking Back, Looking Forward 83

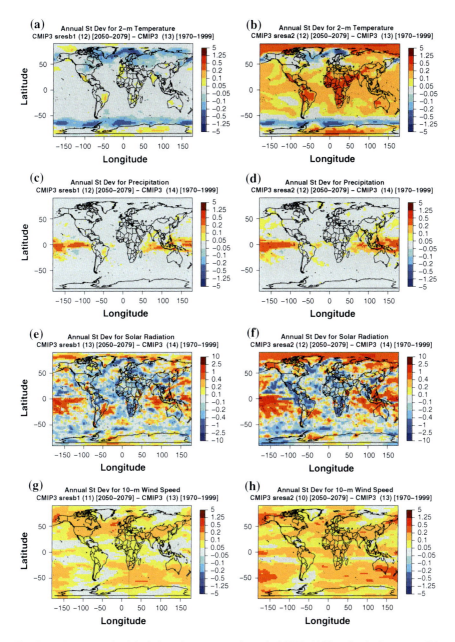

Fig. 7 Projected standard deviation changes over the period 2050–2079 under the low-range B1 emission scenario (*left panels*) and the high-range A2 emission scenario (*right panels*). From the top: 2 m temperature (in °C), precipitation (in mm day^{-1}), solar radiation (in W m^{-2}) and 10 m wind speed (in m s^{-1}). Changes are computed as differences with respect to the 1970–1999 climate run results. Numbers in parenthesis in the title indicate the number of models used for the averaging of each of the two periods considered. *Data source* CMIP3 model output

et al. (2008) conclude that increases in excess of 2 m are physically untenable. Although increases of up to 2 m in this century cannot be ruled out, this does not mean that they are inevitable or even likely. For climate change to produce much more than 1 m of sea level rise, ice sheets would probably have to contribute considerably more to the rise than they do now (Lowe and Gregory 2010).

Accurate projections of global mean sea-level rise require a thorough understanding of the cause of recent, widespread and intensifying glacier acceleration along Antarctic ice-sheet coastal margins as well as in terrestrial water storage. These are currently two of the more poorly understood processes. Indeed, the extent and magnitude of ice-shelf thickness change, the underlying causes of such change, and its link to glacier flow rate are so poorly understood that its future impact on the ice sheets cannot yet be predicted (Pritchard et al. 2012). In addition, Pokhrel et al. (2012) estimated that 42 % of the observed sea-level rise is due to terrestrial water storage, which is affected by activities such as reservoir operation and irrigation. Specifically this component was not properly accounted for in the IPCC AR4.

Global climate models are currently not capable of producing mean sea-level projections. Hence, these are obtained by means of statistical approaches which are loosely based on an understanding of physical processes. The general assumption is that the relationship between sea level rise and temperature (or forcing) will hold in the future and for a much greater range of warming than occurred during the period from which it was calibrated. Therefore, there is no guarantee that such an assumption will hold in the future (Lowe and Gregory 2010).

It is also important to emphasise that the projected changes are given as global means. As noted earlier (Fig. 5), changes in sea level have a regional character. Thus, while most parts of the globe have been experiencing an increase in mean sea level, there are also areas displaying a negative trend (e.g. many parts of the western coast of Central and South America). Other parts are also affected by subsidence and hence the current observed mean sea-level increases may be due to this phenomenon. There is little doubt that sea level rise is likely to continue, but that the rise by the year 2100 is almost certain to be below 2 m and that there is currently very little evidence to suggest that increases at the top of this range are likely (Lowe and Gregory 2010).

4.4 Projected Changes in Permafrost

As shown in Fig. 6, the surface temperature in the Siberian/Arctic region is projected to increase by several degrees during the twenty-first century. Such increase may have serious implications for the evolution of the permafrost extent, particularly that currently found in near-surface soils. Indeed, this portion is most vulnerable to climate change and its degradation has the potential to initiate a number of feedbacks, mostly positive, in the Arctic and in the global climate system (McGuire et al. 2006).

Using the Community Climate System Model (CCSM3), one of the CMIP3 models, Lawrence and Slater (2005) showed that the projected strong Arctic warming could drive severe degradation of near-surface permafrost during the twenty-first century. From the simulated present-day permafrost spatial extent of 10.5 million km^2, the model projected a reduction by an order of magnitude by 2100, namely as little as 1.0 million km^2 of near-surface permafrost would remain. However, in a new version of the land model component the rate of near-surface permafrost degradation, in response to the strong Arctic warming (+7.5 °C vs. land), is slower particularly during the early twenty-first century (81,000 versus 111,000 km^2 yr^{-1}, Lawrence et al. 2008). Even though the rates of degradation are initially depressed, the strong Arctic warming is enough to substantially reduce the total area containing near-surface permafrost in CCSM3 by 2100. Permafrost degradation of this magnitude is likely to invoke a number of hydrological, biogeochemical and ecological feedbacks in the Arctic system (Lawrence et al. 2008), including potential release of considerable amount of methane (CH$_4$) which would provide a positive feedback to global warming. However, it is worth emphasising that such processes are still highly uncertain, not least because of the scarcity of observations (Krey et al. 2009).

5 Extreme Weather Events Relevant to Energy

Climatic extremes might not necessarily coincide with extreme events of significance to the energy sector. It is therefore essential to focus on the energy sector as the target when considering weather/climate-driven extremes of importance to this sector. Perhaps because of the wide relevance of meteorology to many societal sectors, definition of extreme weather/climate events tend to be meteorologically focussed, as in the case of the IPCC AR4, in which an extreme event is defined as an event rare or rarer than the 10th or 90th percentile of the observed probability density function at a particular space and time of year. In the perceptions of many an 'extreme' or 'severe' weather event is one that causes inconvenience in some manner, a perspective that bears no necessary relationship to 10th and 90th percentiles of a specific meteorological variable. This assertion is especially valid when anthropogenic factors contribute to increase the potential vulnerability, such as in the concreting of flood plains or deforestation of mountains.

In attempting to overcome the ambiguity in this definition of extreme events, at the NATO Advanced Research Workshop for Weather/Climate Risk Management for the Energy Sector it was suggested an alternative definition: 'an event on any timescale that contributes to stresses beyond the operating parameters of a sector' (Troccoli et al. 2010). This definition formally accepts that severe events cannot be defined in hydro-meteorological/climate terms alone, and therefore, as implied in the IPCC definition, they cannot be characterised uniquely using climate models, as location and economic activities at that location also play roles. As a consequence, characterisation of future extreme weather events of relevance for the energy industry are even more challenging.

There are, however, certain guidelines that might be offered, also based on the results discussed in this chapter, and assuming that the climate does not pass any tipping points for rapid change:

- Increasing warmth is almost certain to reduce heating demands but increase cooling demands overall. However, interannual variability might become more pronounced and, as a consequence, cold periods will not disappear. Temperature tolerances of energy sector infrastructure may be tested more regularly, as may those of cultivated biofuels. Infrastructure on permafrost may also be affected.
- Flooding and droughts will continue, the former certainly being affected additionally by human activities; contingencies for increased intensities and frequencies of both should be included in risk management, even if it remains difficult to establish impacts at any specific location and time. Impacts on infrastructure (including silting of reservoirs), on demand, on the production of biofuels and on hydro-generation should be considered.
- Mean sea-level rise appears inevitable, and might be accompanied by increased risk of coastal storm damage even without storms intensifying. Potential issues include risks to coastal generation and offshore infrastructure, including production platforms and wave and tidal generators.
- Increased intensities of some thunderstorms might be expected in a warmer climate, although the objective scientific evidence for this supposition is weak. Were this to happen then there might be more frequent localised wind storms, intense rainfall and more hail. Energy infrastructure therefore requires consideration, as in the case of solar power generators.
- While hurricanes destructiveness has been reported to be increasing (Emmanuel 2005; Webster et al. 2005), a later study found that their global frequency may be decreasing (Emmanuel et al. 2008). However, Pielke et al. (2008) showed that the normalised damage associated with U.S. main-land hurricane landfalls highlights the tremendous importance of societal factors in shaping trends in damage related to hurricanes. As people continue to flock to the nation's coasts and bring with them ever more personal wealth, losses will continue to increase.

6 Summary

Climate risk management requires insights into how the current climate impacts the energy sector as well as projected climate changes in the next decades. In this chapter, a selection of climatic variables and some their potential impacts on the energy sector have been highlighted.

As discussed, there is large uncertainty in climate projections. Likewise, in poorly observed regions like parts of Africa or Siberia it is difficult to characterise the current climatic variability. So, how should this information be used? In brief, these assessments provide an initial input for the next step in the impact and risk management chain, particularly in terms of identifying regions/energy sector segments more vulnerable to potential impacts. Higher spatial and temporal resolution analyses would need to be carried out for the identified regions/energy sector segments. Specific examples of specialised use of weather and climate information for energy systems are presented in following chapters.

More specifically, from a meteorological point of view, weather and climate information could provide more readily available input into the energy sector (and therefore reduce uncertainty in impact estimation) by adopting risk management options such as:

- Identifying areas requiring specialised monitoring to overcome the deficiencies/discrepancies in our current knowledge (e.g. that provided by reanalyses) so as to be able to provide the underpinning meteorological information needed for energy planning and operations;
- Complementing the information derived from reanalyses and climate models with meso-scale or micro-scale models (via the so-called dynamical down-scaling) or by means of statistical down-scaling models so as to provide climate characteristics at the location and time resolution needed for specific power generation facilities, transmission/distribution networks, energy demand, etc.;
- Facilitating the engagement between weather and climate information providers and energy users (possibly at regional level) with the aim to provide early warning and advisory weather and climate services.

Acknowledgments ECMWF ERA-Interim data used in this study have been obtained from the ECMWF data server (http://data-portal.ecmwf.int/). Sea-level data from the TOPEX/Poseidon, Jason-1 and Jason-2 satellite missions have been downloaded from CSIRO's site http://www.cmar.csiro.au/sealevel/sl_data_cmar.html. The Coupled Model Intercomparison Project 3 (CMIP3) climate model outputs have been obtained from (https://esg.llnl.gov/).

References

Bonelli P, Lacavalla M (2010) Trends in snow deposition on overhead electric lines: using synoptic data to investigate the relationship black-out risk/climate change. In: Troccoli A (ed) Management of weather and climate risk in the energy industry. NATO Science Series. Springer Academic Publisher, Dordrecht, pp 305–314

Botzen WJW, Bouwer LM, van den Bergh JCJM (2010) Climate change and hailstorm damage: Empirical evidence and implications for agriculture and insurance. Resour Energy Econ 32(3):341–362

Cazenave A, Llovel W (2010) Contemporary sea level rise. Annu Rev Mar Sci 2:145–173

Davy R, Troccoli A (2012) Interannual variability of solar electricity generation in Australia. Sol Energy. doi:10.1016/j.solener.2011.12.004

Dee DP et al. (2011) The ERA-Interim reanalysis: configuration and performance of the data assimilation system. Quart J R Meteorol Soc 137:553–597. doi:10.1002/qj.828

Ebinger J, Vergara W (eds) (2011) Climate impacts on energy systems: key issues for energy sector adaptation, World Bank Publication, Herndon

Emmanuel K (2005) Increasing destructiveness of tropical cyclones over the past 30 years. Nature 436:686–688

Emanuel K, Sundararajan R, Williams J (2008) Hurricanes and global warming: results from downscaling IPCC AR4 simulations. Bull Amer Meteor Soc 89:347–367. doi:10.1175/BAMS-89-3-347

Fikke S, Ronsten G, Heimo A, Kunz H, Ostrozlik M, Personn P-E, Sabata J, Wareing B, Wichura B, Chum J, Laakso T, Santti K, Makkonen L (2007) Atmospheric icing on structures measurements and data collection on icing: state of art—COST 727. MeteoSwiss, Switzerland, 75, pp 110

Johnston PC, Gomez JF, Laplante B (2012) Climate risk and adaptation in the electric power sector. Asian development bank publication, Philippines. Available at: http://www.iadb.org/intal/intalcdi/PE/2012/12152.pdf

Krey V, Canadell JG, Nakicenovic N et al (2009) Gas Hydrates: entrance to a methane age or climate threat? Environ Res Lett 4:1–6. doi:10.1088/1748-9326/4/3/034007

Lawrence DM, Slater AG (2005) A projection of severe near- surface permafrost degradation during the 21st century. Geophys Res Lett 32:L24401. doi:10.1029/2005GL025080

Lawrence DM, Slater AG, Romanovsky VE, Nicolsky DJ (2008) Sensitivity of a model projection of near-surface permafrost degradation to soil column depth and representation of soil organic matter. J Geophys Res 113:F02011. doi:10.1029/2007JF000883

Leblond A, Farzaneh M et al (2006) Guidelines for meteorological icing models, statistical methods and topographical effects. Electra 225:116

Lowe JA, Gregory JM (2010) A sea of uncertainty. Nat Rep Clim Change http://dx.doi.org/10.1038/climate.2010.30

Makkonen L (1998) Modeling power line icing in freezing precipitation. Atmos Res 46:131–142

McGuire AD, Chapin FS, Walsh JE, Wirth C (2006) Integrated regional changes in Arctic climate feedbacks: implications for the Global Climate System. Annu Rev Environ Resour 31:61–91. doi:10.1146/annurev.energy.31.020105.100253

McVicar TR, Roderick ML, Donohue RJ, Li LT, Van Niel TG, Axel T, Jürgen G, Deepak J, Youcef H, Mahowald NM, Mescherskaya AV, Kruger AC, Shafiqur R, Yagob D (2012) Global review and synthesis of trends in observed terrestrial near-surface wind speeds: implications for evaporation. J Hydrol 416–417:182–205

Meehl GA, Stocker TF, Collins WD, Friedlingstein P, Gaye AT, Gregory JM, Kitoh A, Knutti R, Murphy JM, Noda A, Raper SCB, Watterson IG, Weaver AJ, Zhao Z-C (2007) Global Climate Projections. In: Solomon S, Qin D, Manning M, Chen Z, Marquis M, Averyt KB, Tignor M, Miller HL (eds) Climate Change 2007: The Physical Science Basis. Contribution of Working Group I to the Fourth Assessment Report of the Intergovernmental Panel on Climate Change. Cambridge University Press, Cambridge, United Kingdom and New York, NY, USA

Niall S, Walsh k (2005) The impact of climate change on hailstorms in southeastern Australia. Int J Climatol, 25(14):1933–1952

Pielke RA Jr, Gratz J, Landsea CW, Collins D, Saunders MA, Musulin R (2008) Normalized Hurricane Damage in the United States: 1900–2005. Nat Hazards Rev 9(1):29–42. doi:10.1061/ASCE1527-6988

Pokhrel YN, Hanasaki N, Yeh PJF, Yamada TJ, Kanae S, Oki T (2012) Model estimates of sea-level change due to anthropogenic impacts on terrestrial water storage. Nat Geosci. doi:10.1038/NGEO1476

Pfeffer WT, Harper JT, O'Neel S (2008) Kinematic constraints on glacier contributions to twenty-first-century sea-level rise. Science 321:1340–1343. doi:10.1126/science.1159099

Pritchard HD, Ligtenberg SRM, Fricker HA, Vaughan DG, van den Broeke MR, Padman L (2012) Antarctic ice-sheet loss driven by basal melting of ice shelves. Nature 484:502–505. doi:10.1038/nature10968

Schaeffer R, Szklo A, Frossard Pereira de Lucena A, Soares Moreira Cesar Borba B, Pinheiro Pupo Nogueira L, Pereira Fleming F, Troccoli A, Harrison A, Boulahya MS (2012) Energy sector vulnerability to climate change: a review. The Int Energy J 38:1–12. doi:10.1016/j.energy.2011.11.056

Solomon S, Qin D, Manning M, Chen Z, Marquis M, Averyt KB, Tignor M, Miller HL (eds) 2007. Contribution of Working Group I to the Fourth Assessment Report of the Intergovernmental Panel on Climate Change. Cambridge University Press, Cambridge, United Kingdom, and New York, USA

Schuur EAG, Bockheim J, Canadell JG et al (2008) Vulnerability of permafrost carbon to climate change: implications for the global carbon cycle. Bioscience 58:701–714

Trenberth KE, Jones PD, Ambenje P, Bojariu R, Easterling D, Klein Tank A, Parker D, Rahimzadeh F, Renwick JA, Rusticucci M, Soden B, Zhai P (2007) Observations: surface and atmospheric climate change. In: Solomon S, Qin D, Manning M, Chen Z, Marquis M, Averyt KB, Tignor M, Miller HL (eds) Climate change 2007: the physical science basis. Contribution of Working Group I to the Fourth Assessment Report of the Intergovernmental Panel on Climate Change. Cambridge University Press, Cambridge

Tribbia J, Troccoli A (2008) Getting the coupled model ready at the starting blocks. In: Troccoli A, Harrison M, Anderson DLT, Mason SJ (eds) Seasonal climate: forecasting and managing risk, NATO Science Series. Springer Academic Publishers, Dordrecht, pp 91–126

Troccoli A, Muller K, Coppin P, Davy R, Russell C, Hirsch AL (2011a) "Long-term wind speed trends over Australia", J Clim, 25:170–183, doi:10.1175/2011JCLI4198.1

Troccoli A, Zambon F, Hodges K, Marani M (2011b) "Storm surge frequency reduction in Venice under climate change", Clim Change, doi:10.1007/s10584-011-0093-x

Troccoli A et al (2010) Weather/climate risk management for the energy sector: workshop recommendations. In: Troccoli A (ed) management of weather and climate risk in the energy industry, NATO Science Series. Springer Academic Publisher, Dordrecht, pp 327–332

Vautard R, Cattiaux J, Yiou P, Thépaut J-N, Ciais P (2010) Northern hemisphere atmospheric stilling partly attributed to an increase in surface roughness. Nat Geosci 3:756–761. doi:10.1038/ngeo979

Vlasova L, Rakitina GS (2010) Natural risks management in the gas transmission system (GTS) of Russia and contribution to climate services under global climate change. In: Troccoli A (ed) Management of weather and climate risk in the energy industry, NATO Science Series. Springer Academic Publisher, Dordrecht, pp 315–325

Wang XM, Chen D, Ren ZG (2010) Assessment of climate change impact on residential building heating and cooling energy requirement in Australia. Build Environ 45(7):1663–1682

Webster PJ, Holland GJ, Curry JA, Chang HR (2005) Changes in tropical cyclone number, duration and intensity in a warming environment. Science 309:1844–1846

Renewable Energy and Climate Change Mitigation: An Overview of the IPCC Special Report

Ralph E. H. Sims

Abstract Renewable energy systems currently meet only around 7–8 % of the total global heating, cooling, electricity and transport end-use energy demands (Traditional biomass provides around 6.3 % of global primary energy and all other renewables around 6.7 %, but end-use energy is a more useful statistic used in this context). However, rapid growth in renewable energy has been apparent in recent years as a result of improved technology performance efficiencies and lower costs being demonstrated. Given appropriate support policies, renewables have the economic potential to significantly increase their share of total energy supply over the next few decades. The IPCC Special Report on this subject released in May 2011, covered cost trends, opportunities and barriers. This chapter summarises the findings of that report (The author, who was a Co-ordinating Lead Author for "Integration" of this IPCC report and co-author of the Summary for Policy Makers (SPM) and Technical Summary, acknowledges the inputs from around 150 co-authors and staff of the IPCC Technical Support Unit who contributed to writing the report and producing the SPM on which this chapter is largely based. See http://srren.org for the full report, list of authors, and extensive list of references that support the assessment). Most renewable energy resources are dependent on the local climate so there is a risk that they may be impacted by climate change. The size of the technical potentials of renewable energy resources and their geographic distribution could be affected, but there remains much uncertainty. The potential of renewable energy resources, even if significantly reduced, will still far exceed the projected global demand for primary energy, at least out to 2050. All mitigation scenarios show that renewable energy could provide a large share of energy demand in all regions.

R. E. H. Sims (✉)
Centre for Energy Research, Massey University, Palmerston North, New Zealand
e-mail: R.E.Sims@massey.ac.nz

1 Introduction

An assessment was made by the Intergovernmental Panel on Climate Change (IPCC) on the scientific, technological, environmental, economic and social aspects of renewable energy (RE) to climate change mitigation (IPCC 2011). The Special Report (known as "SRREN") had six chapters on each of the main technologies (bioenergy, solar, hydro, geothermal, ocean and wind[1]) as well as cross-cutting chapters on mitigation potentials and costs, integration into present and future energy systems, sustainable development, and policies and financing.

It is fully appreciated that as well as RE, there are several other options for reducing greenhouse gas (GHG) emissions arising from the energy system while still satisfying the demand for energy services. These include energy efficiency and conservation, fuel switching (for example, between coal and gas), nuclear electricity and carbon dioxide capture and storage options. In addition, a number of geo-engineering technologies have been proposed to either remove GHGs from the atmosphere or manage the incoming solar radiation levels. Each of these mitigation options can be evaluated in order to assess and compare their mitigation potential, associated risks, costs and contribution to sustainable development. The IPCC report findings as presented here focused on RE mitigation opportunities within a portfolio of mitigation options.[2]

A wide range of evaluations have consistently found that the global technical potential[3] for RE far exceeds the global energy demand. Solar energy has the highest potential but all six major RE sources have substantial potential (Table 1). Even in regions where relatively low levels of technical potential have been identified using local resources, significant opportunities are normally present for increased deployment compared to current RE supply levels. In the longer term, there are limits to some RE sources once their technical potential has been fully realised (for example, when all the high wind speed sites in a district have been covered in wind turbines). Factors such as sustainability concerns, public acceptance, economic factors and infrastructure constraints may limit future deployment of some RE technologies. In such cases their socio-economic potential is lower than their technical potential.

Since the mid-nineteenth century, modern societies have increasingly depended on the combustion of oil, gas and coal to provide heat, mobility and electricity. The ever-increasing demand for energy services is projected to continue in order to

---

[1] In order of Chaps. 2 to 7 of the SRREN.

[2] An extensive literature list up to early 2011 is given in each chapter of the IPCC report should further details be required to substantiate any of the points addressed in this chapter.

[3] The technical potential is the amount of useful energy that can be obtained by the full implementation over time when using known technologies and practices relating to that specific RE resource. It excludes any impacts that competing costs, barriers and policies might have, though obvious practical constraints (such as the need for solar heated water to be produced in close proximity to the hot water demand) are normally included.

Table 1 Indicative minimum and maximum technical potentials of primary energy for the main RE resources can be compared with global primary energy demand in 2008 of 492 EJ

	Biomass	Direct solar	Geothermal	Hydropower	Ocean	Windpower
Minimum (EJ/yr)	50	1,500	250	50	10	100
Maximum (EJ/yr)	500	50,000	2,500	100	500	600

Note The global end-use heat demand in 2008 was 164 EJ, and electricity demand was 61 EJ
All energy data quoted here and in the SRREN is based on the direct equivalent accounting method which differs slightly from the physical accounting method as used by the IEA

meet the basic human needs for welfare, comfort, lighting, cooking, mobility, communication and health of a growing global population, the aspirations for development of developing countries, and to drive productive processes. Should this growing demand continue to be met mainly by fossil fuels, there are concerns of increasing environmental impacts, both local air pollution and increases in atmospheric concentrations of GHG, as well as future energy security issues. The transition to a low-carbon or 'clean' energy economy is therefore imperative—but the promise is difficult to fulfil. The International Energy Agency, in its 2012 Energy Technology Perspectives report (IEA 2012) stated that although rapid technology change is possible from deploying some renewable technologies, notably solar PV with 42 % annual average growth over the past decade and on-shore wind at 27 %, most clean energy technologies will require additional support if they are to provide a more secure energy system and make the required contribution to reducing GHG emissions.

Current policies will not result in a sufficient increase in global RE shares to help meet the GHG mitigation ambitions necessary to achieve the 2 °C maximum global mean temperature rise above pre-industrial levels agreed by all national negotiating teams at the fifteenth and sixteenth sessions of the Conference of Parties (held in Copenhagen in 2010 and Cancun in 2011) of the UNFCCC. The additional benefits of RE should also be considered by policy makers when assessing their future potentials. If implemented properly, RE deployment can also contribute to social and economic development, security of energy supplies, improvements in environmental impacts and health, and provide energy access to those currently without modern energy services.

2 Trends and Future Scenarios

Renewable energy accounted for almost 13 % of global primary energy in 2008 (IEA 2010) of which around half was traditional solid biomass fuels used, as well as some coal, for cooking and heating by 2.7 billion people in developing countries (Fig. 1).

The share of RE is projected to increase significantly by most mitigation scenario analyses (as exemplified by a typical one shown in Fig. 2). This will require policies (Sect. 7) to stimulate changes in the energy system by improving energy

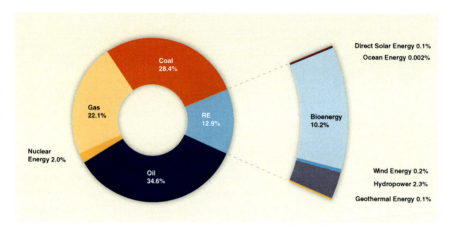

Fig. 1 Shares of total global primary energy (492 EJ) in 2008 with renewable energy at 12.9 % of which traditional biomass is almost half

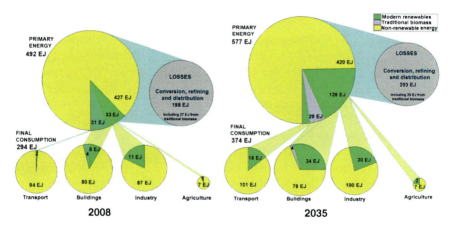

Fig. 2 RE shares (*green*) of primary and final consumption energy in the transport, buildings (including traditional biomass for cooking and heating), industry and agriculture sectors in 2008 and an indication of the projected increased RE shares needed by 2035 in order to be consistent with a 450 ppm CO_2eq stabilization target. *Notes* Areas of *circles* are approximately to scale. Energy system losses occur during the conversion, refining and distribution of primary energy sources to produce energy services for final consumption. 'Non-renewable' energy (*yellow*) includes coal, oil, natural gas (with and without carbon dioxide capture and storage (CCS) by 2035) and nuclear. Data based upon IEA World Energy Outlook 2010 (IEA 2010). Energy efficiency improvements above the 2008 baseline are included in the 2035 projection

efficiency in all sectors; attracting the necessary increases in investments in technologies and infrastructure; and reducing the current dependence on relatively cheap fossil fuels. This will involve removal of subsidies in many countries which, in 2010, totalled around $409 billion (IEA 2011).

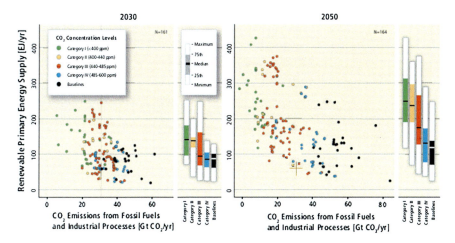

Fig. 3 Global RE primary energy supply projections of 164 scenarios in 2030 and 2050. *Notes* Colour-coding depicts baselines and categories I–IV of atmospheric carbon dioxide stabilisation concentration levels. Crosses depict the relationship in 2007. *Vertical bars* show the RE deployment level range for each category

In the SRREN, 164 mitigation scenarios were assessed. More than 50 % projected that the levels of global RE deployment by 2050 would more than triple the 2008 level of around 33 EJ of primary energy supply (Fig. 2). The share of RE could rise from the current 12.9 % (including traditional biomass) to 17 % in 2030 and to 27 % in 2050. The highest RE shares, as projected by one scenario, were 43 % in 2030 and 77 % in 2050.

The lower GHG stabilisation level scenarios tend to have higher RE deployment compared with their baselines (Fig. 3). As the RE shares increase, CO_2 emissions from fossil fuels and industry decline, leading to lower stabilisation levels. Most baseline scenarios assume continuing demand growth for energy services this century and show a doubling of the RE share by 2030 due to constrained future fossil fuel resource availability and the costs of RE continuing to decline as a result of improved performance and experience.

The scenarios confirmed that there are many low-carbon supply combinations possible linked with end-use energy efficiency improvements to produce low GHG concentration levels, but that in the majority of pathways, RE becomes the dominant technology alongside nuclear and carbon dioxide capture and storage (CCS). The range of assumptions made concerning future population growth, rate of economic development, fossil fuel reserves, policy approaches, carbon prices and the ability to integrate renewables into energy supply systems (see Sect. 5), has led to a wide range of RE deployment projections but with the majority increasing above the current level.

There were good indications that growth in RE will be widespread, and although distribution of RE deployment between regions varied between the scenarios, higher growth in the long term is more likely in developing (non-Annex 1)

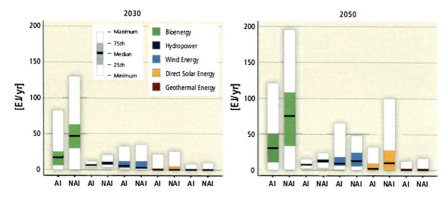

Fig. 4 Ranges of RE primary energy supply in 2030 and 2050 for bioenergy, hydropower, wind power, direct solar and geothermal technologies taken from 164 long-term mitigation scenarios. *Note A1* Annex 1 countries. *NA1* non-Annex 1 countries

countries than in OECD (Annex 1) countries, particularly for bioenergy and direct solar energy technologies (Fig. 4).

The primary energy shares between each of the RE technologies varies across the scenarios, and between regions. No single technology is totally dominant, though geothermal and hydropower gave lower contributions to the total potential than wind, solar and bioenergy. All of the 164 scenarios confirmed that the technical potential of RE technologies is not a limiting factor to expanding global deployment. Based on four selected illustrative scenarios, only 2.5 % of the overall global RE technical potential would need to be developed by 2050.

The four illustrative scenarios were selected to represent the range. Global cumulative savings in CO_2 emissions were around 70–90 % of the 1,530 Gt CO_2 from fossil fuels and industry, as extrapolated from the IEA Reference scenario (IEA 2010) out to 2050. The attribution of this CO_2 reduction to RE technology uptake varies with assumptions made in each of the scenarios, including the energy sources displaced by RE, and complex system behaviour.

3 Climate Change Impacts on Renewable Resources

The technical potentials of RE resources, their size and geographic distribution, could, in many regions, be impacted as climate change becomes more evident. Research into the possible effects is being undertaken but there remains much uncertainty. The precise magnitude and nature of such impacts is not well understood. RE resources are often dependent on the local climate so that future changes could affect that resource base, whether, for example, it results from an increase or decrease of wind speeds or precipitation levels.

- *Biomass resources.* Impacts of climate change on biomass resources are more difficult to assess than for other RE resources due to the large number of feedback mechanisms involved. Particularly for purpose-grown energy crops or crop residues, climate change can impact on crop productivity due to changes in soil condition and moisture contents, solar irradiation, intensification of the hydrological cycle and changes in seasonal distribution such as snow melt. While positive ecophysiological effects on plant growth may result from elevated CO_2 concentrations,[4] the overall magnitude and pattern of impacts of climate change on plant growth remain uncertain. Some studies show that a doubling of atmospheric CO_2 concentration levels could increase plant growth by up to 25 %, but other studies dispute this as, in the long term, the plants may adapt to accommodate higher levels. In some locations, crop yields might increase, and in others they might decrease. While positive effects might occur, detrimental effects that offset these cannot be ruled out.

Changes in precipitation levels can result in extremes (floods or droughts) that impact on crop production. At warmer temperatures, and particularly under higher CO_2 concentration levels, higher transpiration levels usually occur but, for some plant species, can be partially offset by greater water use efficiency (by increased stomatal closure). Semi-arid lands in some regions may therefore be able to support energy crop production due to this improved water-use efficiency.

Overall, it is considered that, if kept below 2 °C, a global mean temperature increase would have only a small effect on the technical potential of biomass production on a global basis. Agricultural management, plant breeding and the land area available are likely to have a greater impact. However, regional variations are expected and in tropical areas in particular, temperature increases above 2 °C may well result in declining crop productivity.

- *Solar resources.* Changes in the distribution and variability of atmospheric water vapour, cloud cover, rainfall and turbidity are likely to impact on solar systems in some locations. Some climate models have shown that, in some regions, variation in global monthly mean solar irradiance levels at the Earth's surface do not vary by more than 1 % due to anthropogenic forcing. There is no current observation or other evidence to indicate a substantial impact of climate change on global solar resources. Some research on global dimming and global brightening indicates probable impacts on solar irradiation but to date there is no evidence. If some regions receive lower solar radiation levels than at present, they may possibly benefit from a resulting lower demand for cooling services. The overall effect on the technical potential for solar energy is expected to be small.
- *Hydro resources.* The resource potential depends upon the volume, variability and seasonal distribution of water availability (run-off) which varies with topography and size of water catchment. Run-off could be affected by climate change due to changes in river-flows as a result of variations in precipitation; seasonal variations in rainfall and snow melt; glaciers receding; extreme

---

[4] Greenhouse growers often use "CO_2 fertilisation" whereby they artificially increase the CO_2 levels in the enclosed atmosphere using bottled CO_2 or un-flued combustion heating plants, to enhance crop productivity.

weather events causing floods and risks to dams; extreme droughts giving risks to generation system reliability; and sediment loads increasing due to changing hydrology and extreme events. This could lead to dams silting up, thereby reducing water storage volumes, and cause turbine blades to wear out faster from increased abrasion.

Models have projected large-scale changes in water run-off for hydropower plants this century with probable increases in high latitudes and the wet tropics, decreases in mid-latitudes and some drier regions of the tropics, uncertainties in desert regions, and uncertain variations in the monsoon regions at lower latitudes. Many discrepancies exist when comparing outputs from the models which leads to high uncertainty. Significant seasonal variations are also expected between regions. An increase in climate variability, even with no change in average run-off, can lead to reduced hydropower generation unless more storage capacity is built or the operation of the plant modified to suit the new hydrology. Indirect effects of water availability on hydropower generation may result from increased competition for irrigation and drinking due to projected fresh water deficiencies, but the impacts of this are difficult to assess.

Overall, the impacts through changes to annual precipitation volumes and run-off are expected to be slightly positive. As one indicator, the total output from all hydropower plants that were operating worldwide in 2005 (when they generated 2,931 TWh) is likely to be around 1 % higher by 2050. By that time however, there will probably be substantial regional variations in changes to precipitation, with most of the increases probably occurring in Asia, decreases expected in Europe and few other major changes expected elsewhere. Local effects and variations may be significant, so some hydropower plants may be impacted more than others, even within the same national borders.

- *Wind energy.* The distribution, annual variability and quality of the wind resource could be impacted by climate change but climate models are unable to confirm the extent (IPCC 2007). Some research suggests that multi-year mean annual wind speed averages will change by no more than ±25 % this century over most of North America and Europe, although other studies have shown this could be as high as ±50 % in northern Europe. The west coast of South America could experience an increase in mean annual wind speeds of around 15 %. Few other regions have been analysed and overall, wind resource change projections for a given location remain uncertain. Wind turbines and their operation could be impacted by extreme weather effects where more frequent, extremely high, gale force winds could cause increased stresses and ultimately damage may occur from structural fatigue.

Changes in seasonal and decadal variations in long-term average wind speeds could occur. At present, for example, northern Europe and the north-east Atlantic have higher wind speeds in the winter than the summer. Climate models are, as yet, unable to reproduce these conditions so future projections cannot be made with any degree of accuracy. Similarly, any historic changes in near-surface wind speeds cannot be attributed specifically to climate change as other factors may have played a role.

Overall, the global technical potential of wind resources used for power generation is unlikely to be greatly impacted by climatic change, but changes in the regional distribution of the wind energy resource may occur. In addition, sea level rise could impact on the foundation and anchoring costs of future designs of deep-water, off-shore wind turbines.
- *Ocean energy.* Several prototype wave technologies have already experienced damage during extreme weather events. Therefore designing devices that have greater reliability under such extreme conditions in an already challenging environment is critical, but they can add significantly to the capital costs. Future sea level rise should be accounted for when designing tidal range systems or anchored wave devices. Tidal ranges, ocean currents and ocean thermal systems are less likely to be impacted by climate change but no detailed analysis has been conducted to date.
- *Geothermal systems.* These are unlikely to be affected by climate change impacts, other than through the possible temperature rise of river water used for cooling at a power generation plant.

Water availability can influence choice of RE technology as is the similar case for water-cooled thermal power and nuclear power plants. Such plants could become vulnerable to changing water supply conditions should climate change add to water scarcity or average temperatures. Dry cooling systems are already used in areas where water scarcity is already a concern, so technical solutions are feasible. Hydropower and bioenergy depend on good water supply availability as do solar PV and concentrating solar power (CSP) systems for cleaning the panels.

Life cycle analysis (LCA) is a useful tool for assessing the full climate change mitigation potential of an RE technology. A full LCA also includes other indicators such as water use, materials consumption, levels of local air pollution, etc. For electricity generation, most LCAs indicate that the GHG emissions from RE technologies are usually lower than those associated with fossil fuel use, even when associated with CCS (Fig. 5). Under present climatic conditions, typical average values for all RE systems range from around 5–50 g $CO_{2\text{-eq}}$/kWh (excluding land-use change emissions—see Sect. 6) compared with fossil fuel thermal power generation ranging from 450 to 1,000 g $CO_{2\text{-eq}}$/kWh (gas-fired being the lower range and coal-fired the higher). Future changes in climatic conditions could impact on GHG emission levels of RE technologies, but analyses are yet to be undertaken.

4 Costs

Several scenario studies have shown that if RE deployment is constrained, achieving targets for low GHG concentrations will be more difficult and the mitigation costs will increase. By what amount it varies with the assumptions made in the analyses is uncertain.

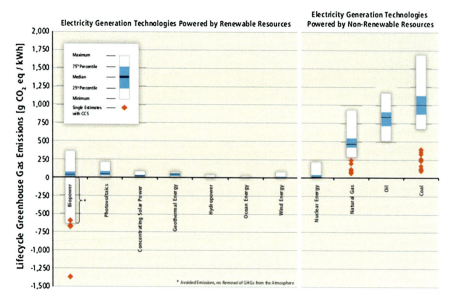

Fig. 5 Estimates of life cycle GHG emissions for a range of electricity generation technologies, including biopower, natural gas-fired and coal-fired power generation plants linked with CCS. *Notes* Biopower and hydropower exclude land-use impacts. Negative GHG emissions from biopower are due to avoided emissions from organic wastes and landfills, not from GHG removal from the atmosphere. Linking biopower with CCS can remove CO_2 from the atmosphere—see Sect. 6

For most RE technologies, their *levelised cost of energy (LCOE)*[5] largely depends on the availability of local RE resources but often tends to be higher than the equivalent cost of conventional energy sources (electricity, natural gas, gasoline, diesel, LPG, etc.). Energy prices vary widely between countries and states as do RE prices, so only general comparisons can be made here.

Analysis is needed at the local level to assess whether or not RE options are cost competitive. Certainly, there are many examples where RE is competitive with current market energy prices (Fig. 6), usually in regions with favourable RE resource conditions or where infrastructure is lacking for conventional energy systems to be supplied (such as national electricity grids not reaching rural areas). In most regions of the world, policy measures are still essential to ensure rapid deployment of most RE systems. However, there are exceptions such as New Zealand where hydro, geothermal, bioenergy and wind produce about 75 % of total electricity generation with no policy supporting measures in place. Due to

---

[5] The LCOE represents the cost of an energy generation system over its assumed lifetime. It is calculated as the price per unit at which energy from a specific source must be generated in order to break-even over the project lifetime. It excludes the cost of delivery to the final customer, any costs of integration, subsidies, tax credits or external environmental costs.

Renewable Energy and Climate Change Mitigation

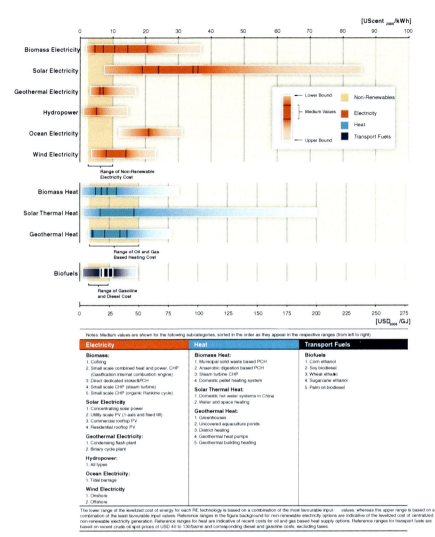

Fig. 6 Ranges of levelised costs of energy for selected RE technologies compared with typical costs of non-renewable electricity, heating using oil or natural gas, and gasoline and diesel transport fuels. (The wide ranging cost bars result from including the full range of sub-technologies (e.g. both on- and off-shore wind) and also projects located where poorer RE sources exist (such as solar thermal water heating in high latitude regions where low solar radiation levels exist). The lower cost ends of the ranges are based on the most favourable situations, whereas the upper ends of the ranges are based on a combination of least favourable inputs.) *Note* Range of typical non-renewable electricity costs are based on the LCOE of centralised generation plants. Heating, gasoline and diesel costs are based on oil spot prices between US$ 40–130/barrel

very good natural resources, RE power-generation can compete in the free market with gas and coal-fired generation options, even though using national supplies of natural gas and coal.

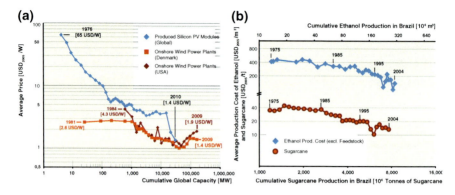

Fig. 7 Selected experience curves showing: (**a**) price reductions (USD/W of installed capacity) as installed total capacity increases over three decades or so for silicon solar PV modules, and for onshore wind turbines in Denmark and USA; and (**b**) the production cost trends over three decades for sugarcane (USD/t) and ethanol (USD/m^3) in Brazil. (Reductions in the price (or cost) of a technology per unit of capacity tend to understate the reductions in the LCOE when performance improvements occur.)

Possible technological advances may result in lower prices over the next decade or two, for example from:

- enhanced geothermal systems (originally termed "hot dry rocks"),
- improved delivery logistics for biomass feedstocks,
- increased energy crop yields,
- advanced biofuels produced using new processes,
- multi-product bio-refineries,
- advanced solar PV manufacturing technologies,
- CSP systems,
- several emerging ocean energy technologies, and
- off-shore wind deep-water foundations and turbines.

Hydropower technologies are relatively mature but opportunities exist to upgrade existing turbines with improved designs that have improved technical performance and to further develop low-head, mini-systems that can operate in a wide range of locations.

The price of most RE technologies has declined over the last few decades (Fig. 7) and this trend is expected to continue as a result of greater project experience, lower labour input intensity, increased mass production, higher market competition and improved performance from continuing RD&D. Periods of rising prices, partly due to demand increasing above supply (e.g. wind turbines) or shortages of materials (e.g. silicon for solar PV cells), have been experienced for a few technologies, but normally only over short periods.

In addition to technology costs, increased deployment of RE technologies often requires investment in new infrastructure. Total investment in RE systems to meet a future low-carbon economy (targeting 450 ppm CO_2 atmospheric stabilisation

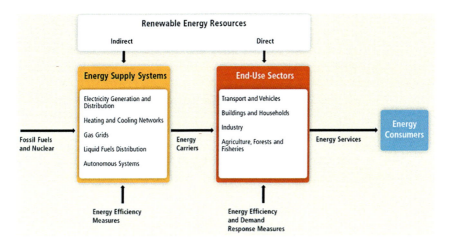

Fig. 8 In order to contribute towards providing energy services to consumers, RE integration can be either indirect into existing energy supply systems or direct when used on-site by the end-use sectors

levels), as estimated from the four scenarios that were selected to cover the wide range, will need to reach up to around US $5,100 billion between 2011 and 2020, with an additional US $7,180 billion needed over the following decade. The annual average of these investments equates to less than 1 % of the world's gross domestic product (GDP). Taking the electricity generation sector as an example, the current global annual investment will need to be increased five times on average during this 20-year period.

The externality costs of conventional energy systems should be monetized in order to make RE systems more competitive and for energy supply comparisons to then be based on what is often termed a 'level playing field'. Increased market prices also lead to more competitive RE systems. The LCOE is not the only determinant of the value of RE technologies. Their attractiveness also depends on broad environmental and social aspects such as their contribution to meeting specific energy services, helping meet peak electricity demands and any ancillary costs imposed on the overall energy supply system, such as the additional costs relating to maintaining system stability in order to integrate high shares of variable RE electricity generation systems (wind, solar PV, waves).

5 Integration

Various RE resources have already been successfully integrated into some energy supply systems and end-use sectors at relatively high shares (Fig. 8). Integrating RE into most existing energy supply systems and end-use sectors is technologically feasible, even at a greatly accelerated rate above the present rate. At higher shares of RE, additional challenges will result, whether for including integration directly into the end-use sectors or indirectly into supply systems. Whether RE is used for electricity, heating, cooling, gaseous fuels or liquid fuels, the integration challenges are contextual, site specific and can include the need for adjustments to existing energy supply systems.

The characteristics of a specific RE resource can influence the degree of difficulty for successful integration. Although most RE resources are widely distributed geographically, others, such as large hydro or geothermal, can be more centralized so that integration options are constrained by geographic location. Wind, wave and solar resources are variable with limited predictability. Biomass resources tend to have lower physical energy densities and their technical specifications differ from fossil fuels. These sorts of characteristics can constrain the ease of integration of RE and result in additional costs of the system, particularly when reaching higher shares of RE. The costs associated with RE integration are often difficult to determine. They may include costs for additional network infrastructure investment, system operation, and losses and other adjustments to the existing energy supply systems as needed. Research on the costs of RE integration has been sparse and accurate cost estimates are lacking.

The costs and challenges of integrating increasing shares of RE into an existing energy supply system depend on the current share of RE, the availability and characteristics of RE resources, the system characteristics and how the system might evolve and develop in the future.

Electricity systems. RE can be integrated into all system types, ranging from large, inter-connected continental-scale grids down to small, stand-alone autonomous systems on individual buildings. System characteristics of relevance include the generation mix and its flexibility, network infrastructure, energy market designs and institutional rules, location of energy demand, demand profiles, and control and communication capability. Wind, solar PV and CSP without storage can be more difficult to integrate than hydropower, bioenergy or CSP with storage and geothermal energy that are deemed to be dispatchable.[6]

As the shares of variable RE sources increase, maintaining system reliability may become more challenging for the electricity supply system operator and also involve additional supply costs. Means of reducing the risks and costs of variable RE integration include:

- having a portfolio of RE technologies;
- developing more flexible operation of existing systems;
- building complementary flexible generation;

---

[6] Power generation plants, such as reservoir-hydro, geothermal or biopower that can be scheduled to generate electricity as and when required are classed as dispatchable. CSP plants are classified as dispatchable only when some of the heat is stored for use at night or during periods of low sunshine. Variable RE technologies such as wind, solar or wavepower are deemed as partially dispatchable since generation can occur only when the RE resource is available.

- improving short-term weather forecasting;
- having better operation of the system using advanced planning tools;
- identifying electricity demands that can respond to supply availability;
- developing energy storage technologies (including pumped hydro and reservoir-based hydropower);
- and modifying institutional arrangements.

Strengthening and extending the electricity network transmission and/or distribution infrastructure (including inter-connections between large systems) may be necessary because of the fixed geographical distribution and remote locations of many RE resources.

District heating. Such systems can use low-temperature, RE heat inputs from solar thermal, geothermal and biomass combustion, including biomass sources with few competing uses such as refuse-derived fuels and municipal solid waste (MSW). The provision of thermal storage capability and flexible combined heat and power cogeneration can help to overcome the challenges of supply and demand variability for heating systems as well as provide a demand response service for electricity systems.

District cooling. Not yet commonly deployed, this technology can make use of any local, cold natural waterway. Where a district heating network exists, this same network can be used to distribute the cold water.

Gas distribution grids. Biomethane (resulting from cleaning biogas or landfill gas) and, in the future, RE-derived hydrogen and synthetic natural gas, can be injected into natural gas pipelines and then used for a range of applications. Successful integration requires that appropriate gas quality standards are met and pipelines upgraded where necessary.

Liquid fuel systems. Transport fuel distribution systems already successfully integrate biofuels for transport applications. Pure (100 %) biofuels, or more usually, biofuels blended at lower proportions with petroleum-based fuels, usually need to meet technical standards consistent with vehicle engine fuel specifications. Some liquid biofuels can also be used for cooking and heating applications, for example, in the form of ethanol gels.

The ease of RE integration across end-use sectors depends on the region, the desired RE characteristics specific to each sector and the technology. There are multiple pathways for increasing the shares of RE in a sector.

Transport sector. Liquid and gaseous biofuels are already commonly integrated into the fuel supply systems in a growing number of countries and this is expected to continue. Other integration options for transport fuels include centralized production of RE hydrogen for fuel cell vehicles and RE electricity for rail and electric vehicles. Future progress of these options will depend on the required infrastructure and vehicle technology developments. Decentralized, on-site production of biofuels, hydrogen or electricity for local transport vehicle use may be feasible. Future uptake of electric vehicles could support the development of flexible, distributed, electricity generation systems by providing energy storage services in the batteries.

Building sector. RE technologies can be integrated into both new and existing building structures to produce electricity, heating and cooling services for the building users at both the commercial and domestic scales. Production of energy surplus to demand may be possible, particularly for energy efficient building designs, and the surplus can then be exported to the grid and sold to provide revenue. In developing countries, the integration of RE supply systems is feasible for even modest dwellings.

Industry sector. Agriculture, as well as food and fibre process industries including wood processing, often use biomass to meet direct heat and power demands on-site. They can also be net exporters of surplus fuels, heat and electricity to adjacent supply systems (FAO 2011). Increasing the integration of RE for use by industries is an option in several sub-sectors, for example through electro-thermal technologies or, in the longer term, by using RE hydrogen.

Energy systems will need to evolve and be adapted in order to accommodate high RE shares. Long-term integration efforts could include investment in enabling infrastructure; modification of institutional and governance frameworks; attention to social aspects, markets and planning; and capacity building in anticipation of RE growth. Integration of less mature technologies, including advanced biofuels, fuels generated from solar energy, solar cooling, ocean energy technologies, fuel cells and electric vehicles, will require continuing investments in RD&D as well as capacity building and other supporting measures.

Electricity is expected to attain higher shares of RE earlier than either the heat or transport fuel sectors at the global level. Parallel developments in electric vehicles, increased heating and cooling using electrical heat pumps, flexible demand response services including the use of smart meters, energy storage and other technologies could become associated with this trend.

Overall, in spite of the complexities, in locations where suitable RE resources exist or can be supplied and infrastructure can be developed, there are few, if any, fundamental technological limitations to integrating a portfolio of RE technologies into an energy supply system to meet a majority share of total energy demand. The actual rate of integration and the resulting shares of RE will be influenced by future costs, policies, environmental issues and social aspects.

6 Sustainable Development

Economic development has historically depended on fossil fuel consumption leading to high GHG emissions. RE has the potential to reduce this dependence and contribute to future sustainable development. For developing countries, this will help the achievement of the Millennium Development Goals (although RE is not specifically mentioned).

In poor rural areas whose residents lack access to centralised energy systems, small RE systems can help provide heat and electricity cost-effectively. This can be in the simple forms of:

- domestic biogas plants using human, animal and crop wastes;
- ethanol gels used as cooking fuels to displace fuelwood and dung;
- windmills for water pumping or solar PV-powered water pumps;
- solar hot water technologies;
- solar crop-drying; and
- solar PV and small wind home power generation systems.

At the larger scale, solar, wind, hydro or biogas can provide electricity for community-scale mini-grids, and liquid biofuels can also be produced commercially at the small- to medium-scales for local use to enhance mobility. This could have positive impacts on job creation, improved access to markets and be a stimulus for capacity building, skills development and education.

RE deployment can also help reduce expenditure on imported fuels, produce greater energy security by diversifying energy resources, and reduce vulnerability to supply disruptions. To ensure electricity supply reliability however, the variability of wind and solar systems requires technical and institutional measures appropriate to local conditions. In OECD countries, the expectations are that electricity will always be available at the flick of a switch and transport fuels will always be available to purchase from a service station 24 h each and every day. In many non-OECD countries, especially in rural areas, such expectations do not exist and residents accept that energy services are supplied with a far lower degree of reliability. Indeed, for the 1.4 billion without access to electricity, as well as the additional 1.3 billion with some modern energy access but still dependent on traditional biomass for heating and cooking, any means of improving their current energy access is welcomed.

Bioenergy deployment can lead to lowering GHG emissions resulting from organic waste disposal in landfills when the gas produced (mainly methane) is collected and used, often in combined heat and power plants. Carbon dioxide is produced during combustion of the methane gas, but this has a far lower greenhouse potential than the methane that would otherwise have been released into the atmosphere.

A combination of bioenergy projects combined with CCS technologies could lead to reductions in atmospheric levels of CO_2. Where energy crops are sustainably grown, harvested, combusted and linked with CCS, if the crops are replanted after each harvest, then the crop plants continue to absorb CO_2 during the photosynthetic process and therefore act as a "pump" to move carbon from the atmosphere to a permanent store (Fig. 9). In order to stabilise atmospheric levels of GHG around 450 ppm, models show that annual global GHG emissions will need to become negative by the end of this century. Therefore bioenergy/CCS will increase in R&D priorities since the process is one of the few practical ways of atmospheric CO_2 removal, though the economics have yet to be determined.

The sustainability of biomass production is influenced by land use and management practices. However, the GHG implications of land-use management and land-use changes in carbon stocks are not yet well understood. Changes in productive land and forest use or management can have direct or indirect effects on

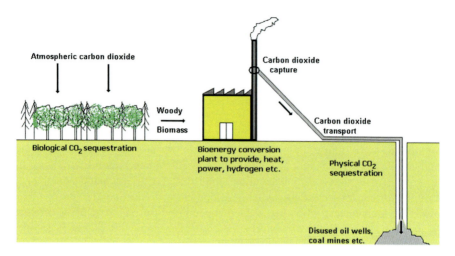

Fig. 9 Energy crops, if grown sustainably and replanted after every harvest, can continually absorb CO_2 from the atmosphere which can then be sequestered after conversion of the biomass to useful bioenergy carriers, thereby, over time, lowering the atmospheric CO_2 concentration level

terrestrial carbon stocks that can be either negative or positive. Examples of negative impacts include the deforestation of peat lands to produce oil palm crops and the increase of the land area under corn crops for biofuels resulting in other crops such as soybean having to be grown elsewhere, at times necessitating further deforestation. Where marginal land is utilised for energy crop production, or biochar (IBI 2012) is incorporated into the crop production cycle, soil carbon contents can be increased and carbon, as part of the carbon cycle, is removed from the atmosphere. This concept could become a key mitigation option, although since it is difficult to measure or model soil carbon uptake and retention over a long term, a large degree of uncertainty remains.

Policies can be implemented for land use, zoning, choice of biomass production systems and minimising any adverse impacts that biomass production might make on biodiversity. They can be designed to ensure that rural development, improvements to agricultural management and climate change mitigation can be realised, though their effectiveness in this regard has not been evaluated.

7 Co-Benefits, Barriers and Policies

In addition to climate change mitigation, RE technologies can provide other benefits. Maximising these benefits depends on the specific technology and site characteristics for each specific RE project. RE systems, in both OECD and non-OECD countries, can contribute to social and economic development, energy access, energy security and a reduction of negative impacts on health and the natural environment, so long as their deployment is carefully implemented. For most countries, increasing the present share of RE in the total energy mix will require policies to stimulate changes in the conventional energy system and also to attract the necessary increases in investment for both technologies and the related infrastructure.

Other than higher costs, a variety of technology-specific challenges are barriers to the significant up-scaling of RE, thereby constraining its potential contribution to climate change mitigation. Some examples are given below.

- Sustainable production and use of biomass requires the proper design, implementation and monitoring of frameworks to minimise the negative impacts. This can also help to maximise the benefits relating to social, economic and environmental issues.
- Regulatory and institutional barriers can impede the deployment of solar PV systems as can power supply system integration and electricity transmission and distribution issues.
- Solar thermal systems for heating water can also be constrained by local planning regulations governing building consents needed for any renovations.
- Geothermal systems, including enhanced systems, need careful design in some locations to avoid land subsidence or even small earthquakes.
- Newly proposed large hydropower projects have been widely constrained in recent years due to ecological and social impacts, cross boundary water supply restrictions as well as remoteness from load requiring costly transmission lines to be constructed. Power lines in themselves usually require planning consents to be approved. These barriers tend to be site specific. They could possibly be overcome by sustainability assessment tools and by early collaboration between multiple parties, such as all water-users along a waterway. Local developments of small, mini- and micro-projects may circumvent some of these barriers.
- Emerging ocean energy technologies require higher investments in R&D as well as demonstrations than is presently the case, with dedicated policies and regulations needed to encourage early developments of these immature technologies.
- Wind energy projects often need to overcome technical and institutional solutions to transmission constraints and allay concerns about integration into electricity systems when high shares (around 10–20 % of total system generation or above) are reached. Public acceptance has also proved to be an issue, especially as installed capacities increase in a locality.

8 Conclusions

There is strong agreement and much evidence that RE systems will have a greater role to play in meeting the ever-growing demands for energy services in the future. This will be in association with energy efficiency measures, as well as in parallel with other low-carbon technology options such as nuclear power and fossil fuels linked with CCS.

The more recent Global Energy Assessment (Johansson et al. 2012) confirmed the findings of the SRREN by stating that 'renewable energies are abundant, widely available, and increasingly cost effective'. This report also concluded that most RE technologies are already commercially available and will be necessary in order to meet a range of pathway objectives that aim to give the desired energy transition to support a sustainable future.

The many proven co-benefits of RE deployment will help to encourage a faster deployment than has been the case to date, even though growth of some technologies has been high. This will be aided by on-going cost reductions for RE technologies that are now more competitive with conventional energy systems than a decade ago. Both centralised scales of large electricity and heat generation plants and decentralised community and domestic scales with numerous small generating plants, can be economically viable in locations where RE resources are good.

The risk of the present RE resource availability declining in the future as a result of climate change is worth considering when developing a specific project in a given location. However, based on the present knowledge of possible regional climate change impacts, overall global RE resources will be affected to a minimal degree, if at all.

References

FAO (2011) Energy-smart food for people and climate, Issue paper, UN Food and Agricultural Organisation, Roma. http://www.fao.org/docrep/014/i2454e/i2454e00.pdf

IBI (2012) International biochar initiative. http://www.biochar-international.org/

IEA (2010) World energy outlook 2010. International Energy Agency, OECD/IEA, Paris. www.iea.org

IEA (2011) World energy outlook. International energy agency, OECD/IEA, Paris. www.iea.org

IEA (2012) Energy technology perspectives. International energy agency, OECD/IEA, Paris. www.iea.org

IPCC (2007) Climate change the physical science basis. Contribution of working group I to the fourth assessment report of the Intergovernmental Panel on Climate Change (IPCC), ISBN 978 0521 88009-1 Hardback; 978 0521 70596-7 Paperback. https://www.ipcc-wg1.unibe.ch/publications/wg1-ar4/wg1-ar4.html

IPCC (2011) Special report on renewable energy sources and climate change mitigation. Prepared by working group III of the intergovernmental panel on climate change. In: Edenhofer O, Pichs-Madruga R, Sokona Y, Seyboth K, Matschoss P, Kadner S, Zwickel T, Eickemeier P, Hansen G, Schlömer S, von Stechow C(eds). Cambridge University Press, Cambridge, United Kingdom and New York, p 1075. http://srren.org

Johansson TB, Nakicenovic N, Patwardhan A, Gomez-Echeverri L, Banerjee R, Benson S, Bouille D, Brew-Hammond A, Cherp A, Coelho ST, Emberson L, Figueroa MJ, Grubler A, He K, Jaccard M, Kahn-Ribeiro S, Karekezi S, Larson ED, Li Z, McDade S, Mytelka LK, Pachauri S, O'Maley M, Riahi K, Rockström J, Rogner H-H, Roy J, Schock RN, Sims REH, Smith KR, Turkenburg WC, Ürge-Vorsatz D, von Hippel F and Yeager K (2012) Global energy assessment. Cambridge University Press, Cambridge, United Kingdom and New York, NY, USA, p 1700. http://www.iiasa.ac.at/Research/ENE/GEA/index_gea.html

Part II
How is the Energy Industry Meteorology-Proofing Itself?

Improving Resilience Challenges and Linkages of the Energy Industry in Changing Climate

Shanti Majithia

Abstract Over the next 50 years changes to temperatures, rainfall patterns, sea levels and more extreme weather are expected globally. The UK energy industry must prepare its infrastructure for the challenges ahead. The Government is investing GBP 200 billion over the next 5–10 years to build a low carbon society. The challenge lies in using this funding to build a climate resilient infrastructure through national strategic planning. This will ensure best value for adaptation and long-term sustainability that supports the transition. The UK Energy companies have taken a sector lead, combining existing and new infrastructure assets with long operational lifetimes to increase the sector's resilience to climate change.

1 Introduction

The UK Government's vision is to implement: (HM Government 2012) *An infrastructure network that is resilient to today's natural hazards and prepare for the future changing climate.*

Evidence that the Earth's climate is warming is unequivocal and some of these recent changes have been attributed to the influences of man-made greenhouse gases. Even if all man-made emissions ceased immediately, the influence of gases already in the atmosphere is likely to result in increasing temperatures, changing rainfall patterns, rising sea levels and more extreme weather throughout the world over the next 50 years. The UK energy industry needs to ensure that its infrastructure can meet the challenges ahead.

A large part of the energy industry infrastructure is highly interconnected with transport, water and information and communication technologies (ICT). These

---

S. Majithia (✉)
ECAS, Berkshire, UK
e-mail: shanti.majithia@hotmail.com

interdependencies could lead to much greater impacts due to a potential cascade of failures. There is, therefore, an urgent need to ensure that the interdependencies are fully understood and considered as part of a business risk process, and that planning strategy is at the top of the companies' risk register. The UK energy sector is already at an advanced stage in dealing with some of the issues by working in partnership with UK climate scientists.

Any future resilience planning must be considered in tandem with UK Government policy addressing both mitigation and adaptation to climate change. Government policies relating to the energy sector are summarised below.

- The *Stern Review on the economics of climate change* (HM Treasury Cabinet Office 2006) presents an authoritative report on the economic consequences of climate change and the need for adaptation.
- The Climate Change Act 2008 (Committee on Climate Change 2012)—The UK is unique in being the only country in the world that has introduced a long-term, legally binding framework to tackle the dangers of climate change. This Act provides a legal framework for ensuring that the Government both mitigates and adapts to climate change. The Act requires that emissions are reduced by at least 80 % by 2050, compared to 1990 levels and by at least 34 % by 2020. The Act also introduces legally binding carbon budgets, which sets a ceiling on the levels of greenhouse gases that can be emitted into the atmosphere to deliver the emissions reductions required to achieve the 2020 and 2050 targets.
- In addition, the Climate Change Act 2008 sets the legal framework for building the UK's ability to adapt to climate change by carrying out a UK-wide Climate Change Risk Assessment (CCRA) (Department for Environment Food and Rural Health 2012a) every 5 years and requiring statutory undertakers such as utility companies to report on how they have assessed the risks of climate change to their work, and what they are doing to address these risks under the Adaptation Reporting Power (Department for Environment Food and Rural Health 2012b).
- A national strategy to deliver emissions cuts was delivered in the UK Carbon Plan 2011, which sets out the transition of the UK to a low carbon economy. Key targets that impact the energy sector include a 15 % target for renewable energy contribution towards total energy demand for 2020, a move to zero-carbon new homes by 2016 and transforming transport by sourcing 10 % of UK transport energy from sustainable renewable sources by 2020. The policy framework for transforming the energy sector to meet these targets is set out in the Energy Act 2008, 2010 and 2011 (Department of Energy and Climate Change 2008), the Planning Act 2008 and Climate Change Act 2008 ensuring that the legislation underpins a long-term energy and climate change strategy of sustainability, affordability and security of supply to meet the low carbon economy.

Preparedness is the key to ensuring that energy infrastructure can meet the challenges that lie ahead. Energy infrastructure assets have long operational lifetimes (up to 60–80 years). The impact of climate change presents a long-term problem and both existing and new infrastructure will need to adapt to ensure the industry is prepared for this challenge. Assets are sensitive not only to the existing climate at the time of their construction, but also to climate variations over the decades. For example, a substantial proportion of infrastructure built in the next 10–20 years will still be in use long after 2050/2080. The energy companies in collaboration with the UK climate scientists have started an evolutionary knowledge-sharing journey to increase the resilience of both new and existing infrastructure.

This chapter describes some significant steps that have been taken to date to achieve a more climate resilient infrastructure. This chapter will also cover the initiatives that the energy sector has made in this climate change journey to address some of the key resilience issues for infrastructure in the coming years. A summary is also provided of the current results from ground-breaking work with UK scientists and the challenges that the energy industry faces to decarbonise.

2 Energy

The energy sector in the UK comprises a number of infrastructure components divided into electricity, gas, oil, renewable and nuclear. The National Infrastructure Plan (NIP) (HM Treasury and Infrastructure UK National Infrastructure Plan 2011) 2011 sets out the investment and strategy needed to fund and replace current ageing infrastructure and states the reasons why resilient infrastructure is imperative for the UK economy:

> The standard and resilience of infrastructure in the UK has a direct relationship to the growth and competitiveness of UK economy, our quality of life and our ability to meet the climate change objectives and commitments.

The current reliability of the energy sector is high and major investments are anticipated in generation, transmission and distribution equipment in order to meet UK greenhouse gas (GHG) and energy reduction targets while maintaining reliability.

The energy industry in the UK has pooled resources to assess the risk of climate change to its sector and incorporate this information into adaptation plans. The risk assessment has been undertaken using UKCP09 (Met Office UK 2012) (the most up-to-date set of climate projections for the UK) and also from tailored collaborative research carried out with UK climate scientists.

This risk assessment has indicated that overall energy industry assets and processes are resilient to the climate change that is projected to occur. Within this assessment there are some assets which require further analysis using more refined data. This is an ongoing process which is incorporated in industry risk management assessments.

It is important to note that even where an asset is at a potential risk in this worst-case scenario model, the risk is localised to the asset and the process it supports.

Future investment to upgrade the energy systems is driven by predicted growth in demand and the development of so-called 'smart'' networks.

"Smart" networks present a parallel challenge for electricity network companies over the coming decades. These initiatives are planned to support the requirement that renewable distributed generation and low carbon (high efficiency) loads can be connected to networks in large numbers, as part of the programme to meet the 2020/2050 carbon reduction targets (National Grid 2011) whilst still maintaining supplies to customers in a cost-effective and reliable manner (more details in Sect. 7).

Overall, this transformation of the network will provide an opportunity to accommodate adaptation to climate change as well.

The energy industry acknowledges that climate adaptation is an evolving science and it is envisaged that the flexible approach that has been adopted will allow for risks to be reassessed as further information becomes available. Gaps in climate science have been identified as part of this study and are described in Sect. 6.

Energy infrastructure (National Grid, Climate Change Adaptation Report 2010) is designed to international standards and the same standards allow infrastructure to operate around the world in varying climatic conditions, including the projected climate conditions for the UK. Climate adaptation reports such as EP1 (Energy Phase 1), EP2 (Energy Phase 2) and ETR 138 (Department for Environment Food and Rural Health 2011) found that there is currently no justification to support adjusting network or asset design standards and except for potential consideration of thermal ratings of equipment and apparatus. The risks posed by fluvial (river) and tidal flooding are well understood and managed on an ongoing basis. The issue of thermal de-rating with increasing temperature needs further investigation. However, it is important to view any potential de-ratings against the response to growth of electricity demand on the transmission network, anticipated to be 0.2 % p.a. (1.4 % p.a. High Growth Scenario) until 2016/2017 (National Grid 7-year statement 2010).

2.1 Energy Delivery in the UK

In general in the UK, energy is transported from the generator or gas import terminal through electricity or gas transmission networks to regional electricity or gas distribution networks that then delivered to energy consumers on behalf of suppliers. This is shown in Fig. 1. Certain end users, primarily large industrial consumers, receive electricity or gas directly from the relevant transmission network, rather than through a distribution network (not shown in diagram).

National Grid Electricity Transmission plc is the owner and operator of the high voltage electricity transmission network in England and Wales; operator, but not

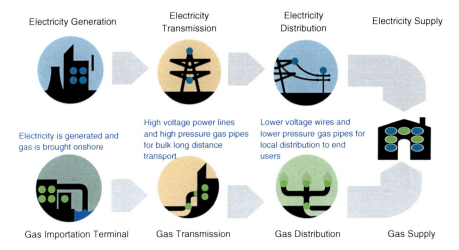

Fig. 1 How the energy industry operates

owner, of the two electricity transmission networks in Scotland; and owner and operator of the national gas transmission system and of four of the eight regional gas distribution networks in Great Britain. National Grid charges electricity and gas suppliers, electricity generators and gas shippers for its services. There are 14 electricity distribution networks in the UK, owned by seven different companies.

2.2 Details of the Electricity Network

The electricity network comprises a mixture of overhead lines and underground cables. In addition there are points on the system, called substations, where voltage transformation takes place and switching and control equipment are located. The interface between the electricity transmission network and the Distribution Network Operators (DNOs) takes place within these grid substations normally at 132 kV as shown in Fig. 2.

These networks are designed to achieve the right balance between cost and reliability. Network planning and design also takes account of normal load growth which has historically been around 1.5–2 % per annum.

3 Infrastructure Components

The individual infrastructure components and their connectivity to the Grid system are described below.

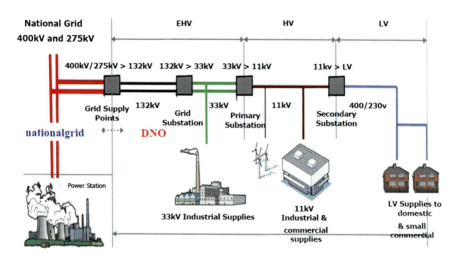

Fig. 2 Electricity transportation process and voltage transformation

3.1 Overhead Lines

(OHL) in the UK transmission networks are constructed using steel towers to support the conductors; these are often referred to as ("pylons")—see Fig. 3. Conductors are suspended from transmission towers using insulators. The conductors are usually aluminium with steel cores and sometimes copper. Different conductor sizes provide different current carrying capabilities. These OHLs normally connect one large substation to another, with no intermediate connections, and are referred to as 'routes'. These routes connect the network together from generators at grid supply points to the grid supply substations, then to the electricity distribution networks which in turn, at lower voltage, radiate out feeding individual customers or small communities/businesses along the route.

3.2 Cables

In the UK, electricity cables are installed and operated at all the common voltages used on the electricity network from low voltage (400/230 v) to high voltage (400 kV).

Fig. 3 Overhead line transmission tower

3.3 Substations

A transmission substation connects two or more transmission lines. The simplest case is where all transmission lines have the same voltage. In such cases, the substation contains high-voltage switches that allow lines to be connected or isolated for fault clearance or maintenance.

3.4 Transformers

Transformers are used to transform voltage from one level to another. Within the transmission systems the most common transformation steps are 400–275 kV and 275–132 kV, which then supplies the distribution networks for further voltage reduction for customer usage.

A large part of the electricity infrastructure is exposed to weather conditions as shown in Fig. 3.

4 Potential Risk to the Infrastructure

There are some significant long-term risks for infrastructure vulnerability (URS 2010) both from extreme weather and gradual climate change in the UK and worldwide. We have already seen major disruption to infrastructure associated with climate events from extreme weather with significant economic consequences [see flooding example Fig. 8]. Some of the key weather and climate sensitivities for the energy industry and their potential impacts are described below. The energy industry experiences a more significant impact from the combination of different weather parameters:

Temperature/heat waves—impact on equipment ratings (e.g. switchgear and transformers) at substations, tunnels, underground cable routes, cable bridges, overhead power lines and towers; overhead lines affected by reduced ground clearance; underground cables affected by increased ground temperatures; heat effects on transformers exacerbated by the urban heat island effect, potential impact on the equipment ratings of control centre systems; increased summer loads reducing the opportunity for planned outages and network reinforcement to enable maintenance—all potentially leading to a possible reduction in the flexibility and capacity of the network; increased air conditioning demand. The load carrying capability of the transformer is primarily dictated by the maximum temperature at which the windings and insulation can be operated without causing damage or faults. The greater the external ambient temperature, the less heating can be permitted from the windings, and consequently the lower the rating. The pattern of demand loading during the day also has an impact because of heat storage, referred to as thermal inertia.

- **Summer drought**—overhead line structures and underground cable systems affected by summer drought and consequent ground movement; reduction in the maximum current rating of cables as thermal conductivity of the ground is reduced.
- **Flooding and heavy rainfall/extreme events**—pluvial and fluvial flooding running the risk that due to extreme flooding a site may be lost or unable to function leading to reduced system security of supply and resilience; flooding of cable bridges; inundation or washaway of overhead towers or routes; risk to substations from flash flooding due to severe rainfall; risk to substations from dam bursts.
- **Sea level rise/tidal surges**—Combined with higher wind speeds, more coastal assets will be exposed to saltwater spray, which due to corrosion will lower asset lifetime. Flooding and inundation leading to the risk of sites being unable to function or lost as a result of extreme flooding; potentially leading to a loss of system resilience or a loss of supply; risk to communications and control infrastructure.
- **Coastal erosion**—risk to infrastructure, including substations, overhead lines and towers, cable bridges, tunnels and underground cable routes.
- **River erosion**—towers, cable bridges and cable routes are at risk of failure if foundations are exposed or weakened, or soil stability is reduced, with routes potentially becoming nonoperational, leading to a loss of system resilience or supply.
- **Vegetation changes**—overhead lines are affected by interference from vegetation, with the risk potentially increasing if growing seasons increase in

the future and there are changes in the growth of vegetation species that are sensitive to climate change.
- **Lightning**—overhead lines and transformers are affected by lightning activity (causing power outages).
- **Atmospheric moisture**—there is the potential for increased rime-icing events if atmospheric moisture increases.
- **Wind storms**—can result in widespread damage caused by trees and wind-blown debris or the exceedance of mechanical load capacity causing tower failures and disruption.
- **Ice storms**—ice accretion potentially causing sag/damage to overhead lines and can cause longer disruption than wind storms.
- **Combined events**—transformers affected by the urban heat island effect coinciding with an increase in air conditioning demand could lead to over-loading in summer months; combinations of adverse weather, such as severe wind and lightning events; conditions suitable for ice loading on lines coinciding with higher wind speeds.

5 Business Preparedness: Infrastructure and Corporate Resilience

The Pitt Review (UK Environment Agency 2009) following the flooding in 2007 stated that

> the driver for business continuity and wider organisational resilience should be in the long-term interests of stakeholders and all those who depend on the organisation in some way.

This has progressed to the Infrastructure and Corporate Resilience Programme (Cabinet Office Infrastructure and Corporate Resilience 2012), led by the Civil Contingencies Secretariat, established in March 2011 to enable public and private sector organisations, of all sizes, to build the resilience of their infrastructure, supply and distribution systems to disruption from all risks (hazards and threats).

The resulting guide has been drawn up in consultation with government departments, agencies, infrastructure owners and operators, trade and professional associations and regulators. It provides a model of resilience that does not depend on additional regulation or standard-setting, but shares best practice and advice to enable owners and operators of the UK's infrastructure to improve the security and resilience of their assets, with support from the regulators where relevant.

5.1 UK Energy Partnership

Energy companies in the UK have worked alongside climate scientists at the Met Office Hadley Centre over the past 5 years on research to better understand and

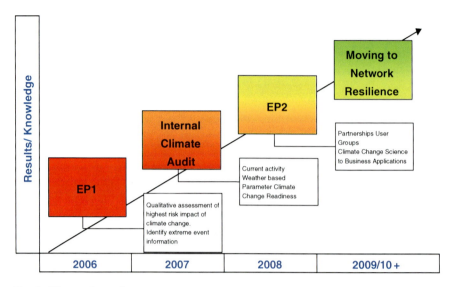

Fig. 4 Climate change journey

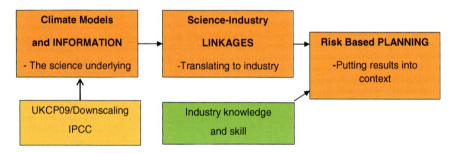

Fig. 5 Partnership approach

prepare for the wider impacts of climate change on UK electricity and gas assets and business operations. This engagement and a number of initiatives related to climate change impacts and the journey taken so far have been summarised in Fig. 4.

5.2 Energy Phases 1 and 2

The project was split into Energy Phase 1 (EP1)—customer requirements and scoping, which facilitated Energy Phase 2 (EP2)—assessing the risk of climate change on the UK's energy industry's infrastructure and business. EP2 was carried out with industry experts working in partnership (see Fig. 5 below) with a view to bring the science closer to practical business application for planning and investment purposes.

EP2 was vital for the industry because of the need to be able to plan 15–40 years ahead for any current and future asset investment. The EP2 project has given the energy industry some of the tools needed for future planning. Energy infrastructure is costly and can have a lifespan of 60–80 years, therefore, the energy industry took the expert advice of the Met Office on climate change. This has helped anticipate the potential impacts of climate change and allowed the industry to future climate proof what it builds in the coming years.

EP2 investigated a number of issues including soil conditions and their impact on cables; how the urban heat island effect might change so the distribution industry can plan city infrastructure; relationship between electricity network resilience and weather; creating a tool to predict sea surges at sites of interest and wind projections.

Potential changes in demand are another key factor as higher temperatures could increase electricity requirements during the summer, as air-conditioning units become more widely used. However, this is balanced by expected lower demand in the winter.

The project covered the following areas:

- Developed innovative new techniques that apply climate models to energy systems applications so that the industry is better placed to adapt to climate change;
- Investigated future wind resources, enabling the industry to understand the continued uncertainty of future wind power. This will assist risk management and investment decisions;
- Modelled future soil conditions and their impact on cables so that companies can understand the cost and benefits of installing cables for a more resilient future network;
- Deliver a tool to enable the energy industry to assess if rising sea levels should be considered in more detail;
- Investigated how the urban heat island effect may change in the future so that networks can develop plans for their infrastructure in cities;
- Produced guidance to help make best use of public domain information on climate change such as the United Kingdom Climate Impacts Programme new scenarios of climate change (UKCP09);
- Delivered new site-specific climatologies of temperature, wind speed and solar radiation that account for climate change so that decisions can be based on realistic climate expectations;
- Examined the relationship between historic weather patterns and network fault performance with a view to developing a tool to predict future network resilience.

The EP2 project found that because of climate change:

- With a few exceptions, such as the thermal ratings of equipment and apparatus, there is currently no evidence to support adjusting network design standards. For example, existing design standards for overhead line conductors do not require change;
- The risk profile for distribution transformers will be affected. Design thresholds of temperature will be exceeded more often and there will be more hot nights in the cities;
- Soil conditions will change; higher temperatures and seasonal differences in soil moisture are expected. Future conditions could be included in cable rating studies by increasing average summer soil temperatures in the models by approximately 0.5 °C per decade;
- Wind resource is uncertain and understanding future resource represents a significant challenge for both investors in wind generation to understand how Climate Change affect their investments (20-year asset should not be affected much) and more extreme wind events impact on electricity network assets which have a much longer time horizon.

5.3 Climate Risk Assessment on Future Network Resilience

Following on from EP2, further work was carried out to carry out a climate risk assessment of the UK electricity networks (McColl et al. 2012). Transmission and distribution networks are susceptible to localised faults caused by weather such as lightning, snow and high winds. The objective of this project was to investigate whether the networks' risk to weather-related faults may change in the future as a result of climate change. Present-day resilience of the networks to weather was assessed by examining past data for weather-related faults. These assessments were then used in conjunction with the Met Office's regional climate model to assess the potential hazard to the network from weather-related faults in the future. The vulnerability of the network to these types of faults is likely to be affected by a number of factors: the age of equipment, the demand on the system and the timing of maintenance schedules. Regional differences in vulnerability are also apparent and can be attributed to many factors; for example, the length of the network in a licence area, the number of customers and the ratio of underground to overground equipment are all likely to contribute to these differences, the age of equipment, timing of maintenance schedules and demand on the network. Detailed information to analyse these differences was not available, but alternatively the average number of customer interruptions caused by a type of weather-related fault is known. This metric is used as an indicator of vulnerability as it summarised the impact of a fault and is likely to be influenced by the factors described above. The headline results were:

Improving Resilience Challenges and Linkages of the Energy Industry 125

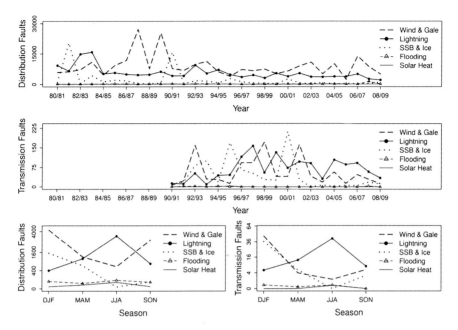

Fig. 6 Annual totals of weather-related fault types on both the distribution (*top*) and transmission (*middle*) networks. The *bottom plots* are of the average number of weather-related faults to occur each season for the distribution (*left*) and transmission (*right*) networks

- In the majority of areas the network is currently at greatest risk from lightning, wind and gale followed by snow. This is because although wind and gale faults occur more frequently, the network is more vulnerable to lightning faults.
- The risk of wind and gale may remain the same, or increase/decrease by a small amount.

Combining projections of future faults with the vulnerability metric, it was found that:

- Wind and gale faults occur frequently; due to the uncertainty associated with future climate projections of wind, we cannot quantify how this risk may change in the future. This is shown for distribution and transmission in the four charts of Fig. 6.
- The risk of lightning faults to the network is projected to increase in the future by up to 40 % as a consequence of more days of higher convection. This percentage changes in the lightning faults relative to the baseline (1990–2009) for each 10-year period up to 2080 in one licence area is shown in Fig. 7 below.
- Snow faults may decrease in the future due to fewer days of snow, but when it does snow the intensity of the event may be the same or may increase.

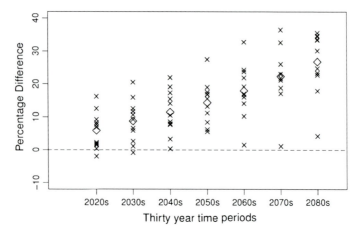

Fig. 7 The percentage changes in future to 2080 in the lightning faults. The black crosses represent each of the 11 model variants and the diamond the mean

Fig. 8 Example of flooding in Gloucestershire, UK 400 kV substation

5.4 Resilience of Flooding of Grid and Primary Substations

The serious incidents of flooding in UK (in the Carlisle, South Midlands and South Yorkshire, Gloucestershire areas) during the summer of 2005, 2007 and 2008 highlighted the potential vulnerability of electricity substations to major flood incidents from current levels of flooding, see Fig. 8 of a substation being flooded.

In the absence of any specific guidance on the level of acceptable flood risk or regulatory impact assessment, it was recognised that the extent of the responsibility has been unclear. This facilitated the development of an industry Engineering Technical Report, ETR 138, setting out a common approach to the assessment of flood risk (inclusive of allowances for climate change and sea level rise) and the development of target mitigation levels that are subject to cost benefit assessment. Some parts of the energy industry are preparing:

- a scheme to undertake flood mitigation work at all sites at risk of a 1:100 year fluvial and tidal flood event
- undertake the next phase of detailed site surveys for sites indicated at risk from a 1:200 year flood event inclusive of climate change
- undertake assessments to understand the risks posed to substations from other sources of flooding, e.g. pluvial (extreme rainfall), and erosion risks.

As a consequence of large flooding events, the electricity transmission and distribution industry has set out target levels (standards) of resilience for different assets within their sector, which includes a risk-based target of the 1 in 1,000 (0.1 %) annual probability flood for the highest priority assets defined as "Critical National Infrastructure". Other measures to improve resilience include the capacity to reconnect or provide an alternative energy supply to consumers. This model of co-operation in the development of standards is being rolled out further to evaluate other hazards in the energy sector.

6 Information Gap Analysis in Climate Science

The energy industry has identified gaps in the climate science knowledge to enable better assessments of the risks going forward. Specifically, the following limitations have been identified which the climate community will need to address in the near future:

- There is no information on future changes in frequency and intensity of wind and gales, including the combined probability of low wind speed (dead calm) events with high ambient temperatures. There are probabilistic projections of wind speed, but they are associated with large uncertainty.
- There is no information on future changes in the frequency and intensity of lightning;
- There is no information on future changes in frequency and intensity of snow, sleet, blizzard, ice and freezing fog.

For the energy industry to better manage its business risks, the scientific community will need to fill the knowledge gaps noted above.

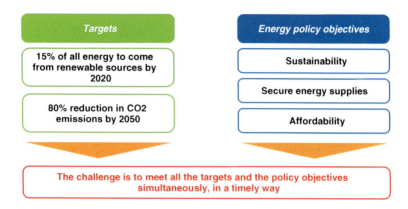

Fig. 9 Targets and energy policy objectives

7 Market Mitigation 2020/2050

There is a parallel challenge for electricity network companies, over the coming decades, in the "Smart" networks in addition to climate change. This initiative is planned to support the requirement that renewable distributed generation and low carbon loads can be connected to networks in large numbers, as part of the programme to meet the 2020/2050 Carbon Reduction targets mentioned at the beginning of this chapter, whilst still maintaining supplies to customers in a cost-effective and reliable manner. This will mean that networks are likely to undergo considerable change at the same time that work may need to be carried out to improve resilience to climate change impacts.

The scale of the change to "smart" networks is likely to be very large—the resultant upgrade may be far larger than required to accommodate potential adaptation requirements. Therefore, although it is essential to research fully the potential effects of climate change in order to understand the potential impacts and mitigations, it is probable that the scale of any network upgrades will be dictated by the drive to low carbon networks.

The energy industry, as part of Electricity Network Strategy Report (ENSG) (Energy and Utilities 2012), has developed pathways for electricity generation and demand scenarios consistent with the EU targets for 15 % of UK energy to be produced from renewable sources by 2020 and greenhouse emissions targets for 2020 and beyond. The challenge is to meet targets and energy objectives in a timely manner as set out in Fig. 9. This figure illustrates the importance of synchronising government targets and energy policy objectives, in a timely manner, whilst making best use of up-to-date skills in climate science.

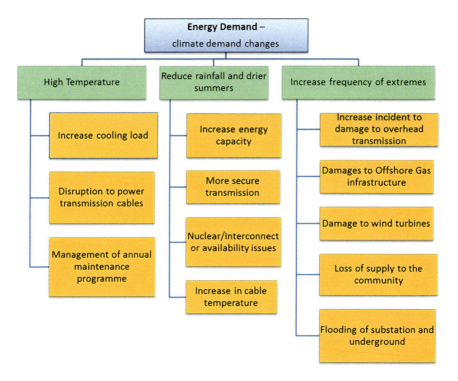

Fig. 10 Meteorology, climate and energy linkages

8 Linking Meteorology, Climate and Energy

As has been seen throughout this chapter, there is a strong link between meteorology, climate and energy which is demonstrated in the diagram in Fig. 10. It shows how temperature, rainfall and extreme events could impact energy usage, infrastructure and transmission, and how these are linked. This linkage is vital for both the scientific and engineering community to understand and work together in the future to produce practical solutions to cope with the challenges of a changing future climate.

9 Conclusion

This chapter has given a high-level view of how the UK is keeping the climate change challenge agenda in the forefront for the energy sector by embedding the regulatory requirements in its national legislative framework.

The progress so far has been demonstrated by working in partnership with UK climate scientists. Through this work and other related case studies the energy industry has been able to assess the challenges posed by climate change and identify how best to climate proof and develop resilient infrastructure to meet those as well as related technology changes. The importance of filling the gaps in climate science knowledge to meet energy industry requirements is key to progress. Some practical examples have shared the potential of future work to aid this progress as well as providing tools and processes for other sectors that may not yet have embarked upon their climate resilient process.

Acknowledgments I am grateful to:

National Grid (UK), Energy companies in UK and the UK Met Office (UK) for allowing me to use some of the material in this paper.

Professor Paul Hardaker, FRMetS CMet CEnv—Chief Executive, Royal Meteorological Society UK and Dr. Lynsey McColl, Climate Change Consultant, Met Office Hadley Centre UK for their professional review of this chapter.

References

Cabinet Office Infrastructure and Corporate Resilience (2012) http://www.defra.gov.uk/publications/files/climate-resilient-infrastructure-full.pdf. Accessed March 2012

Committee on Climate Change (CCC) (2012) Independent advisors to the UK government on tackling and preparing for climate change. http://www.theccc.org.uk/about-the-ccc/climate-change-act. Accessed March 2012

Department for Environment Food and Rural Health (Defra)—UK Climate Change Risk Assessment(2012a) http://www.defra.gov.uk/environment/climate/government/risk-assessment/. Accessed April 2012a

Department for Environment Food and Rural Health (Defra)—Adaptation Reporting Power received reports (2012b) http://www.defra.gov.uk/environment/climate/sectors/reporting-authorities/reporting-authorities-reports/. Accessed April 2012b

Department of Energy and Climate Change 2008—Energy Act (2008) http://www.decc.gov.uk/en/content/cms/legislation/energy_act_08/energy_act_08.aspx. Accessed March 2012

Department for Environment Food and Rural Health (Defra) (2011) http://www.ena-eng.org/ENA-Docs/EADocs.asp?WCI=DocumentDetail&DocumentID=8021. Accessed Dec 2011

Energy and Utilities—Stakeholders and decision makers in the energy industry (2012) http://www.energyandutilities.org.uk/industry_news/content/1586/. Accessed April 2012

HM Government (2011) Climate Resilient Infrastructure: Preparing for a Changing Climate, May 2011 http://www.defra.gov.uk/publications/files/climate-resilient-infrastructure-full.pdf. Accessed March 2012

HM Treasury Cabinet Office (2006) Stern Review on the economics of climate change http://webarchive.nationalarchives.gov.uk/+/http:/www.hm-treasury.gov.uk/sternreview_index.htm. Accessed March 2012

HM Treasury and Infrastructure UK National Infrastructure Plan (2011) http://www.hm-treasury.gov.uk/national_infrastructure_plan2011.htm. Accessed March 2012

McColl L, Palin E, Thornton H, Sexton D, Betts R, Mylne K (2012) Assessing the potential impacts of climate change on the UK's electricity network. Clim Change 115:821–835

Met Office UK (2012) UKCP09: Gridded observation data sets. http://www.metoffice.gov.uk/climatechange/science/monitoring/ukcp09/. Accessed March 2012

National Grid (2011) Transportation of British energy 2011 Development of energy scenarios, July 2011 http://www.nationalgrid.com/NR/rdonlyres/C934D455-1438-4949-A7CD-3FE781FE39E8/47855/DevelopmentofEnergyScenariosTBE2011.pdf. Accessed March 2012

National Grid (2010) Climate Change Adaptation Report—National Grid Electricity Transmission plc. http://www.nationalgrid.com/corporate/Our+Responsibility/Resources/Publications+and+Speeches/pubs/ccapadtationelec.htm. Accessed March 2012

National Grid seven year statement (2010) http://www.nationalgrid.com/uk/Electricity/SYS/. Accessed March 2012

URS: Adapting Energy (2010) Transport and Water Infrastructure to the Long-term Impacts of Climate Change. Ref. No RMP/5456. http://archive.defra.gov.uk/environment/climate/documents/infrastructure-full-report.pdf. Accessed March 2012

UK Environment Agency (2009) The Pitt review—lessons learned from the 2007 summer floods. http://www.environment-agency.gov.uk/research/library/publications/33889.aspx. Accessed March 2012

Combining Meteorological and Electrical Engineering Expertise to Solve Energy Management Problems

Giovanni Pirovano, Paola Faggian, Paolo Bonelli, Matteo Lacavalla, Pietro Marcacci and Dario Ronzio

Abstract This chapter describes some activities carried out in Italy dealing with the impact of climatic and meteorological variability on the Italian Electric System. Every item presented attends to a real problem coming from the interaction between atmosphere and energy system. In particular the activities described are relevant to the evaluation of climate change impact on the energy system, the prediction of solar and wind energy production, the monitoring and forecasting of risks for the power system coming from the weather due to insulators surface pollution, ice and snow loads on overhead lines, storms and lightning phenomena.

1 Introduction

Meteorological events and climate changes can have serious impacts over energy sources. Hydropower generation, for example, is the energy source that is likely to be most directly affected by meteorological events in a wide range of time and space scales. Moreover, drought periods may have serious impacts even on thermal power plants operations where water is normally used as coolant fluid.

A different problem comes from the variability over short time periods (hourly and sub-hourly) of solar and wind energy sources that affects the grid management and the energy market, when the percentage of these renewable resources is significant. On the other hand, the knowledge of variability over longer time horizons (inter- and intra-annual variability) of the same sources is used for planning energy investments. For these reasons, in many countries, electrical utilities involved in power production and grid management ask for meteorological products to their local National Meteorological Services or private weather companies.

G. Pirovano (✉) · P. Faggian · P. Bonelli · M. Lacavalla · P. Marcacci · D. Ronzio
RSE S.p.A.—Ricerca Sistema Energetico public company, Milano, Italy
e-mail: giovanni.pirovano@rse-web.it

This chapter describes some cross-disciplinary activities carried out in RSE[1] dealing with the impacts of climatic and meteorological variability on the Italian Electric System. Every item presented in this paper is linked to a real problem coming from the interaction between atmospheric events and the power system. The proposed solutions originate from the dialogue and cooperation between different expertises working in the same company.

2 The Meteorological Demand from Energy Community

An important issue, influencing data and expertise sharing between Meteorological and Energy community, is the kind of relationship that can arise among different enterprises operating in these fields. This issue was one of the main arguments discussed in the ICEM 2011 (Troccoli et al. 2012).

Public and private meteorological services strive to develop new products addressed to the energy companies, but the dialogue among them is not always easy and synergetic. As an example, the energy data availability for meteorologists sometimes is a crucial problem: i.e., energy companies are often stingy in providing their data and weather forecast services are restricted in customizing applications.

Furthermore, only major meteorological institutions (National Weather Services and Universities) can have sufficient public funds for the research activity directed to new weather tools and application procedures. In European Union, public funds are allocated for Energy-Meteorology research, treated by consortiums of universities and private companies. In this frame, some important projects have been carried out for the short-term wind power forecast, as ANEMOS,[2] or for the solar radiance retrieval from satellite, as HELIOSAT.[3]

In the context of renewable energy production, the need for short range weather forecast at local scale represents a new challenge for scientific institutes and meteorological services.

At RSE, a public research company funded by the Government, different expertises are engaged for dealing with interdisciplinary projects like those of energy-meteorology. In this company, atmospheric scientists work together with electrical engineers in order to study the interactions of weather with the energy sector and develop tools for an optimized management of the electrical system. In

---

[1] RSE Spa—Ricerca sul Sistema Energetico—is a joint stock company, whose unique shareholder is GSE SpA, which develops research in electro-energy, with particular focus to the strategic national projects of general public interest, financed by the Research Fund for the Italian Electrical System under the Contract Agreement between RSE and the Ministry of Economic Development.

[2] ANEMOS-PLUS Project, (http://www.anemos-plus.eu/).

[3] Heliosat—3 Project, (http://cordis.europa.eu/search/index.cfm?fuseaction=proj.printdocument&PJ_RCN=4938509).

such a way, some problems, dealing with data availability or electric technical aspects, can be overcome. Some items, carried out at RSE, are described in the following.

3 Climate Change Impact on the Electric System

The Italian energy system is strongly linked to local climatic conditions. Temperature and precipitation changes are likely to affect both energy use and energy production. Some of their possible impacts are rather obvious: in warming conditions less heating will be needed for industrial, commercial, and residential buildings, while cooling demand will increase, with changes varying by region and by season. Climate change could affect the energy sector in regions depending on water supplies for hydropower and/or thermal power plant if the availability of cooling water was to be reduced at some locations because of climate-related decrease or seasonal shifts in river runoff. Moreover higher maximum temperature may have negative effects on electric line efficiency.

The climate system is global and it is characterized by a broad range of spatial and time scales. Global climate models (GCMs) are fundamental research tools for the understanding of climate. GCMs can effectively address large-scale climate features such as general circulation of the atmosphere and ocean and large-scale patterns of temperature and precipitation but they cannot capture regional and local climate aspects, because of their low resolution (about 100 km for the best), limited by computational costs.

As energy demand and production require information about climate changes at regional/local scales, *downscaling* methods are required to generate information below the grid scale of GCMs. There are two main approaches, *statistical* and *dynamical*. The first involves finding robust statistical relationships between large-scale climate variables (typically mean sea level pressure) and local ones (as temperature and precipitation). The second uses regional circulation models (RCMs) nested in GCMs: the boundary conditions, coming from a GCM, "drive" the RCM simulation over a limited area that, because of its limited dimension, may have a high spatial resolution, without overly increasing computational costs. High resolution improves the representation of landscape feature such as mountains, valleys, lakes, as well as other surface morphologies. For this reason, it is possible to describe local and regional circulation and precipitation features as well as temperature patterns, wind structures, and so on.

Several regional models have been developed in different international research projects. In particular, in the framework of the European Project ENSEMBLES (http://www.ensembles-eu.org) an ensemble prediction system for climate change based on state-of-art, high resolution, global and regional Earth system models has been developed for Europe and validated against a quality-controlled, high resolution gridded dataset E-OBS (http://eca.knmi.nl). E-OBS is a daily observational dataset for Europe developed for precipitation, minimum, mean, and maximum

Table 1 List of RCMs used in this study, ordered according to the Research Institute

Institute	Driving GCM	RCM	Run	
CNRM[a]	ARPEGE_RM5.1	Aladin	CNRM-RM5.1_ARPEGE	(C&A)
DMI[b]	ARPEGE	HIRHAM5	DMI-HIRHAM5_ARPEGE	(D&A)
DMI	BCM	HIRHAM5	DMI-HIRHAM5_BCM	(D&B)
ETHZ[c]	HadCM3Q0	CLM	ETHZ-CLM_HadCM3Q0	(E&H)
ICTP[d]	ECHAM5-r3	ICTP-REGCM3	ICTP-REGCM3_ECHAM5	(I&E)
KNMI[e]	ECHAM5-r3	RACMO	KNMI-RACMO2_ECHAM5	(K&E)
HC[f]	HadCM3Q0	HadRM3Q0	METO-HC_HadCM3Q0	(M&H)
SMHI[g]	HadCM3Q0	RCA	SMHIRCA_HadCM3Q3	(S&H)
SMHI	ECHAM5-r3	RCA	SMHIRCA_ECHAM5	(S&E)
SMHI	BCM	RCA	SMHIRCA_BCM	(S&B)

[a] Météo-France (CNRM)
[b] Danish Meteorological Institute (DMI)
[c] Swiss Institute of Technology (ETHZ)
[d] The Abdus- Salam International Centre for Theoretical Physics (ICTP)
[e] The Royal Netherlands Meteorological Institute (KNMI)
[f] UK Met Office, Hadley Centre for Climate Prediction and Research (HC)
[g] Swedish Meteorological and Hydrological Institute (SMHI)

surface temperature for the period from 1950 and it is updated constantly (Haylock et al. 2008).

Using observation data from E-OBS together with ten RCMs selected from ENSEMBLES dataset, a modeling analysis has been done to elaborate climate change projections in the next few decades over Italy, to support policy makers dealing with Italian energetic sector.

The models used are listed in Table 1.

The ENSEMBLES simulations from 1961 to 2050 have been analyzed at horizontal resolution of 25 km, at daily and monthly scale, under the SRES A1B emission scenario (Nakicenovic et al. 2000) over the area shown in Fig. 1.

Considering as reference the 30 years period 1961–1990, climate projections have been done for the period 2011–2040.

The performance analysis has been focused on some subareas of the domain, particularly interesting for energy demand and production in Italy. They are: the Po Basin, two Alpine Regions, and the Po Valley (Fig. 2).

Climate changes at regional scale are very difficult to predict because of the great spread among models: Fig. 3 shows RCMs estimates of annual surface air temperature (*tas*) in comparison with E-OBS data for the Po Valley.

As the maximum values of energy consumption in the mediterranean region are related to cold winter conditions (for heating) and hot summer conditions (for cooling), a preliminary investigation about the performances of the ten RCMs in describing the present climate has been done at seasonal scale.

The model performances in the period 1961–2010 have been evaluated through the bias index defined as follows:

$$BIAS_{RCM} = VAL_{RCM} - VAL_{EOBS}$$

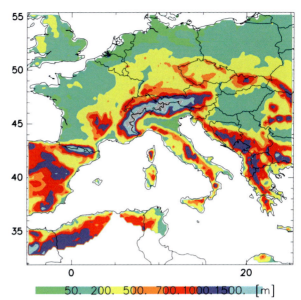

Fig. 1 Analysis Domain and Orography description

where VAL_{EOBS} is the mean seasonal value of the meteorological variable gathered from E-OBS data-set, VAL_{RCM} is the seasonal estimate from the RCM. (Tables 2 and 3) report season bias errors respectively for temperature and precipitation for each sub-area.

International scientific papers recommend the use of *ensemble means* to reduce the uncertainty and to give a variability range in the projections (Christensen et al. 2007; Prein et al. 2011; Kjellström et al. 2011; Faggian and Giorgi 2009). Following this suggestion, future climate scenarios have been elaborated computing an ensemble mean (ENS), in which only the models with the best performances in the description of past century climate characteristics over Italian Region have been used. They are: ICTP-REGCM3_ECHAM5 (I&E), KNMI-RACMO2_ECHAM5 (K&E), CNRM-RM5.1_ARPEGE (C&A); METO-HC_HadCM3Q0 (M&H). Though this ensemble mean gives a better description of the current climate, significant bias still remain if compared with observations.

To reduce errors, *bias correction* techniques have been implemented, according to international methodologies used in impact studies (Dosio and Paruolo 2011; Piani et al. 2010).

Considering the period 1961–1990, the probability distribution function (PDF), for each grid point, of RCMs daily temperature simulation values has been analyzed in comparison with the PDF of E-OBS data. Then transfer functions were derived to fit the PDF model data to the reference PDF.

Averaging the model PDFs, a new temperature distribution (COR PDF) was obtained and the accordance with the observed data was verified for the 20-year period 1991–2010 (not shown).

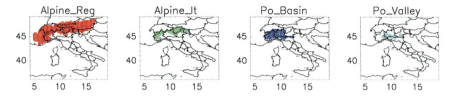

Fig. 2 Sub-regions of particular interest for energetic sector in Italy

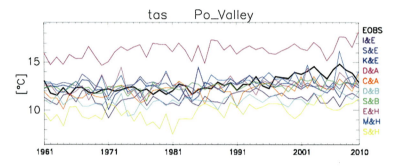

Fig. 3 Annual temperature values from 1961 to 2010 gathered from E-OBS data-set (*black line*) and RCMs outputs (*coloured lines*)

Table 2 Tas: seasonal mean E-OBS values and RCM Bias for the period 1961–2010, in green the smallest errors, in red the worse ones

BIAS		EOBSval	I&E	C&A	D&A	D&B	E&H	K&E	M&H	S&E	S&H	S&B	ENS
Po Basin	DJF	1.5	-0.6	-2.4	0.5	-3.1	-3.9	-0.5	-2.2	-2.2	-5.9	-2.6	1.4
	MAM	9.1	-3.0	-2.4	-0.4	-3.6	-4.9	-1.2	-2.4	-2.6	-4.1	-2.8	-2.2
	JJA	18.6	-3.1	0.1	3.6	-3.8	-0.7	-0.1	1.7	-2.6	-4.0	-3.0	-0.3
	SON	10.6	-2.4	-2.3	1.3	-3.3	-3.1	-1.0	-1.1	-2.1	-4.3	-2.0	-1.7
AlpineReg	DJF	-0.5	-0.5	-2.4	-0.4	-3.7	-3.7	-0.7	-2.5	-1.8	-4.7	-1.9	-1.5
	MAM	6.5	-2.6	-2.6	-0.9	-4.1	-4.6	-1.6	-2.8	-1.7	-3.0	-1.6	-2.4
	JJA	15.5	-2.7	0.2	3.1	-3.4	-0.6	-1.0	0.9	-2.3	-2.9	-1.9	-0.7
	SON	8.0	-1.8	-2.4	0.9	-3.4	-2.8	-1.1	-1.2	-1.6	-3.0	-1.1	-1.6
ItalyAlps	DJF	0.2	-0.8	-2.6	-1.2	-4.8	-5.7	-1.4	-3.3	-4.4	-7.5	-4.8	-2.0
	MAM	7.5	-3.0	-2.6	-2.1	-5.4	-6.8	-2.3	-3.4	-3.9	-5.4	-4.0	-2.8
	JJA	16.9	-3.0	0.0	1.6	-5.1	-2.7	-1.5	0.5	-3.9	-4.9	-3.8	-1.0
	SON	9.1	-2.4	-2.5	-0.5	-4.8	-4.6	-1.9	-1.9	-3.4	-5.1	-3.1	-2.2
Po Valley	DJF	3.4	0.4	-1.1	2.1	-1.5	-1.7	-0.2	-0.8	0.8	-3.7	0.7	-0.4
	MAM	12.1	-2.3	-1.8	0.8	-2.2	-2.8	-1.2	-1.5	-1.1	-2.7	-1.3	-1.7
	JJA	21.9	-2.5	0.9	5.1	-2.7	1.2	-0.1	2.7	-1.8	-3.5	-2.4	0.2
	SON	13.3	-1.8	-1.6	2.8	-2.0	-1.6	-1.1	-0.4	-1.0	-3.8	-1.0	-1.2

These methodologies were applied to give some information about future change in energy demand. The energy demand depends closely on surface temperature variability, with maximum energy requirements correlated to temperature extreme values. According to Giannakopulus et al. (2009), the temperature impacts on energy demand have been analyzed using two degree-day indexes, to characterize the heating (*hdd*) and cooling (*cdd*) energy request:

Table 3 Pr: seasonal mean E-OBS values and RCM Bias (%) for the period 1961–2010, in green the smallest errors, in red the worse ones

BIAS		EOBSval	I&E	C&A	D&A	D&B	E&H	K&E	M&H	S&E	S&H	S&B	ENS
Po Basin	DJF	2.0	57.0	-3.0	22.0	42.0	10.0	41.0	26.0	81.0	43.0	117.0	41.0
	MAM	3.0	48.0	42.0	19.0	78.0	28.0	25.0	52.0	73.0	64.0	109.0	43.0
	JJA	3.0	-6.0	30.0	-58.0	67.0	-2.0	-40.0	0.0	15.0	91.0	111.0	24.0
	SON	3.0	35.0	1.0	6.0	48.0	24.0	12.0	17.0	43.0	35.0	88.0	35.0
AlpineReg	DJF	3.0	77.0	13.0	70.0	104.0	39.0	48.0	62.0	71.0	39.0	99.0	51.0
	MAM	3.0	54.0	39.0	36.0	82.0	22.0	37.0	56.0	73.0	57.0	82.0	46.0
	JJA	4.0	4.0	7.0	-53.0	35.0	-12.0	-19.0	9.0	30.0	51.0	62.0	13.0
	SON	3.0	41.0	2.0	10.0	70.0	15.0	23.0	21.0	53.0	25.0	73.0	28.0
ItalianAlps	DJF	2.0	92.0	2.0	50.0	77.0	20.0	67.0	45.0	102.0	49.0	152.0	57.0
	MAM	3.0	70.0	57.0	51.0	120.0	39.0	51.0	75.0	97.0	81.0	129.0	63.0
	JJA	3.0	-8.0	29.0	-44.0	73.0	17.0	-28.0	13.0	40.0	107.0	117.0	20.0
	SON	3.0	53.0	10.0	28.0	79.0	30.0	30.0	32.0	60.0	40.0	107.0	40.0
Po Valley	DJF	2.0	21.0	-2.0	-10.0	23.0	1.0	23.0	-8.0	40.0	28.0	74.0	23.0
	MAM	2.0	17.0	30.0	-31.0	13.0	11.0	0.0	18.0	27.0	28.0	49.0	29.0
	JJA	2.0	-12.0	16.0	-80.0	45.0	-42.0	-57.0	-33.0	-26.0	37.0	57.0	12.0
	SON	3.0	-4.0	-15.0	-35.0	0.0	3.0	-10.0	-17.0	5.0	0.0	32.0	12.0

$$\text{heating degree day} \quad hdd = \max(T^* - T_i, 0.)$$
$$\text{cooling degree days} \quad cdd = \max(T_i - T^{**}, 0.)$$

where $T^* = 15\ °C$, $T^{**} = 25\ °C$ are respectively the *base temperature*[4] for *hdd* and *cdd*.

Focusing over *cdd* patterns in summer season (Fig. 4), a good accordance between reference values (MOD E-OBS) and computed values (COR PDF) was obtained in the reference period. The future scenarios are characterized by an increase of about 1 °C (last diagram of Fig. 4) and by an increasing of 20–25 % of hot cases (not shown). As a *cold bias* of about 0.5 °C was found over Po Valley in the verification period 1991–2000 (not shown), the future projections likely underestimate the actual increase in peak energy demand for cooling in the next decades over Italy.

4 Forecasting "Weather Energy"

The analysis of climate scenarios for the next decades provides fundamental information in planning solar and wind farms. Over the short-term, weather forecasting might help utilities in integrating into the grid the power produced by the farms. The most popular renewable energies, solar and wind, are fluctuating energies, in the sense that they are characterized by intermittency due basically to local weather conditions. The skill to harness these energy sources is a challenging effort and this can be performed by using empirical models, such as autoregressive, neural networks, or advanced Numerical Weather Prediction (NWP) models, or a mix of them. The formers may be used for few hours ahead, because their predictive capabilities rely on the presence of persistent trends, whilst the latter ones

---

[4] The temperature where the consumption is at its minimum

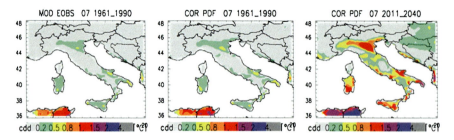

Fig. 4 cdd gathered from E-OBS data in reference period (*left diagram*), cdd estimated from COR PDF in reference period (*central diagram*) and cdd projections for the scenario 2011–2040 (*right diagram*)

allow forecasts up to three days ahead. Forecasts have to be tailored for use by utilities, and more and more accurate predictions are continuously requested (see also George 2013; Renné 2013; Gryning et al. 2013; Haupt et al. 2013; Katzfey 2013; Dubus 2013; Kay et al. 2013; Lorenz et al. 2013).

Despite of well-known astronomical parameters, solar resource depends on microphysics and aerosols content in the atmosphere nearby the site of interest. Performing a good forecast of solar irradiation by means of a NWP implies an efficient parameterization of all the processes involving clouds both in space and in time.

Different surface irradiation components, evaluated by means of a NWP model, are fundamental information required by different typologies of solar power plants. In particular, photovoltaic plants need global irradiation, while concentrating solar systems are very sensitive to the Direct Normal Irradiation. In the former situation, it is possible to extract some energy even in cloudy conditions, because of the diffuse component scattered by clouds. In the latter a small cloud may reduce sharply the power production; for this reason reliable higher frequency forecasts would be desirable.

It is possible either to use directly the NWP radiative outputs or an off-line Radiative Transfer Model (RTM). The convenience of using an off-line model lies in the potentiality of managing different radiative schemes or including some site-depending parameterizations, but usually these models don't describe microphysics processes, rather they depend on a non-complete description of clouds furnished by the NWP.

Forecasts have to be verified against measurements. In RSE experiments, the NWP Regional Atmospheric Modeling System (RAMS) (Pielke et al. 1992) is applied with horizontal resolution of 5 km and its output feeds a RTM that is a modified version of the Harrington scheme (Gabriel et al. 1997). Measurements come from three certified solar stations located in Northern, Central, and Southern Italy. In this manner, it is possible to examine the RTM or NWP behavior in different climatic conditions. The forecast duration is 3 days ahead and generally there is a little degradation in the performances with the forecast horizon (about 5 %). In Fig. 5 hourly forecasted and measured global, diffuse and direct normal

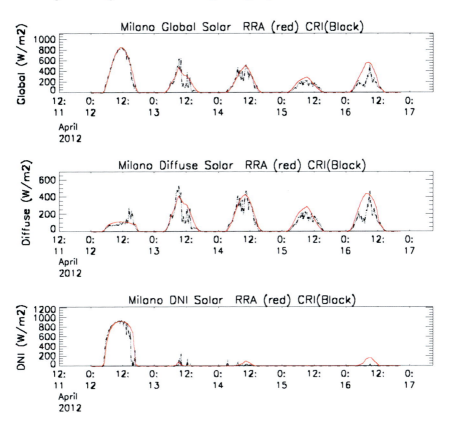

Fig. 5 Hourly solar global (*top*), diffuse (*middle*) and direct normal (*bottom*) irradiance forecast during a clear (first day) and cloudy days in Milan (12–16th April 2012); *black dots* ground measurements, *red lines* RAMS since 12–16th April 2011

irradiation in Milan are shown for a 4-days period. In clear sky conditions, the accordance between model evaluations and measurements is very high, especially for the global irradiation. Regarding the hourly data in Milan for the whole year 2011, the correlation is about 0.82 (Fig. 6).

The wind power forecast system used at RSE is based on the output of RAMS, but a Model Output Statistic (MOS) correction is applied to reduce systematic errors caused, for instance, by a wrong model representation of topography or surface roughness. The MOS technique is based on a Neural Network, and so measurements for the training are needed. Further research is focused on the ECMWF Ensemble Prediction System (EPS) in order to predict the accuracy of a deterministic forecast. Positive results have been obtained in calibrating the forecasted wind spread with local observations

Fig. 6 Forecasted versus measured horizontal global irradiance in Milan using only cloudy hourly data for the whole 2011

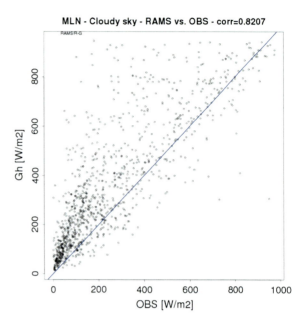

5 Weather Risks for the Power System

Some extreme weather events threaten the continuity of power supply. Floods, hurricanes, thunderstorms, and snowstorm put a strain on power grids at a time when the energy supply is fundamental to the operation of relief, telecommunications, hospitals, and there is the need to limit the damage and casualties. The effects of natural disasters increase their severity in the most densely populated and urbanized regions of the world, regardless to the development's level of the country. In Italy, the electricity distribution and the transmission systems are vulnerable to natural events such as thunderstorms, salt deposits, and heavy snowfalls. Appropriate forecasting systems allow optimal dispatching solutions aimed at containing the effects of such events.

5.1 Pollution Deposit on Insulator

Pollution deposits (marine salt, industrial and agricultural pollutants), in the presence of elevated humidity levels, greatly increase the surface conductivity of lines and substations insulators and can lead to surface discharges and electrical short-circuits. The persistence of these conditions poses a threat to the continuity of service of the electrical grid and may produce extended blackouts in large areas of the network (the last blackout phenomena due to pollution interested the whole

Fig. 7 View of the AMICO system and of its insulator probe

island of Sardinia, and required an extensive washing of the insulators of lines and substations before restoring the electrical grid into service).

In order to know in advance the possible occurrence of critical contamination deposits conditions and to properly plan washing operations of the insulators, a system for monitoring pollution insulator level was planned and realized.

The acronym of the system is AMICO (**A**rtificially **M**oistened **I**nsulator for **C**leaning **O**rganization) (Rizzi et al. 2007) and it carries out the measurement of the degree of pollution of insulators by means of a probe insulator placed in the monitoring zone and subject to the same pollution as the insulators that need to be monitored (Fig. 7).

The level of contamination is detected by applying a voltage to the insulator probe and measuring its conductivity. The particularity of the operation is that the measurement is carried out under controlled humidified surface conditions: the temperature of the probe is in fact taken below the dew point by a coolant liquid inside the insulator. In this way, it is possible to monitor the pollution deposits also during dry periods, in which the ambient humidity is not able to produce any conductivity of the contamination layer, but in which salt, carried usually by strong winds coming from the sea, may accumulate at a high rate on the insulators.

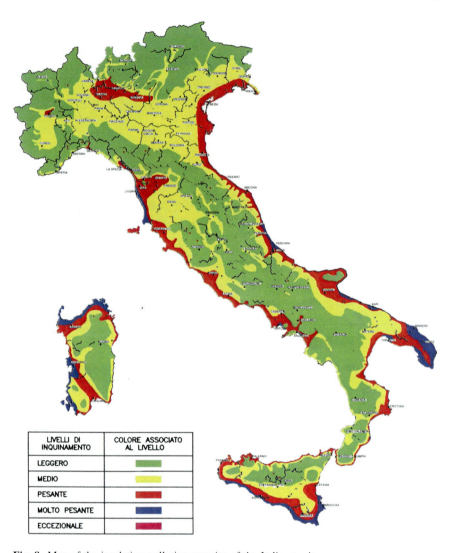

Fig. 8 Map of the insulation pollution severity of the Italian territory

Five AMICO units are installed inside TERNA (the Italian TSO[5]) substations, chosen among those subject to higher pollution levels. The places of installations are indicated in the map (Fig. 8) that reports the insulation pollution severity over the Italian territory.

---

[5] Transmission System Operator.

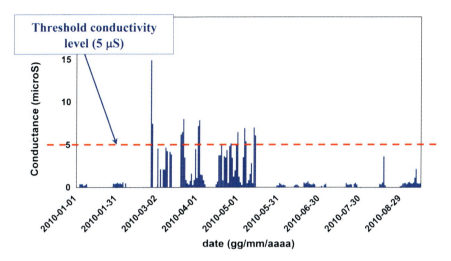

Fig. 9 Graph with site conductivity measurements

When the measured conductivity values are greater than given threshold values (Fig. 9) an alert information is sent to the TSO to allow the adoption of adequate operations.

The AMICO systems are also equipped with a meteorological station for the acquisition of wind speed and direction, temperature, relative humidity, and rainfall.

Analyses of meteorological conditions associated with pollution events are regularly carried out: as above indicated critical conditions often coincide with long periods with the absence of rain and with strong winds coming from the sea.

Another system for monitoring insulator pollution (always by measuring surface conductivity) is illustrated in Fig. 10 (Pirovano et al. 2011). In this case, the system is installed directly on High Voltage lines insulators chains.

The system (called ILCMS[6]) performs measurements of leakage current of insulator under line voltage and under actual meteorological conditions, so in this case we measure the current when there actually is humidification and pollution. The two systems perform complementary measurements.

One AMICO and one ILCMS units are also installed in the experimental station for natural pollution of insulators of EDF R&D in Martigues (close to Marseille), a location characterized by exceptional values of pollution levels.

In this station, measurements are carried out on insulators of different types (glass, silicon, glass covered by silicon) in order to test the material and make performance comparison and evaluation of expected useful life. The direct comparison of different insulator solutions installed in a very severe natural

---

[6] Insulator Leakage Current Monitor System.

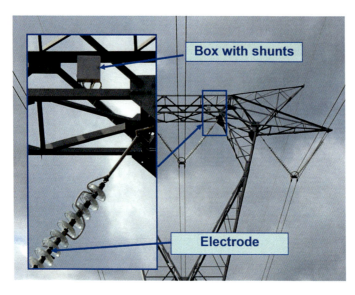

Fig. 10 System for monitoring pollution on overhead line insulators

environment, is of particular importance for the selection of the insulators to install on overhead lines and substations.

In addition to insulator conductivity monitoring, the level of pollution of the environment is also evaluated by the measurement of soluble (mainly marine salt) and nonsoluble pollutants, contained in samples of deposits removed from different types of insulators, installed on suitable poles placed in the vicinity of the AMICO systems (Fig. 11).

These measurements, together with other from tests carried out in laboratory, where insulator pollution can be reproduced (by applying salt fog or by humidifying a layer of salt and kaolin on the insulators), are used for the definition of threshold values for the different types of insulators of the Italian grid.

In order to verify the possibility to extend the pollution monitoring to other places than those of AMICO systems, and to experiment the feasibility of forecasting the saline contamination, studies were carried out for setting up an empirical model of contamination levels using meteorological variables.

Taking into account air humidity, wind strength, and direction, precipitation amount measured on a coastal site, it was possible to set up a threshold model based on the number of dry days, length of windy periods from the sea and the final humidity conditions. Preliminary positive verifications of this model have been carried out with reference to periods characterized by meteorological conditions favorable to salt contamination and to relevant conductivity measurements (from AMICO systems) for some sites along the Italian shores. Feeding the empirical model with the above-mentioned variables, produced by a numerical

Fig. 11 View of high voltage insulators for salt deposit measurements and for laboratory tests

meteorological model, will make it possible to forecast the critical contamination days and give warnings to grid people.

5.2 Ice and Snow Accretion on Overhead Lines

The overload of snow on OverHead Lines (OHL) leads to a decrease of the quality of transmission service and, in some cases, it can also cause serious security problems in the performance of that activity. Where anti-icing and de-icing technologies cannot be applied, as in Italy, dispatching solution based on particular weather prediction systems remains the only approach applicable to the entire High Voltage network. The phenomenon of icing formation on power lines as shown in Fig. 12, is known internationally as "Wet-snow accretion on OHL" (Sakamoto 2000). The overload of ice on conductors may cause the breakage of the wire itself or even the fall of the lattice.

The impacts of these phenomena on various human activities have been treated by the International scientific community through the action COST 727 (Fikke et al. 2007), funded by the EU, which had the objective to exchange experiences between experts from various European countries. About the effects of icing on power lines, there are the recommendations of International Electrotechnical Commission (IEC 1997) and International Council on Large Electric Systems (CIGRE 2005) dealing with methods of measurement and modeling the phenomena. The IWAIS (International Workshop on Atmospheric Icing of Structures) is another important initiative.

An activity research has been undertaken in order to investigate the physics of "wet snow", consisting either in the analysis of historical meteorological series or in new observational activities. In fact, the lack of experimental data, due to the difficulties in reproducing wet-snow in the laboratory, and the interest in the

Fig. 12 Wet-snow accretion on overhead line during a heavy snowfall in the Alps

weather conditions at the base of these events, justify carrying out experimental campaigns necessary to collect unconventional weather data and ice accretion parameters.

5.3 Atmospheric Icing Phenomenon

Investigations on the winter electrical outbreaks in Italy show that they are mainly caused by the particular characteristics of snowfall. In the international literature (Admirat and Sakamoto 1988; Farzaneh 2008), this phenomenon is called "**wet-snow icing**" and it is different from the other phenomenon "**in-cloud icing**" more frequent in Northern Europe and Canada. Wet-snow is a precipitation icing that occurs when the surface air temperature is close to the freezing point. There is not an exact temperature range for wet-snow conditions, but the most well-known interval, supported by many observations in different countries, is within the range of 0–2 °C. In this atmospheric layer with positive temperature, the snowflakes can partially melt and increase their liquid water content. Under these conditions, the snowflakes settle on the conductor and join together, not only by the mechanism of collision, but also because of the strong coalescence due to the presence of liquid water promoting the growth of a sleeve around the cable. The typical duration of wet snow events is 18–24 h, producing snow sleeves till 15 cm in diameter and causing an extra load on conductors up to 8–10 kg/m. This overload can produce serious damages to Overhead Lines, exceeding the mechanical resistance of lattice tower supports. In some cases, the conductor undergoes an extra load due to the intense wind blowing after the accretion event.

5.4 WOLF (Wet-Snow Overload aLert and Forecasting)

At RSE a forecast and an alert system, named **WOLF** (**W**et snow **O**verload a**L**ert and **F**orecasting), has been developed and tested with the aims to identify every day the forecasted weather conditions favorable to the overload of snow over the entire national high voltage network and to provide an estimate of the mass of ice on conductors. WOLF can automatically send alert messages via SMS to those responsible for network maintenance who can take appropriate solutions. The system integrates a wet-snow accretion model (Makkonen 2000; ISO 12494-2000), an interface for the forecasted data and a GIS-display system. The forecast weather data, coming from the nonhydrostatic model RAMS,[7] are filtered with appropriate selection criteria for the wet-snow identification. Every day RAMS provides the principal meteorological variables on a regular grid with a mesh of about 5 km, with a forecast horizon of 78 h. The outputs of the NWP feed the "Makkonen model," assuming a conservative growth of the sleeve on a cylindrical conductor. The grid of the meteorological model has been intersected with the path of the Italian transmission lines to obtain for each point of the model, the characteristics of power lines.

The WOLF output, presented to the user on a WEB-GIS, allows to display the alert points on several specific thematic layers (Fig. 13). WOLF provides, for each grid-point, a chart where the time evolution of the principal weather forecasted variables (as air temperature, precipitation amount, wind intensity, and direction) are shown. The ice-sleeve growth on conductors, in terms of ice load and thickness, are depicted too. The WOLF system has been verified considering an archive of electric interruption messages reported by the Italian Transmission System Operator, mainly due to the load of ice and snow on the components occurred between October 2010 and April 2011.

5.5 Storms and Lighting Phenomena

Severe thunderstorms may become a serious risk during warm seasons in Italy. In particular, the region between the Po Valley and the Alps experiences one of the highest thunderstorm frequencies in Italy. Most events are caused by high intensity convective cells that may develop and vanish in a few hours producing their effects in a small area, less than 10 km wide. The associated deep convection causes severe weather such as hail, gusty winds, tornadoes, and heavy rain.

Due to the high urbanization of Lombardy (about 450 inhab. per sq. km.), any severe thunderstorm can damage house roofs, scaffolds, and old trees; road signs often fly away breaking down electric overhead lines; large hailstones damage cars

---

[7] "Regional Atmospheric Modelling System", developed by Colorado State University, division: ASTER Research Corporation Mission and ATMET.

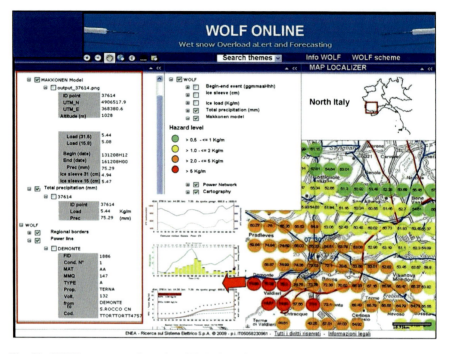

Fig. 13 WOLF outputs on the active-GIS desktop for the Piedmont wet-snow event

and windows. Furthermore, short but heavy rain causes flooding of underpasses, flash floods or landslides.

Overhead power lines undergo these severe weather situations without the possibility to avoid damages and electric outages. In fact, the lines damages are often caused by objects thrown on them by the strong wind or by lightning falling close to them. Electric companies may only repair the damage as soon as possible, in order to re-enable the power supply. In this context, a thunderstorm alert system based on radar and satellite monitoring can help company rescue teams to work better and faster.

Since 1993 the Mount Lema radar, managed by MeteoSwiss (Switzerland Meteorological Service) and located close to the Italian boundary, provides useful data on the whole Lombardy region.

With the project ST-AR (STorm—ARchive), an archive has been built in which the ground effects of severe convective storms that happened in the Po Basin have been assessed: the severity of a thunderstorm has been classified by computing the index "Probability of Damage" (PD) (Collino et al. 2009), the radar and satellite data are stored together with damages reports from newspapers.

Combining Meteorological and Electrical Engineering Expertise 151

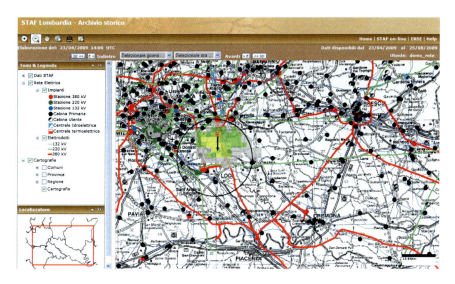

Fig. 14 Web GIS interface of STAF with an overhead power lines map

5.6 STAF (Storm Track Alert and Forecast)

Since 2006, in the frame of the research on the interaction between severe weather and the electrical grid, RSE is collaborating with Lombardy Region Authority of Civil Protection Office to develop and test a severe thunderstorms detection and nowcasting system, called STAF (Bonelli et al. 2011).

STAF is an information system able to locate and track severe storms cells, to calculate their dangerousness, the vector motion, to forecast their position in half an hour. The system is able to send alert messages to user's service by using GSM-SMS network when heavy thunderstorm comes into a target area. STAF is provided by a WEB GIS Interface (Fig. 14) that allows to track cells in real time over different GIS layers, such as overhead power lines map, physical and road maps, geo-hydrological maps with different boundaries. Through the same interface, it is possible to reanalyse past events for diagnostic purposes.

Every 5 min, the system acquires and elaborates a volumetric scan of radar reflectivity and infrared radiances, detected from the geostationary satellite METEOSAT, in order to obtain the cloud top temperature (Fig. 15). The volumetric scan of reflectivity is composed by 12 layers, from 1 to 12 km height, and it is used to reconstruct the thunderstorm area, through the vertical projection of the maximum reflectivity over each radar pixel (VMI—Vertical Maximum Intensity). In this projection, radar information changes from 3D to 2D making easier the trajectory analysis. The thunderstorm intensity is derived from the number of points

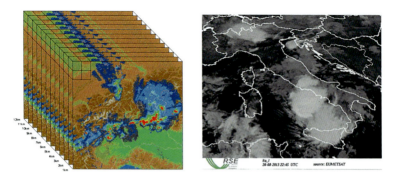

Fig. 15 Radar volumetric reflectivity and thermal infrared 10.8 μ MSG channel

over the vertical of each pixel having reflectivity greater than 44 dBZ.[8] This value is mentioned in literature as marker of the deep convection and presence of hail (Hering et al. 2007).

STAF may process lightning strikes too, in order to associate a number of strikes to every thunderstorm cell, even if it is not verified that they are closely related to its intensity.

STAF provides information about the storms intensity, motion, and the associated damage probability (PD) to the urbanized terrain. The PD index has been verified during summer 2009, when STAF was applied in a field test, involving a group of Civil Protection observers and users in order to register independent observations of damage collected during the test.

6 Conclusion

This chapter describes the activities carried out in Italy within the frame of grid management and risk reduction associated to climate changes and adverse meteorological events.

Renewable energies, especially wind and solar, are strongly dependent on highly variable weather processes and the increased penetration of these sources leads to strong fluctuations in power flows on the electrical grid. Accurate power forecasting, intelligent grid management, and increased flexible storage capacity are mandatory for the efficient development of the future energy policies in Europe.

Moreover the setting up of warning systems, based on weather analyses and forecast, are fundamental for a timely response in case of adverse weather

---

[8] Decibel of equivalent reflectivity.

phenomena such as strong snowfall, thunderstorms and wind gusts, which may endanger the safety and availability of the electric grid.

In order to increase the resilience of the energy system, it is also necessary to take into account climate change impacts by the introduction of more adequate design and construction criteria for new electric infrastructures.

Renewable energy production, reliability of overhead power lines, and definition of future climate scenarios are some of the items covered by the multiannual research program currently underway at RSE, where an effective collaboration has been established in order to guarantee a good match between the meteorological knowledge and the energy industry needs.

References

Admirat P, Sakamoto Y (1988) Wet snow on overhead lines: a state of the art. In: Proceedings of the 4th international workshop on atmospheric icing of structures, Paris, pp 7–13

Bonelli P, Marcacci P, Bertolotti E, Collino E, Stella G (2011) Nowcasting and assessing thunderstorm risk on the Lomabrdy region (Italy). Atmos Res 100(4) ISSN 0169-8095

Christensen JH, Carter TR, Rummakainen M, Amanatidis G (2007) Evaluating the performance and utility of regional climate models: the PRUDENCE project. Clim Change 81:1–6

CIGRE, Working Group B2.16, Task Force 03 (2005) Guidelines for meteorological icing models, statistical methods and topographical effects

Collino E, Bonelli P, Gilli L (2009) ST-AR (STorm—ARchive): a project developed to assess the ground effects of severe convective storms in the Po Valley. Atmos Res 93:483–489

Dosio A, Paruolo P (2011) Bias correction of the ENSEMBLES high-resolution climate change projections for use by impact models: evaluation on the present climate. J Geophys Res 116:D16106. doi:10.1029/2011JD015934

Dubus L (2013) Weather and Climate and the Power Sector: Needs, Recent Developments and Challenges. Weather Matters for Energy. Springer, New York, USA

Faggian P, Giorgi F (2009) An analysis of global model projections over Italy, with particular attention to the Italian Greater Alpine Region (GAR). Clim Change 96:239–258

Farzaneh M (2008) Atmospheric icing of power networks. In: Farzaneh M (ed) Springer, Berlin

Fikke S, Ronsten G, Heimo A, Kunz H, Ostrozlik M, Personn P-E, Sabata J, Wareing B, Wichura B, Chum J, Laakso T, Santti K, Makkonen L (2007) Atmospheric icing on structures measurements and data collection on icing: state of art—COST 727. MeteoSwiss, Switzerland, 75:110 pp

Gabriel P, Harrington JY, Stephens GL, Schneider T (1997) Adjoint perturbation method applied to two-stream radiative transfer. J Quant Spectrosc Radiat Transf 59:1–24

George T (2013) Weather and Climate Impacts on Australia's National Electricity. Weather Matters for Energy. Springer, New York, USA

Giannakopoulos C, Hadjinicolaou P, Zerefos C, Demosthenous G (2009) Changing energy requirements in the mediterranean under changing climatic conditons. Energies 2:805–815

Gryning S-E, Badger J, Hahmann AN and Batchvarova E (2013) Current Status and Challenges in Wind Energy Assessment. Weather Matters for Energy. Springer, New York, USA

Haupt SE, Mahoney WP, Parks K (2013) Wind Power Forecasting. Weather Matters for Energy. Springer, New York, USA

Haylock MR, Hofstra N, Klein Tank AMG, Klok EJ, Jones PD, New M (2008) A European daily high-resolution gridded dataset of surface temperature and precipitation. J Geophys Res (Atm) 113:D20119. doi:10.1029/2008JD10201

Hering AM, Ambrosetti P, Germann U, Sénési S (2007) Operational nowcasting of thunderstorms in the Alpine area. In: 4th European conference on severe storms, Trieste—Italy

IEC (1997) Overhead lines—Meteorological data for assessing climatic loads. International electrotechnical commission technical report 61774

Katzfey J(2013) Regional Climate Modelling for the Energy Sector. Weather Matters for Energy. Springer, New York, USA

Kay M and MacGill I (2013) Improving NWP Forecasts for the Wind Energy Sector. Weather Matters for Energy. Springer, New York, USA

Kjellström E, Nikulin G, Hansson U, Strandeberg G, Ullerstig A (2011) 21st century changes in the European climate: uncertainties derived from an ensemble of regional climate model simulation. Tellus 63A:24–40

Lorenz E, Kühnert J, Heinemann D (2013) Overview of Irradiance and Photovoltaic Power Prediction. Weather Matters for Energy. Springer, New York, USA

Makkonen L (2000) Models for the growth of rime, glaze, icicles and wet snow on structures. Philos Trans R Soc 358:2913–2939

Nakicenovic N et al (2000) Special report on emissions scenarios: a special report of working group III in the intergovernmental panel on climate change. Cambridge University Press, Cambridge, p 600

Piani C, Weedon G, Best M, Gomes S, Viterbo P, Hagemann S, Haerter J (2010) Statistical bias correction of global simulated daily precipitation and temperature for the application of hydrological models. J Hydrol 395:199–215. doi:10.1016/j.jhydrol.2010.10.024

Pielke RA, Cotton WR, Walko RL, Tremback CJ, Lyons W, Grasso LD, Nicholls ME, Moran MD, Wesley DA, Lee TJ, Copeland JH (1992) A comprehensive meteorological modeling system RAMS. Meteorol Atmos Phys 49:69–91

Pirovano G, Chiarello S, Mannelli ED, Omodeo GP (2011) Sistemi di monitoraggio degli isolamenti in aria di linee e stazioni. Convegno Nazionale AEIT Prospettive Economiche e Strategie Industriali, Milano, 27–29 Giugno 2011

Prein AF, Gobiet A, Truhwetz H (2011) Analysis of uncertainty in large scale climate change projections over Europe. Meteorol Z 20, N°4:383–395

Renné DS (2013) Emerging Meteorological Requirements to Support High Penetrations of Variable Renewable Energy Sources: Solar Energy. Weather Matters for Energy. Springer, New York, USA

Rizzi G, Omodeo P, Panzeri E, Valagussa C (2007) Evaluation of surface pollution as trigger for cleaning operation. CIRED – Vienna 21-24 may 2007, paper 0510

Sakamoto Y (2000) Snow accretion on overhead wires. Phil Trans R Soc London 358(1776):2941–2970

Troccoli A et al. (2012) Promoting new link between energy and meteorology. Bull Am Meteorol Soc. e-View: doi: http://dx.doi.org/10.1175/BAMS-D-12-00061.1

Weather and Climate Impacts on Australia's National Electricity Market (NEM)

Tim George and Magnus Hindsberger

Abstract This chapter presents information on weather and climate effects in the Australian National Electricity Market (NEM). The effects are considered in two classifications: Operational effects and impacts, representing the short- to medium-term effects and responses and how these are managed by power system operators. Planning impacts, representing the longer term considerations that must be given to power system planning and development in order to accommodate increasing renewable energy generation and climatic changes. The main focus is on wind and solar power, both of which are already present in the NEM and both of which are forecast to increase significantly over the coming decade. Solar and wind variability and diversity are key characteristics for both operational and planning purposes and existing and proposed approaches are discussed.

1 Introduction

With the recent growth in renewable generation, the Australian energy markets are becoming increasingly influenced by weather. As a result, forecasting capability, taking weather into account, is becoming more important for the energy industry and is an area of increased research.

The Australian Energy Market Operator (AEMO) plays a key role as it is the operator of the electricity system and market operator for the gas and electricity markets in the Eastern and Southern states of Australia. AEMO is also the national transmission planner, a function that requires it to take a long-term view on developments within the industry.

T. George (✉)
DIgSILENT Pacific, Brisbane, Australia
e-mail: tim.george@digsilent.com.au

M. Hindsberger
Australian Energy Market Operator, Brisbane, Australia

This chapter deals with:

- Operational aspects, which tend to be weather driven and
- Planning aspects that are influenced by longer term climate considerations.

In an operational context, the most significant impact of weather has historically been on electrical and gas demands, because of the dependency on energy-intensive air conditioning (electricity) and space heating (both gas and electricity). Operational demand forecasting tools have been developed over time to improve intra-day and short-term demand forecasting. Over the past decade, however, development of wind generation has added to the weather impacts on the power system and led to the development of the Australian Wind Energy Forecasting System (AWEFS). This national forecasting system plays an important role in mitigating the effects of intermittency in the NEM. An increase in solar generation will require a similar focus on forecasting to facilitate operation of the power system.

In the planning context, the energy markets will need to adapt to a range of climate-related issues and emerging weather-dependent generation technologies, among others, in order to deliver a secure and reliable energy supply.

On the technology side, AEMO must:

- Plan for managing the increasing level of wind generation that is expected as a result of the incentives provided by the Australian Government's Large-scale Renewable Energy Target (LRET).
- Monitor and plan for large-scale introduction of solar technologies including increased uptake of distributed or large-scale photo-voltaic (PV) and solar thermal generation.

In relation to climate change, AEMO should take into account the potential impacts on:

- Demand for electricity and gas due to changes in average and peak temperatures.
- Generation of electricity from wind-, solar- and hydro-based resources.
- Transmission capability due to changes in temperatures (line ratings).
- System risks due to more extreme and widespread events such as bushfires, floods and storms.

The chapter will discuss the key aspects of power system operation and planning as affected by weather and climate.

This chapter first considers the drivers affecting the power system that make considerations of weather and climate more important. This is followed by an overview of the characteristics of weather-dependent generation technologies.

The next two sections then discuss, respectively, the considerations made by AEMO in the operational horizon (real-time to 2 years) and planning horizon (beyond 2 years). Lastly, the key findings will be summarised.

2 Setting the Scene

Traditionally, the major impact from weather and climate on the energy system has been on demand. The temperature-sensitive component of the electrical demand is expected to continue to grow, as reliance on air conditioners increases. There is also the possibility of fuel switching in the colder states like Victoria and possibly South Australia as air conditioners may increasingly be used for winter heating in preference to gas.

But increasingly, weather and climate is affecting power generation. The drought in the first decade of the millennium both led to reduced hydro generation levels in the NEM but also reduced output from some large thermal units due to restricted supply of cooling water. Increasing wind generation, particularly in South Australia and Tasmania, have changed generation patterns in those states. Looking into the future, apart from wind power, the impact of solar generation is expected to be significant.

In its 2010 National Transmission Network Development Plan (NTNDP), the AEMO identified the need for investment in generation of $40–120 billion over the next 20 years and a further investment of some $4–9 billion in the transmission networks (AEMO 2010). The studies considered five market development scenarios and, for each scenario, two different carbon price trajectories, giving a total of 10 cases that were studied. The carbon price trajectories were selected from a set of four, namely zero, low, medium and high (over $100/t in 2030), with the last three based broadly on estimates published in 2008 by the Commonwealth of Australia Treasury (Australian Government 2008).

AEMO's analysis of the scenarios comprised modelling of market outcomes and the development of a generation expansion plan. For each of the five scenarios, and each of the two carbon price outcomes in each scenario, a generation investment schedule was developed. Based on this schedule, transmission network requirements were assessed and options developed.

Key outcomes of relevance to this chapter include the impact of the Australian Government's LRET policy, which seeks to provide incentives for investment in renewable energy. AEMO's analysis found that in all scenarios with a carbon price, the energy targets in the LRET were met. Critically, this resulted in the investment of, in some cases, more than 6,000 MW of new wind generation in the NEM before 2020.

Separately, almost 1,000 MW of rooftop solar was installed in the NEM power system by the end of 2011 in response to various state and Commonwealth government incentive schemes (AEMO 2012c). Although many of these schemes have been modified to reduce the incentives, media coverage suggests further investment is expected as the cost of PV solar energy is decreasing rapidly and could break even with retail energy costs within a small number of years.

There is also considerable development in the solar thermal arena with 1,100 MW installed globally by 2011 (ITP 2012). The Commonwealth Government's 'Solar Flagships' programme may see the first solar thermal technologies introduced to the NEM. The key attraction of solar thermal technologies is the ability to store energy (sometimes in the form of molten salts) and have the capability to continue to generate energy, at least for a while, when the sun is not shining. Furthermore, it may be possible for these energy storage systems to mitigate variability in wind generation output and generation from solar technologies without storage.

3 Characteristics of Weather-Dependent Generation

This section describes how wind, solar and hydro generation are affected by weather and climate in Australia.

3.1 Wind Generation Characteristics

South Australia has one of the highest levels of wind generation in the world and, as a region, is second to the Iberian Peninsula and similar to Ireland (ECAR Energy 2011). Wind energy now provides 20 % of the annual energy production in the state. As at the end of 2011, there was 1,150 MW of installed wind generation capacity with several thousand more MW being considered in a number of projects (AEMO 2011a). The maximum demand (MD) reached in 2011 in South Australia was 3,433 MW when the temperature was 43 °C. At the time of the peak demand, the output of all of the wind generation was only ∼60 MW. So wind contributed with just ∼5 % of its installed capacity. But, on the contrary, wind generation at times of low demand can be significant and prices in South Australia have been observed to go negative at times of high wind and low demand. This market signal indicates an oversupply of generation and encourages generation to reduce or disconnect.

Broadly speaking, the wind generation in South Australia:

- Is affected by the same weather systems—when the wind does blow, all the generators in the state operate, increasing supply and dropping South Australian prices. The wind in South Australia is also quite closely correlated with that in neighbouring Victoria, which tends to put further downward pressure on interstate prices.
- Achieves capacity factors in the 30–40 % range[1] (AEMO 2011a), although these can be quite variable year on year.

---

[1] The lower number is typical for earlier wind farms, in particular those with environmental constraints or transmission bottlenecks. Newer wind farms at favourable sites can achieve capacity factors of around 40 %.

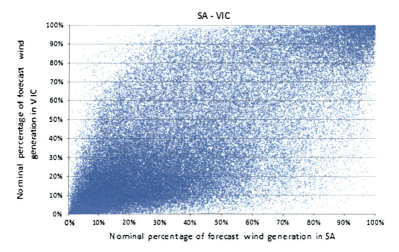

Fig. 1 Hourly wind diversity between South Australia and Victoria

- May only have limited availability during high demand periods as some wind turbines shut off because they exceed their design operating temperatures. Also, while the hottest days are often associated with a relative strong northerly breeze, very hot days with almost no wind occasionally occurs.
- Is potentially limited from significant further expansion because remote, wind-rich regions are not well connected to the transmission system, nor is there much 'spare' capacity to transmit, at times, all of the South Australia generation into the rest of the NEM.
- Is becoming adversely affected by marginal loss factors, which devalue energy produced in generation rich and remote areas because of the higher losses associated with transfers to load centres.

There is significant interest in wind generation in Victoria, which had almost 500 MW of committed capacity at the end of 2011 with investor interest in several thousand more MW (AEMO 2011a). Generation in the western parts of Victoria is exposed to similar weather patterns to those of South Australia. Recent changes to the development approvals framework, which restricts where wind turbines can be located, will make some of the planned projects less likely to proceed in this state.

An important aspect of wind integration is the degree to which there is diversity between wind patterns across the geography of the NEM. The higher the diversity, the less will be the overall variability. To illustrate this, wind patterns across South Australia, Victoria and New South Wales have been analysed over the period 2002–2008 based on synthetic wind generation data.

Figure 1 shows the correlation of South Australian and Victorian wind generation. Figure 2 shows the correlation between South Australia and New South Wales. Closer alignment to the diagonal from the bottom left to the top right indicates high correlation, and hence lower diversity. As seen, wind generation in

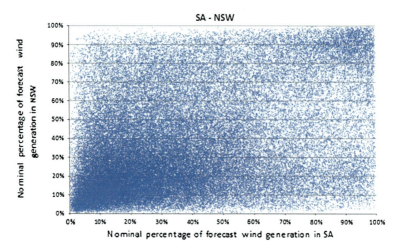

Fig. 2 Hourly wind diversity between South Australia and New South Wales

Table 1 Wind contribution factors

Region	Average capacity factor (2010–2011) (%)	Contribution factor Summer MD (%)	Contribution factor Winter MD (%)
New South Wales	25.6	9.2	0.4
Victoria	29.2	7.7	3.9
South Australia	32.6	5.0	3.5
Tasmania	39.2	1.0	1.0

South Australia is more correlated with wind generation in Victoria compared with New South Wales wind generation.

The clustering of points in the lower left of the diagrams represents periods where the wind is low in both locations, something that must be taken into consideration when operating and planning the power system.

The correlation of wind across Australian states is discussed further in Sect. 5.2.

Table 1 gives the AEMO calculated contributions from wind generation at the time of peak demand (MW) for each of the NEM states except Queensland, where insufficient data was available (AEMO 2011a). The table refers to the maximum demand (MD), which is the highest demand in MW averaged over a half-hour period in the season indicated.

A longer term picture can be estimated using the synthetic 2002–2008 data and assessing wind strengths at times of peak demand. This provides potential contributions factors over a longer period of time but caution is still required in relying on these synthetic figures as they are not using real wind farm data and will not capture reduced output as result of network congestion and tripping of turbines due to extreme heat. Also, this assumes a dispersed development of wind generation

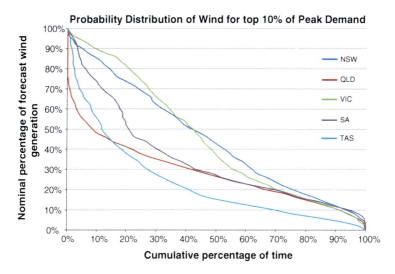

Fig. 3 Estimated state-based wind contributions to peak demand periods using 2002–2008 wind hourly data (synthetic)

within the state, which increases the diversity of wind output compared to historical development, which is currently concentrated in a few areas. The resulting wind contribution factors to summer MD are shown in Fig. 3 (AEMO 2012a).

The key message from the contribution data is that at times of peak demand, the output from wind farms can realistically only achieve around 10 % of the installed capacity. This provides a need for operational strategies or new technologies to manage peak demands. Increased flexibility on the demand side and storage technologies are examples of possible mitigation strategies.

3.2 Characteristics of Solar Generation

The two key solar technologies are photovoltaic (PV) and concentrated solar thermal (CST). The rapid uptake of domestic PV and the evident reduction in costs of this technology are stimulating interest in how to manage even larger market share of PV, including with utility scale projects.

CST, particularly if coupled with salt or other thermal storage facilities, is better understood and less subject to rapid variation as insolation changes with cloud cover (or nightfall!).

PV has a very rapid response to changes in insolation but, on a large scale, this may not cause the same degrees of variability as wind generation because of the widely distributed nature of the installations (except for utility size systems concentrated in a specific area). At a more local level (within an area fed by one distribution substation), the effects may be more pronounced and could require

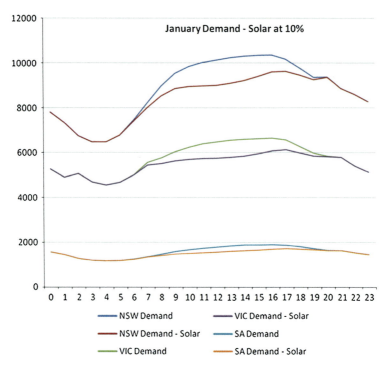

Fig. 4 Indicative impact of 10 % solar PV penetration by state on a sunny summer day (*source* AEMO)

investment in additional plant to allow for the rapid variations in power flow as rooftop PV systems are affected by cloud shadows.

Figure 4 shows the assumed contribution of rooftop PV to the demand a typical summer and winter day for some of the states. The analysis assumes PV capacity for each state is installed to a level equivalent to 10 % of the peak demand. Curves are then plotted for each state to show the effect of this level of solar penetration compared to the case without any solar. The contribution from the installed solar can then be seen at various times of day for both winter and summer. These curves show how solar is quite closely correlated with high (summer) demand periods and can actually reduce the capacity required to meet peak demands reliably.

Figure 5 shows the less rosy picture of rooftop PV contributions to demand on a sunny winter day. In the NEM, states like New South Wales and Queensland do not have the same degree of difference between winter and summer peak demands as the southern (mainland) states where gas space heating is prevalent. It is conceivable, therefore, that winter peak demands may grow relative to summer peaks and that some states might even revert to winter peaking.

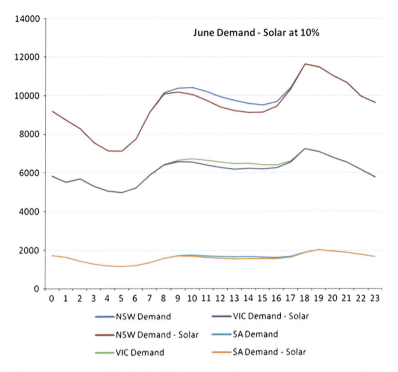

Fig. 5 Indicative impact of 10 % solar PV penetration by state on a sunny winter day (*source* AEMO)

3.3 Hydro Generation Characteristics

There are three classes of hydro generation that are of interest:

- Storage hydro with dams that can store significant amounts of water (or energy).
- Run of river hydro where output is dependent on river flow or a cascade of power stations on a river system.
- Pumped storage hydro where energy can be pumped into a reservoir at times when energy prices are low and used to generate energy when prices are high.

In terms of climate impact, increasing periods of low rainfall can restrict hydro potential and cooling water supply to thermal generators and this is an obvious climate change concern. This particularly affects storage and run of river hydro and should be taken into account in the longer term system planning.

Pumped storage hydro is an important means of mitigating wind variability but is costly and dependent on the availability of suitable sites. Prolonged droughts can also affect the availability of pumped storage (as can floods!).

4 Operational Considerations

The operational outlook is typically from real-time to 2 years. It is unlikely that any significant investment can be achieved within the 2-year period; so operational strategies are developed to manage security and reliability with existing and committed assets.

4.1 Operational Forecasting of Demand

As previously discussed, weather—in particular temperature—is a key driver of electricity demand given the reliance of electricity for both cooling and heating.

For operational use, AEMO uses a state-of-the-art Demand Forecasting System (DFS)[2] that uses statistical models to automatically generate short-term forecasts every half-hour for all NEM regions including sub-regions that can have a significant impact on transmission network constraints.

The major factors considered in the statistical models are;

- Temperature profile.
- Weather season.
- Week day/weekend.
- Unusual conditions (school/public holidays/daylight savings).

4.2 Operational Treatment of Wind Generation

Operationally there are a number of issues that must be managed as the degree of wind penetration increases. Focusing only on the weather related issues, we have:

- Intermittency or variability—supplies must be dispatched to exactly meet demand and variability must be accounted for by other generation or the modulation of demand. Important metrics are:
  - The size of the maximum change over a variety of operations forecast periods, ranging from 5 min to 40 h.
  - The peak output and the minimum output.
- The need to forecast wind speed, and through an energy conversion model, the generated output.
- The need to prevent, where possible, rapid changes in wind generation output (see discussion on 'semi-scheduled' generation in the next section).

---

[2] This system was introduced in late 2011, see AEMO press release: http://www.aemo.com.au/electricityops/0140-0055.html.

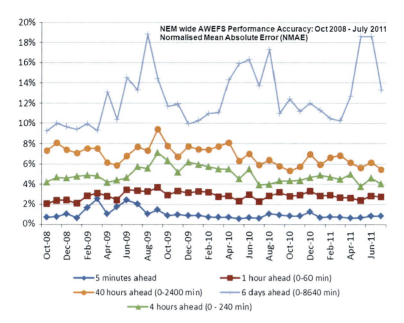

Fig. 6 Indicative forecast performance, in MW error (forecast vs. actual), of AWEFS over a number of time frames (*source* AEMO)

4.2.1 Market Design

The design of the NEM is well suited to integrate generation from intermittent resources such as wind power. One reason is the 5-min dispatch instructions where other markets tend to have half-hourly or hourly dispatch decisions. Another advantage comes from the late market gate closure in the NEM where market participants can change volumes (but not prices) right up to the dispatch interval for physical delivery. As the forecasting uncertainty increases the further out you forecast (see discussion around Fig. 6 in Sect. 4.2.2), this is much preferred over day-ahead type markets, where bids and offers are typically fixed 12–36 h before physical delivery.

Another interesting market design characteristic of the NEM is the classification of 'semi-scheduled' status to wind generators. This means that the output of a wind farm through the central dispatch system can be constrained down to prevent transmission networks becoming overloaded because wind output is too high (or demand too low, etc.). Of course, being semi-scheduled does not help rapid declines in output because the wind drops. This type of change, however, can be far better accommodated if it can be forecast in advance, even by as little as 5 min.

4.2.2 The Australian Wind Energy Forecasting System

The AWEFS project was established in response to the growth in intermittent generation in the NEM and the increasing impact this growth is having on NEM forecasting process. The project was financed through the Australian Commonwealth Department of Resources, Energy and Tourism (DRET). This project provides a vehicle for improving the research, development and application for forecasting of intermittent energy.

AWEFS integrates seamlessly with the DFS discussed previously. The essential characteristics of the AWEFS system are:

- Each wind farm provides current wind information to AWEFS.
- The European ANEMOS system provides wide area wind forecasts.

The system will produce generation forecasts for all NEM wind farms (≥ 30 MW) for all NEM forecasting time frames as follows:

- Dispatch (5 min ahead)
- 5-Minute pre-dispatch (5 min resolution, 1 h ahead)
- Pre-dispatch (30 min resolution, up to 40 h ahead)
- ST PASA[3] (30 min resolution, up to 7 days ahead)
- MT PASA[4] (daily resolution, 2 up to years ahead).

Other generators, based on coal, gas, hydro, etc., can view these forecasts and review the quantities they will offer into the market at what prices.

Market efficiency will improve if forecasts (both of wind and demand) are accurate.

The accuracy of the AWEFS system over various time frames is shown in Fig. 6, which shows the normalised mean absolute error of the forecast wind generation output against the actual observed output. The impact of variability can be greatly mitigated by predictability of the changes. This is clearly illustrated by the ease with which power systems can manage large, rapid and sustained changes in demand, primarily because the changes are predicted by forecasts and operators and market participants have sufficient time to plan for the changes. As seen, the forecasting accuracy of the AWEFS system is generally very good 5 min ahead (similar to the market gate closure), but also reasonable 4 h ahead giving operators of thermal plants a good indication of whether they need to ramp up or down the output from these plants.

---

[3] Short-Term Projected Assessment of System Adequacy—a measure of the available generation to meet demand, covering the next 7 days.

[4] Medium-Term Projected Assessment of System Adequacy, covering a 2-year outlook period.

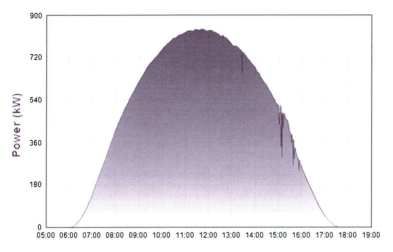

Fig. 7 Solar generation during a sunny day—7 April 2012 (*source* UQ)

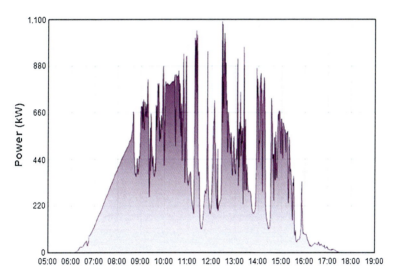

Fig. 8 Solar generation during a cloudy day—11 April 2012 (*source* UQ)

4.3 Operational Considerations of Solar Power

As was the case for wind, it is important for the market to have good estimates of solar generation for the full range of operational time frames. In the case of PV, the key variables are location, time of day and cloud cover. The impact of cloud cover in particular is shown in Figs. 7 and 8, which are based on solar generation data available from a 1.2 MW solar PV installation at University of Queensland.[5]

As seen, the variability of output from single systems can be very high especially on partly cloudy days where shadows from clouds will move over the PV panels and cause rapid reductions in output. The temperature de-rating effect can also be seen from the peak output on a cloudy day (lower temperature) against a hot sunny day.

A dispersed system made up of many, smaller systems, such as residential rooftop solar PV systems, will have lower variability on partly cloudy days because the output is averaged over a wide area. This aggregation over distributed systems will reduce the impacts of rapid changes but the power system will still be affected by variability caused by large-scale weather events such as storm fronts. Solar eclipses are rare and predictable but could also have major impacts.

Some thought is being given to developing a forecasting system equivalent to AWEFS, or an add-on component that could possibly be integrated into AWEFS, to provide solar forecasts (ASI 2012).

With respect to CST with storage, the forecasting system is likely to be based on the same principles but the energy conversion model will need to consider the specifics of the plant, with the heat input effectively being integrated over the day and drawn down when the solar incidence reduces below the aggregate output. The slower response time of CST and the use of synchronous generators as the energy conversion device make these systems easier to integrate into the power system than the asynchronous PV (and wind) systems, which effectively have little or no inertia.

4.4 Operational Consideration of Rainfall

To assess system security under different rainfall scenarios, AEMO in each quarter is undertaking a probabilistic (Monte-Carlo) assessment of energy adequacy under different rainfall scenarios. This study, with the name Energy Adequacy Assessment Projection (EAAP) has an outlook horizon of 2 years.[6]

5 Planning for the Longer Term

Within the 2–20 years planning horizon, predicting weather outcomes become impossible and the key issue is to understand and utilise information about weather behaviour—such as averages and not least distribution of outcomes.

---

[5] Data available from: http://www.uqld.smartersoft.com.au/user/reportEnergy.php.

[6] See: http://www.aemo.com.au/AEMO%20Home/Electricity/Resources/Reports%20and%20Documents/EAAP.

5.1 Demand Forecasting within the Planning Horizon

An important future consideration for the characteristics of the demand side arises from new technologies. These technologies include:

- The ability to control demand by switching off non-essential loads using advanced communication and metering systems (some of these technologies and techniques fall into the so-called SMART grid basket).
- The take-up of new technologies like electric vehicles and fuel switching that will act to increase demands.
- Impacts of energy efficiency measures that will reduce demands, such as improved electric motor efficiencies and better thermal insulation.
- The further development of distributed generation systems, like solar PV on rooftops, that can materially change demand patterns during the day.
- Development of small-scale energy storage systems, such as the use of batteries from electric vehicles, that will allow local demand to better match output variable energy sources like wind and solar.

One thing is certain: new technologies and increasing costs of energy will change consumption characteristics (quantity, time of day). These changes must be considered by power system planners and investors alike.

AEMO is undertaking significant work in the area of longer term forecasting. In particular, AEMO is working on creating consistent forecasts for all states across the NEM from 2012. Previously the state forecasts were done by different organisations and collated by AEMO. This is known as the National Electricity Forecasting Report (NEFR).[7] Also, AEMO (2011b) discussed how the load shape might change in the future due to uptake of Solar PV and electric vehicles.

5.2 Considering Wind Power within the Planning Horizon

Adding over 6,000 MW of wind generation over the next 7 years accentuates the operational efforts that will be required to predict the output from NEM wind farms. Further developments of wind generator technology are seeing lower wind strength sites becoming more attractive than previously. For example, in 2010, AEMO was aware of some 2,500 MW of interest in wind farm developments in New South Wales. In 2011, this rose to 6,500 MW (AEMO 2011a).

The siting of wind farms depends on a range of factors but, as part of its planning, AEMO must be able to characterise individual and aggregated contributions from all wind farms, as the dispatch process is national and considers all generators in the NEM.

---

[7] Further information is available from: http://aemo.com.au/forecasting/forecasting.html.

As part of the wind integration work summarised in AEMO (2011b), AEMO has obtained wind data for a period of 8 years and matched this with demand data across the NEM. This has allowed the analysis of diversity of wind energy within and between NEM regions. Figures 3 and 4 were based on this analysis. Table 2 summarises the correlation between NEM states.

A high degree of diversity is desirable, as this typically would mean that there would be, at a given time, a high probability of some wind in some of the states. Similarly, it would reduce the risk of spilling surplus wind power, as it would be unlikely that all wind farms generate at maximum at one point in time in a high diversity system. However, if there are transmission limitations, this may prevent the sharing of wind energy across the NEM, reducing the potential benefits of diversity.

It takes 5–10 years to build a transmission line and trends are for this lead-time to increase. Transmission infrastructure is costly and investors and consumers who have to pay for transmission investments are reluctant to invest on the off-chance that generation investors will turn up after a line is built. Similarly, generation investors are reluctant to build billion dollar projects where there is inadequate (or no) transmission capacity to get their product to market. Building a wind farm is also considerably quicker than building a transmission line. Transmission planners and policy makers in the energy industry hence need to address the disparities in build-times and the need for some degree of certainty around the transmission capacity available to new generation investments.

Technology is also playing a role in where and when investors choose to locate their power stations. Improvements in lower speed turbines have seen new locations, close to transmission lines, become favoured development sites. There may well be cumulative inefficiencies that arise from this trend but, in a competitive generation market, it is not a trivial matter to capture benefits of scale across several competing investments.

Understanding what drives investors is critical for transmission planners to provide adequate transmission capacity. In the past few years it appears that investors have changed their focus from raw wind strength to factors such as:

- Knowledge of the wind conditions.
- Correlation of wind strength with high demand periods.
- Congestion on the transmission network.
- Marginal loss factors.

The first two bullet points are directly relevant to forecasters while the latter are likely to have a strong influence on the selection of preferred development sites.

Internationally, several market and transmission system operators have been incentivised to build transmission into areas considered having attractive resources ahead of commitments by generators to invest at those locations. In essence, this transfers the asset stranding risk to transmission customers but provides an upside that potentially more efficient generation investments may occur, providing downward pressure on energy costs. Examples include the Competitive Renewable

Table 2 Correlation in wind energy production between states

State	SA (%)	QLD (%)	NSW (%)	TAS (%)
QLD	0			
NSW	38	7		
TAS	32	1	35	
VIC	69	0	50	64

Energy Zones (CREZ) in Texas[8] and transmission developments in New Zealand (Transpower 2010), where new transmission has been justified on the basis that it opens up (selected) resource rich areas.

Another key consideration is just how much wind generation can be accommodated on a power system. Modern wind generators are asynchronous and can operate over a fairly broad range of power frequency. Conventional generation is synchronous, and all synchronous generators in an AC power system must operate at exactly the same average speed.

While the level of wind penetration is relatively low, the power system behaves in a predictable way (synchronously, like it has in the past). As the level of wind generation increases, the characteristics of the power system change and new operational issues are introduced. In the Irish power system, for example, it is considered that the instantaneous wind penetration should be restricted to 50 % of the system demand with the existing structure of the power system for reliability reasons (Ecar 2011). Beyond this level of penetration, there may be additional investments required (in control systems, rotating plant, demand flexibility, storage, etc.) for the power system to remain secure and reliable and it is planned to increase this limit as the various countermeasures are deployed. Studies suggest (DigSILENT 2010) an upper achievable limit of 75 %.

In NEM terms, the 50 % threshold alone would allow for up to around 20,000 MW of wind generation! Should this level of wind generation be achieved, the importance of understanding and forecasting variables, asynchronous generation, such as wind and solar, cannot be overstated.

5.3 Considering Impacts of Solar Generation within the Planning Horizon

Due to increasing solar generation, particularly distributed rooftop systems, the summer peak demands may be reduced and shortened in duration. Typically, in New South Wales the peak summer demand occurs around 16:30 each day—a time when rooftop solar could be generating ∼ 20–30 % of its rated output (as per

---

[8] More information on CREZ is available from: http://www.texascrezprojects.com/ or see Ecar Energy (2011).

Fig. 4). These impacts, coupled with the possibility of fuel switching, could result in peak demands switching from summer to winter in some states.

The reduced duration of peak demands as a result of PV contributions is also expected to reduce the need for network investment, provided an economic approach is adopted. Under an economic planning framework, investments only proceed when benefits exceed the costs. Shorter duration peaks will reduce benefits from investments and thus defer the timing of investments. This represents material savings in network investments.

As result, understanding the impact from rooftop solar PV developments on the electricity demand from households will be very important from a planning perspective. AEMO released its first NEM solar PV uptake forecast in May 2012 as part of its National Electricity Forecasting Report discussed in Sect. 5.1.

5.4 Taking Climate Change into Account

One of the major factors to consider is **temperature**.

Most power system equipment is fundamentally limited by absolute temperatures. For example, a power line must not be operated above its design temperature. Many transmission lines are constrained by legacy design decisions that limit conductor temperature to as low as 50 °C. If the outdoor temperature is 40 °C, much less power can be sent through the lines before they reach 50 °C. If a line is operated above its design temperature, the lines start to sag and this may cause minimum clearance distances to be violated, causing public safety risk from flashovers—and on hot day this could trigger bush fires.

An upper end temperature for conventional conductors is of the order of 120 °C and new transmission line designs tend to have their limits much closer to that maximum. Operating a line above its maximum conductor temperature can irreversibly damage the conductor through annealing.

As a second example, transformers have insulation that degrades under high temperatures. Output is limited to avoid the hottest part of the transformer from exceeding design temperatures.

For both lines and transformers, the operating temperature is the result of an equilibrium of the heat producing and heat dissipating elements of the transformers or lines. The heat producing elements are (mainly) the resistive losses while the heat dissipation is due to flow of coolant (including air). Higher ambient temperatures reduce the cooling ability and, therefore, reduce the amount of heat that can be generated by energy transfers. Effectively, higher ambient temperatures reduce the rating of the assets.

With all mainland NEM states summer peaking (driven by air conditioners to a large degree), the transmission assets must carry peak loads at the time when their rating is at its lowest.

Transmission planning considers the likely maximum temperatures as part of the design and rating process. The data used in these processes are based on historical observations. If global and regional temperatures increase, and particularly if heat waves become more severe or prolonged, some of the underlying analysis supporting transmission line ratings will need revisiting. Lower ratings of existing transmission assets will drive the need for earlier new investments to maintain the ability to meet demands. These investments need not be new transmission assets but could be, for example, in control systems to reduce peak flows following contingencies or in demand side participation to lower peak demand.

Coupled with the reduction of capability as temperatures rise is the temperature sensitivity of demand. Planning is currently based on 10 % probabilities of exceedance (POE)—those temperature and load conditions that are expected to be achieved 1 year in 10. Higher temperatures—say a 5 % POE—will not only result in lower asset ratings but also higher than forecast peak demands.

Other **extreme weather** impacts should also be considered.

If climate change drives more extreme weather events—storms, wind gusts, droughts—the security of the power system may be affected through physical impacts. Many aspects of the power system design are weather dependent and based on observed historical extremes.

Transmission lines, for example, have towers designed to withstand extreme wind gusts and are shielded and insulated to provide levels of protection from lightning strikes. Significant increases in these parameters could result in increased mechanical and electrical faults, reducing, for periods, the transmission capacity available to supply consumers or connect generators to load centres.

Some of these risk factors can be mitigated—for example, by reinforcing towers with more steel—but others, like lightning shielding, are perhaps too costly to address.

Bushfires and flooding are also potentially significant influencers of reliability and security. As an example, Fig. 9 shows Kerang Substation in Victoria which was close to being flooded in January 2011. Only extraordinary effort by volunteers building sandbag walls prevented this.

Remediation is possibly too costly for existing infrastructure but, from a planning perspective, analysis of the consequences and probabilities of high-impact low-probability events are increasingly required to determine whether additional costs to provide physical diversity across infrastructure are justified. Examples include having different routes for transmission lines rather than concentrating many assets in a single corridor.

A key aspect of climate change and the ability to accommodate its effects in infrastructure planning is getting a good understanding of the probabilities of occurrence, and for that a key input will be from meteorologists.

Defensive approaches that AEMO is already adopting include:

- Consideration in some analyses of 5 % POE demands and temperature conditions (for asset ratings).
- Improved techniques to assess benefits arising from the avoidance of high-impact low-probability (HILP) events, so that greater diversity can be provided where economically justified.

Fig. 9 Aerial photo of Kerang Substation in Victoria in January 2011

- Adoption of real-time line rating systems that make use of direct weather measurements to calculate conductor temperature, avoiding inadvertent overloading or damage to lines.

6 Conclusions

The nature of power systems, including the NEM, is changing materially because of the introduction of increasing levels of renewable energy, particularly wind and solar, and the changing characteristics of the demand for electrical energy.

Weather, particularly wind and solar insolation, affects renewable energy generation and also has a material impact on the electrical demand through energy-intensive devices such as air conditioners.

Power systems are currently robust enough to handle large and rapid changes in demand as a result of good demand forecasting and the nature of synchronous power systems. In order to maintain similar levels of reliability to those experienced at present, additional tools will be required to forecast and manage both the supply and the demand side. Many of the forecasting tools will need to incorporate a greater degree of meteorological forecasting to account for weather effects.

Increasing levels of variable generation and changing demand characteristics are revising the planner's paradigm and introducing significantly higher levels of uncertainty, and therefore risk, than previously observed. An important means of mitigating these risks will be the development of improved forecasting tools for both the supply and the demand side.

In addition to this, climate change has the potential to reduce the rating of existing assets as temperatures rise and may also affect the availability of assets if the severity of wind gusts and storm intensities, for example, increase.

Overall, the planning processes require, ultimately, a trade-off between the environmental footprint of generation (increasingly favouring renewables), reliability and cost.

Increased penetration of weather-dependent technologies and climate change impacts are affecting:

- Reliability, which is affected by:
  - Increased reliance on variable renewable energy sources like wind and solar, which may not be available to generate when needed;
  - Reduced network ratings as temperatures increase and
  - Plant availability as affected by storms and other weather events.

- Costs, which are affected by:
  - Higher capital costs of (most) renewable energy generators;
  - Storage or other support systems to manage variability of (some) renewable technologies and
  - Reduced availability of transmission network equipment during large environmental events.

Power system engineers and meteorologists clearly need to work together!

References

AEMO (2010) 2010 National Transmission Network Development Plan. Australian Energy Market Operator. http://aemo.com.au/planning/ntndp.html. Accessed Dec 2010

AEMO (2011a) 2011 Electricity statement of opportunities. Australian Energy Market Operator. http://aemo.com.au/planning/esoo2011.html. Accessed Aug 2011

AEMO (2011b) 2011 National Transmission Network Development Plan. Australian Energy Market Operator. http://aemo.com.au/planning/ntndp2011.html. Accessed Dec 2011

AEMO (2012a) Wind integration in electricity grids, work package 3—simulation using historical wind data. Australian Energy Market Operator. http://aemo.com.au/planning/0400-0056.pdf. Accessed Jan 2012

AEMO (2012c) Rooftop PV information paper—National Electricity Forecasting. Australian Energy Market Operator. http://aemo.com.au/electricity/~/media/Files/Other/Forecasting/Rooftop_PV_Information_Paper.ashx. Accessed June 2012

ASI (2012) A review of solar energy forecasting requirements and a proposed approach for the development of an Australian SOLAR Energy Forecasting System (ASEFS). CSIRO for Australian Solar Institute. http://www.australiansolarinstitute.com.au/SiteFiles/australiansolarinstitutecomau/ASI_Solar_Energy_Forecasting_Final_Report_-_Public_1.1.pdf. Accessed March 2012

Australian Government (2008) Australia's low pollution future: the economics of climate change mitigation. Australian Government, Canberra. http://archive.treasury.gov.au/lowpollution future/report/downloads/ALPF_report_consolidated.pdf

DigSILENT (2010) All island TSO facilitation of renewable studies. DigSILENT for EirGrid, Dublin, Ireland. http://www.uwig.org/Faciltiation_of_Renwables_WP3_Final_Report.pdf. Accessed June 2010

Ecar Energy (2011) Work package 1—wind integration in electricity grids: international practice and experience. Ecar Energy for Australian Energy Market Operator. http://aemo.com.au/planning/0400-0049.pdf. Accessed Oct 2011

ITP (2012) Realising the potential of concentrating solar power in Australia. IT Power (Australia) PTY LTD for the Australian Solar Institute. Accessed May 2012

Transpower (2010) Final approval for lower south island facilitating renewables project. Press release, Transpower. http://www.transpower.co.nz/n3717.html. Accessed 9 Aug 2010

Bioenergy, Weather and Climate Change in Africa: Leading Issues and Policy Options

Mersie Ejigu

Abstract The impact of weather on the generation, transmission, distribution and use of energy is wide and severe in Africa, where 80 % of the population relies on traditional biomass (solid wood, twigs, and cow dung) energy. Achieving quick transition from traditional to modern energy sources and increasing access to clean and affordable energy services are among the key objectives of energy strategies of many African countries and the New Partnership for Africa's Development (NEPAD)—the blueprint for Africa's development. In this endeavor, modern bioenergy, particularly the production and processing of its liquid form, biofuels, has ascended to the top of the energy development agenda. This chapter explores the two way connection between bioenergy and weather/climate change to shed light on the extent of vulnerabilities of the bioenergy sector to weather and how climate change is impacted by developments in the sector. The chapter caps its analysis by proposing measures to better adapt to and mitigate climate change while improving energy production and efficiency.

1 Introduction

Africa has the lowest energy consumption level in the world and accounts for 3.5 % of global consumption in 2009 (IEA 2010). Of the total energy Africa consumes, about 58 % is derived from traditional biomass (solid wood, twigs, and cow dung), electricity 9 %, petroleum 25 %, and coal and gas each 4 % (IEA 2010). The consumption pattern shows two extremes: a rural sector highly dependent on traditional biomass energy and a modern sector largely dependent on fossil fuels.

M. Ejigu (✉)
PAES, Addis Ababa, Ethiopia
e-mail: mejigu@paes.org

Over the past decade, interest in bioenergy has been gathering momentum. Among the key drivers are: growing interest in energy security; the desire of countries to produce own energy, and improve energy access; and priority importance given to the transition from traditional biomass to modern sources of energy (UNECA and AU 2011). The EU targets for replacing fossil fuels by biofuels have also triggered large investments in African countries.

Bioenergy, as used here, refers to modern bioenergy. It represents the sustainable and more efficient production of energy derived from plants and agricultural crops, which almost all countries can easily grow. It has, thus, the potential to empower countries to produce own energy and reduce dependence on the highly volatile and expensive fossil fuels. Bioenergy helps also reduce greenhouse gas emissions as it burns off less carbon with negligible emissions of sulfur dioxide and nitrate, which are urban pollutants. The carbon dioxide emitted during burning is absorbed by the new plants and recycled, rather than fully released into the atmosphere like fossil fuels.

There is a wide range of plants and crops that can be used as feedstocks. Much of the bioenergy feedstock research, however, is largely work in progress. With its tropical climate and soil conditions, Africa has the potential to grow almost all types of feedstocks. With advances in second and third generation bioenergy technologies that tend to favor Africa, the extensive development of bioenergy is both necessary and inevitable.

The biofuels investment momentum has cooled down in recent years. Among the reasons mentioned are study findings and wide media reports that show: (i) adverse socio-economic impacts including: higher food prices; (ii) widespread land grabs and crowding out of small farmers by biofuels large-scale investments; and (iii) monocultures that result in net increase of GHG emissions. While these predicaments are undisputed, the economic, social, and environmental benefits and costs depend, however, on the type of feedstock used, how and where it is produced and processed. In turn, the how and where these feedstocks are produced, are largely determined by weather.

The following sections discuss briefly Africa's energy profile, vulnerability to climate change, the benefits and risks of bioenergy and the way forward to manage the impact of weather.

2 Africa's Energy Profile and Vulnerability to Climate Change

2.1 Energy Production, Use, and Efficiency

Today, over seven out of ten Africans (57 % urban and 88 % rural population) have no access to electricity (IEA IEA 2010). With a projected annual growth of over 20 % in Africa's electrification rate, there will be 698 million people without

Table 1 Total final consumption of energy (Mtoe)

	1990	2007	Share (%) 2007	2015	2030
Biomass & waste	169	261	56	287	
Electricity	21	43	9	57	
Other renewables	0	0	0	0	
Oil	70	112	24	122	
Gas	9	29	6	34	
Coal	19	17	4	16	
Heat	0	0	0	0	
Total final consumption	289	463	100	516	
Pop. dependent on biomass (million)		608			765

Source IEA 2010

access to electricity by 2030, an increase of 111 million in 2030 from the level today (IEA 2010), Table 1.

In addition to demographic pressure, higher economic growth rates of recent years have significantly increased the demand for electricity. Economic growth in Sub-Saharan Africa (SSA) in 2011 was higher than in other regions estimated at 4.9 % GDP growth; closer to the pre-crisis (2003–2008) level of 5 %" (World Bank 2012).

Frequent power outages are common in many African cities. To cope with this, containerized mobile diesel units have been introduced for emergency power generation at "a cost of about US$0.35/KWh, with lease payment absorbing more than 1 % of GDP in many cases" (UNECA and AU 2011). This is in addition to a loss of 0.1 % in per capita income growth equivalent to a 1.9 % GDP growth attributed to Africa's deficient power infrastructure. It is estimated that Africa loses 10–40 % of its primary energy input (Kirai 2006).

Indoor air pollution causes an estimated 2 million excess death per year, or 5 % of the global burden of disease (IEA 2010). In Asia, such exposure accounts for between half and one million excess deaths every year. In sub-Saharan Africa the estimate is 300,000–500,000 excess deaths (WHO 2012). As a cause of death and illness, indoor air pollution is a larger problem than tuberculosis, AIDS, or malaria. The economic burden of this pollution is estimated at 0.5–2.5 % of the world GNP, some $150–750 billion per year (EIA 2000).

Improving efficiency and access to affordable and clean energy are among the leading objectives of energy policies of countries as well as NEPAD. To meet NEPAD goals, for example, the African power sector requires a total of USD 40.8 billion a year in investments. Hydropower claims most of this investment. Among Africa's planned large-scale hydropower investments are: the Grand Inga Dam[1] in the Democratic Republic of Congo (39,000 MW) and the Grand

---

[1] Africa: The Grand Inga Dam—Can It Really Happen? By Simon Allison, 16 November 2011 http://allafrica.com/stories/201111160521.html.

Ethiopian Renaissance Dam (5,250 MWs)[2] While the Arab Maghreb Union (UMA) countries are grouped within COMELEC (Comité Maghrebin de l'Electricité), countries south of the Sahara (Eastern, Southern, West, and Central Africa) have established sub-regional power pools to facilitate electricity power sharing, promote trade, which are expected to contribute to regional stability and integration. Clearly, weather plays a critical role in the success of the hydropower sector (see discussion in Sect. 2.2 below).

Africa's total investments in renewable energy, which modern bioenergy is a part of, rose from $750 million in 2004 to $3.6 billion in 2011,[3] although the forecast for 2012 is way below 1 billion dollars. Apart from "Morocco's 1.12 billion USD, Africa's investment in renewable energy, excluding large hydropower, remains low compared to the rest of the world" (UNEP 2012).

2.2 Vulnerability to Weather and Climate Change

A majority of the African population are highly vulnerable to climate risks. Among the factors behind this vulnerability are: heavy dependence on subsistence rain-fed agriculture, geography (where the physical effects of climate variability and change are very severe), extensive pastoral-based livestock farming, excessive reliance on traditional biomass energy, low adaptive capacity and weak governance. The most recent drought crisis (2011) in the Horn of Africa, for example, threatened the lives of about 13 million people (AADP August 2012). Traditional sources of energy are getting scarce too requiring women, who bear the primary responsibility for household energy to walk long distances, leaving their children unattended and increasingly rely on cow dung depriving the soil fertility.

Reading of long-term weather variables and community perceptions indicate: (i) increasing mean temperatures (the climate is warming up); (ii) downward trends in long-term rainfall; (iii) increased year-to-year rainfall variability but more on the downside; (iv) poor temporal rainfall distribution within a year—early or late onset of rainfall and dry spells in critical crop growth cycle (seasonal distribution); and (v) recurrence of droughts has increased over time notably in arid and semi-arid areas but also extending to the most productive highlands of the country.

For example, meteorological observations in Eastern Africa and the Sahel, see below, show extreme variability help explain the degree of exposure to climate stress and risk Charts 1, 2, 3, and 4.

The impact of climate change and variability is felt in all development sectors and at all levels. For example, if we consider three levels for analytical purposes:

---

[2] http://grandmillenniumdam.net/
[3] http://allafrica.com/stories/201112230448.html

Bioenergy, Weather and Climate Change in Africa 181

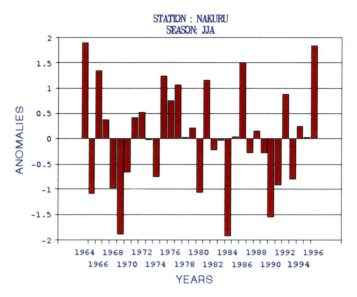

Chart 1 Rainfall anomalies (mms) Nakuru, Kenya Station. *Source* Ogallo (2009)

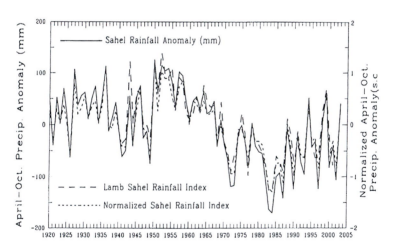

Chart 2 Sahel rainfall (mms). *Source* Ogallo (2009)

national or aggregate; market level; and household/farm level, the likely impacts can be summarized as follows.

National or aggregate level impacts of climate change include: lower economic growth and increased poverty; declining labor productivity and worsening food insecurity often leading to increased human and livestock mortality rates; women walking long distances for fuel wood, thus reducing care for their children; lower water availability, declining water tables, drying of hydropower dams, hydropower

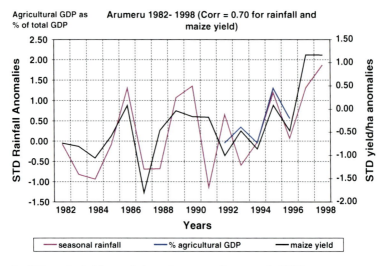

Chart 3 GDP and rainfall correlations. *Source* von Braun et al. (1998)

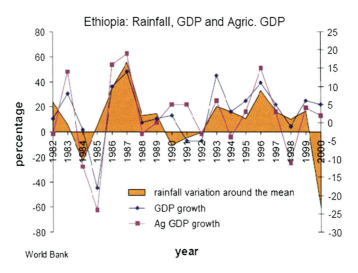

Chart 4 Rainfall, GDP, and agricultural GDP correlation: the case of Ethiopia. *Source* World Bank (2005, 2006)

stations operating below capacity, and long-term reduction of hydropower potential; heightened ethnic tensions and political instability. For example, UNEP's reports show a 30 % decline of rainfall in Darfur during the past 80 years, with a projected decline of 70 % in the years to come (UNEP 2007). The graphs below show strong correlation between rainfall and maize yield as well as between rainfall and GDP.

The impact of rainfall variability on crop production as well as sectoral and aggregate output (GDP) is severe and wide spread. Rainfall explains 94 % of the variance in GDP (IMF 2004).

Market level impacts of climate change include: disintegration of markets (especially remote local markets), with increased power outages due to reduced capacity of hydropower, shift from electricity to charcoal in urban areas with upward pressure on charcoal prices, which in turn aggravates deforestation, increased use of diesel generators, hence increase in GHG emissions, volatility of agricultural production and exports resulting in low foreign exchange earnings.

Household/farm level impacts of climate change include: decreased pasture, water, and fuel wood; changes in production and employment—distorted traditional planting season and disrupted crop cycle; inadequate moisture for crop growing and heightened moisture stress in traditionally good crop growing and grazing areas, and lowering food consumption—declining level and changes in diet and food culture. The level and magnitude of household vulnerability to climate change varies from one country to the other as the map below shows.

The significance of bioenergy and the impacts on weather need to be seen in relation to other renewable energy resources. The development of the bioenergy sector is neither a standalone initiative nor is perceived as a solution to Africa's energy problems. Opportunities and challenges in other sectors, e.g., hydropower, geothermal, solar and wind greatly influence the magnitude of investment and priority accorded to bioenergy, hence a brief discussion below.

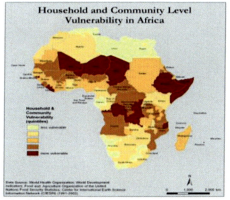

Source Busby et al. (2010)

3 Bioenergy Vis-à-Vis Other Renewables: The Significance of Weather

Africa is hugely endowed with hydropower, solar, wind and geothermal resources. In the area of hydropower, which is and continues to be the primary source of electricity, Africa has developed only less than 10 % of the potential. The sector, however, is adversely impacted by: (i) recurrent drought and consequent drying of dams; (ii) high up-front investment requirements; and, (iii) high risks (technical, economic, commercial, environmental, and social) arising from the investment and management of large hydropower investments, in particular, including population displacement, biodiversity loss, ecosystem disturbance). These constraints highlight the need to work out the optimum energy mix on short-, medium-, and long-term basis and the potential role of bioenergy in this mix.

Because of its proximity to the Equator, Africa has the world's highest yearly average amount of solar radiation with about 95 % of the daily global sunshine above 6.5 kWh/m^2 falling on Africa during winter (UNECA and AU 2011). High initial investment costs and the lack of capacity to maintain solar panels at the village level have impeded wider adoption. Today, solar energy use in Africa is mostly at the household level and small-scale applications.

Despite Africa's enormous unexploited wind energy resources, particularly along some coastal and specific inland areas, wind energy development is at its infancy and is not widely used. Africa's total wind turbines installed capacity stood at 906 MW in 2010 (0.5 % of the worldwide capacity) (UNECA Biofuels Technology Options 2011). Of this total, 169 MW were added in the year 2009), in three countries, Egypt, Morocco and South Africa with Egypt, Morocco, and Tunisia accounting for 890 MW out of Africa's total of 906 MW.

Geothermal power is the other key renewable energy source estimated at 14,000 MW, but only 0.6 % of this potential has been commercially used with Kenya's (127 MW) and Ethiopia (7 MW) (UNIDO 2009). The entire rift valley, stretching from northeastern Ethiopia to Mozambique, shows strong indications of geothermal potential. East Africa alone has the potential to generate 2,500 MW of geothermal power. High initial investment cost is a major constraint.

4 Bioenergy: Climate Opportunities and Risks

4.1 Bioenergy Forms and Uses

Bioenergy refers to renewable energy derived from biomass (biological materials) and can be used for household purposes, i.e., cooking, heating, lighting as well as for industrial or manufacturing processes. It is produced in solid, liquid and gas forms: (i) solid fuelschips, pellets, briquettes, logs used for household uses including cooking, heating, lighting as well as for industrial or manufacturing processes; (ii) ethanol gel—used for cooking in traditional African cooking stoves; (iii) conventional ethanol: produced by the conversion of starchy and sugar crops (sugarcane, wheat, cassava, sorghum, maize, etc.) into alcohol through fermentation and distillation; (iv) conventional biodiesel: produced from oilseeds (rapeseed, oil palm, soybean, sunflower seed, coconut, linseed, cotton seed, ground nuts, castor, sesame seed, corn, jatropha as well as animal fats: beef tallow, pig lard, poultry fats, used cooking oils and oil extracted from algae; (v) biogas: involves converting biomass (plants, wood including waste) into biogas through anaerobic digestion; (vi) producer gas: generated when wood, charcoal or coal is gasified with a limited supply of air; (vi) bio-hydrogen: hydrogen (H2) obtained from biomass (plants and organic waste) through biological process for use by industries (many refineries, coal gasification industries, diesel desulfurization) with huge potential to run transport engines; (vii) lignocellulosic ethanol production: involves extracting fermentable sugar, which can then be converted into alcohol from the lignocellulosic material found in plant stalks and waste seed husks through biological enzymatic process; (viii) algal biodiesel and ethanol—often referred to as the third generation biofuel, algal fuel (biodiesel and ethanol derived from algae) and involves strain selection, culture media, algae growing, and harvesting.

Currently, however, modern bioenergy is dominated by the production of conventional bioethanol and biodiesel dominated by two countries: Brazil (sugarcane) and the United States (corn). Sugar cane remains the primary feedstock for bioethanol while rape seeds and palm are the primary feedstock for biodiesel.

4.2 The Benefits: GHG Emissions Reduction and Beyond

Bioenergy has wide-ranging economic, social, and environmental benefits, which include: (i) improved household and commercial access to cleaner fuel; (ii) enhancing the development of the agriculture sector by offering opportunities for investment and infrastructure development; (ii) diversification of renewable energy sources; (iii) broad-based development and greater multiplier effects as bioenergy creates the possibility to engage a large segment of the farming population in feedstock production; and (iv) reducing carbon dioxide emissions as new plants absorb the carbon dioxide released upon burning thus recycling it without any net carbon dioxide released into the atmosphere.

The realization of the benefits stated above depends on the type of feedstock used, how and where the bioenergy is produced. Nevertheless, given its climatic conditions, in particular tropical climate suitable for fast plant growth, Africa has the potential to grow all types of bioenergy feedstocks including second and third generation cellulosic biofuels.

4.3 Bioenergy Risks

One of the main risks of bioenergy, biofuels in particular, is competition with food production. Most bioenergy feedstock crops (e.g., sweet sorghum, corn, and rapeseeds), used today, are staple food for a majority of the African population. Any use of these crops to produce biofuels may reduce food availability production and raise prices. In countries where these crops are used either directly or indirectly as animal feed, the livestock sector would be adversely affected too. Given the current low level of agricultural technologies, land tenure problems, and poor agricultural management practices, it may also be hard for farmers to produce food and fuel simultaneously.

Depending upon the biofuels feedstock, greenhouse gas emissions is also an issue of grave concern. The main biofuels feedstocks, sugarcane and oil palm, in particular, grow well and perform best in high rainfall and warm areas—the same areas which host Africa's remaining tropical forests. If the production of the feedstock involves the conversion of natural habitats and ecosystems such as peat lands, forests, and grasslands, the climate benefits that accrue to bioenergy will not only be eroded but result in net emissions "as much as 10 times more carbon dioxide than conventional fuel, depending on the type of land used… palm oil produced on converted rainforest land produces 55 times more carbon emissions than palm oil produced on previously cleared land" (Hileman et al. 2010).

In addition to the adverse environmental impacts of monocultures, which is widely documented, there is also the risk of accelerated rainforest clearance and ecosystem destruction as investors hunt for good soils and rainfall. Biofuels investment has seriously threatened Africa's rather small forest reserve, the continent's richest concentrations of biodiversity. Investment firms continue to insist on grabbing the best and most fertile land. In a continent where environmental governance is weak, the likelihood of increasing destruction of rainforest and biodiversity loss is high.

Large land acquisitions and consequent crowding out of farmers to give space to big business is the other major risk. Most bioenergy crops are produced in large commercial farms while many African households remain on small farms. These large commercial farms either bring new land into production or convert degraded, abandoned land, or land that is considered marginal to productive use. Although much of the biofuels' promise was the latter, the former has been the dominant strategy. Further, the notion of huge uncultivated land resonates in many political circles with many people espousing bioenergy that offers opportunities to convert the uncultivated area into energy wealth. Nevertheless, current "estimates greatly exaggerate the land available, by over-estimating cultivable land, under-estimating present cultivation, and failing to take sufficient account of other essential uses for land" (Young 1999). The notion of cultivable land includes mountains, forests, bushes, lakes, wetlands, gorges, national parks, and protected areas. Thus, the actual cultivable area is much smaller than the figures suggest. Further, "in Africa 16 % of all soils are classified as having low nutrient reserves while in Asia the equivalent figure is only 4 %."

As Table 2 shows, Sudan has given out the largest land area of close to 4 million hectares. A recent publication, shows that Ethiopia has placed 3,589,678 hectares for lease under the Federal Land Bank in five regions of the country: Amhara 420,000 (not yet confirmed); Afar 409,678; BeniShangul (691,984); Gambella (829,199); Oromia (1,057,866), and SNNP (180,625) (Dessalegn 2011). The same publication shows that the land is being leased out at

Table 2 Land acquired for biofuels production in selected African countries

Country	Projects	Area ('000) hectare	Median size (hectare)	Domestic share %
Ethiopia	406	1,190	700	49
Liberia	17	1,602	59,324	7
Mozambique	405	2,670	2,225	53
Nigeria	115	793	1,500	97
Sudan	132	3,965	7,980	78

Source Deininger et al. (2011); World Bank (2011)

"ridiculously" low rent that ranges between 14.1 birr (less than one US dollar) and 135 birr (about USD 8) (Dessalegn 2011).

Information is lacking to analyze the social and environmental implications of these investments. As Table 2 above shows, in Sudan 78 % of the investment is locally owned (domestic share) compared to Liberia's 7 %. However, there is no empirical evidence suggesting that local investors will be more socially and environmentally responsible than foreign investors.

5 The Direct and Indirect Impacts of Weather/Climate Change

Climate change and variability impact the choice of feedstock, level of production, costs, where it is produced, and how it is produced (rain fed or irrigated). These impacts take several forms:

(i) **Impact on Africa's Bioenergy Potential**: The availability of adequate moisture is an important factor for plant growth. For example, the main biofuels feedstocks: sugarcane and oil palm are high water requiring and growing them under rainfed conditions can only be done in Central Africa and some parts of West Africa which threatens Africa's small remaining tropical forest belt. Other feedstocks like sweet sorghum maize, jatropha, and soybean do not provide optimum yields under conditions of moisture stress. The amount of rainfall in Africa varies considerably from 0 mm of annual rainfall in the Sahara and Kalahari deserts to 4,500 mm in Central Africa. In general, Africa's rainfall is characterized by the following:

- Uneven rainfall distribution. Central Africa gets most of Africa's rainfall. The concentration of precipitation in a rather small part of the continent means less potential for large, rain-fed commercial farms. Further, Central Africa includes what remains of Africa's tropical rainforest, which must be conserved to maintain the ecosystem integrity of the entire continent.
- Rainfall variability within the same region. In addition to the rainfall variability between, for example, Southern Africa and Central Africa, or between the Sahel countries (Mali, Niger, etc.) and the Gulf of Guinea (Liberia, Sierra Leone, etc.),

there is a marked difference in the amount of rainfall within the same region. Extreme variability of the local rainfall would require adoption of wide-ranging varieties of bioenergy crops, which could limit the size of commercial farming.
- Recurrent drought and flooding. Since the 1970s, the Horn of Africa, the Sahel countries, and Southern Africa have experienced recurrent drought much more frequently than previously observed. The Horn of Africa in particular has been hit hard with a resulting famine of apocalyptic proportions. In Ethiopia and Somalia, the severe drought in 2006 was followed by flooding.
- Erratic rainfall. Several African countries receive annual rainfall during certain months of the year, with some months being completely dry. This rainfall pattern reduces the potential for growing crops with a long growing season.

(ii) **Choice of Feedstock, Productivity, and Cost**: The range of plants and crops used as biofuels feedstocks is rather limited. Table 3 shows energy yields of the main feedstocks. Accordingly, sugarcane is the most efficient crop for the production of bioethanol and the highest energy-yielding crop. Sweet sorghum is second, but provides only 56 % of the ethanol that sugarcane produces. The biofuel yield of maize is far lower than sweet sorghum.

On the biodiesel side, palm oil is the best performer in terms of biofuel and energy yield. Jatropha, with a biofuel yield of 700 l per hectare is rather low, although it stands as having the second best potential after palm oil.

Based on current technologies, below is a brief analysis of each crop:

- **Sugarcane**: Sugarcane requires adequate annual rainfall (1,400–1,800 mm) and warm temperature (22–38 °C). The plant performs well under a long growing season (15–16 months) on soils with pH in the range of 5–8.5. Sugarcane productivity is highly influenced by climatic conditions and productivity ranges from 50–100 t/ha (weight of wet stem). Productivity in some African countries, particularly Zambia and Zimbabwe, reaching 140 t/ha in place (UNECA and AU 2011). Today, much of Africa's sugar is produced under irrigation. For rain-fed sugarcane production, which is least cost, availability of adequate and well-distributed rainfall throughout the growing period is necessary for maximum yields. The minimum temperature for active growth is approximately 20 °C. Accordingly, the most suitable areas for rain-fed sugarcane are Central Africa, some parts of West Africa and Madagascar, which house Africa's remaining small tropical forest. Meteorology has, thus, severely limited Africa's potential in rain-fed sugarcane production.
- **Maize/Corn**: Maize is widely grown in Africa under rain-fed conditions. It is a staple food crop in many countries and a backyard crop. It is used also for making traditional alcoholic beverages. As a result, its use as a biofuels feedstock is banned in many countries, including South Africa.
- **Sweet Sorghum**: Widely grown in many parts of Africa under rain-fed conditions, sweet sorghum is known for its resistance to drought and tolerance for heat. It is a staple food crop for many Africans and is also used to produce alcoholic beverages.

Table 3 Comparative analysis of performance of biofuels feedstocks

Crop	Seed yield (t/ha)	Crop yield (tons/ha)	Biofuel yield (l/ha)	Energy yield (GJ/ha)	Annual rainfall range in mm	Altitude/temperature range for optimum growth
Sugarcane (juice)	–	100	7,500	157.5	1,400–1,800	0–1,000 m/22–38 °C
Palm oil	9,800	70	3,000	105.0	>2,000	<400 m/22–32 °C
Sweet sorghum	–	60	4,200	88.2	500–800—has no flooding tolerance	Variable, but does not handle cold well
Maize	–	7	2,500	52.5	500–800	Above 15 °C
Jatropha	740	2–12.5, depending on rainfall	700	24.5	300–1,000	0–500 m/ above 20 °C
Soybean	480	990 kg/hec	500	17.5	450–700	10–30 °C, depending on stage of germination, ideal is between 21–27 °C
Cassava	–	–	1,537	61	–	–
Sugar beets	–	–	–	–	450–960	20–35 °C
Sunflower seed	–	–	–	–	600–1,000	–
Algae	–	–	30,000	–	–	–
Rapeseed	–	–	544	–	–	−6–4 °C
Castor beans	–	–	1,600	54.3	–	–

1 gigajoule (GJ) = 278 kWh
Source: (first 6 rows) Francis X. Johnson, Stockholm Environment Institute, 2006
Palm oil http://www.newcrops.uq.edu.au/newslett/ncn10214.htm
Sweet sorghum http://www.tropicalforages.info/key/Forages/Media/Html/Sorghum_(annual).htm
Rainfall for maize, soybean, wheat http://www.fao.org/docrep/U3160E/u3160e04.htm#2.1%20water%20requirements%20of%20crops
Temperature for Maize http://www.fao.org/ag/AGL/AGLW/cropwater/maize.stm
Jatropha http://www.jatrophaworld.org/9.html
Soybean http://www.cgiar.org/impact/research/soybean.html; http://www.nsrl.uiuc.edu/news/nsrl_pubs/insectbooks/guidelines/introduction.pdf
Sugar Beets http://www.tmau.ac.in/tech/swc/sugarbeet.pdf; http://www.agr.hr/jcea/issues/jcea4-2/pdf/jcea42-3.pdf; http://www.wg-crop.icidonline.org/37doc.pdf; http://www.biodieseltechnologiesindia.com/biodieselsources.html
Castor beans http://www.hort.purdue.edu/newcrop/duke_energy/Ricinus_communis.html
Cassava http://www.mekarn.org/msc2003-05/theses05/phallaabs.pdf; http://gristmill.grist.org/story/2006/2/7/12145/81957

- **Palm Oil:** Palm is an important energy crop, the most efficient among the biodiesel crops. The Central Africa region, which is also a tropical rainforest area, is most suitable for rain-fed oil palm. Indeed, large-scale oil palm plantations are almost all established in large forest areas and involve forest clearance. Oil palm monocultures are also associated with soil nutrient depletion and erosion. Oil palm production, unlike other crops, is unamenable for smallholder agricultural scheme. Therefore, in many parts of Africa the benefits of saving Africa's rainforest and maintaining hydrological functions and services provided by ecosystems far outweighs energy benefits that accrue from palm-based biodiesel. Large-scale production of rain-fed oil palm appears limited or will result in high climate debt.
- **Jatropha**: Jatropha grows well in low-altitude areas with annual rainfall between 300 and 1,000 mm per year. This makes many parts of Africa suitable for jatropha production. Jatropha allows intercropping with yams, pulses, grain, or legumes, while the oil cake is used as an organic fertilizer. It, thus, contributes to increasing food production while reducing soil nutrient loss. Moreover, jatropha can be harvested three times a year. It can also do without much irrigation and does not require much pesticides or fertilizers. Jatropha gives about one-fourth of the biofuel yield of palm oil, but the climate benefits could outweigh energy benefits. However, knowledge about this plant is generally limited to corroborate the view that it is a miracle crop.
- **Soybean**: Soybean is one of the most promising food and energy crops. It is a soil-enriching crop that performs well with annual rainfall between 450 and 700 mm.[4] Although its biofuel yield is about 500 l/ha, it requires far less fertilizer and pesticides compared to, for example, maize. Soybean grows well in many parts of Africa and its contribution to greenhouse gas reduction is more substantial compared to maize.
- **Castor Oil**: Castor grows throughout Africa, but generally as a wild plant. Castor is believed to have enormous potential for biodiesel production and appears to be superior given its economic and ecological benefits, which include: no competition with the food sector; belongs to the bean family—soil-enriching not depleting like maize and palm; castor oil maintains its fluidity at extremely high and low temperatures; and, variety of medicinal and other values too. Further, perhaps more importantly, the castor oil plant is easy to grow and drought resistant, which makes it an ideal crop for the semi-arid and arid regions of Africa.
- **Cassava**: Cassava is common in many African countries and is a staple food crop. It grows under a variety of moisture and soil conditions. Cassava is produced manually at the small-scale or household level. Cassava has "one of the highest rates of CO_2 fixation and sucrose synthesis for any C_3 plant." Cassava performs well in pockets of tropical, high rainfall, and humid areas of Africa, where expansion will be constrained because of the high priority given to conservation of tropical forests.

---

[4] http://www.fao.org/docrep/u3160e/u3160e04.htm

(iii) **Length of Growing Period**: For plant growth, particularly crops like sugarcane that have a long growing period (LPG), it is not the amount of rain but its distribution that is more important. LPG is defined as "the period (in days) during the year when soil moisture supply is greater than half potential evapotranspiration (PET)" (www.fao.org/ landsuitability maps). Accordingly, FAO puts LPG into four categories: (i) arid: LGP less than 75 days; (ii) semi-arid: LGP in the range 75–180 days—agriculture areas compete with livestock; (iii) sub-humid: LGP in the range 180–270 days; and (iv) humid: LGP greater than 270 days. Any area that has a growing period of less than 75 days is considered unreliable and unsuitable for rain-fed agriculture, although in the southern part of the Sahara desert, the eastern part of the Horn of Africa, and southwest Africa bordering the Kalahari desert, there could be livestock and limited farming activity through the use of boreholes and artificial dams. Here again, Central Africa and the surrounding countries have a longer growing period consistent with the amount of rainfall they receive, limiting Africa's potential in the production of the main biofuels feedstock: sugarcane and oil palm.

In sum, weather has considerably diminished Africa's potential in conventional biofuels feedstock's (sugar cane, oil palm, etc.) in addition to the limited availability of suitable land. Still, with rapid advances in research in developed countries that brought new energy crops into production, lligocellulosic and algae technologies, Africa continues to possess huge potential given its tropical climate suitable for fast growth of almost all types of crops and plants all year round.

6 Main Policy Issues and Options

Climatic conditions determine what, where, and how bioenergy is produced and processed. In addition, macroeconomic policies and developments in the agriculture, natural resources, industry, water resources sectors influence bioenergy. Bioenergy development, in turn, impacts food, environment, natural resource, industry, and infrastructure sectors. A bioenergy policy should, thus, be seen as an integral part of a country's national development strategies that requires consideration of economic, social, environmental, political, and cultural factors as well as social organization, sub-regional and global trade and investment relations.

In Africa, bioenergy policy development is at an early stage. Today, only about ten African countries have some kind of national policy related to biofuels (UNECA and AU 2011). The experience of these countries and others in the processing of developing a bioenergy policy is characterized by: (i) the tendency to overlook potential negative impacts of biofuels and a rush toward concluding investment deals without carefully examining environmental and social impacts; (ii) the promotion of foreign investment without analyzing strong backward and forward linkages to the economy; and (iii) limited consideration of the entire life cycle of bioenergy production and process of the investment (Ejigu 2011).

Among the critical policy issues are:

(a) Land use planning. Agroecological zonation and land use planning are vital for the sustainable production of bioenergy feedstocks. The land use plan considers the type and location of the plant species to be grown for biofuel feedstock, socio-economic and environmental viability, farming and harvesting systems involved in their production, transportation to refineries, the type and location of the production facilities, and transportation of the fuel to market. Decisions regarding how crops are grown and managed will determine their effects on carbon sequestration, native plant diversity, competition with food crops, greenhouse gas emissions, water quantity, and water and air quality.

What is woefully inadequate is local level weather data over a long period required for planning and policy development. Meteorological stations in many countries are few and too widely spread to capture the wide variation in local climate, particularly in arid and semi-arid areas. It is also hard to access weather-related data from Met offices in almost all cases, even if the data exists because weather data is often seen as a national security issue. Thus, there is a need to strengthen the capacity of Met offices in a holistic manner. With the growing importance of climate issue in national development policy, there is need for sector wide recognition that the old model of Met offices—a modest scientific weather bureau—would not enable them meet today's expectation in the context of climate change. There is need to set up a properly budgeted, staffed, and trained public information section.

(b) Choice of feedstock supply and use. The type of feedstock used, how and where it is produced determine the extent of economic, social, and environmental benefits of bioenergy. The life cycle assessment (LCA) of different biofuel crops reveals large differences in yields, climatic requirements, energy balances, and water and carbon footprints. Under existing production patterns and technological conditions, the use of staple food crops for bioenergy should be avoided. The opening up of a new window of consumption, i.e., energy for these crops, will drive food prices up, which will make food expensive and inaccessible. Even if a government introduces food subsidies, which actually is unwarranted and imposes financial burden beyond the capacity of most African countries to bear, the higher prices could trigger land use changes as investors acquire large tracts of for feedstock production. The supply of the displaced food and feed commodities subsequently decline, leading to higher prices for those commodities. The land use conversion may result in undesirable social and environmental changes that need to be factored in.

(c) Scale of operations. The scale of operations, i.e., small producers, medium, and/or large-scale plantations matters in the choice of feedstocks. Most feedstock crops can be economically viable and more so environmentally sustainable in small-scale and community-based production and processing schemes. Large-, medium-, and small-scale production and processing can be complementary and have different impacts on development. There are areas where large-scale production of bioenergy could have high returns and advantages of economic scale, while small- and medium-scale enterprises have greater potential to create backward, forward, and lateral linkages to the economy. Further, large-scale and small producers can target different markets. For example, while large-scale producers aim at producing, say high quality bioethanol for the transport sector, small- and medium-scale producers can cater for household uses, i.e., heating, cooking, and lighting. A sustainable bioenergy policy should be designed in a manner that will make small producers and low income groups (which constitute a large segment of the population) central (both as producers and consumers) to transition toward higher agricultural income growth and agro-industrial processing.

(d) Subsidies and government support. One of the distinguishing features of the energy sector is subsidies to both nonrenewable and renewable energy production and consumption. IEA estimates subsidies at the global level at 37 billion for electricity and 20 billion for biofuels in 2009 (IEA 2010). Indeed, the bioenergy development in the now successful countries (e.g., Brazil, U.S., Europe, China, and India is characterized by heavy government subsidy. For example, EU subsidizes farmers at the rate of Euros 45 per hectare while the U.S. subsidizes ethanol production by 51 cents per gallon (Harder, Science News 2006). Current policies designed to promote bioenergy development include production subsidies and incentives for local processing subsidies as well as tariff and non-tariff barriers with the view to encouraging individual and corporate investors to move into production with minimal risks and also give them time to establish the industries. While such policies could be justified on economic, social, and environmental, and even in some cases, national security grounds, they have the potential to distort national, and in some cases global, markets. However, the pace and viability of the bioenergy sector will continue to be determined by dynamics in the petroleum industry, which is also a highly subsidized sector (IEA 2010). Thus, a well-strategized government support is critical to bioenergy development and inevitably. While government subsidy is unlikely in the African context, subsidies given to biofuels producers in countries outside Africa is bound to put heavy pressure on production and processing of feedstock in Africa, which African countries have to respond to make biofuels competitive in the global market.

(e) Production and processing. With its largely tropical climate suitable for fast growth and diverse ecological conditions, Africa has the potential to grow almost all types of feedstock in a cost-effective manner. However, Africa has little to gain as a raw material producer. Much of the value addition and backward and forward linkages to the rest of the economy are realized at the processing stage. As feedstock suppliers, countries will forego considerable economic benefits (higher prices per unit of output, markets created for inputs at the processing stage, transfer of skills, etc.), social benefits (employment opportunities both at the installation and operational stages), and even environmental benefits (use of the waste products as fertilizers). Further, meeting the energy requirements of the local population and supplying fuel to local markets including meeting blending targets presents huge investment opportunity. Thus, there is need to shift the bioenergy decision-making process from a supply push (an effort to accommodate investors) to a demand-driven program that aims to meet a country's energy demand based on own feedstock and in-country processing (refining) capacity. A sustainable bioenergy policy requires bioenergy investments to combine feedstock production and processing (refineries) and guarantees local markets.

(f) Price competitiveness with oil. Bioenergy costs and benefits are always expressed in relation to petroleum prices. When petroleum prices rise, bioenergy investments become lucrative. But when petroleum prices fall, bioenergy becomes a losing undertaking. The breakeven price for biofuels, today, is believed to be between $35 and $60 per barrel of oil equivalent, with Brazil's ethanol estimated to break even at $35 compared to around $45–55 in US and EU, which reflects the high costs of biofuels production. Biofuels will continue to be expensive, hence the need for subsidies in the short to the medium term, which will be a heavy burden to governments, particularly in the African setting. To enhance the competitiveness of bioenergy, bioenergy development should be grounded on a large production base that embraces small holder production and processing scheme, environmental and social benefits, in addition to the backward, vertical, and lateral linkages to the wider economy. This will certainly help lay the foundation for rural transformation and a country's industrialization.

(g) Managing the Food, Fuel, and Feed Competition. The biofuel feedstocks currently commonly used (e.g., corn, rapeseeds, lentils, etc.) are staple food/feed crops to most Africans. While it is possible for technological changes in fuel crop production to give impetus to food production growth with the net result being higher food supply, social, political, and environmental factors, even economics, are dictating the shift from food crops to non-food crops and from large-scale plantations to greater involvement of small-scale producers. Even here, although the decision of how much to produce for fuel or food is likely to be based on the household's economic and social needs, higher prices of feedstock are bound to heavily influence these decisions in their favor. However, there is wide range of crops and plants that can be used for bioenergy; and with technological advances that made it possible to grow energy crops in areas deemed unsafe for consumable crops, such as beside roads, next to polluting industry or on contaminated land, or being irrigated by treated waste water, or even on marginal lands assuming economics hold, it is possible to produce enough food while increasing the bioenergy supply. A sustainable bioenergy policy is founded on an integrated and balanced pursuit of economic, social, and environmental objectives and avoiding the food, feed, and fuel competition.

(h) Land tenure policy and property rights. Tenure insecurity that prevails in many African countries has become a constraint to increasing agricultural production, conserving soil and water resources, and planting trees. Investment in energy crops also requires long-term commitment and secured land holdings.

The Framework and Guidelines on Land Policy in Africa (AU, UNECA and ADfB 2010), while calling for clarification of property rights in agriculture and sustainable land management, highlights "serious concerns about land needs" for energy development and questions "the capacity of many countries to meet their internal agricultural requirements as land is taken out and the ecological trade-offs involved in the scramble by foreign investors for land" to grow biofuels feedstock. While some countries have already established programs and policies to improve land tenure, other countries that have not done so need to critically consider land tenure and property rights issues improvements in their bioenergy policy formulation process.

(i) Bioenergy and biodiversity. Halting environmental/land degradation while expanding bioenergy production is a critical element of a sustainable bioenergy policy. Very few African countries can take pride in their environmental policies of the past decade. Deforestation and land degradation have continued unabated. The rate of replanting has been woefully inadequate to offset the effects of deforestation. As current biofuels feedstocks, sugarcane, palm oil, corn, etc. are high moisture requiring and soil fertility depleting, sustainable management of natural resources under fast-growing bioenergy production is a daunting task. Moreover, and much seriously, planting some of the highly productive biofuels feedstocks, such as palm oil, required clearing tropical forest opening wide the scope for a large-scale, massive deforestation. One source of hope is the collective position taken and political commitment made to fighting climate change through effective climate adaptation and mitigation. Through strategic choices of biofuel feedstocks, gradually developing those that enrich soils and do not require much water to grow and further moving to lignocellulosic and algae based biofuels, a sustainable bioenergy policy ensures the protection of the environment including Africa's fast dwindling forest resources as well as and the maintenance of ecosystem integrity and diversity.

(j) Bioenergy markets and trade. Bioenergy trade is an issue of global interest and perhaps one of the fastest growing sectors worldwide. Today, there are many foreign investors interested in the production and processing of certain bioenergy products with the view to supplying a foreign market. Issues of economic, social, and environmental sustainability, subsidies, tariff and non-tariff barriers, fair-trade practices, and certification are key agenda items of an ongoing debate.

(k) Bioenergy governance: Governance, here, refers to institutions, policies, customs, relational networks, laws and regulations, property rights, stakeholders' participation in policy development, access to knowledge, finance, information, and education that foster the sustainable development of bioenergy. Indeed, bioenergy development is a multisectoral and multilevel undertaking that requires the active engagement of the government, the private sector, civil society, and institutions of higher learning.

There are four possible areas of intervention: (i) knowledge development and capacity building of Met Offices. Filling existing climate knowledge gap on CC impact through generating sub-national climate data and systematic assessment of vulnerable areas and populations and threatened ecosystems strengthening the capacity of Met offices in a comprehensive manner; (ii) Preparing disaster preparedness and risk management action plans as integral part of a national development strategy; (iii) Promote and expand collaborative, adaptive, and multidisciplinary research through, among others, twinning arrangements between developed and African countries; (iv) Improved access to research results on new biofuels feedstocks including perennials, lignocellulosic and microalgae.

7 Conclusion

The connection between bioenergy and weather/climate is strong, multifaceted, and nonlinear. The sustainable development of bioenergy has the potential to contribute substantially to improving access to affordable and clean energy and also to raising living standards. However, how bioenergy development is designed, the kind of feedstock used, how it is produced, and where it is processed are critical elements of a sustainable bioenergy program. There is clearly strong need for vigorous investment in developing the local climate knowledge base and also improving the coverage, reliability, and timeliness of meteorological data.

References

Busby J et al (2010) Locating climate insecurity: where are the vulnerable places in Africa?

Deininger K and Byerlee D with Lindsay J, Norton A, Selod H, and Stickler M (2011) Rising global interest in farmland: can it yield sustainable and equitable benefits? The World Bank, Washington, DC

Dessalegn R (2011) LAND TO INVESTORS: large-scale land transfers in ethiopia, forum for social studies, Land governance for equitable and sustainable development, IS Academia, Addis Ababa

Ejigu M (2008) Toward energy and livelihoods security in Africa: smallholder production and processing of bioenergy as a strategy. Nat Res Forum UN Sustain Dev J 32(2):152–162

Ejigu M (2011) Sustainable bioenergy policy framework in Africa: toward energy and livelihoods security, first draft, prepared for the food security and sustainable development division, UNECA

Energy Information Administration (EIA) (2000) International energy outlook, energy information administration, office of integrated analysis and forecasting, U.S. Department of Energy

International Energy Agency (IEA) (2010) World Energy Outlook 2009. http://www.iea.org

International Monetary Fund (IMF) (2004) The federal democratic republic of Ethiopia: Ex post assessment of long term fund engagement. Staff Country reports, 18 Aug 2004

Kirai P (2006) Energizing Africa through energy efficiency. Presentation at the 6th global forum for sustainable energy, GEF-KAM ENERGY EFFICIENCY PROJECT, Vienna, Austria, 29 Nov–1 Dec 2006

Mitchell Donald (2011) Biofuels in Africa: opportunities, prospects, and challenges. The World Bank, Washington, DC

Ogallo LA (2009) Climate change and sustainable development of IGAD region. Presentation at the IGAD climate change workshop, Djibouti

Stratton R, Wong HM, Hileman J (2010) Lifecycle greenhouse gas emissions from alternative jet fuels. MIT: partnership for air transportation noise and emissions reduction, Project 28, June 2010, Version 1.2, http://web.mit.edu

UNECA AU, ADfB (2010) Framework and guidelines on land policy in africa, Addis Ababa, Ethiopia

UNECA and AU (2011) Economic report on Africa 2011: governing development in Africa—the role of the state in economic transformation. Addis Ababa, Ethiopia

UNEP (2007) Sudan post-conflict environmental assessment, Nairobi, Kenya

UNEP—Frankfurt School Collaborating Centre for Climate Change & Sustainable Energy (2012) Finance global trends in renewable energy investment 2012

UNIDO (2009) Scaling up renewable energy in Africa. http://www.uncclearn.org/sites/www.uncclearn.org/files/unido11.pdf

World Bank (2005) Water resources, growth and development :a working paper for discussion. Prepared by the World Bank for the panel of finance ministers, The U.N. Commission on Sustainable Development

World Bank (2006) Ethiopia—managing water resources to maximize sustainable growth: a World Bank water resources assistance strategy for Ethiopia

World Bank (2012) Africa pulse update, February 2012. Outlook for Sub-Saharan Africa is positive, but downside risks loom large. http://web.worldbank.org

World Health Organization (WHO) (2012) WHO, fact sheet air quality and health. http://www.who.int/mediacentre/factsheets/fs313/en/

Young A (1999) Is there really a spare land? a critique of available land estimates in developing countries. Environ Develop Sustain 1:3–18

Part III
What can Meteorology Offer to the Energy Industry?

Weather and Climate Information Delivery within National and International Frameworks

John W. Zillman

Abstract Weather and climate information plays an important role in decision-making in the energy sector. Energy-specific meteorological services are based heavily on the national and international meteorological infrastructure put in place to provide information, forecasting, warning and advisory services to all weather- and climate-sensitive sectors of society. Recent advances in atmospheric, oceanic and hydrological observation, data collection, modelling, forecasting and warning systems, and in national and international arrangements for service provision, offer many opportunities for increased benefit for the energy industry from effective application of meteorological and related science and services.

1 Introduction

Meteorology serves the energy industry through research into atmospheric influences on energy demand, generation, distribution, supply and use and through provision of a wide range of weather and climate services to inform decision-making in the energy sector (Maunder 1986; Price-Budgen 1990; World Meteorological Organization 2007).

Most of the weather and climate services used by the energy sector are purpose- and user-specific but essentially all such energy-focussed services are based, ultimately, on the integrated national and international infrastructure for meteorological and related observation, research, modelling and service provision operated by countries to meet the information needs of all weather- and climate-sensitive sectors of their national economies (Zillman 1999, 2003a).

Meteorological services deliver public and private benefit by enabling improved decision-making which leads to greater (financial and other) gains in virtually all

J. W. Zillman (✉)
Melbourne, VIC, Australia
e-mail: J.Zillman@bom.gov.au

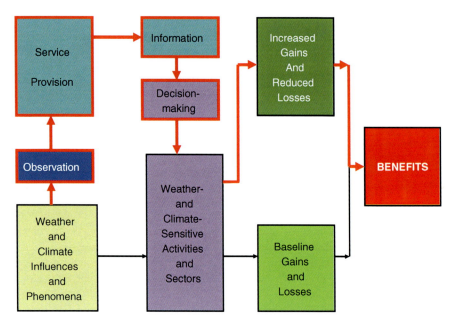

Fig. 1 The role of meteorological observation and service provision and application (*bold red boxes*) in delivering benefits to weather- and climate- sensitive activities and sectors by providing forecast, warning and other information which enables better decisions than would otherwise be possible resulting in increased gains and reduced losses relative to the baseline impacts and outcomes (*non-bold black boxes and arrows*) in the sectors concerned

sectors of the economy and reduced loss of life, livelihood and property from hazardous meteorological phenomena (Fig. 1). Energy is one of the most weather- and climate-sensitive sectors of society and the energy industry can benefit greatly from informed use of meteorological information (Page 1990; Dubus and Parry 2012).

By way of introduction to the many established and emerging sector-specific meteorological services for the energy industry, this chapter briefly:

- explains the nature and mechanisms of weather and climate;
- describes the origin and scope of meteorological and related services;
- introduces the concept of a national meteorological service system;
- outlines the international framework through which nations work together in the provision of meteorological services; and
- identifies some recent international initiatives which will influence the future of weather and climate information delivery for the energy sector.

2 The Nature of Weather and Climate

Almost every aspect of human activity is influenced by the state and behaviour of the atmosphere around us—its temperature, humidity, cloudiness, wind, rainfall and so on. All of these are determined, more or less directly, by the incoming energy

Weather and Climate Information Delivery

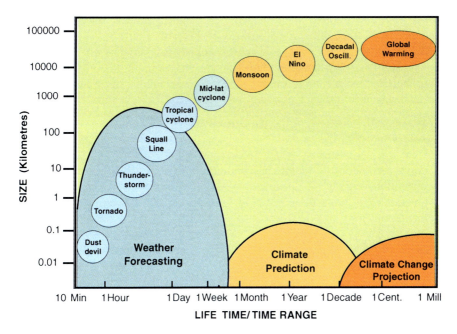

Fig. 2 Characteristic time and space scales of some weather and climate phenomena (*in circles*) and, very schematically, the time range (*horizontal scale*) and relative skill/reliability (*vertical scale*) of weather forecasting (*blue*), climate prediction (*light orange*) and climate change projection (*dark orange*)

from the sun incident on the spherical rotating earth driving the global circulation of the atmosphere and ocean and generating cyclones and anticyclones and all the other familiar phenomena of weather and climate (Bureau of Meteorology 2003).

Meteorology is the science of weather and climate. Weather is a description of the state of the atmosphere over minutes, hours, days and weeks. Climate is a statistical description (in terms of means, variability and extremes) of its behaviour over longer timescales, usually years, decades or centuries, in one sense 'average' weather. The characteristic time and spacescales of some of the important phenomena of weather and climate are shown schematically in Fig. 2.

In order to map the patterns of weather and climate over the earth's surface, to build up a historical description of their behaviour in the past and to provide the starting point for prediction of how they will evolve in the coming days, weeks, months, years and decades, it is essential to maintain long-term in situ and space-based observing networks and systems over the globe. The total annual cost of global weather and climate observing networks, including the expensive meteorological satellite systems, is around US$5–7 billion (Zillman 1999; Global Climate Observing System 2010).

Because the behaviour of the atmosphere and ocean are governed by the well-established laws of physics, it is possible to simulate the processes and phenomena shown in Fig. 2 through sophisticated mathematical models run on powerful computers with three-dimensional grids over the earth's surface.

These numerical models enable meteorologists to assimilate all the observations into an internally consistent three-dimensional description of the state of the atmosphere and ocean and to run the model forward in time much faster than the real atmosphere to provide:

- analyses and reanalyses of the patterns of weather and climate over the globe;
- weather forecasts out to the 2 weeks (approx.) limit of predictability of individual weather systems;
- climate predictions for months and seasons out to a few years with experimental predictions for a decade or more; and
- climate change projections resulting from human interference with the global climate system.

There are many different ways of characterising the 'predictability' of future weather and climate and the level of confidence attached to projections of human-induced climate change including the use of 'ensembles' of model runs whose scatter provides an indication of the probability of a range of possible outcomes. The approximate limits of predictability of weather and climate and the level of scientific confidence in model projections of human-induced climate change are summarised schematically in the lower part of Fig. 2.

3 Meteorological and Related Services

Meteorological service provision has a long history (Davies 1990). In the broadest sense, humans have made use of weather and climate information to guide their decisions on where to live, what to grow, when to harvest and how to organise their daily lives for thousands, and probably tens of thousands, of years. But it is only over the last few hundred years that systematic meteorological observations have provided a reliable description of past and present weather and climate and dramatic progress in understanding of the atmosphere and ocean has provided a scientific basis for societally useful forecasts of their future evolution. Nowadays, meteorological services are routinely used in decision-making in almost every sector of society (World Meteorological Organization 2007).

Historically, meteorological services were regarded as including the provision of basic data and information on the state of the land and ocean surface as well as on all aspects of the atmosphere but, with the emergence of operational oceanography and hydrology, it has become appropriate to recognise the existence of separate categories of oceanographic and hydrological services and to view meteorological services as the weather, climate and air quality subset of a broad suite of Earth system science services providing information on:

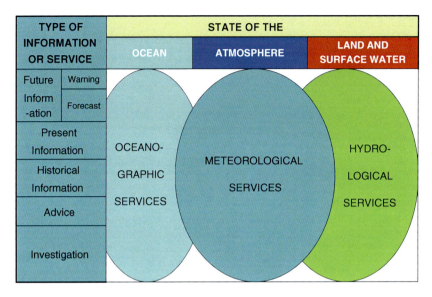

Fig. 3 The scope of meteorological and related services in terms of Earth system domain (ocean, atmosphere, land) and the type of information or other (eg advisory or investigative) service provided

- the past behaviour of the various domains (ocean, atmosphere, land) of the Earth system;
- the current (physical, chemical and biological) state of the atmosphere, ocean and land surface;
- their expected future behaviour, including both forecasts of the various environmental variables and warnings of specific hazardous phenomena and conditions; as well as
- advice on the implications and applications of oceanic, atmospheric and terrestrial science; and
- conduct of investigations of specific oceanographic, meteorological and hydrological problems;as shown schematically in Fig. 3.

The five basic types of meteorological service shown in Fig. 3 can be further categorised in many ways, including in terms of:

- the meteorological variables on which information is provided;
- the phenomenon (especially hazardous phenomena) involved;
- the method of service delivery;
- the economic characteristics of the service provided; and, most importantly:
- the user communities served.

Almost every meteorological element impacts on society in some way and there are specific meteorological services focussed on the provision of information on past, present and future values, including both categorical and probabilistically forecast values, of all the standard *meteorological variables* including:

- solar (direct and diffuse) and infrared radiation;
- temperature (mean, maximum and minimum surface air temperatures as well as vertical temperature profiles and soil and ocean temperatures) and humidity;
- wind speed and direction near the surface and their variation with height;
- cloudiness (including type, height and thickness of cloud layers), rainfall and evaporation; and
- atmospheric composition, turbidity and visibility;as well as such derived quantities as moisture and heat fluxes and atmospheric stability, all of which can be determined from measurements or calculated from the output of numerical models.

Meteorological services are also often described and organised in terms of the specific *weather or climate phenomenon* involved. Highest priority is usually given to detection, forecasting and warning of dangerous phenomena and conditions. These embrace a wide range of meteorological and related oceanographic and hydrological hazards and give rise to specific services focussed on warning of:

- severe weather phenomena (thunderstorms, hail, lightning, tornadoes, tropical cyclones, gales, blizzards,…);
- other hazardous weather conditions (heat waves, fire weather conditions, cold snaps, fog, icing, freezing rain, …);
- air quality hazards (urban air pollution, smoke haze, acid rain, volcanic ash,..);
- climatic hazards (drought, pest infestations, disease outbreaks, crop failures,…);
- marine hazards (waves, currents, storm surges, tsunamis, icebergs, marine pollution, algal blooms …); and
- hydrological hazards (riverine and flash flooding, erosion, landslides, …).

Service categorisation by *method of delivery* recognises the vital role of the various print, broadcast and other electronic communications media as well as the more personally interactive forms of service provision, giving rise, for example, to source-based meteorological service categories such as:

- TV weather services (including both weather segments on news and other channels and dedicated services such as 'The Weather Channel');
- Radio weather services (providing local temperatures and forecasts and immediate broadcasts of severe weather warnings and the like);
- Newspaper weather services (synoptic maps, satellite photos, rainfall lists, prognoses and city, town and district forecasts);
- Recorded telephone services (latest temperatures, forecasts, warnings) and SMS messaging systems (e.g. for storm, hail and lightning warning);
- Internet-based weather services (everything);
- Publications (observational data, scientific papers, reports, bulletins); and
- Consultative services (usually involving personal interaction and report preparation).

It is important in the design of both general and sector-specific meteorological services and in determining the appropriate basis for their provision to take account of the *economic characteristics* of the services, according to their rivalry in consumption and excludability. This leads to the important categorisation of meteorological services either as public goods which, in general, are most efficiently provided by government through taxation, or private goods which are most efficiently provided through the operation of market processes (Gunasekera 2004). In some countries, especially those in which cost recovery and commercialisation policies have been introduced into meteorological service provision, it has proved useful to recognise a public versus private goods-based distinction between what are referred to as 'basic' and 'special' services (Zillman and Freebairn 2001) and to view national meteorological service provision, in the broadest sense, as involving the provision of:

- *basic infrastructure, data and products*: the basic national meteorological infrastructure which underpins the provision of the full range of services and may itself be recognised as the provision of a 'basic service' to present and future generations;
- *basic service*: those services provided at public expense to discharge the governments' sovereign responsibilities for protection of life and property, for the general safety and well-being of the national community and for provision for the essential information needs of future generations; and
- *special services*: those services beyond the basic service aimed at meeting the needs of specific users and user groups and which may include provision of specialised data and publications, their interpretation, distribution and dissemination.

The users of meteorological services include virtually every person on the planet. The total user community is often broken down into the 'general public' embracing individuals, households and a wide range of government and non-government organisations on the one hand; and specific users and user sectors, on the other. The main specialist *user communities* for whom specific user-focussed services are provided in most parts of the world include:

- agriculture;
- aviation;
- defence;
- construction;
- emergency management;
- energy;
- health;
- natural resource management (including especially water resources);
- shipping; and
- tourism.

The energy sector (planning, construction, generation, trading, transmission, distribution and use) is one of the major users and beneficiaries of meteorological services in both developed and developing countries and, in some countries, it is the most significant user (Dutton 2002; Spiegler 2007). The information needed by the energy sector includes:

- climatological information and studies of wind profiles, plume dispersal and the like for conventional electricity generation plant siting and design;
- daily weather information, especially temperature, wind and cloudiness, both forecast and actual, local and distant, for generation and grid load management, financial decision-making and the like;
- the full range of agrometeorological (weather and climate) services in support of biomass-based fuel production;
- seasonal and longer term outlooks for demand assessment, hydropower operational planning, financial risk management (e.g. through weather derivatives) and so on;
- climate change assessments and projections for input to long-term energy policy and greenhouse gas abatement strategy development;
- detailed measurement and modelling of solar, wind, wave and tidal energy potential for renewable energy system planning; and
- scientific and socio-economic data for investigation of new energy supply and transmission systems.

The renewable energy sector is especially dependent on reliable meteorological information, as elaborated in following chapters, including information for:

- variable resource assessment;
- wind and solar power forecasting, both for day-ahead trading and for real-time grid integration to improve the economics and reliability of using renewables in the energy mix;
- planning, operation and maintenance of wind farms and solar arrays; and
- designing wind turbines and solar collectors that can withstand the environments in which they are sited.

Energy generators, traders, suppliers and consumers make use of both widely available basic weather and climate services accessed through the media (including especially meteorological websites) and special services provided either externally or internally, focussed on both short-term and long-term operational and business decision-making.

4 The National Meteorological Service System

The provision of meteorological services, such as those described and categorised above, within individual countries depends on the establishment and operation of an integrated national meteorological service system.

Almost every country has some form of 'end-to-end' national meteorological service system involving four basic components: a national observation network; a

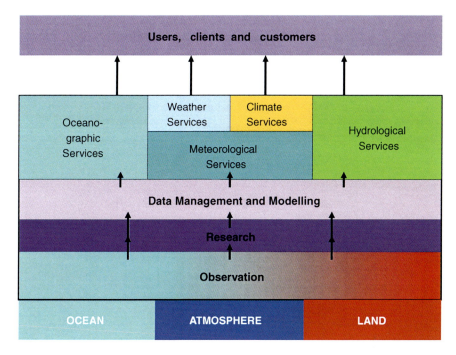

Fig. 4 The four main components (observation, research, data management, service provision) of an integrated end-to-end (in this representation, bottom-to-top) national meteorological service system

research and development effort; a data management and modelling/prediction/archival centre; and a service provision system; with each of these focussed primarily on the atmospheric domain but including, also, to varying degree, the ocean and land surface, especially the lakes and rivers (including river flow forecasting, flood warning and other hydrological service provision), as illustrated schematically in Fig. 4.

There is usually substantial overlap between the national meteorological service system and the (normally, far less well developed) national oceanographic and hydrological service systems. In many countries, all four components of the system, but especially the service provision arrangements, reflect the now somewhat arbitrary historical separation of meteorological services into 'weather' and 'climate', as shown in the upper part of Fig. 4.

The national meteorological service system in essentially all countries is based on the operations of the National Meteorological Service (NMS), the agency, usually primarily government funded, responsible for the operation of the basic national meteorological observation and data processing infrastructure, provision of the basic service and fulfilment of the government's obligation for international meteorological cooperation through the World Meteorological Organization (WMO).

The NMS is an essential component of the basic infrastructure of every country (Zillman 1999). Though the detailed roles and responsibilities of NMSs vary from country to country, most are charged with a five-fold mission of:

- observation and monitoring of meteorological and related variables over their national territory;
- research aimed at improved understanding of their countries' weather and climate;
- assembly and custodianship of the national climate record and modelling and prediction of the national weather and climate;
- service provision, including especially public forecast and warning, to their national communities; and
- meeting their governments' responsibilities for international cooperation in meteorological data collection, exchange and service provision.

The ultimate goals of the NMS also vary from country to country but, in most countries, they focus primarily on:

- protection of life and property;
- reduction of the impact of natural disasters;
- safety and efficiency of transportation;
- national economic development;
- national security;
- community health, welfare and quality of life;
- environmental quality;
- advancement of knowledge;
- efficient planning and management of government and community affairs; and
- meeting the environmental information needs of future generations.

An overview of the role and operation of NMSs is provided in a series of statements from the Executive Council of WMO (e.g. World Meteorological Organization 1999). Zillman (2003b) provides a summary of the early twenty-first century state of NMSs around the world and a recent WMO Executive Council Statement (World Meteorological Organization 2011a) provides a detailed listing of contemporary NMS functions and services.

The four other major players in the national meteorological service system whose roles complement that of the NMS are:

- the *mass media* who play a key role in the delivery of services, especially the basic service, to the community at large (e.g. Leep 1996);
- *private sector service providers* who provide a range of special services, usually on a commercial basis, to individual clients, customers and user groups (e.g. Spiegler 2007);
- the *academic/research sector* who contribute to the advancement of the science and, to varying degree, engage in data collection and certain niche categories of service provision (e.g. National Research Council 2003); and

- *other service providers*, often government agencies or other publicly funded operations, focussed on specific government functions or user communities, and including in-house services provided in some major public sector organizations and business enterprises (e.g. National Research Council 2010).

5 The Global Framework for Meteorological Service Provision

The essential overall international framework for provision of meteorological services is provided by the World Meteorological Organization (WMO).

The World Meteorological Organization is a Specialised Agency of the United Nations with 189 Member States and Territories and a Convention which provides for global cooperation in;

- establishment of observing networks and systems and collection and exchange of meteorological and related data;
- meteorological and related research;
- establishment of world/global and regional centres for the provision of meteorological products and services;
- application of meteorological information and services for the benefit of international shipping and aviation and of all sectors of the national communities of its Members; and
- training and capacity building in meteorological science and services.

Although NMSs usually take the lead in their governments' implementation of the WMO Convention, WMO is an organisation of countries, not of NMSs, and WMO programmes are intended to embrace the total global effort in meteorology, including that of the media and the private and academic meteorological sectors of Member countries—not just the work of the NMSs. In doing this, WMO works closely, at the international institutional level, with other UN organisations with related responsibilities such as the Intergovernmental Oceanographic Commission (IOC) of UNESCO for the ocean and the non-governmental International Council for Science (ICSU) on meteorological and related research.

The World Meteorological Organization has traditionally carried out its role through its Members' contribution to a set of Major Programmes corresponding to its major responsibilities under the Convention. Over recent decades, it has carried out its work through:

- the World Weather Watch Programme as the global framework for meteorological observation, data collection and processing, and product preparation in support of the full range of weather and climate services at global, regional and national levels. Historically, the World Weather Watch was regarded as consisting of the Global Observing System (GOS), Global Telecommunications

System (GTS) and Global Data Processing and Forecasting System (GDPFS). In recent years, these have been substantially integrated into what are now known as the WMO Integrated Global Observing System (WIGOS) and the WMO Information System (WIS);
- the World Weather Research Programme (WWRP) and World Climate Research Programme (WCRP), the latter in collaboration with ICSU and the IOC, to advance the underpinning science of weather and climate prediction;
- a set of service programmes (grouped together as the Applications of Meteorology Programme (AMP)) focussed both on the community at large (Public Weather Services Programme (PWS)) and several individual major sectors (especially aviation, agriculture and marine) and including both general information and forecasting services and a range of specific warning services (e.g. for tropical cyclones); and
- the World Climate Programme (WCP), especially its components focussed on data and monitoring (WCDMP) and applications and services (WCASP) which was established following the 1979 First World Climate Conference and successively restructured following the 1990 Second World Climate Conference and the 2009 Third World Climate Conference to provide an overall international framework for assisting countries in using climate science and services for the benefit of their national communities;along with a number of supporting capacity building and related programmes which, together, provide the overall global framework for meteorological service provision as shown schematically in Fig. 5.

It is important to stress that, just as at the national level, the operation of the overall global framework for meteorological service provision is totally dependent on the underpinning observing and research programmes and the global operational infrastructure for data processing and modelling provided through a network of World/Global and Regional/Specialised Meteorological Centres. While, historically, the WWW GOS, GTS and GDPFS were envisaged as supporting the full spectrum of 'weather' and 'climate' services, the 1979 establishment of the WCP (Zillman 1980) sought to address the hitherto somewhat under-emphasised challenges of climate variability and change and the burgeoning societal need for climate information and prediction services.

Within the international planning and coordination mechanisms of WMO, as well as at the national level within its Member countries, the weather and climate service and applications programmes include specific components focussed on the energy sector. The initial focus of the WCP and many of its achievements over its early decades were concentrated in four priority applications sectors, including energy (Bruce 1991), as follows: WCP—Energy; WCP—Food; WCP—Water; and WCP—Urban and Building.

WMO's mechanisms for international capacity building, information sharing and, where necessary, coordination amongst countries, involve both formal intergovernmental specialised bodies (e.g. expert intergovernmental 'technical commissions') and less formal scientific and operational conferences, workshops

Weather and Climate Information Delivery 213

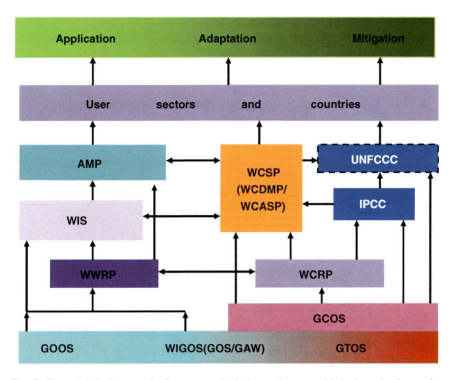

Fig. 5 The global framework for meteorological services provided through the major programmes of WMO and its international partner organisations. In addition to the acronyms explained in the text: GOOS is the IOC-led Global Ocean Observing System; WIGOS is the WMO Integrated Global Observing System encompassing both GOS and GAW and WMO-related parts of the co-sponsored GOOS and GTOS; GOS is the World Weather Watch Global Observing System; GAW is the WMO Global Atmosphere Watch (for atmospheric chemistry); GTOS is the FAO (Food and Agriculture Organization)-led Global Terrestrial Observing System; GCOS is the WMO-IOC-UNEP-ICSU Global Climate Observing System which is built on GOOS, WIGOS and GTOS; IPCC is the Intergovernmental Panel on Climate Change; UNFCCC is the UN Framework Convention on Climate Change; WCSP is the World Climate Services Programme based on a consolidation of the former WCDMP and WCASP; WIS is the WMO Information System; AMP is the WMO Applications of Meteorology Programme

and groups of experts. Integrated international planning and guidance over the period 1983–2007 was provided through a detailed programme-based 'WMO Long-term Plan' which provided guidance both to the WMO Secretary General in the use of the WMO Regular Budget in support of programme implementation and to individual NMSs for the planning and development of the corresponding activities at the national level. More recently, the overall international planning framework has been provided by a WMO Congress-approved 'WMO Strategic Plan 2012–2015' focussed on five 'Strategic Thrusts' (World Meteorological Organization 2011a):

- Improving service quality and service delivery;
- Advancing scientific research and applications as well as development and implementation of technology;
- Strengthening capacity building;
- Building and enhancing partnerships of cooperation; and
- Strengthening good governance.

Much of the coordination and information flow within the international meteorological system is facilitated by the NMSs whose Directors normally serve as Permanent Representatives of their countries with WMO but important roles are also played by international coordination mechanisms such as the International Association of Broadcast Meteorology (IABM) and similar groups of equipment providers (Association of Hydrometeorological Equipment Industry (HMEI)) and commercial meteorological service providers.

6 Future Directions

Five important recent initiatives on the international meteorological scene warrant special mention in terms of their potential influence on future arrangements for weather and climate information delivery to the energy sector. They are:

- the Intergovernmental Panel on Climate Change (IPCC);
- the Global Earth Observation System of Systems (GEOSS);
- the Madrid Action Plan on social and economic benefits of weather, climate and water services;
- the Third World Climate Conference and the Global Framework for Climate Services (GFCS); and
- the WMO Strategy for Service Delivery.

The 1988 establishment of the IPCC by WMO and the United Nations Environment Programme (UNEP) set in train a remarkable international scientific assessment process with profound consequences for international and national energy policy (Houghton 2005). The successive IPCC Assessment Reports and Special Reports including its recent Special Report on renewable energy (Intergovernmental Panel on Climate Change 2012), have provided authoritative advice to the global community generally on the threat of human-induced climate change resulting from carbon dioxide emissions, especially from the stationary energy sector, and on a range of other issues at the interface of energy and human-induced climate change. The IPCC assessment process and the interpretation and application of the findings in its assessment reports provide the most authoritative source of advice on most aspects of the science and implications of human-induced climate change. The international response to the findings of the 2014 IPCC Fifth Assessment Report must be expected to have major implications for the energy

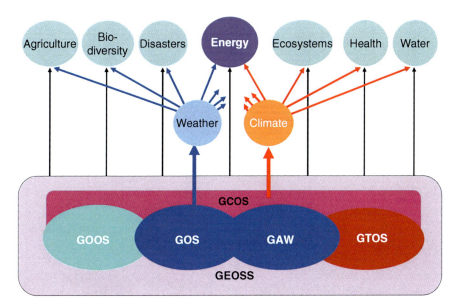

Fig. 6 The provision of weather and climate observations and information to the energy (and other) Societal Benefit Areas (SBAs) within the framework of GEOSS. The acronyms for the various domain-based global observing systems contributing to GCOS and GEOSS are as given in the caption to Fig. 5.

industry worldwide and require extensive collaboration between the climate and energy communities.

In 2005, an initially ad hoc intergovernmental Group on Earth Observations (GEO) agreed on a 10-year Implementation Plan for GEOSS aimed at providing all the in situ and space-based earth observations needed for effective application in nine key Societal Benefit Areas (SBAs), three of which were weather, climate and energy (Zillman 2005). Though initially focussed on enhancing collaboration amongst the various national agencies and international programmes and organisations engaged in in situ and space-based observation of the different Earth system domains, the 88 countries and 64 'participating organisations' now engaged in the GEO mechanism have increasingly focussed their attention on delivery of user-focused information to the various SBAs. It is envisaged that GEOSS outcomes in the energy sector will support environmentally responsible and equitable energy management, better matching of energy supply and demand, reduction of risks to energy infrastructure, more accurate inventories of greenhouse gases and pollutants and a better understanding of renewable energy potential. The overall architecture of GEOSS is still evolving. But it is clear that the major part of the GEOSS contribution to the energy (and other) sectors will continue to be based on the established weather and climate observation and service provision framework of the UN System as shown schematically in Fig. 6.

The 2007 Madrid Conference on 'Secure and Sustainable Living' (World Meteorological Organization 2009a) provided a detailed assessment of the nature and value of meteorological and related services in six major weather- and climate-sensitive sectors including 'Energy, transport and communications'. It reviewed the energy sector's requirements for services and the extent to which the information available is capable of meeting current and expected future needs. The Conference developed a 'Madrid Action Plan', subsequently endorsed by the 2007 WMO Congress, which set out a 15-point agenda for accelerated action on improving the quality and user value of weather, climate and water services. Through its 'Forum on Social and Economic Applications and Benefits of Weather, Climate and Water Services' and other mechanisms, WMO is progressing many of the specific proposals from the Madrid Conference including the development of new energy-needs-specific observing, data collection and modelling systems in support of new wind, solar, wave, tidal and other alternative energy generation systems (World Meteorological Organization 2011a).

The Ministerial Third World Climate Conference (WCC-3) held in Geneva in August–September 2009, undertook a detailed assessment of the achievements of the World Climate Programme (including WCP-Energy) over the previous three decades and agreed to establish a new GFCS to improve the provision and application of climate services in all the key climate-sensitive sectors worldwide. The Expert Segment of the Conference issued a detailed Conference Statement which provided a set of specific conclusions and recommendations for each individual sector. For the energy sector, the climate and energy experts at WCC-3 concluded (World Meteorological Organization 2009b) that:

> Climate information is essential for ensuring the most efficient production and consumption of essentially all traditional forms of energy including coal and gas-fired generation and the distribution and utilization of electricity, and is especially important for the design and operation of infrastructure and facilities for renewable energy sources such as hydro, wind, solar, tidal and bio-energy. Season to multi-decadal climate variations give rise to changes in energy demand but also in energy availability and supply. Primary energy is traded globally and often delivered within complex energy grids. In particular, the generation of renewable energies is often itself climate dependent. Energy prices may also be affected by climate variations. The discussions on sustainable energy highlighted the climate information that is currently available, the extent to which it is already being used and the current and future needs of climate information from the energy sector".

The WCC-3 energy experts went on to stress the need for:

- *Historical and quality observations.* Historical and high quality weather and climate observations are needed for the energy sector especially in developing countries;
- *Seamless predictions.* Seamless predictions from global climate models (monthly to seasonal to decadal timescales) with much improved resolution are needed;
- *Updated reanalysis.* There is need for quality reanalysis of meteorological data that is regularly updated;

- *Reliable access.* Reliable access to climate information using readily available servers and grid technology is important;
- *Joint partnerships.* Establishment of joint partnerships between the energy sector and climate service providers is desirable;
- *Mainstreaming climate information.* It is vital to mainstream climate information into long-tem development plans, in particular for the energy sector;
- *Vulnerability assessments.* Assessments of the vulnerability to severe weather and extreme climatic events are needed for energy infrastructures including generation, transmission, transformation, processing, distribution and extraction;
- *Strengthening partnerships.* Partnerships should be strengthened between the energy sector and the climate service community;
- *Active participation.* Active participation by civil society is needed to improve decision-making in issues linking climate services and energy; and
- *Capacity-building and technical cooperation.* The transfer of energy and climate technology between developed and developing countries requires capacity building and technical cooperation.

Following the WCC-3 decision to establish the GFCS, a major intergovernmental and interagency process was put in place to advance its design. A High-level Task Force (World Meteorological Organization 2011b) agreed on an overall structure for the GFCS which served as the basis for development of a detailed implementation plan for intergovernmental approval in October 2012. Many individual climate services enhancement initiatives are already underway in support of the GFCS, including the Climate Services Partnership (CSP) established by the October 2011 First International Conference on Climate Services (ICCS-1) at Columbia University, New York City.

At the same time as it approved its overall strategy for 2012–2015 and the development of a GFCS Implementation Plan, the 2011 WMO Congress, (the four-yearly intergovernmental session of Principal Delegates from its 189 Member countries) approved a detailed 'Strategy for Service Delivery' (World Meteorological Organization 2011a) aimed at guiding all the players in both the international and national meteorological service provision system on ways of improving the effectiveness of service delivery and application by end users. The six 'strategy elements' which provide the overall framework for the Strategy are:

- Evaluate user needs and decisions;
- Link service development and delivery to user needs;
- Evaluate and monitor service performance and outcomes;
- Sustain improved service delivery;
- Develop skills needed to sustain service delivery; and
- Share best practice and knowledge.

The WMO Service Delivery Strategy is seen as an important initiative in support of the implementation of the Madrid Action Plan, the WMO Strategic Plan and the GFCS Implementation Plan and can be expected to reinforce the international focus on new and enhanced meteorological services for the energy industry over the coming decades.

7 Conclusion

The various national and international institutions of meteorology recognise the importance of further strengthening the energy-meteorology partnership in implementing the many new and innovative weather and climate services that will be needed to support major enhancements in the global energy system over the coming decades. This will require:

- New and improved meteorological observing networks and models that are better focussed on the needs of the energy sector;
- Better training of the meteorological community in the provision of weather and climate information for the energy sector;
- Continued research into all aspects of the use of meteorological information for improved energy efficiency;
- Stronger mechanisms for ensuring essential data and information exchange between the energy and meteorological communities; and
- An enhanced energy focus in the further planning and implementation of the GFCS.

Most of all, however, it will require increased professional interaction and dialogue between the energy and meteorology communities around the world. There is a long history of collaboration to provide the basis for confidence that the global meteorological service provider community will embrace the opportunity to play their part.

References

Bruce JP (1991) The world climate programmes' achievements and challenges. In: Proceedings of the 2nd world climate conference. Cambridge University Press, Cambridge, 149–155
Bureau of Meteorology, (2003) The greenhouse effect and climate change. Bureau of Meteorology, Melbourne, pp 74
Davies DA (1990) Forty years of progress and achievement. Hist Rev WMO. WMO, Geneva, pp 205
Dubus L, S Parry (2012) Developing climate services: the role of the energy sector. In: Climate Exchange, World Meteorological Organization/Tudor Rose, Leicester, 156–158
Dutton JA (2002) Opportunities and priorities in a new era for weather and climate services. Bull Amer Met Soc 83:1303–1311
Global Climate Observing System (2010) Implementation plan for the global observing system for climate in support of the UNFCCC. GCOS-138. Secretariat of the World Meteorological Organization, Geneva, pp 180

Gunasekera D (2004) Economic issues relating to meteorological services provision. BMRC Research Report No. 102. Bureau of Meteorology, Melbourne

Houghton JT (2005) Climate change and sustainable energy. Weather, 60(7):179–185

Intergovernmental Panel on Climate Change (2012) Renewable energy sources and climate change mitigation: Special report of the intergovernmental panel on climate change. Cambridge University Press, Cambridge, pp 1075

Leep R (1996) The American Meteorological Society and the development of broadcast meteorology. In: Historical essays in meteorology 1919–1995. American Meteorological Society, Boston, 481–507

Maunder WJ (1986) The uncertainty business: Risks and opportunities in weather and climate. Methuen, London, pp 420

National Research Council (2003) Fair weather: Effective partnerships in weather and climate services. The National Academies Press, Washington DC, pp 220

National Research Council (2010) When weather matters: Science and services to meet critical societal needs. The National Academies Press, Washington DC, pp 181

Page JK (1990) Operational aspects of using meteorology for energy purposes. In: Using meteorological information and products. Ellis Harwood, New York, 211–235

Price-Budgen A (ed) (1990) Using meteorological information and products. Ellis Harwood, New York, pp 491

Spiegler DB (2007) The private sector in meteorology: an update. Bull Amer Met Soc, 88:1272–1275

World Meteorological Organization (1999) The national meteorological service and alternative service delivery. A statement by the executive council of the world meteorological organization on the future role and operation of meteorological services. Secretariat of the World Meteorological Organization, Geneva

World Meteorological Organization (2007) Weather, climate and water services for everyone. WMO No. 1024. Secretariat of the World Meteorological Organization, Geneva, pp 70

World Meteorological Organization (2009a) Secure and sustainable living. The findings of the international conference on secure and sustainable living: social and economic benefits of weather, climate and water services. WMO No. 1034, Geneva

World Meteorological Organization (2009b) Working together towards a global framework for climate services. Report of the world climate conference-3. WMO No. 1048, WMO Secretariat, Geneva, pp 80

World Meteorological Organization (2011a) Sixteenth Congress: Abridged final report with resolutions. Secretariat of the World Meteorological Organization, Geneva

World Meteorological Organization (2011b) Climate knowledge for action: a global framework for climate services—empowering the most vulnerable. The report of the high-level task force for the global framework for climate services. WMO No. 1065. Secretariat of the World Meteorological Organization, Geneva. pp 240

Zillman JW (1980) The world climate programme. Search, 11:108–111

Zillman JW (1999) The National Meteorological Service. WMO Bulletin, 48:129–159

Zillman JW, Freebairn JW (2001) Economic framework for the provision of meteorological services. WMO Bulletin, 50(3):206–215

Zillman JW (2003a) Demands on meteorology. EMS Publications Series 1, European Meteorological Society, Berlin, 5–14

Zillman JW (2003b) The state of National Meteorological Services around the world. WMO Bulletin, 52(4):360–365

Zillman JW (2005) GEOSS—a new framework for international cooperation in earth observation. ATSE Focus, 136:8–12

Meteorology and the Energy Sector

Geoff Love, Neil Plummer, Ian Muirhead, Ian Grant and Clinton Rakich

Abstract The energy sector has a diverse requirement for meteorological services to support decision-making for both day-to-day operations and for longer term strategic planning. This requirement is driven in part by the natural climate variability (including extreme weather events) and increasingly by climate change as manifested through the physical climate and through policy responses to the issue. The meteorological services required for decision-making in this sector can be broadly categorised into two ways: (i) those that support decision-making concerning the implementation and operation of new technologies for energy production, and (ii) those that support decision-making for maintaining service and reducing emissions by existing energy sector infrastructure. This chapter focuses on the electricity production sector and examines the types of services that are currently available, and also those that are likely to be needed in the future. The chapter concludes with a discussion of the likely climate and weather service provision mechanisms that will best meet the energy sector's needs, and the role that the Global Framework for Climate Services could be expected to play in meeting these needs.

1 Introduction

Meteorology is the science of the atmosphere, and as such it encompasses the science of both weather and climate. This chapter addresses the issues surrounding the availability, supply and use of meteorological services in the energy sector, and so as to limit the scope of the chapter, to that part of the sector responsible for the production of electricity and, in particular, examines the challenges and opportunities meteorological services will increasingly assist in addressing.

G. Love
World Meteorological Organization, Geneva, Switzerland

N. Plummer · I. Muirhead (✉) · I. Grant · C. Rakich
Australian Bureau of Meteorology, Melbourne, Australia
e-mail: i.muirhead@bom.gov.au

2 Factors Affecting Electricity Production

The energy sector's requirement for meteorological services clearly needs to be addressed within the context of climate change, but this is certainly not the only, and possibly not even the dominant, driver of change for the energy sector in the decades ahead (see "Improving Resilience Challenges and Linkages of the Energy Industry in Changing Climate" by Majithia, "Combining Meteorological and Electrical Engineering Expertise to Solve Energy Management Problems" by Pirovano et al., "Weather and climate impacts on Australia's National Electricity Market (NEM)" by George and Hindsberger and "Weather & climate and the power sector: Needs, recent developments and challenges" by Dubus). The economics of the industry will always be fundamental, and evolution and revolution in the cost of the various energy production technologies, whether determined through government policy or by new scientific and technological developments, will be the key driver. However, the demographics of the market (size and location of the energy consumers) for energy will continue to be an important driver in the overall market.

Starting with the last of these three, demographics, it is clear that rising population, particularly in the urban centres is increasing the demand for energy, with peak demand in many countries now occurring in summer as the total capacity of installed air-conditioning systems continues to grow. Added to this is the demand for reliable, high quality electricity supply as an ever increasing number of business and homes rely on, or at least expect continuity of access to computing resources. The aggregate sensitivity of demand for electricity to the meteorological conditions by nations can be reflected by the use of the measure of degree-days.[1] A relatively benign climate is one for which the sum of the annual average number of Heating degree-days and Cooling degree-days is relatively small (say less than 2,000), but of course national averages may hide important regional variations for large countries (Baumert and Selman 2003). It must also be noted that the energy industry must be scaled to accommodate the extremes, not just the annual mean values of heating and cooling degree-days, now and in the future. Another way of quantifying the dependence of power consumption on temperature is by defining a temperature gradient per megawatt, as in Dubus (see "Weather and Climate and the Power Sector: Needs, Recent Developments and Challenges" in this volume).

Technological change is also occurring at a rapid pace. For countries with access to reserves of coal, it is clear that large-scale, coal-fired plants are the most cost-effective way to generate electricity, particularly if the infrastructure to deliver this electricity has been established and is a "sunk cost" already met by the electricity users. However, the price competitiveness of energy production systems whose efficiency is affected by meteorological conditions, such as wind and solar, is increasing.

---

[1] Unit for estimating the demand for energy required for heating or cooling. In the US, the typical standard indoor temperature is 65 °F (18.3 °C). For each 1 °F decrease or increase from this standard in the average outside temperature for each day this occurs, one heating or cooling degree-day is recorded. *Source* http://www.BusinessDictionary.com

Fig. 1 A flooded open-cut coal mine at Babinda, Queensland, Australia, January 2011 (Lyndon Mechielson, Newspix)

Table 1 Wind field for an 80,000 km² test area, from Coppin et al. (2003), in the vicinity of the Great Dividing Range in Eastern NSW, Australia, showing the percentage of the non-woodland area, and actual area, where specified mean wind thresholds are exceeded

Lower threshold of mean annual wind speed (m·s^{-1})	Percentage of land area at or above threshold mean wind speed (excluding woodland) (%)	Land area (km²)
9	0.02	19
8.5	0.08	71
8	0.16	134
7.5	0.54	460
7	3.07	2,635
6.5	12.13	10,396
6	28.60	24,500

Meteorological conditions not only affect demand for electricity but also production and distribution. Extreme events such as the floods that occurred in Queensland, Australia in December 2010 through to early 2011 caused widespread disruption as open-cut coal mines in that State were flooded (Fig. 1) and electricity distribution interrupted (QFCOI 2012). For the producers of hydroelectricity the occurrence or otherwise of rainfall is a major determinant on their ability to generate electricity. For solar producers the key is sunshine hours while for wind farms the variable used to assess the productivity of sites is the mean wind speed at the height of the hub of the wind generating propeller, with a mean wind speed exceeding 8 m·s^{-1} considered excellent sites and 7 m·s^{-1} good. From Table 1 it can be seen that the area of land suitable for wind farms increases exponentially as the efficiency of the wind generation technologies improves, making them economically viable at lower speeds. Regardless of technology efficiency, however, the number of viable sites will always be limited by competing land-use requirements.

3 The Energy Sector's Requirement for Meteorological Services

In this section, an attempt is made to take a user perspective on the demand for meteorological services to the sector of the economy responsible for electricity generation and distribution. To further limit the scope of the analysis, the chapter considers a subset of five classes of decision-makers who might conceivably use meteorological information to make decisions relating to the electricity supply sector. These are: the public, electricity grid managers, policy makers, energy sector investors and energy traders (see Table 2). The information requirements of these different classes of decision-makers differ markedly, and have the potential to conflict in some areas.

3.1 The Public

The public generally expects their electricity supply to work effectively day-in, day-out. The times when meteorological factors work towards threatening supply will most likely be those when severe weather is likely, including those times of sustained hot and cold spells. If a severe storm or bushfire during a heat wave threatens supply, the public has a reasonable expectation of forewarning. Similarly, if high demand for electricity during a heat wave is likely to lead to "brown outs" or electricity rationing, again the public expects forewarning. In the lead up to, and during such events, products and services that combine weather information from the meteorological community integrated with information relating to infrastructure vulnerability and likely demand are required from the electricity sector. There may also be interest in the climatology, or likelihood of the occurrence, of such events. Public meteorological services have been found to be most effective when simple products are delivered on a routine basis.

3.2 Electricity Grid Management

Electricity providers manage the grid supplying consumers (households and industry) by manipulating the mix of electricity sources and distribution of electricity between regions. The variability in consumer demand will be driven by a

Table 2 The demand for a number of different types of meteorological information for a subset of five "classes" of decision-makers with an interest in the electricity segment of the energy sector

Class of decision maker	Demand for weather data and information	Demand for climate data and information	Demand for complex products and analyses
The public	Real time data of temperature, wind, rainfall (including real-time radar observations of rainfall) along with forecasts of these elements, particularly for their area of residence.	Information concerning averages and extremes of temperature, wind, sunshine duration and rainfall (and related severe weather events) for their area of residence.	Simple analyses, particularly of extreme events likely to impact on supply and pricing. Straightforward analysis of possible climate change impacts on supply.
Grid managers	Real time data of temperature, wind, solar radiation and rainfall, along with forecasts of these for all areas relevant to the elements of grid they manage, and for the grids that affect theirs. Seasonal outlooks for the likely weather assist in scheduling maintenance.	Information concerning extremes of wind, temperature and rainfall and related severe weather events (return periods, etc) for all areas relevant to their grid management responsibilities.	Rapid, accurate (real-time and detailed) weather and scientifically validated climate timescale inputs to complex supply and demand models – will be tailored to meet the needs of grid management. Output not publicly disseminated.
Policy makers	Rapid advice on forecasts of extremes that may have short- to medium-term impacts on infrastructure within their policy domain.	Information concerning extremes of rainfall, wind, and temperature, along with disasters that might affect supply and demand (floods, bushfires, tropical cyclones, drought, etc).	Sophisticated modelling to underpin the short- and medium-, and long-term forecasts used in policy making. These would be a mixture of publicly available and "In-Confidence" products.
Investors	Advice on short-term (1 hour to 7 days) forecasts of extremes that may have short- to medium-term impacts on actual and potential investment	Information concerning extremes and variability of solar radiation, temperature, wind and rainfall (including flood and bushfire potential) for all areas of actual and potential investment.	Climate timescale projections, as an element in a multi-disciplinary analysis of the likely impact of climate change on energy supply (by technology) and regional demand.
Traders	Real time data of temperature, wind, solar radiation and rainfall, along with forecasts of these for all areas relevant to trades.	Information concerning extremes and variability of temperature, wind and rainfall for all areas affecting trades.	Sophisticated modelling to underpin the short- and medium-term forecasts used in trades; these would be Commercial-in-Confidence products

Importance to decision maker: High ▇ Medium ▇ Low ▇

Coloured shading provides a subjective assessment of the importance of three types of meteorological products

number of factors; for example, is it a weekend, public holiday or normal working day, and will it be a high heating or cooling degree-day for some part of the grid? Variability of supply is also important; has rainfall been high enough to support a good level of peak hydroelectricity supply, will the peak demand have to be accommodated with a high level of electricity from gas generation, and, as the newer sources such as wind and solar become significant, how will their availability impact supply? Of growing importance in a number of countries is the need to manage the increase in generation capacity of distributed small photovoltaic power systems. To help assess these and other factors the electricity grid manager needs real-time temperature, wind, solar radiation and rainfall data as inputs to relatively sophisticated supply and demand models, the output from which is an input to the overall decision-making process. It is likely that historical climatological information would not be highly important to the grid manager for day-to-day operations. However, multi-week forecasts and seasonal outlooks can provide valuable guidance when assessing future load demands or scheduling maintenance outages.

3.3 Policy Makers

Policy makers have two timescales of interest. They are keenly interested in extreme weather events that impact negatively on supply of electricity (and consequently the consumer). They will expect detailed warnings of the likelihood of these events that are integrated with information describing the vulnerability of the supply to the event. During extreme weather events policy makers require frequent updates during the lifetimes of such events. The policy makers will also be critically interested in the timescales of the investments in infrastructure in the industry, typically 40–50 years. On the longer timescales, they would expect climatological scenarios reflecting the multi-decadal variability of the climate to be integrated with sophisticated modelling of energy demand. Policy makers would expect to be supplied with a mixture of "in-confidence" advice from the model analysis as well as products and information that was tailored for public release.

3.4 Energy Sector Investors

Investors (as distinct from energy traders) are likely to be less interested in the short-term outlook for the energy sector than that of the medium to long term on the scale of years. It is considered here that the investor has the intention of investing in some aspect of electricity generation for the medium to long term through the commitment of funds to electricity by way of equity or debt mechanisms. The investor will most probably require access to sophisticated modelling that would underpin decision-making if the scale of investment were to be significant. For example, at the level of state and national governments multidisciplinary analysis bringing together the known climatology of extremes, some assessment of their possible changes under climate change, and evolutions in electricity generating technologies, changing population and industry demographics would all need to be brought together in an integrated assessment of the financial viability of the investment. Smaller scale investment, and one-off investment decision would be unlikely to command such detailed (and expensive) analyses.

3.5 Traders

From the meteorological sector, energy traders require good forecasts of likely disruptions to supply along with forecasts of heating and cooling degree-days, preferably ones that are not available to their fellow traders until they have made their trades. To obtain a market edge high volume, traders would likely seek priority access to a model—often developed "in-house"—that uses real-time meteorological inputs and provides the trader with forecasts of likely demand, together with wind and solar energy production where markets exist for these.

4 Overview: Uses of Meteorological Information in the Energy Sector

The provision of basic real-time meteorological data, including observations of temperature, wind velocity, rainfall, solar radiation, radar and satellite imagery is a valuable service needed to underpin decision-making in this sector. The users of climate services also require high quality forecasts of these parameters (where scientifically possible) and, if possible, access to the climatological norms for key parameters used for decision-making. The real-time observations and short-term forecasts need to be authoritative, quality controlled and reliably and routinely available if decision-making is to be properly informed. In the slightly longer time frame, seasonal outlooks (Fig. 2) can provide guidance to assist with scheduling construction and maintenance activities. Efficient markets for electricity consumption, and for investment in supply, must provide all decision-makers access, at least in principle, to the same data. That is, for the market for electricity to be efficient the buyers and sellers in that market place must have access to meteorological data services.

It should be emphasized that real-time, routine short-term forecasts of temperature, wind and rain are generated by sophisticated numerical weather prediction systems, with the highest quality forecasts produced by a small number of very advanced forecasting centres operating massive supercomputer installations

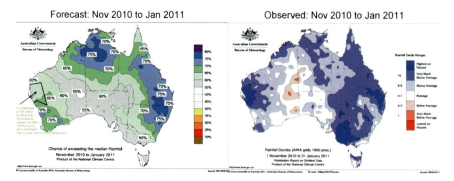

Fig. 2 Seasonal climate outlooks can provide valuable guidance on the likely conditions to occur in the months ahead. This information can be used to provide advanced estimates of generating capacity, energy demand and the likelihood of favourable conditions for construction or maintenance. In this example, the regions identified as having a greater than 50 % chance of exceeding median rainfall correlate well with the observed rainfall pattern.*Source* Australian Bureau of Meteorology

and supported by hundreds of technical experts. The data outputs from these centres are freely exchanged around the world by the meteorological community and are available for integration into bespoke energy sector forecast systems. It is clear that the greatest return on the investment in these systems, in global terms, is achieved by the most widespread use of the resultant data in decision-making processes across a wide variety of industry sectors and in the maximum number of countries.

With the possible exception of the public, which requires access to robust but simple analyses and forecasts, all user sectors have a high demand for integrated, multidisciplinary analyses that make the most effective use of meteorological information relevant to their decision-making. The integration of different data types into sophisticated, user specific analyses and forecast systems is a specialist skill requiring multidisciplinary inputs, including from climate scientists. Even in developed countries the pool of such expertise is relatively small; one of the challenges in the provision of improving climate services is to build this skill pool. A second challenge is that while the basic data and forecasts are clearly a public good, the bespoke forecasting systems serving a specific user are clearly in the commercial domain.

To assist the private investor and companies investing in solar or wind electricity generation technologies, governments in developed countries may produce national or regional climatologies of parameters such as measures of the available solar energy for domestic hot water production or conversion to electricity for the country (Fig. 3). Governments may also collaborate with the private sector to generate more detailed products. An example of such a product is the regional scale maps of mean wind speed at a typical wind generator hub height (Fig. 4), where the location of transmission lines available to collect electricity generated using wind power have been added to assist with site "prospecting".

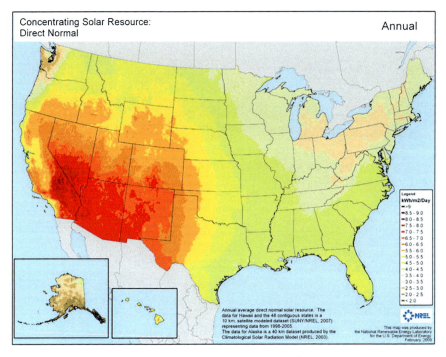

Fig. 3 Concentrating solar potential for the USA (Units kWh/m^2/day). *Source* US EPA—http://www.nrel.gov/gis/images/map_csp_us_10km_annual_feb2009.jpg. The U.S. Department of Energy's National Renewable Energy Laboratory (NREL) developed the Concentrating Solar Resource model. Specific information about this model can be found in Maxwell et al. (1998) and George and Maxwell (1999)

5 An Energy Sector Case Study: Solar Resource Assessment

Assessment of the solar resource available for solar power generation is an example of a growing application that is driving creation of new data services related to weather and climate. This industry needs data on the variation of solar radiation in space and time, both on climate timescales for planning and weather timescales for operation. Ground-based and satellite-based measurements provide complementary historical and present observations, and numerical weather prediction (NWP) has a role in solar forecasting.

Historical solar data are needed to support the planning and financing of solar power stations. The data required at a potential location for a power station project include not only mean radiation amounts but also the variability at a range of timescales from minutes to years, together with knowledge of the extremes. Moreover, to have the "bankability" needed to support financing decisions these data must be accompanied by uncertainty estimates that can be translated into a

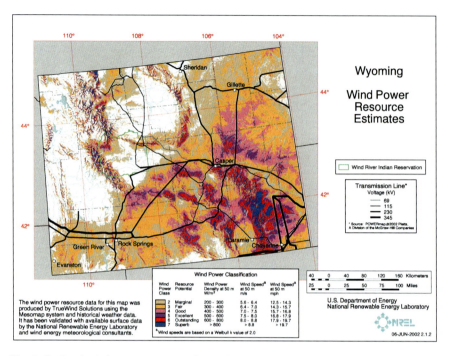

Fig. 4 Regional wind power resources. *Source* http://www.windpoweringamerica.gov/

level of confidence in, for instance, projections of the potential year-to-year variation in project revenues. The accurate assessment of interannual variability and extremes requires data record lengths of at least a decade and preferably several decades. Well-maintained ground-based measurements provide the data with the smallest uncertainties, and a number provide high resolution spectral data to assist with photovoltaic systems design, but current networks of suitable stations are spatially sparse. The imagery from the geostationary meteorological satellites yields detailed spatio-temporal information on the presence and characteristics of the cloud that critically controls radiation amounts; data that can be analysed to estimate surface radiation parameters. These satellite-based estimates, while being indirect and having larger uncertainties and generally lower temporal resolution than ground-based measurements, have the advantages of being spatially complete with a multi-decadal observation record.

Composite observing systems can combine surface and satellite observations to draw on the best of both platforms: the spatially extensive satellite data can be calibrated and validated by the sparse but more accurate surface data. Public good services include the support of long-term surface observation networks and provision of basic data products, such as spatial products derived from combining satellite and surface data, for preliminary solar prospecting. An example of such a public project is the Solar Energy Resource Mapping Project being undertaken by the Australian Bureau of Meteorology and Geoscience Australia

(Geoscience Australia 2013). Customised high-value products, such as the collection of short-term surface observations at a location of interest and their integration with longer term, more spatially comprehensive satellite data, are often produced as a commercial activity by private meteorological services.

To maximise their value solar resource data need to be easily accessible. Meteorological Data Services must provide data at a resolution and in formats that suit users with a variety of needs and levels of data handling capability. For instance, the Australian Bureau of Meteorology makes the data from its surface radiation network available at daily, half-hourly and one-minute resolutions. Accessibility becomes more challenging with the large data volumes associated with time series of spatial data such as are derived from satellite observations. Advanced data service technologies such as Web Map Services and server-side processing, customisable by the user, promise a capability in the near future to meet user requirements. Further, the need for meteorological and related data other than solar radiation to support solar power plant operation, such as air temperature and water availability, will motivate the development of data services that package diverse data streams.

International coordination in the production and dissemination of solar resource data is important to maximise impact, minimise the burden of provision and access, and minimise the confusion of users when presented with diverse and fragmented sources of data. An example of coordination is the International Energy Agency's Solar Heating and Cooling program's Task 36 on solar resource knowledge management. This five-year programme brought together technical experts to, amongst other objectives, standardise the formatting and benchmarking of international solar resource datasets to ensure worldwide intercompatibility and improve accessibility. One outcome was the associated European MESoR project to harmonise pre-existing solar resource datasets for Europe in a single, open source, prototype Web portal (http://project.mesor.net/web/guest/unifying). Figure 5 shows an example of time series solar resource data extracted for a point location through the Web map interface.

The demand for solar forecasting services will increase as the deployment of solar electricity generation increases. Forecasts over the full weather range of minutes to days are increasingly needed in a variety of applications including optimising the operation of large power stations, balancing the electricity grid in the face of variable supply from rooftop photovoltaics and operating the energy market. On the climate timescale, seasonal and yearly outlook time frames assist in scheduling maintenance and assessing cycles in the demand for energy. Longer term trends in the solar resource are needed for planning the performance and financing of solar power stations that may have a lifetime of several decades.

Numerical weather prediction is the key solar forecasting tool on timescales of hours to days, although the cloud and radiation schemes in these models typically are not currently adequate for solar energy applications. Satellite-based solar forecasting, whereby clouds are tracked in image sequences and their motion projected forward in time, has an important contribution to make on the scale of minutes to hours. The latest generation of geostationary meteorological satellites is

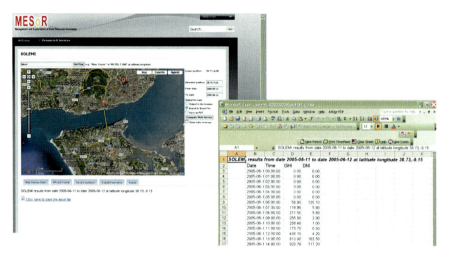

Fig. 5 The MESoR prototype Web portal, shown here giving access to the Solar Energy Mining (SOLEMI) satellite-based solar radiation dataset and a spreadsheet showing data extracted at a location. *Source* MESoR web site

contributing to improvements in forecasting and monitoring, with an imaging period of 10 or 15 min or less, and a greater number of spectral bands targeted to detailed characterisation of the state of the atmosphere and clouds.

6 A New Global Information Service

There are many potential weather and climate services that will be of great use for the electricity production and distribution sector. The UN System, with leadership from the WMO, is now engaged in putting in place a Global Framework for Climate Services (WMO 2011). This Framework is aimed at making available globally the best possible climate information, which will inevitably also provide access to a growing array of real-time (weather timescale) data, structured in a way that will feed decision support systems.

While many countries have an effective climate service assisting with reducing risks and maximising opportunities dependant on the climate, there more broadly remains a significant gap between the supply of climate services and the needs of users. This is particularly the case in developing and least developed countries, which are also the most vulnerable to the impacts of climate variability and change.

To be useful, climate information must be tailored to meet the needs of users and be derived from high quality observations across the entire range of climate parameters and of relevant socio-economic variables. It must also be supported by basic and applied research to improve our understanding of, and ability to

Meteorology and the Energy Sector

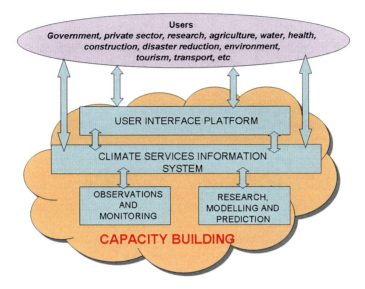

Fig. 6 A schematic of the components of the Global Framework for Climate Services with capacity building occurring within, and between all other components

communicate information about climate and weather systems. Further effort is also needed by governments and others to overcome the currently significant restrictions concerning sharing of, and access to, climate and other relevant data.

Efforts to provide effective climate services globally will only be successful if capacity is systematically built to enable all countries to manage climate risk effectively. Current capacity building activities to support climate services need to be scaled up and better coordinated within and between nations. A comprehensive capacity building initiative is needed to strengthen existing capabilities in the areas of governance, management, human resources development, leadership, partnership creation, science communication, service delivery and resource mobilisation.

7 Components of the Global Framework for Climate Services

The components of the Framework that are now being implemented by the WMO and other UN-System partners are as follows (Fig. 6):

1. The User Interface Platform will provide a means for users, user representatives, climate researchers and climate service providers to interact, thereby maximising the usefulness of climate services and helping develop new and improved applications of climate information.

2. The Climate Services Information System is the system needed to protect and distribute climate data and information according to the needs of users and according to the procedures agreed by governments and other data providers.
3. The Observations and Monitoring component will ensure that climate observations are generated at a quality and frequency necessary to meet the needs of climate services.
4. The Research, Modelling and Prediction component will assess and promote the needs of climate services within research agendas.
5. The Capacity Building component will support systematic development of the necessary institutions, infrastructure and human resources to provide effective climate services.

Many of the foundational capabilities and infrastructure that make up these components already exist or are being established, but they require coordination and strengthened focus on user needs. The role of the Framework should, therefore, be to facilitate and strengthen, not to duplicate.

8 Some Concluding Remarks

The energy sector is sensitive to meteorological events on all timescales, particularly in relation to climatological extremes impacting both consumer demand for energy and the capacity and security of transmission infrastructure. While the sector has been utilising meteorological information in its most basic form for many decades, it is only in recent times that it has become part of integrated design and decision support systems. Further, opportunities abound for new and improved meteorological services to the energy sector. Necessary ingredients are good data and good science, combined with close and effective engagement between meteorological service users and providers.

At present, industry relevant meteorological consultants are a rare breed, and there are relatively few operational weather and climate services tailored to maximise the effectiveness of decision-making in the private sector. A key outcome of the implementation of the Global Framework for Climate Services would be the building of this sector throughout the world (WMO 2012).

References

Baumert K, Selman M (2003) Heating and cooling degree days. A report of the World Resource Institute
Coppin PA, Ayotte KA, Steggel N (2003) Wind resource assessment in Australia—a planners guide. A report of the CSIRO Wind Energy Research Unit, CSIRO Land and Water
Dubus, Weather and climate and the power sector needs, recent developments and challenges (this volume)

George R, Maxwell E (1999) High-resolution maps of solar collector performance using a climatological solar radiation model. In: Proceedings of the 1999 annual conference. American Solar Energy Society, Portland, ME

George T, Hindsberger M, Weather and climate impacts on Australia's National Electricity Market (NEM) (this volume)

Geoscience Australia (2013) Solar energy. http://www.ga.gov.au/energy/other-renewable-energy-resources/solar-energy.html. Accessed 24 Jan 2013

Majithia, Improving resilience challenges and linkages of the energy industry in changing climate (this volume)

Maxwell E, George R, Wilcox S (1998) A climatological solar radiation model. In: Proceedings of the 1998 annual conference. American Solar Energy Society, Albuquerque, NM

Pirovano et al, Combining meteorological and electrical engineering expertise to solve energy management problems (this volume)

Queensland Floods Commission of Inquiry (2012) Final report, March 2012, 654 pp

World Meteorological Organization (2011) Climate knowledge for action: a global framework for climate services, WMO No. 1065, 240 pp

World Meteorological Organization (2012) Climate exchange. Tudor Rose, Leicester, 290 pp

Earth Observation in Support of the Energy Sector

Pierre-Philippe Mathieu

Abstract This chapter briefly describes how Earth Observation (EO) from space—in particular from satellite missions of the European Space Agency (ESA)—can support the energy sector by delivering accurate, consistent, and timely information on the state of the environment and natural resources. Some examples are presented of EO demonstration pilot projects performed in partnership with leading industrial players in Oil and Gas and Renewable Energy sector within the framework of the ESA Earth Observation Market Development (EOMD) program. The benefits and limitations of EO-based information services in supporting the whole life cycle of energy production, from technical and investment feasibility study up to the distribution and trading of electricity are highlighted and discussed.

1 The View from Space: A Unique Perspective to Help the Energy Sector

Today, most of the energy processes—ranging from exploration of natural resources up to their exploitation, distribution (supply), and consumption (demand)—are heavily affected by environmental and climatic conditions (e.g., amount of renewable resources, supply infrastructure, and demand).

Hence, reliable information about the state of our environment is increasingly needed by energy managers to assist them in optimizing energy extraction from natural resources while minimizing their cost and impact on the environment (Ebinger and Vergara 2011). This includes measurements of a wealth of meteorological, climate, and environmental parameters, at all scales in space—from local to global—and time—from minutes up to decades (Troccoli et al. 2010, 2013).

P.-P. Mathieu (✉)
European Space Agency, ESA/ESRIN, Via Galileo Galilei, 00044 Frascati, Italy
e-mail: pierre.philippe.mathieu@esa.int

Table 1 Examples of geophysical parameters needed to support decision-making in a variety of energy applications, ranging from exploration, resource analysis, design, siting, operation and distribution

	Some Essential Variables of Interest for the Renewable Energy Sector
Solar energy	Global horizontal irradiance, direct normal irradiance, aerosol optical depth, dust content, cloud cover
Wind energy	Wind speed/wind magnitude (at hub height), surface roughness
Bio energy	Biomass, vegetation indices, net primary production, topography, soil moisture
Geothermal energy	Land surface temperature, surface deformation
Hydropower	Run-off, snow extent, snow water equivalent, rainfall rate, lake height
	Some Essential Variables of Interest for the Oil & Gas Sector
Off-shore exploitation	Sea-level, ocean currents, met-ocean state, bathymetry
On-shore exploitation	Land cover type, topography, ground motion, geological characteristics
Arctic exploration	Sea ice extent, ice thickness, ice motion, iceberg

These parameters can be to some extent measured from space with EO satellites. They represent a limited subset of the needs and no order of priority has been assigned in the list. Traditional meteorological parameters, such as air temperature and surface pressure, are also required but are not detailed here. Requirements on sampling and accuracy are defined in more detail within the GEO energy task documents (sbageotask.larc.nasa.gov) and GEO energy community of practices (www.geoss-ecp.org)

The measurements needed, and their associated requirements—in terms of spatial resolution, temporal sampling, availability, reliability, accuracy, and stability—are heavily dependent on the specific *energy sector* as well as the type of *application* as reflected in Table 1. For example, exploration, assessment, design of power plants generally require "long-term" time series, while operation and trading would need (quasi) "real-time" information and short-term forecast. Also, requirements on accuracy can dramatically change according to whether users need to have a first look at energy potential resources to the region of exploration or analyze accurately the conditions to support design. Not all users always need the highest resolution and accuracy of products, in particular for a quick assessment.

Earth Observation satellites in orbit around the Earth can help energy managers addressing this *measurement challenge*. By remotely sensing radiation, EO satellites are able to infer in a consistent manner some of the parameters needed to support energy activities, ranging from weather and climatic conditions up to vegetation health and chemical composition of the atmosphere. This capability goes well beyond simple pretty pictures (being already very useful) to become a powerful *quantitative* tool. Using the unique vantage point of space, satellites deliver *global* data, covering even in the most remote places where no survey data exist or are possible to obtain. The ability to retrieve historical data from the satellite archive is also a key advantage to allow users to detect changes in the environment. These unique characteristics of wide-area mapping of EO data make

Fig. 1 The ESA Earth Observation Programme

them particularly useful to complement—but not supplement—traditional in situ measurements, which are typically point based, sparsely distributed, or simply completely missing in remote or difficult to access areas (e.g., mountains, polar regions).

Today, a wealth of EO satellites, carrying multiple radar and optical instruments, are continuously monitoring the state of our planet, providing scientists with a continuous stream of data on the state of the ocean, atmosphere, ice sheets, and vegetation. Some of the present and future missions built by the European Space Agency ESA are shown in Fig. 1 and described in more detail in Table 2. For more information on the satellite missions worldwide and their applications, see the recent handbook of the Committee on Earth Observation Satellites (CEOS) prepared for Rio + 20 (eohandbook.com).

International efforts by space agencies worldwide under the auspice of CEOS are also actively ongoing to contribute to the Global Climate Observing System (GCOS) by combining remote sensing information derived from various satellites to derive long-term climate quality data sets of the Essential Climate Variables (ECVs) (GCOS-154 2011) needed by the United Nations Framework Convention on Climate Change (UNFCCC) (article 4.1 g). Given that the majority of satellites contributing to the GCOS have not been designed specifically for monitoring climate, it can create problems of continuity of data sets (e.g., jumps or offset due to changes in observing systems), and is therefore necessary to continuously *re-process* the data archive in the context of new knowledge (e.g., new geoid from gravity missions like the Gravity field and steady-state Ocean Circulation Explorer (GOCE), new algorithms and atmospheric correction) to achieve the climate quality, accuracy, and stability required by GCOS.

The existing fleet of EO missions creates a large volume of data, leading to a series of challenges in terms of access, discovery, distribution, integration, mining and exploitation of this "Big Data." One of the key challenges is to convert the raw data measured by space sensors into information deemed useful for the user, in a similar way as the process of refining crude oil to make it usable. This data-to-information conversion process is not an easy task and depends on the type of

Table 2 Expanding European Observing Capability from Space, describing in more details some of the EO missions built by ESA

The Sentinel Family: Operational monitoring missions supporting the European copernicus initiative

Sentinel-1 mission is series of satellites dedicated to land and ocean services. It will provide Copernicus users with data from imaging radar with high revisit time (1–3 days over Europe). It includes a pair of polar-orbiting, all-weather, day-and-night C-band Synthetic Aperture Radar (SAR) missions operating in several modes with different resolution and coverage, including a main Interferometric Wide Swath Mode (swath of 250 km and ground resolution of 5x20 m), a Wave Mode, a Strip Map Mode and an Extra-Wide-Swath Mode

Sentinel-2 mission is series of satellites dedicated to land monitoring and emergency services. It will address the issue of data continuity for landsat and SPOT-5 data. It includes a pair of polar-orbiting, multispectral (13 bands), high-resolution imaging missions (e.g. 10 m, 20 m, 60 m), monitoring vegetation, soil and water cover, inland waterways, and coastal areas with high revisit (every 5 days at equator for two satellite units)

Sentinel-3 mission is a series of satellites dedicated to land and ocean services. It will address the issue of data continuity of Envisat data. This includes polar-orbiting, multi-instrument mission to measure variables such as sea surface topography, sea and land surface temperature, ocean colour and land vegetation with high-end accuracy and reliability. It includes a variety of sensors including the Ocean and Land Colour Instrument (OLCI), the Sea and Land Surface Temperature Radiometer (SLSTR), the Sentinel-3 Ku/C Radar Altimeter (SRAL), the MicroWave Radiometer (MWR), and the Precise Orbit Determination (POD)

Sentinel-4 mission is dedicated to atmospheric monitoring is a payload that will be embarked upon a Meteosat Third Generation-Sounder (MTG-S) satellite in geostationary orbit

Sentinel-5 mission is dedicated to atmospheric monitoring is a payload that will be embarked on a MetOp Second Generation satellite, also known as Post-EPS. In order to fill the data gap between Envisat Schiamachy and Sentinel-5, a Precursor satellite mission, including the TROPOMI spectrometer will be launched around 2015

The Earth Explorer Research missions dedicated to study the Earth System

GOCE ESA's Gravity field and steady-state Ocean Circulation Explorer (GOCE) mission (launched on 17 March 2009, completed on 11 Nov 2013) dedicated to measuring details in Earth's gravity and modeling the "geoid" — the surface of a hypothetical global ocean at rest — with unprecedented accuracy and spatial resolution. A precise knowledge of the geoid is crucial to advance our understanding of ocean circulation and sea-level change, both of which are influenced by climate. The data will improve our knowledge of processes occurring inside Earth. The satellite employs a state-of-the-art gravity gradiometer incorporating six highly sensitive accelerometers that measure gravity gradients in three dimensions. The mission is determining gravity field variations with an accuracy of 1 mGal (10^{-5} m/s^2) and the geoid with an accuracy of 1–2 cm, both with a spatial resolution of better than 100 km

SMOS (Soil Moisture and Ocean Salinity) is ESA's water mission (launched 2 Nov 2009) dedicated to making global observations of soil moisture over land and salinity over oceans. By consistently mapping these two variables, the mission will not only advance our understanding of the exchange processes between Earth's surface and atmosphere, but will also help to improve weather and climate models. The satellite carries a novel interferometric radiometer operating in the L-band microwave range to capture brightness temperature images

(continued)

Table 2 (continued)

The Sentinel Family: Operational monitoring missions supporting the European copernicus initiative
Cryo Sat-2 is ESA's ice mission (launched 8 April 2010) dedicated to monitoring cm-scale changes in the thickness of ice floating in the oceans and in the thickness of the vast ice sheets that blanket Greenland and Antarctica. Together with satellite information on ice extent, these measurements will show how the volume of Earth's ice is changing and lead to a better understanding of the relationship between ice and climate. Optimized for measuring icy surfaces, the satellite's high-spatial resolution radar altimeter is the first of its kind. It uses different modes to measure the height of sea ice above the waterline and to target the steeply sloping coastal terrain of ice sheets where the most dramatic changes are happening. Reaching latitudes of 88°, CryoSat-2 provides greater polar coverage than earlier missions
Swarm is ESA's magnetic field mission (launched 22 November 2013), comprising a constellation of three satellites, dedicated to precisely measuring the magnetic signals that stem from the magnetosphere, ionosphere, Earth's core, mantle, crust, and the oceans. This sampling, in both space and time, will lead to an improved understanding of the processes that drive Earth's dynamo, which appears to be weakening. Swarm also aims to provide a better insight into Earth's crust and mantle. The Swarm mission takes advantage of a new generation of magnetometers, enabling measurements to be taken over different regions of Earth simultaneously. GPS receivers, combined with an accelerometer, and an electric field instrument will deliver supplementary information to study the interaction of Earth's magnetic field with solar winds, electric currents and radiation, and their effect on the Earth system
ADM-Aeolus is ESA's wind mission will be the first space mission to acquire profiles of the wind on a global scale. These near-realtime observations will improve the accuracy of numerical weather and climate prediction, advancing our understanding of tropical dynamics and processes relevant to climate variability. Aeolus carries an innovative Doppler wind lidar to probe the atmosphere and acquire global wind profiles up to an altitude of 30 km. By demonstrating new laser technology, Aeolus will pave the way for future operational missions to measure wind
Earth CARE is ESA's cloud and aerosol mission dedicated to advance our understanding of the role that clouds and aerosols play in reflecting incident solar radiation back into space and trapping infrared radiation emitted from Earth's surface. The mission will acquire vertical profiles of clouds and aerosols, as well as the radiances at the top of the atmosphere to improve our understanding of Earth's radiative balance. These observations will lead to more reliable climate predictions and better weather forecasts. Developed in cooperation with the Japan Aerospace Exploration Agency (JAXA), the mission carries a high-spectral resolution atmospheric lidar, a radar instrument with Doppler measurement capability to provide cloud profiles, a multispectral imager, and a broadband radiometer
Biomass is ESA's forest mission will provide crucial information about the state of our forests, using for the first time from space P-band SAR measurements to determine the amount of biomass and carbon stored in forests. The data will be used to further our knowledge of the role forests play in the carbon cycle. The mission was selected as ESA's seventh Earth Explorer in May 2013

applications. Many barriers (e.g., lack of awareness from the user about EO technologies and capabilities, poor understanding of real user needs, and requirements by the data providers) and challenges exist to make this conversion valuable, moving toward usable products. Delivering the right information often requires integration of EO data with ancillary data (e.g., traditional knowledge, model output) and in situ data (anyway needed to validate satellite data) into a Geographical Information System (GIS). It also requires putting the information in the right format to be used as input to decision-making process (e.g., input to an energy demand or supply model). This sounds trivial but is often a key hurdle to achieve interoperability and usability.

The transformation of *raw* data into *usable* information is at the heart of "Environmental Information Services." One example is the so-called "Climate Services[1]" developed within the framework of the Global Framework for Climate Services (GFCS) established during the third World Climate Conference (WCC-3) in 2009. A variety of international initiatives contribute to generate the data being the foundation of such services (WMO Bulletin 2011). For example, in Europe, the ESA Climate Change Initiative (CCI) (www.cci-esa.org) aims to generate long-term consistent global satellite data sets of a wealth of ECVs by reprocessing the archive of ESA EO missions providing climate quality and a detailed characterization of the uncertainty. Such climate data sets constitute the *building blocks* and *pre-requisite* to support both climate *science* (e.g., understanding the Earth System) and *services* (e.g., verifying compliance of international treaties, supporting the energy sector). It is interesting to note some strong commonalities between the ECVs required by GCOS and the parameters needed by the energy sector (Table 1). In fact, the ECVs are to some extent simply "Essential Variables," not only useful for climate monitoring, but also supporting a wide spectrum of applications. Other examples of EO-based environmental and climate services are provided within the next section.

2 EO Demonstration Pilot Projects in Support of the Energy Sector

Within this section, we provide a few examples of environmental and climate information services addressing the issue of energy production, exploration, transport, and operations of power plant covering both the "Oil and Gas" and "Renewable Energy" sectors. This (limited) set of EO demonstration pilot projects has been defined in partnership with industrial players such as Shell, BP, AGIP, Vestas, Verbund, BMT, Statkraft, Enel, AMEC Western Geco, and Fugro. The pilot

---

[1] Climate services consist of the generation and provision of a wide range of information on past, present and future climate and its impacts on natural and human systems, and the application of that information for decision-making at all levels of society.

projects have been supported by the ESA Earth Observation Market Development (EOMD) programe (www.vae.esa.int) and make use of EO data in particular from ESA missions Envisat and ERS. The limited set of examples presented here aims to cover the major service domains addressing the Atmosphere, Cryosphere, Ocean, and Solid Earth but is not comprehensive (e.g., monitoring of biomass, wave energy, and transport are not addressed). Some of these prototype information services have now been consolidated further as "pre-commercial" services within the framework of the Copernicus Downstream services (copernicus.eu).

2.1 Quantifying Renewable Energy Resources from Space

Wind energy is experiencing one of the fastest growths across the whole Renewable energy industry. The financial success of wind farms is strongly bound to the wind resources available over the plant lifetime (hence the revenue) but also to other factors affecting the initial investment such as the impact on environment, access to turbines for maintenance, and connection to the grid network for distribution (hence the cost). Quantifying these factors is critical to perform technical and financial feasibility studies of prospective sites but also to secure long-term investment.

The traditional way to assess the potential energy yield of a prospective wind farm is by using data from a meteorological mast, which is very expensive in terms of installation and maintenance. Although this approach is very accurate, it can only provide point-measurement data for a short period of time (typically 1 year), while the wind field is generally highly variable in space and time (Hasager et al. 2006). This issue is further compounded for offshore farms, as the amount of wind offshore is sometimes estimated from on-shore measurements. Using local data can, therefore, be an issue to assess effectively the "bankability" of prospective farms.

In contrast, satellites can indirectly measure wind—but only over sea—through active sensors, like scatterometers, altimeters, and Synthetic Aperture Radar (SAR). EO data thereby provides a more comprehensive and spatially resolved view of the ocean wind climatology and the entire probability distribution (Fig. 2). The new generation of algorithms is now able to extract more information on wind magnitude and direction from the radar Doppler signal (Wergeland et al. 2011; Mouche et al. 2011).

EO missions such as Envisat that carry imaging radar sensor have provided users with information on coastal wind availability, the state of ocean (e.g., wind/wave met-ocean conditions), and land (e.g. roughness, vegetation cover). Within the next decades, the Sentinel missions will ensure continuity of such data streams for several decades, with high revisit time over Europe. This information will be critical to assist decision-making regarding operability and availability of wind turbines.

Other operational user-driven missions, like the Meteosat family in geostationary orbit, are delivering a wealth of information on availability of natural resources including a variety of renewable energy resources such as solar power (Cogliana et al. 2008). For example, MSG satellites deliver global maps of

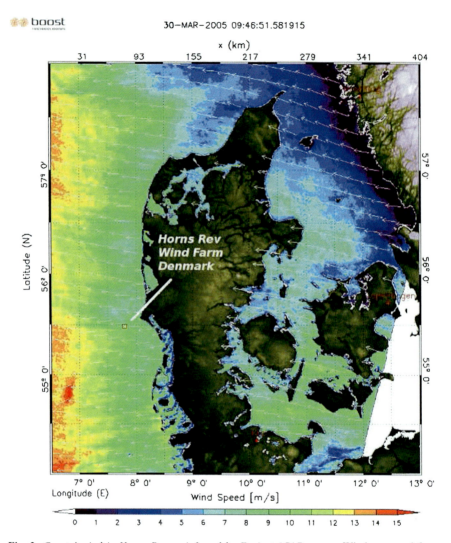

Fig. 2 *Coastal wind in Horns Rev as inferred by Envisat ASAR sensor.* Wind measured from space by the imaging radar instrument ASAR onboard the ENVISAT satellite around Denmark on 14-11-2005 (the location of the Horns Rev wind farm is indicated by a red star). The magnitude and variability of coastal wind resources can be assessed by using complementary EO sources of ocean wind data, including global data from scatterometers at about 25 km resolution, which provide a suitable temporal sampling, and regional data from SAR at about 100 m resolution, which provide the spatial details at the coast. The data set can also be used to study the shadow effect of mountains and the wake created by turbines. Courtesy ESA for ASAR data, BOOST technologies for processing with SARTool

irradiance up to 1 km resolution every 15 min. By combining EO-based irradiance maps with other EO products, such as Digital Elevation Model (DEM) and cloud cover/aerosols maps, it is possible to estimate the solar energy yield expected from a solar energy power plant. The ability to go back in time in the archive of Meteosat data—spanning several decades—provides the long-term time series and statistics of direct/diffuse solar irradiance together with cloud conditions necessary to quantify solar resources. This information forms the basis of a portfolio of climate services supporting solar energy managers in siting, designing, and assessing performances of solar plants, such as the Copernicus downstream services developed within the framework of ENDORSE (www.endorse-fp7.eu). The future generation of meteorological missions such as MTG to be launched in 2020 will further improve these services by providing decision-makers with enhanced sampling capability in space, time, and spectral range.

2.2 Assessing Environmental Impact of Hydropower Plants

The energy of moving water has been harnessed for millennia for a variety of purposes, ranging from powering mills to produce flour from grain or pumping water into irrigation networks. Today, hydropower is mainly used to generate electricity, supplying up to 20 % the global production of the world electricity, mainly through large dams.

One key advantage of hydropower, over other types of "intermittent" Renewable Energy, is its ability to store energy and, therefore, to manage peak load demand. In northern countries and mountainous areas, a large portion of the "potentially available water" is stored in the seasonal snow pack and provides "fuel" to the hydropower reservoir during the snow melt period. One example in Europe is Norway where the vast majority of energy is derived from hydropower. Precise information on the snow reserves, amount of precipitation, and accurate predictions of the onset of snowmelt are, therefore, required to optimize hydropower production. Satellite data—from optical and radar sensors—can assist in this process by providing timely measurements of key hydrological parameters (e.g., precipitation, snow cover, and temperature) and terrain conditions (e.g., DEM, land cover) affecting the river run-off. EO-based climate information services for snow monitoring (Fig. 3) are currently being used by hydropower companies such as Enel, Statkraft, and Verbund to initialize, validate, and constrain run-off models. See Copernicus downstream services CryoLand (www.cryoland.eu).

Another important role of EO for the hydropower sector is to assess the potential and effective environmental impact of dams. Indeed, a major drawback of large dams is often their impact on the environment (e.g., creation of large flooding areas, damaging of aquatic and forest ecosystem, fragmentation of wildlife habitat) and local communities (e.g., displacement of population). This makes the construction of some dams a very controversial issue, as some stakeholders question, whether their positive effects (e.g., electricity, availability of water, control of floods) outweigh their negative social and environmental impacts. To address this question, it is important to assess objectively the integrated impact (negative and positive) of large hydropower infrastructures resulting from the

Fig. 3 *Snow cover measured from space by Envisat MERIS sensor.* Optical remote sensing allows detection of snow extent. By processing the product (*cloud masking*) and combining it with ground measurement of snow thickness, it is possible to estimate the "Snow Water Equivalent" (SWE), which approximates the potential hydropower fuel stored in snow. The new generation of microwave EO missions aims to directly infer the snow water equivalent from space. *Courtesy* ESA

damming of the river and the production of electricity. Traditionally, an Environment Impact Assessment (EIA) study is always performed prior to construction of any hydropower plant, in order to obtain a building permit. However, this is only part of the picture, as dams and their environment are not static, and their real cumulative impact tends to be revealed only several years after the construction. In order to address this issue and need for continuous impact assessment of developments around the dam during operations, a pilot project has been set up in Cana Brava (Brazil) in partnership with GDF Suez, Tractebel Engineering, and Tractebel Energia to design an EO-based tool accurately and objectively measuring the cumulative impact of the Cana Brava dam with auditable and transparent results.

Hence, the idea here is to develop "Sustainability Indicators" which help companies to report on their corporate sustainable development performances along the "triple bottom line" (economic, social, environmental). Such reporting is becoming increasingly important to offer guarantees of transparency and accountability of companies to their stakeholders. Based on a preliminary study gathering user requirements and assessing feasibility of quantifying from space proxy parameters of sustainability, four indicators have been measured addressing issues of land-used change, biodiversity, socio-economic dynamics, and risk of erosion. A map of land-use change has been generated by processing two SPOT5 images from 2004 and 2007, and validated by field measurements. Indicators were then derived from the map by combining different measurements (e.g., fragmentation of habitat, anthropogenic use) associated with land classes. The methodology of derivation of indices—from the quality, radiometry correction of images to photo interpretation and classification—was also clearly described in order to ensure reproducibility of results and transparency to stakeholders. An illustration of the EO data and derived indicator is shown in Fig. 4. These environmental and socio-economic indicators were shown to be quite useful in understanding changes induced by the dam, revealing new regions of economic development (inducing land-use changes) where people have been displaced or where people were attracted. One key advantage of using EO imagery for sustainability reporting for a large group like GDF Suez operating globally is the enhanced objectivity and comparability of indices across different regions of the world. Future Sentinel missions, like Sentinel-2 carrying a high-resolution optical sensor, will provide free high-resolution data on land on an operational basis with a high revisit ensuring continuity of Spot like missions for the next 20 years.

Tony Moens de Hase, Sustainable Development Officer at Tractebel Engineering, part of the Suez group explains: *There are two main reasons why we want to explore the potential of EO in the monitoring of hydroelectricity infrastructure. One is technical accessibility: large reservoirs have a big area of influence and you do not have any other way to survey it all. Second is objective data over time: there is a demand for spatially and temporally homogeneous data on all infrastructures owned by the company to form a sound basis for reporting on sustainable development actions, not influenced by local contingencies.*

2.3 Weather and Marine Forecasting in Support of Offshore Oil and Gas Operations

The ability to perform accurate weather and marine forecasts is of critical importance to achieve cost-effective management of energy activities. For example, small error in weather forecast can lead to bigger errors in load forecast, which can in turn lead to significant economic impact on the price of electricity through trading, and even to enhanced risk of grid-failures (e.g., the Aug 14th 2003

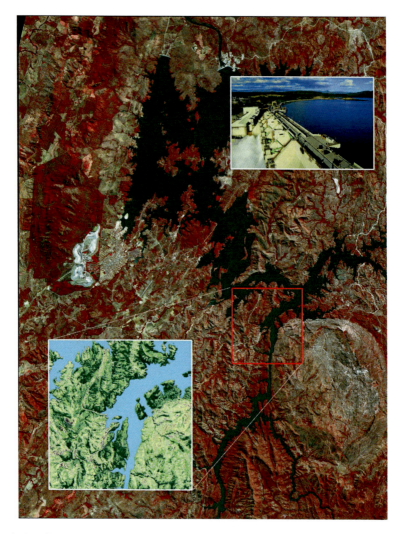

Fig. 4 *Development around the Cana Brava dam in Brazil as seen from space.* Color composite from SPOT5 data with an overlay of map of the biodiversity index illustrating the changes in the site between 2007 and 2003. Service providers are CAP CONSEIL, KEYOBS, and NADAR. *Courtesy* Spot image

Black-out in U.S, see also "Weather and Climate and the Power Sector: Needs, Recent Developments and Challenges" by Dubus and "Weather and Climate Impacts on Australia's National Electricity Market (NEM)" by George and Hindsberger in this book).

Over the last decades, rapid progress in observing technology, modeling, data assimilation, and computing power, have enabled meteorologists to improve significantly their forecasting system, resulting in current 7-day forecasts being as skillful as 5-day forecasts were 20 years ago. By the only virtue of being global, EO data have played a key role in this improvement, filling gaps in the observing system, in particular in remote or difficult to access regions like the Southern Ocean and polar regions. Today, Numerical Weather Prediction (NWP) is performed by routinely assimilating several millions data from EO satellites in their *pure* form as radiances (true observable) in forecast models. For example, at the European Centre for Medium-Range Forecasting (ECMWF), several million radiances from more than 30 satellites representing about 90 % of the total data volume are routinely assimilated within the analysis system. EO has, therefore, now become the single most important component of the global observing network for NWP. Making effective use of EO data, however, took several decades, as early satellite soundings were treated, not as radiances, but as though they were atmospheric profiles from radio-sondes, thereby introducing new sources of errors related to retrieval and contamination (e.g., using the same information as prior for inversion and for initialization of forecast).

Similarly to meteorology, rapid progress in numerical ocean modeling and observations from orbiting, floating, or gliding sensors have enabled oceanographer to better quantify the past, present, and future state of the global ocean. This contributed to the emergence of a new and now mature discipline referred to as "Operational Oceanography" and an international Global Ocean Data Assimilation Experiment (GODAE) community guided by a common vision (www.godae.org). This new discipline has also paved the way toward the foundation of "Marine Services" of immense social value, to support management of maritime resources (e.g., fishery), maritime safety (e.g. oil spill detection), offshore operations, and decadal prediction supporting the climate change adaptation in coastal cities.

High-precision satellite altimetry, such as from Topex, Jason, Envisat, has played a key role in this endeavor as it has provided modelers with synoptic measurements of the height of the ocean along the satellite path with a few centimeters accuracy. This capability is today complemented by ice thickness measurements from the "Cryosat" radar mission, which provides sea ice modelers with unique data to constrain their models. More recently, the Soil Moisture and Ocean Salinity (SMOS) satellite, carrying a novel microwave radiometer measuring "brightness temperature," has provided oceanographers for the first time with global map of salinity, which is key to modeling and understanding of the thermohaline circulation. Also gravity missions like GOCE and GRACE provide unique insight into the geoid, which ultimately improves our understanding and modeling of the global ocean circulation.

Such forecasting capability of weather and marine events provide energy managers with an "Early Warning" system to support operations of power plants, in particular offshore. One example is the exploitation of offshore rig platforms in the Gulf of Mexico, which are subject to the passage of hurricanes, as well as their marine equivalent ocean eddies. These extreme events can significantly disrupt offshore exploration, development construction, and production operations. In 2003, an eddy with strong currents called "Sargassum" (named after the Sargasso

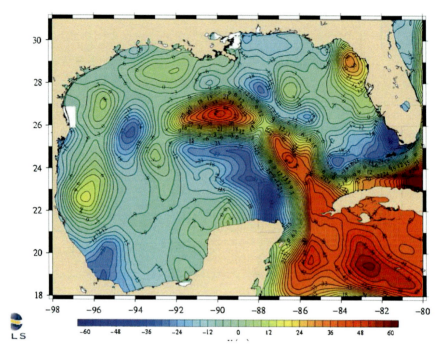

Fig. 5 *Loop current in the Gulf of Mexico as seen from space.* Satellite altimetry is a sophisticated technique that can infer the height of the ocean by measuring the time taken by a radar pulse to travel from the satellite antenna to the water surface and back to the satellite receiver. By combining such information with precise satellite location data and adjusting for the atmospheric effect, it is possible to compute the sea surface height anomalies within millimeters and also derive the associated absolute current velocities. Above is a high-resolution chart of sea surface height obtained by aggregating altimeter measurements from multiple satellite passes (e.g., Jason, ERS, Topex) in order to increase the spatial and temporal resolution. Regions of high gradient correspond to high current velocity. *Source*: Courtesy of CLS and Fugro

Sea), spinning off from a large oceanic river called the "loop" current,[2] traveled through a heavily exploited sector and caused major disruptions (Fig. 5). In 2005, hurricane "Katrina," one of the strongest and most destructive Atlantic hurricane, created a dramatic impact on the US causing severe destruction along the Gulf coast from central Florida to Texas, and also in regions of offshore exploitation. By using the forecasting/nowcasting capability underpinned by sophisticated models and observations—in particular from EO satellites, energy operators are able to anticipate—although not mitigate—some of the impact of these extreme events. For example, SMOS data over Hurricane Igor, which reached category five in the

---

[2] The loop current is formed when warm water from the Caribbean Sea enters the Gulf of Mexico through the Yucatan Straits and "loops" clockwise through the basin before exiting through the Florida Straits to merge with the Gulf Stream.

North Atlantic in 2010 (Reul et al. 2012), have shown that salinity in the surface waters can influence to some extent the strength of hurricane. This is the first time that such changes have been detected from space.

2.4 Supporting Sustainable Mining Operations and Rehabilitation

Due to increasing pressure of stakeholders, many large mining companies are now trying to limit their impact on the environment not only during but also after operations. Hence, these companies now undertake to rehabilitate sites when they are no longer used, either by regenerating habitats of the original site (e.g., growing forest and vegetation) or by transforming the site appropriately to support local communities (e.g., creation of recreational areas or protected park).

In this context, several pilot projects have been performed in partnership with two global mining companies: Rio Tinto Alcan for aluminum production and Lafarge for cement production. The pilot addresses several sites in Africa and Australia and the US operated by the mining groups. For each site, the project aimed to deliver a variety of EO products (e.g., land cover, biodiversity and watershed conditions, topography) in order to (i) assess conformity of mining operations with engagement and environmental rules and (ii) check progress of rehabilitation activities.

The results were quite positive and showed real value for operations and rehabilitation, in particular in regions like Africa where very few *in situ* data were available. The EO products from Radarsat and IKONOS helped companies to manage operation and also to provide data for reporting on sustainability.

Ms. Sharon Lee, Corporate Technical Services of Lafarge (Canada), says: *EO solutions show all-in-one spatial information of impacts and how they are managed. Elevation data and the incorporation of other multidisciplinary information ease interpretation. So, the use of EO solutions is undoubtedly time effective.*

EO also demonstrated its value as a tool of communication to support dialog with stakeholders about sustainable mining operations. For example, 3D virtual rendering of the "current" mining and "rehabilitation plan" at the Awaso site in Ghana based on EO, was used by Rio Tinto Alcan, both internally to support operations of the mine, and externally to foster dialog with local communities.

Mr. Mark Annandale, Community Relations Manager at Rio Tinto Alcan (Australia), says: *EO solutions provide essential tools to share views among multidisciplinary stakeholders and to communicate on social responsibility efforts and achievements. It helps to visualize a site evolution in a very realistic and reliable way and to detect anomalies to be further analyzed on the ground. Through this complementary vision, it complements traditional methods. Also, EO solutions ease stakeholders' dialogue as they allow visualizing of several thematic layers on a sole 3D topography map. Multidisciplinary experts get to exchange on a shared basis all together, to determine the best solution through dialogue* (www.esa.int/eomd).

2.5 Monitoring Ground Motion to Support Carbon Capture and Storage

Carbon Capture and Storage (CCS) refers to the technology attempting to prevent the release of carbon dioxide into the atmosphere released from fossil fuel combustion by capturing, transporting, and ultimately pumping it into underground geologic formations, such as saline aquifers, for long-term storage. Today, CCS is considered a potential means of mitigating the contribution of fossil fuel emissions to global warming. The Intergovernmental Panel on Climate Change (IPCC) estimates that the economic potential of CCS could be between 10 and 55 % of the total carbon mitigation effort until year 2100 but the risks associated with the technique are not well quantified.

Hence, a growing number of regulatory and policy frameworks are put in place to promote and control CCS activities, thereby requiring continuous monitoring of CCS sites. EO can assist such monitoring in different ways. In particular, one way is through use of powerful techniques, such as Interferometry of Synthetic Aperture Radar (InSAR) or Permanent Scatterer Interferometry (PSI), which enable well engineers to remotely monitor ground deformation within sub-centimeter accuracy in some specific conditions (e.g., need for temporal consistency of the target). Such EO-based information is of critical importance to identify subsidence or uplift risks associated with carbon dioxide injection activities, which can in turn lead to leakages.

Traditionally, engineers of wells rely on GPS surveys (using the differential GPS technique) to monitor ground deformation of sequestration sites. However, although very accurate, this method is resource intensive (e.g., data processing) and can only provide an incomplete point-based picture of the ground motion issue. The InSAR technique provides a better synoptic view, which helps engineers to identify "hot spots" where enhanced in situ monitoring is required, providing them with an "optimal sampling" strategy for on-site surveys. It also provides the advantage of being noninvasive and nonvisible and, therefore, does not disturb inhabitants or create environmental concerns.

The method has been tested at different CCS on-shore sites with different levels of performance and applicability. In Algeria, at the In Salah injection site in a deep saline aquifer in Algeria, the PSI method based on high-resolution radar such as TerraSAR-X has been quite successful, capturing the hot spot regions experiencing ground motion associated with carbon injection and sequestration (Mathiesen et al. 2008; Fokker et al. 2011). Combination with ASAR data has also improved the results due to increase PSI density. However, this case was quite ideal as it was

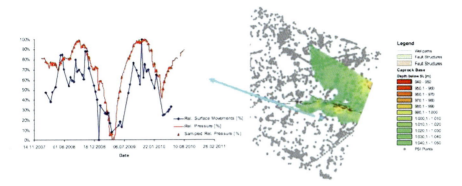

Fig. 6 *Subsidence monitoring in gas site.* Application of radar interferometry with TerraSAR-X in a natural gas storage site in Germany, selected to be representative of future CO2 storage sites in Central Europe. The "Persistent Scatterer Interferometry" (PSI) technique has been applied to quantify subsidence induced by gas pumping. The distribution of 61594 PSI points is displayed along with a time series of displacements (relative) of one specific PSI point versus relative gas storage site pressure (*correlation* = 0.79) illustrating the good correlation. Courtesy Infoterra Gmbh

located in a desert area presenting a strong temporal consistency. The method has also been tested in Southern Germany, where the application of the technique is more challenging due to the presence of heterogeneous land conditions (e.g., forest, agricultural areas, small settlements). The method, however, manages to capture small displacements (Fig. 6) associated with nonlinear movement of the ground (Petrat et al. 2010). Hence, the EO information is a promising piece of information to complement the traditional in situ surveys. The Sentinel missions like Sentinel-1 designed to support interferometry with high revisit frequency will provide a useful tool to quantify ground motion on a weekly basis.

3 Conclusions

EO satellites orbiting hundreds of kilometers above the planet can deliver invaluable information to the energy industry. High-level observational products and the predictions which are based on them, form the foundation of "Information Services," which can help energy managers in a variety of applications, ranging from exploration, extraction, and transport of natural energy resources in a more effective way up to mitigation of hazards, with enormous benefits for our society.

EO data provide the energy sector with timely information that is not available any other way. Satellite data is needed to initialize, validate, and constrain forecast models used within the industry. It also helps engineers to make a "virtual field trip" to any prospective site, even located in the most remote regions, without leaving their office, at a cost much cheaper than traveling to the site. The power of wide-area observation from satellites lies in its ability to assess objectively and consistently the environmental footprint of business activities at the global scale, thereby providing stakeholders with a rapid "third party" check on sustainability of business operations even in the most remote or difficult to access regions. Finally, satellite imagery provides a powerful communication tool to support dialog with stakeholders by putting their environmental issues in a spatial context.

However, EO data are and will remain only *part* of the solution as in situ measurements are and will always be required for validation of remote-sensing data. EO data are often not as accurate as traditional measurements based on expansive field surveys, but what they lack in precision can be made up for in coverage. As such EO data help one to "spatialise" *in situ* data, complementing the traditional approach, and thereby providing a useful additional information input to energy models for demand and load.

The potential of space-based remote sensing to assist decision-making in the energy sector is already substantial and is likely to grow significantly with the new generation of satellite missions providing enhanced spectral, temporal, and spatial capabilities. In particular, the availability of operational services will be improved through the "Sentinel missions" (Table 2) to be launched from 2014 + under the European Copernicus initiative and guided by a full and open data policy. The availability of large volume of EO data is likely to herald a new era of global and timely environmental information. Realizing the full potential of this data, however, remains a challenge requiring further collaboration between users and providers of the technology.

References

Cogliana E, Ricchiaza P, Maccari E (2008) Generation of operational maps of global solar irradiation on horizontal plan and of direct normal irradiation from meteosat imagery by using SOLARMET. Sol Energy 82:556–562

Ebinger J, Vergara W (eds) (2011), Climate impacts on energy systems: key issues for energy sector adaptation, World Bank publication

Fokker PA, Orlic B, van der Meer LGH, Geel CR (2011) Geomechanical modelling of surface uplift around well KB-502 at the In Salah CO2 storage site; 73th EAGE Conference & Exhibition, Vienna

GCOS-154 (2011), Systematic observation requirements for satellite-based products for climate: supplemental details to the satellite-based component of the implementation plan for the global observing system for climate in support of the UNFCCC

Hasager CB, Barthelmie RJ, Christiansen MB, Nielsen M, Pryor SC (2006) Quantifying offshore wind resources from satellite wind maps: study area the north sea. Wind Energy 9(63–74):9

Mathiesen A, Wright I, Roberts D, Ringrose P (2008) Satellite imaging to monitor CO2 movement at Krechba, Algeria, Proceedings of 9th international conference on Greenhouse Gas Control Technologies (GHGT-9), Nov 2008. Washington DC, USA

Mouche, A, Collard F, Chapron B, Dagestad KF, Guitton G, Johannessen JA, Kerbaol V, Hansen MW. In Press. On the use of Doppler shift for sea surface wind retrieval from SAR. IEEE. Transactions on Geoscience and Remote Sensing

Petrat L, Riedmann M, Anderssohn J (2010) CO2 Storage: monitoring of related surface movements from space—potential for central european land cover conditions?, Proceedings of second EAGE CO2 geological storage workshop, Berlin, Germany, 11–12 March 2010

Reul N, Tenerelli J, Chapron B, Vandemark D, Quilfen Y, Kerr Y (2012). SMOS satellite L-band radiometer: a new capability for ocean surface remote sensing in hurricanes. J Geophys Res: Oceans 117

Troccoli A (ed) (2010) Management of weather and climate risk in the energy industry. Springer Academic Publisher, NATO Science Series

Troccoli A, Boulahya MS, Dutton JA, Furlow J, Gurney RJ, Harrison M (2010) Weather and climate risk management in the energy sector. Bull Amer Meteorol Soc 6:785–788. doi:10.1175/2010

Troccoli A et al (2013) Promoting new links between energy and meteorology. Bull Amer Meteor Soc 94:ES36–ES40. http://dx.doi.org/10.1175/BAMS-D-12-00061.1

Wergeland HM, Collard F, Dagestad KF, Johannessen JA, Fabry P, Chapron B (2011) Retrieval of Sea Surface Range Velocities From Envisat ASAR Doppler Centroid Measurements. IEEE Trans Geosci Remote Sens 49(10):3582–3592

WMO Bulletin (2011) Reaching users with climate services, vol 60(2)

Emerging Meteorological Requirements to Support High Penetrations of Variable Renewable Energy Sources: Solar Energy

David S. Renné

Abstract With the emergence of favorable policies and effective financing schemes to allow for large penetrations of variable renewable energy resources tied into the electricity grid, there are increasing R&D efforts underway to understand and predict how this variability can best be managed in grid operations. Weather variability is key to the cause of solar and wind energy output variability; in turn, as variable renewable energy (VRE) resource penetrations increase, their impact on grid reliability, stability, and energy quality also increase. A key challenge for utility and system operators is the ability to *forecast* this variability over time periods of a few minutes to several days in the future. R&D programs on weather-driven energy resource variability and the ability to forecast this variability have been undertaken in the U.S. and elsewhere. These programs include special field measurement campaigns, data processing techniques, and modeling schemes devised to quantify and forecast the characteristics of this variability in the electricity grid. After introducing the challenge of high penetrations of VRE in a transmission or distribution system, this chapter highlights some key studies underway or completed, with special emphasis on characterizing and predicting solar system output variability associated with rapid cloud passages or changing weather patterns. Some of these studies have been undertaken as a significant international collaboration established under the International Energy Agency's (IEA's) Solar Heating and Cooling (SHC) Implementing Agreement's Task 36 "Solar Resource Knowledge Management," and a new Task 46 currently underway titled "Solar Resource Assessment and Forecasting."

D. S. Renné (✉)
NREL (Retired), Boulder, CO, USA
e-mail: drenne@mac.com

1 Introduction

Installation of PV systems around the world has been growing exponentially over the past few years, and a significant percentage of this growth is in the emergence of 1 MW or larger grid-tied "Central Station" PV facilities. Due to changing weather patterns and even to the continuous changes in cloud patterns and amounts, these facilities are seen as variable renewable energy (VRE) resources to the utility industry, and pose unique challenges in planning and operating large penetrations of these systems. Some of these challenges are already being addressed with the large-scale penetration of wind energy into utility systems. However, the emergence of large-scale PV and the potential for large penetrations of Concentrating Solar Power (CSP) projects has created new demands on the solar resource assessment community with requirements for sophisticated solar resource data sets and products that were not envisaged even just a few years ago. This chapter addresses the emergence of new weather data products to address VRE incorporated into a utility system, with a specific emphasis on solar, and especially PV, technologies. After introducing the challenges of VRE to utility and system operators, this chapter provides a review of recent R&D efforts and results, including work by an international collaboration of researchers involved in the International Energy Agency's (IEA's) Solar Heating and Cooling (SHC) Implementing Agreement's Task 36 "Solar Resource Knowledge Management," which was in place from 2005 to 2011, and a follow-up ongoing Task 46 "Solar Resource Assessment and Forecasting."

2 Global Trends in PV Development

At the time of the *International Conference Energy and Meteorology (ICEM)* in the Gold Coast, Australia in November 2011, we reported that a total of 40 GW of PV systems had been installed globally. These were the most recent figures available, and were published in the REN-21 Global Status Report (GSR 2011). Since then REN-21 produced an update (GSR 2012) showing a global total of 70 GW by the end of 2011, and (GSR 2013) shows 100 GW of installed capacity by the end of 2012. Despite the global economic recession, PV production nearly doubled in the year 2010, and increased again by around 75 % in 2011. Overall, by the end of 2012 global PV capacity will be nearly 10-fold higher than it was just 6 years earlier. Of significant importance is that, by the end of 2010, there were nearly 5,000 grid-tied "central station" PV systems installed around the world, each at least 1 MW in size, and this number grew substantially in 2011. In fact, virtually all of the PV installed in the past 2 years has been grid-connected; according to GSR (2012) only 2 % of PV installed in 2011 was for off-grid applications.

The IEA, in its recent Solar PV Technology Roadmap Report (IEA 2010), further indicates that all major applications of PV (commercial and residential as

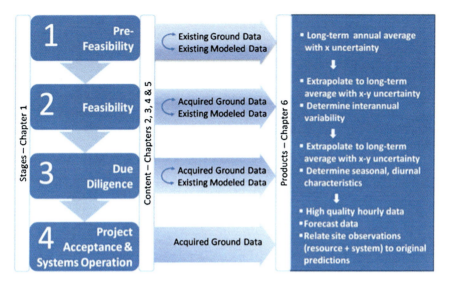

Fig. 1 A typical large-scale project development sequence, showing the four basic phases of a project from concept to operations. Chapters refer to the chapters in the report by Stoffel et al. (2010): *CSP best practices handbook for the collection and use of solar resource data*

well as utility grid-tied systems along with off-grid applications) will continue into the foreseeable future. The report estimates that by 2030 PV could be providing as much as 5 % of the global electricity supply, and 11 % by 2050. A major driver of this growth is the government policy initiatives around the world; GSR (2012) reports that at least 118 countries, many of them developing countries, now have renewable energy targets that include solar.

The scale of these grid-tied projects also increases substantially each year. There is now a >1 GW central station PV facility in Germany, and a nearly 0.5 GW facility under construction in the U.S., along with facilities nearly this size in place or planned in Spain, China, and elsewhere. Especially in China, with the recent creation of a Feed-in Tariff (FiT), its current 5-year plan envisions an installed capacity in the country of 40–50 GW by 2015. Although installed capacity is still a relatively small percentage of the world's total electricity supply (about 1.3 % according to the GSR 2012), PV is now a $100 Billion USD/year industry, and PV continues to remain by far the fastest growing renewable energy technology. Thus, the penetration of PV into many electricity systems, both through central station facilities as well as distributed generation programs such as roof top solar, requires careful operational planning to take maximum advantage of these clean yet variable resources.

The growth of these systems has created increasing demands on the quality of solar resource information available to properly site, size, and operate these systems. Although historically it was possible to use most available solar resource data, either interpolated from existing ground networks or derived through modeling efforts such as the use of satellites to estimate the solar resource over large areas, the scale and cost of current systems requires a new level of sophistication in the quality, completeness, and type of solar resource information. We now often hear the term "data bankability" which implies the degree to which the financing community can rely on available data sets for financing large-scale grid-tied systems. Figure 1, taken from Stoffel et al. (2010) shows how different levels of data accuracy and completeness are needed for different stages of project design, development, and operations. The remainder of this chapter focuses on the type of solar and other weather data requirements that are emerging as a result of this growing industry.

3 Perspectives from the Utility Industry

With the growth of grid-tied PV systems into utility systems, occurring both as central station systems as well as large-scale distributed systems (such as commercial and residential rooftop solar), a number of studies have been launched to better understand how to operate these systems. The goal of many of these studies is to maximize the value of the clean energy benefits offered by VRE resources such as PV and to minimize impacts associated with their variability of output. The IEA, in response to estimates that VRE can comprise anywhere from 18 to 31 % of total electricity supply by 2050, has recently released a report titled "Harnessing Variable Renewables: A Guide to the Balancing Challenge" (IEA 2011). The main theme of the report is to help utility systems identify the degree of flexibility their system has to respond to VRE resources and still meet fluctuating loads without relying significantly on back-up power stations. For example, the report shows that it is easier to predict fluctuations in demand than to predict variations in VRE supply, and that some utilities are better than others to manage large penetrations of VRE.

From the perspective of solar resources, the most critical time frame of concern to utilities in managing VRE is over the minutes to days timescale (Fig. 2, taken from IEA (2011)), where utilities can use their balancing authority to meet fluctuating loads that are further impacted by VRE resources within their system. This is further exemplified in Fig. 3, also from the IEA (2011) report. Figure 3 shows fluctuations for a typical week in total load, or demand (top curve), and that part of the total load or demand that is *not* being met by the wind and solar resources installed in the system (bottom curve). The bottom curve is labeled the "net load," and indicates that other energy resources, which can be ramped up and down rather quickly, or power sources from other balancing regions, must be available to meet that portion of the load not being met by wind and solar. Without this balancing, grid quality, including voltage and frequency fluctuations, will be compromised. From the example in Fig. 3 it is clear that there are even times when the wind and solar resources actually *exceed* the load, and must be curtailed in some way.

Figure 3 also shows the emerging importance of the ability to *forecast* when the solar and wind resources may not be available, or when they may need to be curtailed. For example, the ability to forecast the high wind and solar resources that

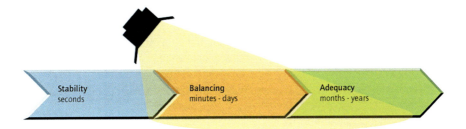

Fig. 2 System integration challenges occur over a spectrum of time frames, but VRE in the minutes to days time frame poses significant new utility balancing challenges (from IEA 2011)

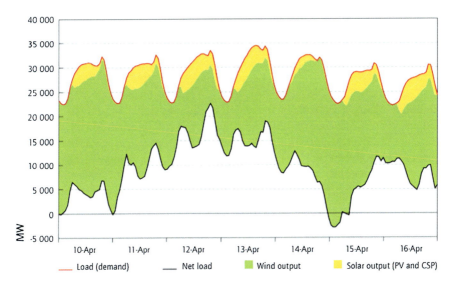

Fig. 3 Variability in demand (*upper line*) and net load (*lower line*) for a challenging week that occasionally showed high penetration of VRE (*wind and solar*) resources. The additional variability in *net load* can pose significant additional challenges to utilities. From IEA (2011), adapted from an earlier report by GE (2010)

occur at a time when the overall demand is dropping, such as seen on 15 April in the figure, allows the utility to curtail other more expensive energy resources that would normally be quickly brought online, and therefore make better use of the cleaner solar and wind resources. On the other hand, there are periods around 12 April where wind and solar resources are relatively low, and yet demand remains high. Knowing this in advance allows the utility to ramp up other mid-merit or peak load power plants quickly to meet the net load, or to exercise its balancing authority to purchase power from adjacent regions to meet the rapid increase in net load that the solar and wind resources are unable to meet.

Principles such as these were applied by GE (2010) to evaluate the capability of the existing transmission and distribution system in the western U.S., to handle large penetrations of VRE (up to 30 % wind and 5 % solar, split between CSP, and PV technologies). The report shows that it is operationally feasible to incorporate this amount of VRE into the western U.S. grid under current conditions, although a number of key steps need to be taken to assure that this level of penetration can be managed, such as increasing the balancing area and enabling coordinated dispatch over wider regions, building new transmission facilities to accommodate renewable energy while also taking more advantage of the existing transmission system, committing additional operating reserves, and increasing the use of sub-hourly scheduling. But among the most important conclusions is to incorporate state-of-the-art resource forecasting at various timescales (sub-hourly, hour-ahead, day-ahead) to effectively dispatch VRE supply as part of the grid operations.

A recent study published by Marquis et al. (2011) took an even closer look at the importance of wind resource forecasting to improve the value of high penetrations of wind in the electricity grid. This article concluded that while VRE resources are currently viewed as "non-dispatchable" from the perspective that wind resources can be curtailed, they are in effect "dispatchable" in the downward direction (that is, when the wind resource drops to the point that the installed wind systems drop off the grid). What is more, improved forecasting can give wind resources some dispatchability in the upward direction (when resources increase, bringing wind power back on-line), and at the same time reduce VRE integration costs by reducing integration charges and curtailments. The article shows that it is important for system operators to have improved forecasts of wind energy ramp events, and in particular "downramps" when other resources will need to be quickly brought online.

These as well as numerous other studies that have come out these past few years indicate the level of sophistication now required of the solar and wind resource assessment communities not only for providing bankable data to support large-scale projects costing (many of which now cost upwards of USD 1 Billion), but also for providing key information on resource variability and predictability that support cost-effective and efficient operation of VRE in a utility system. The rest of this chapter focuses on recent studies and findings addressing some of these resource characterization issues.

4 Challenges for the Solar Resource Community

4.1 Short-Term Resource Variability

There are a multitude of scales of motion and relevant time frames in the atmosphere, as shown in Fig. 4. The key scales of motion that are relevant to utilities in operating VRE in the grid are primarily in the hours-to-days time range. As shown

Emerging Meteorological Requirements to Support High Penetrations 263

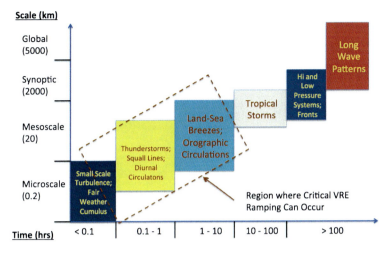

Fig. 4 Examples of atmospheric scales of motion, indicating the region important to understanding VRE impacts on the grid

in the figure, this means that critical atmospheric processes to be considered include minute-by-minute cloud passages over large arrays (microscale), sudden changes in cloud cover associated with mesoscale weather systems such as squall lines and frontal passages (mesoscale), and changes in general cloud characteristics associated with synoptic-scale high- and low-pressure patterns.

An example of the kind of variability that could impact utility operations of large-scale solar arrays is shown in Fig. 5a and b, taken from Kankiewicz et al. (2010). Here we see fair weather cumulus clouds passing across a 25-MW PV array in Central Florida. These conditions result in some portions, or "containers" within the array to shut down momentarily due to the passage of cloud shadows, while other containers are still providing power into the substation. Thus, even though the full array is not producing full power, the ramps observed from the full array output would not be as extreme as would those seen by individual containers. The inset in Fig. 5b demonstrates this phenomenon, showing that portions of the array are always in direct sunlight, which results in the *magnitude* of the 10-s ramp rate becoming smaller and smaller as the array gets larger and larger.

Similar results were obtained from a field study conducted near the Kalealoa general aviation airport on the southwest corner of the island of Oahu (HI). A variable mesh of 17 fast-responding (1-s) solar sensors, all measuring global horizontal irradiance (GHI) with the output of all sensors time-syncronized, was installed near to and within the airport location. An additional Rotating Shadowband Radiometer (RSR) was also installed to provide not only GHI, but also Direct Normal Irradiance (DNI) and Diffuse Irradiance measurements at 3 s intervals. Figure 6 shows the layout of the measurement array.

In a study by Hinkelman (2013), data from the network were analyzed to determine cloud decorrelation factors for various-sized arrays. The data obtained

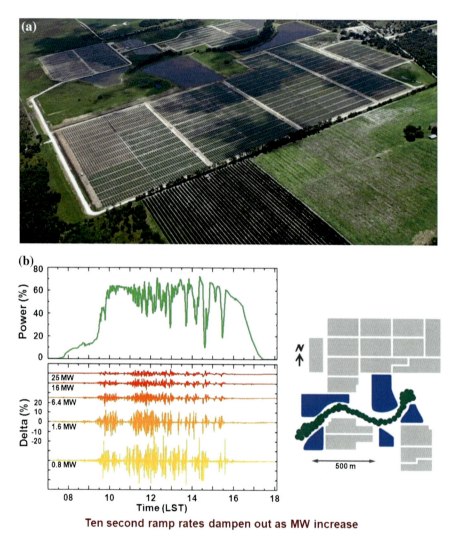

Fig. 5 a Fair weather cumulus passing over a 25 MW PV system in DeSoto, Florida and **b** a case study example of the power output characteristics for various combinations of "strings" over 10 s intervals. From Kankiewicz et al. (2010)

during daylight hours were averaged into 10-, 20-, 30-, 60-, 90-, 180-, and 300 s values, and then the ramps, or differences between two consecutive values were calculated. Finally, the ramp time series for various station pairs, ranging in separations from 100 to 800 m were calculated. In their study, the data were sorted based on along-wind and cross-wind station pairs.

Figure 7 provides the results of 28 days of broken cloud analysis in 2010. The figure shows that the cross-correlation of ramp values drops off significantly over

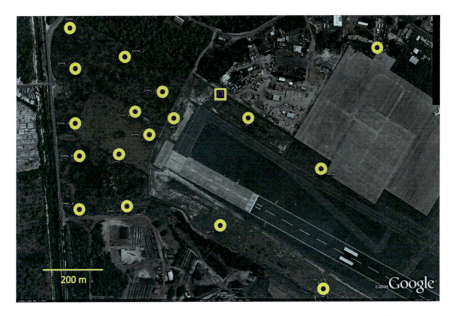

Fig. 6 Pyranometer array at the Kalealoa Airport in southwestern Oahu (Hawaii, USA). *Circles* represent locations of 1 s GHI sensors, and the *square* represents the location of the RSR 3 s sensor. Although the distance between stations is variable, the east-west extent of the array is approximately 900 m, and the north-south extent is approximately 800 m, and the shortest distance between two stations is ~100 m

Fig. 7 Correlation of 60 s ramp rates as a function of distance derived from the Kalealoa Airport (Oahu, Hawaii, USA) high frequency 18-station irradiance network (Hinkelman 2013)

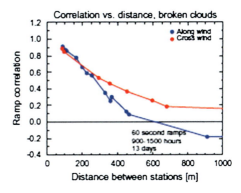

relatively small distances, so that from one end of the array to the other the correlations approach 0. These results suggest that as array sizes increase in conditions of fast-moving broken clouds, the overall ramp rates of the full array should decrease considerable. The studies also show that for longer ramp periods the correlations tend to be higher, so that the smoothing of ramp rates is more noticeable for very short ramp periods compared with the longer averaging periods. The studies by Hinkelman (2013) further show that, for the 60 s ramp data, correlations drop more

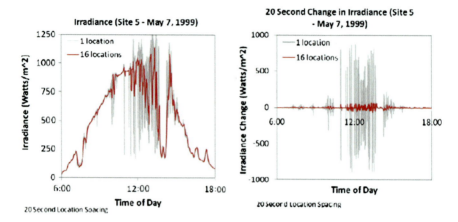

Fig. 8 *Left side* irradiance measured at a single location (*gray line*) and at a virtual network of 16 locations (*red line*). *Right side* 20 s changes in irradiance in a single location (*gray line*) and at a virtual network of 16 locations (*red line*). From Hoff and Perez (2010)

rapidly in the along-wind rather than the cross-wind direction, suggesting that the manner in which arrays are installed and strung together to take advantage of this phenomenon can serve to further mitigate output variability associated with ramps caused by rapid cloud movements across the array.

A number of studies using station pair data obtained from a variety of locations, as well as simulated and actual PV output data for various PV array configurations, have been conducted recently, all with the same general conclusions that variability, especially in the magnitude of the ramp rates, is significantly reduced for larger arrays. A comprehensive survey of these studies is beyond the scope of this chapter. However, one significant study worthy of mention here is one conducted by Hoff and Perez (2010), which undertakes a comprehensive analysis of the reduction in variability obtained from various configurations of fleets of PV systems. The authors adopt the concept of a Dispersion Factor, which is a variable that captures the configuration of the PV fleet, the transit speed of clouds, and the time interval over which variability is calculated. They test this concept with actual irradiance data obtained from the U.S. Department of Energy's Atmospheric Radiation Measurement Program site in central Oklahoma. Using the single station data they create a "Virtual Network" which simulates the irradiance that would be expected over a fleet based on the solar irradiance variability and cloud motion velocities determined at a single station. The reduction in variability that can occur in this virtual network is exemplified in Fig. 8, which compares the irradiance (left side) and 20 s ramps (right side) at a single station with a virtual network consisting of 16 locations.

A key outcome of the work by Hoff and Perez (2010) is that utilities can undertake strategic measures for minimizing variability by understanding local cloud motion conditions and the manner in which different configurations of PV fleets respond to this critical variable. For example, they demonstrate how the output variability decreases as the size of the fleet increases, so that the output variability of a dispersed array of 1,000 MW of 4 kW rooftop solar home systems is a small fraction of the variability that would occur if the 1,000 MW were all installed at a single location. They tested this concept with data from an actual 5 MW PV array in the southwest U.S. to show how variability from the system could have been decreased if the 5 MW had been installed in five 1 MW facilities scattered over the area available to the utility rather than having all 5 MW installed at a single centralized location.

4.2 Forecasting Solar Variability

In order to meet ever-fluctuating load demands, utilities must be able to predict at a high degree of accuracy when and to what extent the loads will change, so that they can take action to increase or decrease the supply of electricity to meet these changes. The bulk of the power supplied to a grid typically comes from base load power plants, traditionally plants that provide large quantities of electricity through steam generation systems or large hydro plants. As the load increases, utilities must bring in additional power sources from facilities that can be started up quickly, such as mid-merit or peaking plants. In some cases, such as where the peaking facilities are not available, or for economic reasons, it may be more cost-effective to purchase excess power from neighboring, or even distant utilities. In addition, for large grid systems, balancing strategies may be required to ensure uniform power flows that match variations in regional loads.

As noted in Sect. 3, introducing VRE into the grid adds a new dimension of complexity to utility and system operators striving to maintain uniform and high quality electricity flows during changing load conditions. VRE resources normally cannot be dispatched in the same way as base load, mid-merit and peaking plants, because VRE resources are only available when weather conditions allow. Furthermore, the VRE supply is constantly changing, but not necessarily in correlation with load changes. In order to make maximum use of these VRE resources, the utilities must also be able to predict very accurately when these resources are available to the grid in time frames that allow the utility to dispatch these resources to meet loads. Thus, accurate solar and wind resource forecasting has become a key topic for energy meteorologists; these forecasts can have a significant impact on how utilities manage VRE in terms of keeping traditional "spinning reserves" available, and in the infrastructure investments required to add VRE to the existing grid system.

Extensive research is currently underway within both the wind and solar communities to develop reliable forecasts over key timescales that can be used by utility and system operators to manage loads with VRE resources on the grid. For example, the NREL Western Wind and Solar Integration Study (GE 2010) highlights a key finding that large penetrations of VRE can be incorporated into the existing western

U.S. grid infrastructure, provided that a number of key steps are taken; one of the most important of these steps is to incorporate the state-of-the-art resource forecasting tools that allow utilities to manage loads more effectively with VRE.

An international group of experts is currently working through the IEA's SHC Program to examine the state-of-the-art forecasting capabilities for various forecasting time frames. This group began its work in 2005 under Task 36 "Solar Resource Knowledge Management," and recently extended this work in a new Task 46 titled "Solar Resource Assessment and Forecasting" (http://task46.iea-shc.org). Within this task a variety of solar resource forecasting methodologies, ranging from on-site all sky camera technologies to satellite-derived cloud motion vectors (CMV) and global and mesoscale numerical weather prediction (NWP) models are being investigated in ways that provide guidance to the industry as to which combination of approaches results in the lowest risk to utilities in terms of relying on these forecasts to manage VRE in their grid. Although the utility ultimately requires forecasts of system output characteristics from minute-ahead to day-ahead time frames, the foundation behind such forecasts is the weather prediction tools that simulate cloud motions and cloud formation and dissipation patterns to provide solar resource forecasts that are used as input to system performance models.

Table 1 is a useful way of summarizing our current knowledge and R&D priorities regarding best practices in solar resource forecasting over various timescales. Very short-term forecasts (<1-h) are important to utilities for load-following; hours- to day-ahead forecasts are required for dispatch and balancing, and longer term forecasts can be useful for longer term scheduling of system operations and system maintenance planning. The next subsections summarize key research underway, and reported through Task 46 activities, in each of the three timescales shown in the Table.

4.2.1 Sub-Hourly Forecasting

Current research efforts for providing sub-hourly solar forecasts, which are needed by utilities to address load-following strategies using VRE resources on the grid, focus on ground-based real-time observations, such as radiometric networks or sky imagers in the vicinity of large-scale PV or CSP systems. By providing observations at a high frequency, trends observed from these systems can be projected forward to alert system operators of impending changes in solar resources. For example, the University of California at San Diego (UCSD) is using both digitized cloud images captured by Total Sky Imagers (TSI) as well as arrays of radiometric observations to develop forecast tools for predicting immediate changes that can be expected in the solar resource at a given location. Ghonima et al. (2012) show successful use of total sky imagers for accurately classifying clear and thick clouds, and for improving detection of thin clouds, which is required for accurate short-term irradiance forecasting. Bosch et al. (2013) have presented two methods for estimating cloud speed from an array of eight radiometric stations, showing

Table 1 Current state of solar forecasting across timescales, and key development challenges

Forecasting timescale	Source of information/ methodology	Development challenges
Sub-hourly	Ground-based observations • Radiometers • TSI	TSI based cloud advection models
	• Visual observations	Statistical models based on high-resolution surface measurements
1–6 h	CMV from satellites Numerical Weather Prediction (NWP) models	NWP model wind profiles for better cloud advection

	• Global
	  - National Digital Forecast Database (NDFD)
	  - GFS
	  - European Centre for Medium-Range Weather Forecasting (ECMWF)
		Statistical cloud formation and dissipation models derived from easily measured parameters (e.g. humidity, temperature etc.)

	• Regional
	  - Rapid Update Cycle (RUC)
	  - GEM
	  - North American Model (NAM)
	• Mesoscale: Weather Research and Forecasting (WRF) model	

1–3 days	Solar Forecasts not directly available, but can be derived from NWPs	Capability to assimilate surface radiation into mesoscale model runs (MOS)
	• NWP plus WRF • HRRR • MOS • ANN	Improved cloud formation and dissipation physics

that the high variability expected under partly cloudy conditions can be detected using this method for use in short-term PV power system forecasting.

MINES ParisTech, located in Sophia Antipolis in southern France, is working in collaboration with the energy company EDF on a prototype of a low-cost fisheye camera developed and set up by EDF's R&D group (Fig. 9). This approach is meant to provide an alternative to total sky imagers for local and very short-term solar forecasting. Their preliminary results demonstrate a very important prerequisite for using fisheye cameras for solar forecasting purposes: establishing a relationship between the hemispheric sky images obtained by the imagers and the different components (global, diffuse, and direct) of the surface solar irradiance. This study has been carried out with high quality radiometric data and hemispheric sky images from an EDF R&D test site on Reunion Island, in the Indian Ocean. This experimental study leads to very conclusive and promising results on sub-hourly estimation of diffuse, direct, and global irradiance from sky images provided by the fisheye camera (Gauchet et al. 2012).

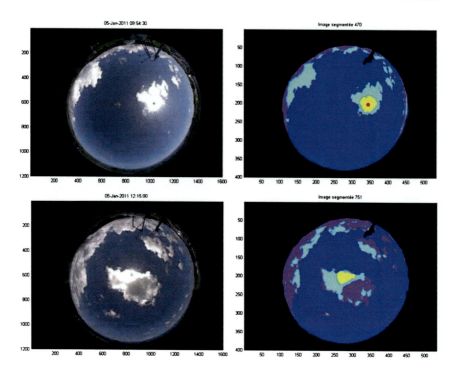

Fig. 9 Examples of clouds/clear sky segmentation result (*right*) for two hemispheric sky images (*left*). From Gauchet et al. 2012

4.2.2 One- to Six-Hour Forecasting

Geostationary weather satellites provide high-resolution visible channel images every 15–30 min that allow users to develop CMVs. CMVs are derived by interpolating the cloud images collected at each observation period down to 1 min values. These CMVs can then be extrapolated or projected out into the future to predict where clouds will occur for up to 6 h or more. These projected images can then be converted into ground-level solar resource forecasts. Descriptions of this approach can be found in Lorenz et al. (2007) and Perez et al. (2010).

4.2.3 One- to Three-Day Ahead Forecasting

For longer term forecasts it is generally found that the accuracy of the CMV scheme drops off to the point where, after 6 h, NWP models are typically more accurate in providing solar resource forecasts. There are several types of NWP models: global models, such as those provided by the European Centre for Medium-Range Weather Forecasts (ECMWF) and the U.S. Global Forecast System (GFS), regional or country-specific models such as the Global Environmental

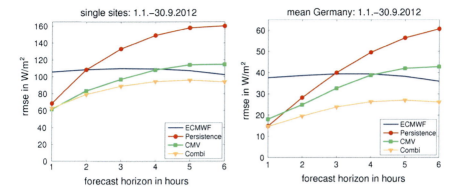

Fig. 10 RMSE of ECMWF based, CMV and combined forecasts in comparison to persistence in dependence on the forecast horizon. For the ECMWF-based forecast, the 12:00 UTC run of the previous day is evaluated independent of the forecast horizon and variations of the RMSE with the forecast horizon are due to the horizon-dependent data sets. *Left* single sites, *right* German mean (from Lorenz et al. 2012)

Multiscale (GEM) model in Canada and the High-Resolution Rapid Refresh (HRRR) Model in the U.S., and Mesoscale Models such as the Weather Research and Forecasting (WRF) model. These models can be run in a variety of configurations, and can also be coupled with ground observational data to produce Model Output Statistics (MOS). Besides these modeling approaches, several research institutions are developing methods using Artificial Neural Networks (ANN) and other machine-learning techniques.

Under IEA/SHC Tasks 36 and 46 a number of collaborative studies have been undertaken by researchers in Germany, Switzerland, Spain, Canada, and the U.S. on the accuracy and validity of forecasts derived from CMVs when compared against longer term NWP models. The general conclusions coming out of these studies is that CMVs can actually provide more reliable forecasts out to about 4–6 h into the future when compared with NWP models. Furthermore, Perez et al. (2010) have shown that the performance of the ECMWF model appears to be somewhat better (in terms of reducing RMSE when compared with ground-based high quality SURFRAD solar measurements in the U.S.) than the GFS for forecasting solar irradiance. Recent work at the University of Oldenburg (Lorenz et al. 2012) further indicates that there is an approximately 10 % improvement (as measured by reduction in RMSE) in solar forecasts when the basic ECMWF model runs are enhanced by a mesoscale numerical model. However, as shown in Fig. 10, which summarizes the reliability of solar forecasts in Germany, even by enhancing global models such as ECMWF with mesoscale models, the CMV approach still appears to be superior out to about 4 h.

5 Summary and Conclusions

The goal of any renewable energy target is to substantially increase the share of renewable energy supply in a transmission system. However, wind and solar energy resources, because of their dependence on weather phenomena, are variable in nature, adding complexities to utility and system operators in optimizing their use in the system. This chapter first demonstrates the challenges that VRE sources can create when grid operators are trying to meet changing load demands. Just as variations in loads themselves must be accurately predicted by system operators to ensure that they are met in a timely manner and in a way that minimizes degradation in energy quality or reliability, the output of VREs operating within the system must also be predicted in order to optimize their use and to avoid disruption of grid operations. Some utility systems have more flexibility than others to manage VRE, depending on the availability of mid-level or peaking power stations that can be quickly brought on-line in case VREs should suddenly drop off.

However, when one investigates the meteorological characteristics associated with large amounts of VRE operating in a grid, several key mitigating factors are observed that help to reduce potential disruptions to grid operations. First, it is well demonstrated that variability (or ramp) characteristics are mitigated with larger or distributed systems, due to the rapid drop off in cross-correlations of resource levels with distance for solar resources. In some circumstances, such as in a subtropical environment characterized by trade wind cumulus clouds, solar ramp rates can be mitigated even for power stations spaced less than 500 m apart, due to the nature of the cloud decorrelation statistics. This is good news for utilities adopting a distributed generation philosophy, or even when operating large, multi-megawatt central station wind and solar plants.

Second, the impacts of VRE on grid operations can be further mitigated if the variability can be accurately predicted over timescales ranging from minutes to a day ahead. Recently, great progress has been made in producing forecasts that could ultimately be of sufficient accuracy to allow utilities to dispatch VRE in order to meet fluctuating loads. In the case of solar forecasting, the use of ground observations (e.g., all sky cameras, total sky imagers, and radiometer networks) appears to be a very promising approach for providing system operators with sub-hourly forecasts. Out to about a 6-h forecasting period, CMV derived from time series of geostationary weather satellite observations in the visible channel appears to be a promising tool. Beyond the 4–6 h forecasting timeline, NWP models are the most promising approach. To further reduce the uncertainty in the forecasts, regional or global-scale NWP models can be downscaled with mesoscale model outputs. Model outputs can be further improved by incorporating ground measurements to produce MOS, or by adopting the use of ANN. These concepts and techniques are explained in more detail in the chapter by Lorenz et al.

Nevertheless, historically solar resources have not been an explicit output of NWP models. Although the technology for using NWPs in wind forecasts has advanced considerably in the past few years, further research is required to improve cloud characterization, and therefore solar radiation estimates, in the solar forecasting tools. These are some of the highest priority R&D efforts underway by energy meteorologists.

References

Bosch JL, Zheng Y, Kleissl J (2013) Deriving cloud velocity from an array of solar radiation measurements. Sol Energy 87:196–203
Gauchet C, Blanc P, Espinar B (2012) Surface solar irradiance estimation with low-cost fish-eye camera. In: COST WIRE workshop on remote sensing measurements for renewable energy, DTU Risoe, Denmark
GE Energy (2010) Western wind and solar integration study. Prepared for the National Renewable Energy Laboratory. 1 River Road, Schenectady, NY. Technical Monitor: Debra Lew. Subcontract No. AAM-8-77557-01. http://www.nrel.gov/docs/fy10osti/47781.pdf
Ghonima MS, Urquhart B, Chow CW, Shields JE, Carzola A, Kleissl J (2012) A method for loud detection and opacity classification based on ground based sky imagery. Atmos Measur Technol Discuss 5:4535–4569 http://www.atmos-meas-tech-discuss.net/5/4535/2012/amtd-5-4535-2012.html
Hinkelman LM (2013) Differences between along-wind and crosswind solar irradiance variability on small spatial scales. Sol Energy 88:192–213
Hoff TE, Perez R (2010) Quantifying PV power output variability. Sol Energy 84(10):1782–1793
International Energy Agency (IEA) (2011) Harnessing variable renewables: a guide to the balancing challenge. http://www.iea.org/publications/freepublications/publication/Harnessing_Variable_Renewables2011.pdf
International Energy Agency (2010) Technology roadmap: solar photovoltaic energy. http://www.iea.org/publications/freepublications/publication/pv_roadmap.pdf
Kankiewicz A, Sengupta M, Mood D (2010) Observed impacts of transient clouds on utility-scale PV fields. Presented at solar 2010, Phoenix, AZ. In: Campbell-Howe R (ed) Proceedings of 39th ASES national solar conference, American Solar Energy Society, Boulder, Colorado
Lorenz E, Heinemann D, Wickramarathne H, Beyer HG, Bofinger S (2007) Forecast of ensemble power production by grid-connected PV Systems. In: Proceedings of 20th European PV conference, Milano
Lorenz E, Kühnert J, Heinemann D (2012) Short term forecasting of solar irradiance by combining satellite data and numerical weather predictions. In: Proceedings of 27th EUPVSEC, Frankfurt, pp 4401–4405, 25–27 Sept 2012
Marquis M, Wilczak J, Ahlstrom M, Sharp J, Stern A, Smith JC, Calvert S (2011) Forecasting the wind to reach significant penetration levels of wind energy. Bull Am Met Soc 92(9):1159–1171
Perez R, Kivalov S, Schlemmer J, Hemker K, Renné D, Hoff TE (2010) Validation of short and medium term operational solar radiation forecasts in the US. Sol Energy 84(12):2161–2172
Global Status Report (2011) Published by the Renewable Energy Policy Network for the 21st Century
Global Status Report (2012) Published by the Renewable Energy Policy Network for the 21st Century. http://www.ren21.net/default.aspx?tabid=5434
Global Status Report (2013) Published by the Renewable Energy Policy Network for the 21st Century.http://www.ren21.net/REN21Activities/GlobalStatusReport.aspx
Stoffel T, Renné D, Myers D, Wilcox S, Sengupta M, George R, Turchi C (2010) Concentration solar power: best practices handbook for the collection and use of solar resource data. NREL/TP-550-46475, National Renewable Energy Laboratory, Golden. http://www.nrel.gov/docs/fy10osti/47465.pdf

Current Status and Challenges in Wind Energy Assessment

Sven-Erik Gryning, Jake Badger, Andrea N. Hahmann and Ekaterina Batchvarova

Abstract Here we discuss the status and challenges in the development of atlases for the assessment of the regional and global wind resources. The text more specifically describes a methodology that is under development at DTU Wind Energy in Denmark. As the wind assessment is based on mesoscale modelling, some of the specific challenges in mesoscale modelling for wind energy purposes are discussed such as wind profiles and long-term statistics of the wind speed time series. Solutions to these challenges will help secure an economic and effective deployment of wind energy.

List of Acronyms

4DVar	Four-dimensional variational
AGL	Above ground level
ASTER	Advanced Spaceborne Thermal Emission and Reflection Radiometer
BouLac	Bougeault-Lacarrère PBL
CFDDA	NCAR Climate four-dimensional data assimilation
CFSR	NOAA Climate forecast system reanalysis
CORINE	Coordination of information on the environment
DEM	Digital Elevation Map
DTU	Technical University of Denmark
EU	European Union
EEA	European Environment Agency
ECMWF	European Centre for Medium-Range Weather Forecasts
ERA Interim	ECMWF Interim Reanalysis

S.-E. Gryning (✉) · J. Badger · A. N. Hahmann · E. Batchvarova
DTU Wind Energy, Technical University of Denmark, Risø Campus,
4000 Roskilde, Denmark
e-mail: sveg@dtu.dk

E. Batchvarova
National Institute of Meteorology and Hydrology, Sofia, Bulgaria

ESA	European Space Agency
FDDA	Four-dimensional data assimilation (nudging)
FINO	German Research Platform in the North and East Sea (Forschungsplattformen in Nord- und Ostsee)
FINO1	FINO platform 1 (German North Sea)
FINO2	FINO platform 2 (German Baltic Sea)
FNL	NCEP Final Analysis
FROGFOOT	Program for WAsP calculations over large areas
GEOS5	NASA Goddard Earth Observing System Model, Version 5
GFS	NOAA Global Forecast System
HRSST	High-resolution SST
LCCS	Land cover classification system
MERRA	Modern Era Retrospective Analysis for Research and Applications
MM5	Mesoscale Model Version 5 (predecessor of WRF)
MYJ	Mellor-Yamada-Janjic PBL scheme
MYNN2	Mellor-Yamada-Nakanishi-Niino Level 2 PBL scheme
MYNN3	Mellor-Yamada-Nakanishi-Niino Level 3 PBL scheme
NASA	National Space Administration (USA)
NCAR	National Center for Atmospheric Research (USA)
NCEP	NOAA National Center for Environmental Prediction (USA)

1 Introduction

This chapter is split into two parts. The first part addresses the pressing need to improve the assessment of the global wind resource. These challenges encompass many aspects of wind energy modelling. The solutions will place wind energy more strategically in the global energy picture for the future. The text describes a methodology for the Global Wind Atlas.

The second part addresses specific challenges in mesoscale modelling for wind energy purposes. This aspect is of particular importance for an accurate assessment of national and site wind resource.

For a general textbook like description of the basics of, e.g. wind structure near the surface, impact of stability, diurnal and spatial variability as well as the effect of averaging time, the reader is referred to Stull (1998) for general and Emeis (2012) for wind energy specific descriptions.

2 Global Wind Resources

The current status and coverage of wind assessment around the world is a collection of more or less *ad hoc* studies, using a broad range of methods, and in turn providing resource products with different specifications and types. The incomplete coverage is natural enough, as wind resource assessments—usually made on a country-wide scale or smaller—have followed needs and motivations on a country-by-country basis, and these are very much dependent on each individual case. The broad range of methods and product types is the result of the number of research centres and companies that are engaged in wind energy assessment, and the rapid development of new methods. Furthermore, the degree to which wind assessment data are open and freely available, as well as the extent to which the methodology is transparent and subject to scrutiny by the wind resource assessment community, is also disparate.

Because of the incomplete assessment of wind resource over the world, policy makers and energy planners have been forced into using coarse resolution global reanalysis data to estimate wind resources. This has a very serious drawback as hills and ridges are better resolved at higher resolutions giving rise to increased wind speeds. Consequently, coarse resolution leads to an erroneous negative bias in the wind resource.

Consequently, the role of wind energy in the future energy mix may be downplayed, with grave implications for modelling approaches to climate change mitigation.

2.1 Wind Resource Assessment

The term, wind resource assessment covers a very broad range of methods and many kinds of data. For example, the assessment can be based on *in situ* measurements and as such pertain to the measurement location and height only, unless some kind of treatment of the measured winds is carried out in order to extend its validity spatially. At the other end of the range, the assessment may be based on modelling, giving wind resource in three dimensions. However, the value of such model-derived assessment is limited without some kind of verification against measurements.

Therefore, the most valued wind resource assessment will feature a combination of measurement and modelling. For example, the European Wind Atlas (Troen and Petersen 1989) developed a pioneering method to analyse *in situ* measurements in such a way that the information obtained from the measurements can be applied away from the measurement location. The analysis is done by modelling the effects due to local changes in terrain elevation, local surface roughness changes, and obstacles, each of which impacts the measured winds. The result of the analysis is a generalised wind climate. To predict the wind resource at a new site requires the application of the same aforementioned models (calculating effects due to local changes in terrain elevation, local surface roughness changes and obstacles) on a generalised wind climate. This method comprises the workings of the Wind Atlas Analysis and Application Program (WAsP, www.wasp.dk) software developed by DTU Wind Energy and now used by over 10,000 users worldwide.

Wind resource assessment of the kind outlined above required a dense network of high quality and long-term measurements. This is because a generalised wind climate is only valid for a limited area. Where good quality measurement data is missing, which is more often the case, numerical wind atlas methodologies are used. The conventional numerical wind atlas uses long-term, but coarse resolution, atmospheric datasets (e.g. reanalysis from NCEP/NCAR, Kalnay et al. 1996) to force mesoscale models, capable of modelling the atmospheric flow at scales ranging approximately from 100 to 5 km. This comprises a so-called downscaling technique. From the mesoscale model simulations, maps of wind resource can be created. In the method developed at DTU Wind Energy, post-processing of the simulations results in a grid of generalised wind climates, which can be used in WAsP. The huge advantage of creating the generalised wind climates is that these can be compared to generalised wind climates derived from measurements in the region of interest. Even if there is a limit to the number of high quality measurements, this comparison of model and measurement derived climate allows for a verification of model results. A proper verification adds tremendous value to a wind resource assessment.

So far, the methods outlined above have been used in numerous locations around the world; most recently in India, northeastern China, and South Africa. However, up to this point, no single unified wind resource assessment has been performed for the whole world, and it is important to note the objective is not to perform a global version of these country-specific studies. A new method is required to generate the Global Wind Atlas, making it efficient to create, and suitable for the needs of the policy makers and energy planners. This is only now becoming a possibility due to developments in global reanalysis datasets, global topographical datasets and microscale modelling tools.

The method underway (within the Danish Energy Agency funded Wind Atlas project) to create the improved global assessment of wind is made up of a chain of processes in which global reanalysis datasets are the meteorological input data and high-resolution wind climate statistics, suitable for analysis and mapping, are the output data.

Global reanalysis datasets with a spatial resolution of around 50 km are now available, see Table 1. These datasets are at a much higher resolution than previously available, compared to, for example NCEP/NCAR reanalysis of $\sim 2.5° \times \sim 2.5°$ (Kalnay et al. 1996), and thus give new possibility for their exploitation for wind resource assessment. A number of reanalysis datasets can be used to investigate the range of wind climates that a set provides. The reanalysis datasets are not wholly independent as the same observational data network is available for assimilation; however, the manner in which assimilation is performed is different, as are the models underlying the reanalysis. For example, there will be differences in the physical parameterizations modelling sub-grid scale processes and surface processes, as well as the description of the surfaces.

Because of their different characteristics, the reanalysis datasets tend to complement each other in providing the best description of the wind climatology within the boundary layer. Because of its higher resolution, CFDDA will probably

Table 1 Description of reanalysis datasets (see also http://reanalyses.org)

Product	Model system	Horizontal resolution	Period covered	Temporal resolution
ECMWF/ERA interim	T255, 60 vertical levels, 4DVar	78 km	1989–present	3/6-hourly
NASA/MERRA	GEOS5 data assimilation system (incremental analysis updates), 72 levels	$0.5° \times 0.67°$	1979–present	3-hourly
NCAR/CFDDA	NCAR/PSU MM5 (regional model) + FDDA	~ 40 km	1985–2005	hourly
NOAA/CFSR	NCEP GFS, ~ 38 km (T358) with 64 vertical levels	$0.5° \times 0.5°$	1979–2010 and updating	hourly

represent the mesoscale processes best. The high vertical resolution of MERRA is a plus for resolving the low-level wind structure, and its long record is a plus to accurately represent inter-annual variations in regional wind climate. Finally, because of the assimilation of a large amount of data, ERA Interim is potentially best in depicting the overall atmospheric circulation. Employing several reanalysis datasets as input can give ensembles of surface wind climates and may lead to a better resource estimates and provide an indicator of uncertainty.

Surface winds given by the reanalysis datasets should not themselves be used directly to estimate global wind resources because the spatial resolution is still too coarse. Spatial variance of wind climates at scales smaller than that resolved in the reanalysis data will contribute significantly to the wind resource. The small-scale spatial variance of wind speeds can be modelled by microscale models. Running microscale models requires that the reanalysis surface winds are treated in such a way to make them generalised winds. Differences in surfaces winds given by three reanalysis datasets can in part be explained by differences in the surface roughness lengths used in each reanalysis model. The objective of generalising the surface winds is to remove the influences of model dependent surface description. Generalised wind climate statistics give the wind conditions for a standard set of heights above the surface and surface roughness lengths. Global generalised wind climate grids will be created, containing sector-wise (*directional*) frequency distribution and sector-wise (*directional*) wind speed distributions.

Microscale modelling over large extents is now possible using the generalised wind climate statistics generated from the reanalysis datasets. A new functionality within the WAsP system, called FROGFOOT, allows resource calculation over a large, high-resolution grid to be performed, see Fig. 1. Each resource calculation uses generalised wind climate statistics from the nearest reanalysis grid points. For WAsP to calculate the local flow at high resolution, high-resolution data of terrain elevation and surface roughness length are also needed.

Fig. 1 Cape Verde numerical wind atlas in FROGFOOT. Each *orange dot* represents a data point with details sector-wise generalised wind climates statistics. FROGFOOT allows wind resources at high resolution to be calculated in WAsP using the coarser grid of generalised wind climate data points. Only part of Cape Verde is shown here

2.2 Topographical Datasets

Modelling of the wind resource on sub-kilometre scales (say, 100–500 m) requires detailed information on the terrain elevation, water body distribution and land cover of the terrain. Several global or near-global datasets are available that could be used for this purpose:

- *Elevation—Shuttle Radar Topography Mission (SRTM), version 2.1, released 2009*: SRTM data are available as grid point spot heights with a resolution of 1 arc-second (continental USA) and three arc-seconds (from 56°S to 60°N, 80 % of the Earth's land surface). Derivative datasets exist, where data voids have been filled. The SRTM datasets are used extensively for wind resource assessment already and they are easy to download, process and transform (e.g. to height contour maps).
- *Elevation—ASTER Global Digital Elevation Model (ASTER GDEM), version 1, released 2009*: ASTER data are available as grid point spot heights with a resolution of one arc-second from 83°S to 83°N (99 % of the Earths land surface). Less experience exists using ASTER data for wind resource assessment but they are in principle easy to download, process and transform. Version 1 of the ASTER GDEM is still viewed as "experimental" or "research grade".

- *Coastline contours—SRTM Water Body Data (SWBD)*: Version 2 of SRTM also contains the vector coastline mask derived by NGA during the editing, called the SWBD, in GIS Shapefile format. SWBD data cover the Earth's surface between 56 °S and 60 °N; the rest of the world is available at lower resolution through the Coastline Extractor hosted by NOAA.
- *Land cover—ESA GlobCover, version 2.1, released 2008:* GlobCover is the highest resolution (300 m) global land cover product ever produced and it is made available to the public by the European Space Agency (ESA). The GlobCover land cover map is compatible with the UN Land Cover Classification System (LCCS).
- *Land cover—regional databases*: Several regional and national land cover datasets exist which may be more detailed and sometimes more readily applicable to microscale flow modelling. As an example, the European Environmental Agency (EEA) has produced vector and raster land cover databases—CORINE—for the 25 EU Member States and other European countries.

2.3 Importance of Resolution

Figure 2 shows the effect of modelling the wind power density at a height of 50 m for a 50 × 50 km area at four different resolutions, namely 10, 5, 2.5 km and 100 m. As the resolution increases features in the terrain become better resolved. Resolved hills and ridges give rise to increased wind speeds. As wind power density is a function of wind speed to the power 3, the impact of the resolved terrain features is significant.

For the 50 × 50 km area, the mean wind power density is estimated to be around 320 Wm^{-2} for the resolutions of 10, 5 and 2.5 km. For the 100 m resolution, the mean power density is around 505 Wm^{-2}, i.e. an increase of 50 % compared to the lower resolution estimates.

The comparison becomes more striking when the distribution of the wind power density is considered. Consider this: we split the 50 × 50 km into two areas, the first area where the wind power density is below the median value and the second area where the wind power density is above the median value. Next, we calculate the mean wind power density in the second higher wind area, we get for the 10, 5, 2.5 km resolution estimates 380 Wm^{-2}, whereas for the 100 m resolution we get 640 Wm^{-2}, an increase of nearly 70 %. The impact gets stronger as we look at the even windier areas. As wind turbines will be deployed at the favourable sites, it is important to be able to capture the distribution of wind power density due to terrain features, and this is only possible by consideration of high-resolution effects.

Even for rather simple terrain, such as in Denmark, the effect of resolution is important. A similar 50 × 50 km area showed an increase of wind power density of around 25 % for the windiest 5-percentile (windiest 1/20th of the area).

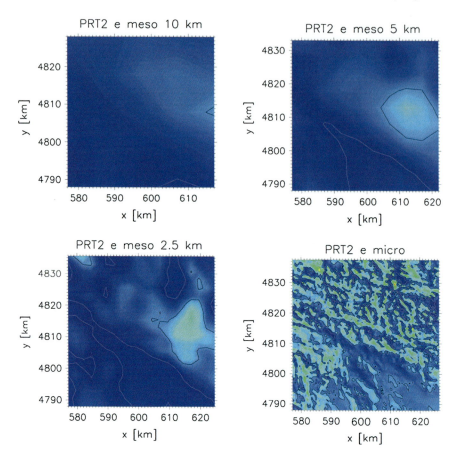

Fig. 2 Wind power density calculated at a height of 50 m for a 50 × 50 km area at four different resolutions. *Top-left* 10 km, *top-right* 5 km, *bottom-left* 2.5 km mesoscale modelling and *bottom-right* microscale modelling at 100 m. The colour scale is the same in all the maps

2.4 Verification

The Global Wind Atlas is not a substitute for the detailed country wind resource assessments. Therefore, it is not expected that the Global Wind Atlas result will correspond exactly to the detailed country studies. However, the Global Wind Atlas result will be a wind resource estimate and an uncertainty estimate, and thus comparison with dedicated country studies should show agreement subject to the margins given by the uncertainty estimates.

Validation of the Global Wind Atlas offshore will be possible via wind data derived by remote sensing techniques applied on satellite images of oceans. Various products are available for this purpose ranging from the now discontinued QuikSCAT, to current Synthetic Aperture Radar (SAR) platforms. Risø DTU has extensive experience in generating wind climate statistics from satellite-derived winds. The manner in which the verification is carried out is open to development, however one possibility is to perform two sets of verification; one global at low resolution using QuikSCAT, the other at selected coastal regions at high resolution using the new satellite image derived wind products. As in the comparison with detailed country wind resource assessments, an agreement within the uncertainty estimate is sought.

3 Dynamical Downscaling for Wind Applications Using Meteorological Models

From the foregoing chapter on the mapping of the global wind energy resources, it is clear that the performance of mesoscale models is the backbone of all wind energy assessment studies. Here, we discuss some aspects of mesoscale modelling of the wind profile and the Weibull distribution of the long-term wind time series, the sensitivity of the driving reanalysis, the Planetary Boundary Layer schemes (PBL) by showing examples of data comparison with simulations performed with limited area Numerical Weather Prediction (NWP) models, which are now being widely used to estimate wind energy resources in areas with inadequate observational coverage. In principle, these models have been optimised to obtain skilful weather prediction of fields such as temperature and precipitation in the short (0–72 h) term. In addition, models such as the Weather Research and Forecasting (WRF) model (Skamarock et al. 2008) contain what appears to be an infinite set of model configuration options that need to be set by the user and which can sometimes alter dramatically the behaviour of the model solution. With this in mind, how skilful these models are in simulating the full spectrum of atmospheric motion necessary to describe the regional wind climate is still a topic of active research. Here, we provide examples of the use of WRF model output for generating wind resource atlases and point out how a few configuration options can alter the end result of the dynamical downscaling. The model simulations are compared to basically two experimental datasets and the corresponding simulations, one deals with offshore and coastal wind profiles up to 100 m and the other is drawn from measurements of the wind profile up to 600 m performed in a coastal site with a long-range wind lidar.

For estimating wind energy resources, the mesoscale model simulations are not conducted in the same way as weather forecasting, and special considerations are required for the model spin-up. The large-scale reanalysis drives the downscaling, we use the mesoscale models to resolve smaller scales not present in the reanalysis. The chosen model set ups are based on these assumptions, which are described in Hahmann et al. (2010).

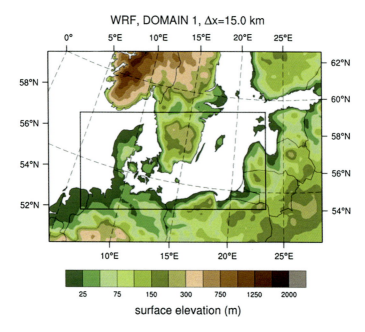

Fig. 3 WRF model configuration and surface elevation (*meters*) used in the simulations. The *inner square* shows the location of domain 2 (*D2*, 5 km grid spacing)

3.1 OffShore and Coastal Wind Profiles, Case Study 1

For this study, measurements are drawn from two offshore research platforms, FINO 1 (N 54° 01′, E 06° 35′) and FINO 2 (N 55° 0′, E 13° 09′) located in the North Sea and Baltic Sea, respectively, as well from a coastal site at Høvsøre (N 56° 3′, E 08° 09′) at the west coast of Denmark.

3.1.1 Model Setup

The model grids are 15 km × 15 km (D1); 5 km × 5 km (D2) grid spacing, and uses WRF version 3.2.1. The model is configured with 41 vertical levels with model top at 50 hPa; 12 of these levels are within 1,000 m of the surface; the first level is located at ∼14 m AGL. The simulations are forced by the USA NOAA reanalysis (CFSR; Saha et al. 2010) at 0.5° × 0.5° horizontal grid spacing; a separate file containing SST and sea-ice fractions is used and comes from the dataset NCEP Version 2.0 global SST (OISST; Reynolds et al. 2002) with 0.25° × 0.25° resolution, and are updated daily. A map of the model setup location is shown in Fig. 3.

The control simulation is 5-years long (2006–2010) and it is run in 11-day long overlapping runs, with the first day discarded. The simulation uses grid nudging which continuously relaxes the model solution towards the gridded reanalysis but this is done only on D1 and above the boundary layer (level 10 from the surface) to allow for the mesoscale processes near the surface to develop freely. We compare next the wind climate generated by various model configurations.

3.1.2 Driving Reanalysis and Sea Surface Temperatures

To examine the wind climate sensitivity to the driving reanalysis, we repeated the control simulation using the ERA Interim reanalysis (Dee et al. 2011). The extra simulation was carried out during 2010. Figure 4 (*top*) shows the relative difference in annual mean wind speed between the two simulations. Differences are small in most areas with the largest of the order of ±3 % especially around the edges of the domain, where the influence of the reanalysis is expected to be the strongest.

The sensitivity to the lower boundary conditions used in the simulations was examined by conducting extra simulations where the OISST is replaced by the daily, high-resolution, real-time, global, sea surface temperature (RTG_SST) available at a horizontal grid spacing of 1/12°. In this area, the largest differences in SST (not shown) are seen along the west coast of Denmark and Sweden, with warmer (up to 2 °C in the annual mean) SSTs in the RTG_SST dataset. These are the same areas where the annual mean winds show the largest differences in the middle and bottom panels of Fig. 4.

Annual wind speeds are larger (negative areas) in the climatology generated from annually averaged warmer SSTs. But, it is interesting to see that the response in the wind speed to the different SSTs depends on the reanalysis used. The differences are larger for the ERA interim compared to the CFSR. In addition, it is expected that in areas where the wind regime is driven by sthermal contrast the sensitivity could be stronger.

3.1.3 Planetary Boundary Layer Parameterisation

Figure 5 shows the wind speed averaged over the period 1–30 June 2009 as a function of height for six PBL schemes available in WRF version 3.2.1 (see Table 2) and the observations at the offshore masts of FINO1 (German North Sea), FINO2 (Baltic sea) and the tall mast at Høvsøre (West coast of Denmark). The spread among PBL schemes is, as expected, lower for the two offshore sites than for Høvsøre, which is located over land. Differences between the model-derived wind speeds and the observations are of the order of 0.5 ms^{-1} at FINO1 and FINO2, but much larger ($\sim$ 1.5 m s^{-1}) at Høvsøre. When considering the "slope" and "shape" of the wind profile simulated by the models and compared to the observations, at FINO1 YSU performs best, at FINO2 MYJ is perhaps the best, and at Høvsøre the MYNN3 seems to outperform the others. Therefore, there is no

Fig. 4 Differences in mean wind speed (*percentage*) at 100 m during 2010 between various model configurations: ERA Interim—CFSR (*top*), HRSST—OISST under CFSR (*middle*) and HRSST—OISST under ERA Interim (*bottom*)

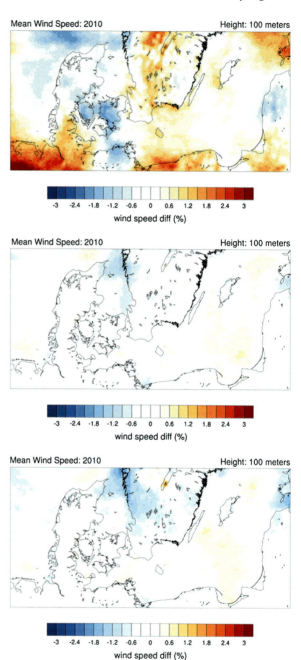

single choice of PBL scheme that is best both at sea and land sites. In addition, when examining other wind-related parameters such as vertical shear between models and the observations, the conclusion might be different.

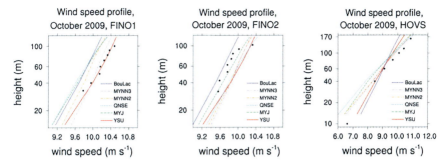

Fig. 5 Vertical profiles of wind speed (ms^{-1}) simulated by WRF using the six PBL parameterisations in Table 2 compared to measurements (*black dots*) at FINO1, FINO2 and Høvsøre. Profiles represent the average for the period 1–30 October 2009. Wind speeds whose directions are affected by wind farm wake (*Høvsøre*) or mast booms (*FINO1 and FINO2*) have been removed from both observations and models

Table 2 Description of the WRF PBL schemes used in the simulations

Scheme	Reference
YSU	Yonsei University scheme
MYJ	Mellor-Yamada-Janjic scheme
QNSE	Quasi-Normal Scale Elimination PBL
MYNN2	Nakanishi and Niino level 2.5
MYNN3	Nakanishi and Niino level 3
BouLac	BouLac PBL

Maps of the differences in wind speed from climatologies using different PBL schemes (not shown) show much larger values than those derived from using different driving reanalysis or SSTs. In particular, since the behaviour is different over land and ocean, systemic and distinct differences are seen over these areas.

3.2 Weibull Distribution at a Coastal Site, Case Study 2

For wind energy assessment, the distribution of the long-term wind speed is very important because the wind power density, approximately, is a function of the wind speed cubed and thus a large fraction of the wind energy will come from the high end of the wind speed distribution, which has a relative low probability. The two parameter Weibull distribution reveals this feature and therefore, is often used for the description of the long-term frequency distribution of the horizontal wind speed. This distribution has received considerable attention in relation to assessment of wind energy from meteorological observations.

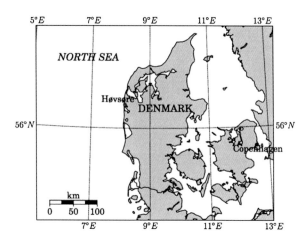

Fig. 6 Geographical location of the Høvsøre site

3.2.1 Site and Measurements

Detailed studies of the wind field at the Høvsøre site have already been reported by Floors et al. (2011, 2013) and Peña et al. (2009). Here, we focus on the statistical distribution of the long-term (annual) wind speed and its Weibull distribution. The measurements were carried out at the Danish National Test Station of Wind Turbines at Høvsøre, which is located at the western coast of Jutland, Fig. 6. Except for the presence of the North Sea to the west, the terrain is flat and homogeneous consisting of grass, various agricultural crops and a few shrubs. The intensively instrumented 116.5 m high reference meteorological mast is located about 1.8 km east of the coastline. Wind speed is measured at 10, 40, 60, 80, 100, 116.5 m with cup anemometers.

In addition, a pulsed wind lidar (WLS70) has been operating near the meteorological mast between April 2010 and March 2011. The wind lidars Doppler shift-based measurements of the wind are available measures from 100 m above the ground and every 50 m up till 1 to 2 km height dependent on the attainable 10-min averaged Carrier to Noise (CNR) ratio.

3.2.2 Numerical Modelling

Ten minutes wind profiles for the period 23 April 2010 to 31 March 2011 were predicted based on the WRF ARW model version 3.2.1. The physical options of model setup include the Noah land surface scheme (Skamarock et al. 2008) and the Thompson microphysics scheme (Thompson et al. 2004). The WRF model calculates the meteorological parameters at 41 vertical levels from the surface to pressure level 100 hPa. Eight of these levels are within the height range of 600 m that is analysed in this study and the first model level is at 14 m.

The model is run in analysis mode. It uses the NCEP Final Analysis (FNL) global boundary conditions that are available every 6 h on a 1° × 1° grid. Two domains with a horizontal grid size of 18 and 6 km, respectively, are used. The simulations are initialised every 10 days at 12:00 and after a spin up of 24 h a time series of 10-min simulated meteorological forecast data from 25 to 264 h is generated. In order to prevent the model from drifting away from the large-scale features of the flow, the model is nudged towards the FNL analysis. Nudging is applied for the wind, temperature and humidity above the 10th model level, which approximately corresponds to 1,400 m, on the outermost model domain during the whole simulation period.

3.2.3 Weibull Distribution

The Weibull distribution of the horizontal wind speed can be expressed as:

$$f(u) = \frac{k}{A}\left(\frac{u}{A}\right)^{k-1} \exp\left(-\left(\frac{u}{A}\right)^{k}\right), \qquad (1)$$

where $f(u)$ is the frequency of occurrence of the wind speed u. In the Weibull distribution the scale parameter A has units of the wind speed. It is proportional to the average wind speed $\langle u \rangle$ for the entire distribution. It is related to the wind speed through:

$$\langle u \rangle = A\Gamma\left(1 + 1/k\right), \qquad (2)$$

where Γ represents the gamma function and k is the shape parameter in the Weibull distribution: for typical wind speed distributions over homogeneous terrain k is in the range 2–3. For decreasing k the mode of the distribution shifts towards lower values of the wind speed at the same time as the probability for higher wind speeds increases.

From the measurements and simulations of the wind speed, the A and k parameters in the Weibull distribution were derived by use of the Climate Analyst which is a part of the Wind Atlas Analysis and Application Program (WAsP).

3.2.4 Scale Parameter

The comparison of the modelled A parameter with measurements shows similarities with the wind speed. Below 60 m the WRF model predicts well the A parameter. Above 60 m the simulated scale parameter underestimates the measured one with the difference increasing with height, Fig. 7.

Fig. 7 Profiles of the scale parameter in the Weibull distribution estimated from measurements and simulations

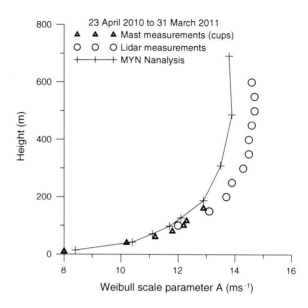

3.2.5 Shape Parameter

Contrary to the scale parameter, which has a rather smooth vertical profile, the shape parameter k has a very distinct form. Investigations over land (Hellmann 1917 and Wieringa 1989) have revealed that k is controlled by two regimes of the atmosphere, the large-scale wind climate and the local boundary layer. This results in a characteristic vertical profile of the shape parameter. It increases from its value near the ground up to a maximum located at around 100–200 m height, depending on the balance between the diurnal variation of the local meteorological conditions and the variability of the synoptic conditions prevailing in the region. The height of the maximum in the k-profile is associated with the height of the stable boundary layer as well as the reversal of the wind regime that occur on stable nights where the diffusion of momentum is inhibited, resulting in low near-surface winds while the wind speed above the stable boundary-layer increases. The range of k is not well investigated; typical values reported in the literature are between 1 and 4. Smaller values of k correspond to higher variability of the wind speed.

It is found that the WRF simulations agree well with the measurements up to 100 m, while above that height the model generally underestimates the k parameter (Fig. 8). The height of the maximum in the k-profile from measurements is about 200 m while it is lower for the model simulations, being about 120 m.

Fig. 8 Profiles of the shape parameter in the Weibull distribution estimated from measurements and simulations

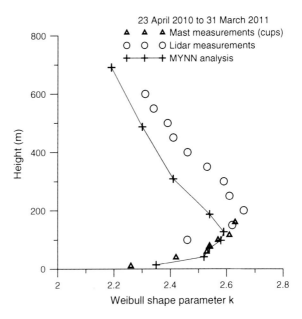

4 Summary and Outlook

The current practice of global energy modellers is to use coarse resolution reanalysis datasets. This has the serious shortcoming that the wind energy resource is underestimated, as small-scale variability of winds is missing. This missing variability is responsible for a large part of the wind resource, which is not captured by the reanalyses. A Global Wind Atlas could provide data at high resolution for the whole world, using a unified and transparent methodology, and an uncertainty estimate based on use of a number of input reanalysis datasets and methodologies. Crucially, the method should employ microscale flow modelling to calculate the small-scale variability of winds, and thus capture the sizable wind resource missed by the current methods.

The performance of the mesoscale models, such as the WRF model, is the backbone of all wind energy assessment studies at different scales. It is shown that the choices of driving reanalysis for the WRF model results in a difference in wind speed of maximum 3 %. Using different lower boundary conditions, results in differences of 1–2 % in the mesoscale wind climate. Using different PBL schemes, results in larger differences in the predicted wind speed compared to the driving and lower boundary conditions. Furthermore, none of the most common parameterisations provides the best solution for both ocean and land conditions. The 1.5 order closure MYNN PBL scheme used in the simulation of the Weibull distribution of the long-term wind speed predicts the general profile of the scale and shape parameters quite well although it underestimates their values.

Acknowledgements The study was supported by the Danish Council for Strategic Research, project number 2104-08-0025 named "Tall Wind", the EC FP7 project NORSEWIND (TREN-FP7EN-219048), and the Danish Energy Agency (grant EUDP 11-II, Globalt Vind Atlas, 64011–0347). The study is a part of activities within the COST Actions ES1002 and ES0702. We would like to thank the Test and Measurements section of DTU Wind Energy for the maintenance of the Høvsøre database.

References

Dee DP, Uppala SM, Simmons AJ, Berrisford P, Poli P, Kobayashi S, Andrae U, Balmaseda MA, Balsamo G, Bauer P, Bechtold P, Beljaars ACM, Berg L van de, Bidlot J, Bormann N, Delsol C, Dragani R, Fuentes M, Geer AJ, Haimberger L, Healy SB, Hersbach H, Hólm EV, Isaksen L, Kållberg P, Köhler M, Matricardi M, McNally AP, Monge-Sanz BM, Morcrette JJ, Park BK, Peubey C, Rosnay P de, Tavolato C, Thépaut JN, Vitart F (2011) The ERA-Interim reanalysis: configuration and performance of the data assimilation system. Q J Roy Meteorol Soc 137:553–597. doi:10.1002/qj.828

Emeis S (2012) Wind energy meteorology. Springer, Berlin pp 196

Floors R, Gryning S, Peña A, Batchvarova E (2011) Analysis of diabatic flow modification in the internal boundary layer. Meteorol Zeitschr 20:649–659

Floors R, Vincent CL, Gryning SE, Peña A, Batchvarova E (2013) Wind profiles in the coastal boundary layer: wind lidar measurements and WRF modelling. Boundary-Layer Meteorol 147:469–491

Hahmann AN, Rostkier-Edelstein D, Warner TT, Vandenberghe F, Liu Y, Babarsky R, Swerdlin SP (2010) A reanalysis system for the generation of mesoscale climatographies. J Appl Meteor Climatol 49:954–972. doi: http://dx.doi.org/10.1175/2009JAMC2351.1

Hellmann G (1917) Über die bewegung der luft in den unteren schichten der atmosphäre. Meteorol Zeitschr 34:273–285

Kalnay E, Kanamitsu M, Kistler R, Collins W, Deaven D, Gandin L, Iredell M, Saha S, White G, Woollen J, Zhu Y, Chelliah M, Ebisuzaki W, Higgins W, Janowiak J, Mo KC, Ropelewski C, Wang J, Leetmaa A, Reynolds R, Jenne Roy, Joseph Dennis (1996) The NCEP/NCAR 40-Year reanalysis project. Bull Amer Meteor Soc 77(3):437–471

Peña A, Hasager CB, Gryning S-E, Courtney M, Antoniou I, Mikkelsen T (2009) Offshore wind profiling using light detection and ranging measurements. Wind Energy 12:105–124

Reynolds RW, Rayner NA, Smith TM, Stokes DC, Wang WQ (2002) An improved *in situ* and satellite SST analysis for climate. J Climate 15:1609–1625

Saha S, Moorthi S, Pan H-U, Wu X, Wang J, Nadiga S, Tripp P, Kistler R, Woollen J, Behringer D, Liu H, Stokes D, Grumbine R, Gayno G, Wang J, Hou Y-T, Chuang H-Y, Juang H.M, Sela J, Iredell, Treadon R, Kleist D, Delst P van, Keyser D, Derber J, Ek M, Meng J, Wei H, Yang R, Lord S, Dool H van den, Kumar A, Wang W, Long C, Chelliah M, Xue Y, Huang B, Schemm J-K, Ebisuzaki W, Lin R, Xie P, Chen M, Zhou S, Higgins W, Zou C-Z, Liu Q, Chen Y, Han Y, Cucurull L, Reynolds R, Rutledge G, Goldberg M (2010) The NCEP climate forecast system reanalysis. Bull Amer Meteor Soc, 91:1015–1057. doi: http://dx.doi.org/10.1175/2010BAMS 3001.1

Skamarock WC, Klemp JB, Dudhia J, Gill DO, Barker DM, Duda MG, Huang X-Y, Wang W, Powers JG (2008) A description of the advanced research. *WRF version 3*. NCAR/TN–475 + STR, NCAR Technical note, Mesoscale and microscale meteorology division, National center for atmospheric research, Boulder, Colorado, USA, p 113

Stull R (1998) An introduction to boundary layer meteorology. Kluwer Academic Publishers, Dordrecht, The Netherlands, p 666

Thompson G, Rasmussen RM, Manning K (2004) Explicit forecasts of winter precipitation using an improved bulk microphysics scheme, part I: description and sensitivity analysis. Mon Weather Rev 132(2):519–542

Troen I, Petersen EL (1989). European wind atlas. Published for the commission of the european communities directorate-general for science, reserach and development, Brussels, Belgium for Risø National Laboratory, p 656

Wieringa J (1989) Shapes of annual frequency distribution of wind speed observed on high meteorological masts. Bound-Layer Meteorol 47:85–110

Wind Power Forecasting

Sue Ellen Haupt, William P. Mahoney and Keith Parks

Abstract The National Center for Atmospheric Research (NCAR) has configured a Wind Power Forecasting System for Xcel Energy that integrates high resolution and ensemble modeling with artificial intelligence methods. This state-of-the-science forecasting system includes specific technologies for short-term detection of wind power ramps, including a Variational Doppler Radar Analysis System and an expert system. This chapter describes this forecasting system and how wind power forecasting can significantly improve grid integration by improving reliability in a manner that can minimize costs. Errors in forecasts become opportunity costs in the energy market; thus, more accurate forecasts have the potential to save substantial amounts of money for the utilities and their ratepayers. As renewable energy expands, it becomes more important to provide high-quality forecasts so that renewable energy can carve out its place in the energy mix.

1 The Need for Renewable Energy Power Forecasts

Because wind and solar resources are highly variable, it is important to be able to forecast them in order to effectively and economically integrate them into the electrical power grid. Although the details of how utilities and power grids integrate these variable resources vary throughout the world, there is a critical need to predict the timing and magnitude of the power output from these variable generation systems. This chapter focuses on wind energy forecasting technologies as well as the benefits to electrical utilities of utilizing advanced forecasting

---

S. E. Haupt (✉) · W. P. Mahoney
National Center for Atmospheric Research, Boulder, CO 80301, USA
e-mail: haupt@ucar.edu

K. Parks
Xcel Energy Systems, Inc., Denver, CO 80202, USA

capabilities. In particular, this chapter describes a collaboration between the U.S. National Center for Atmospheric Research (NCAR) and Xcel Energy Services, Inc. (Xcel Energy) to conduct research to develop and deploy a state-of-the-science wind power forecasting system that was required for Xcel Energy to effectively integrate and balance an increasing capacity of wind power into its operations. The details of the usage of the system by Xcel Energy may be more applicable to utilities and authorities in the United States, but the design of the system can be generalized, based on best practices in the literature, and is applicable over a wide variety of geographical, market, and utility operating scenarios.

A wind power forecasting system must address the utility and independent system operator (ISO) needs to have a high-quality forecast of power generation on several temporal and spatial scales. For wind energy, this means being able to first predict the wind resource, and then being able to convert the wind speed prediction to estimated power output. The problem, of course, is that wind energy availability depends on the current and future state of the atmosphere. Although there are standard prediction methods for atmospheric flow, those predictions are, by nature, uncertain. It is widely accepted in the meteorological community that the atmosphere is inherently chaotic, particularly at the smallest spatial scales, and that a limit exists to predictability, as the expected accuracy degrades in time (Lorenz 1963; Wyngaard 2010). This does not imply that the power output cannot be predicted, but rather that in addition to providing a best deterministic estimate, it is also important to quantify the uncertainty of the forecast (Buizza et al. 2009).

The motivation for Xcel Energy to become a leader in implementing wind power forecasting is that it was the first utility in the United States to push its wind energy capacity past 10 % of total operating capacity and has plans to more than double that amount. As the wind capacity of a utility grows, it becomes more difficult to effectively integrate this variable resource into the power mix. Xcel Energy had reached the wind capacity threshold (~ 8 %) where it was economically beneficial to invest in developing advanced wind energy forecasting technologies to reduce the forecast error, and hence, lower the cost of integrating wind into the power grid. Wind power forecasting can significantly improve that integration by predicting the expected available power, which improves system reliability in a manner that can also minimize costs (Ela and Kemper 2009). An accurate wind energy forecasting system enhances the ability of a utility to predict the wind resource, supporting day-ahead unit commitment, and power trading, as well as decision-making related to load balancing, selection of alternative generation units, and transmission capacity.

The meteorological research community has developed numerical weather prediction (NWP) systems and statistical forecasting technologies to benefit the public good, concentrating primarily on optimizing the prediction of temperature and precipitation. Those NWP forecasts provide useful input for wind power prediction but those models are not usually tuned to optimize wind forecasts in the lower atmospheric boundary layer, which represents the lowest kilometer or so during the day and collapsing down to around a couple 100 m at night. Thus, it is important for wind energy prediction to incorporate additional technologies designed to optimize wind speed.

The remainder of this chapter provides a basic overview of how these technologies have been blended into the NCAR/Xcel Energy wind prediction system. A more technical description of the system appears in Mahoney et al. (2012) and Parks et al. (2011). Although discussion focuses on wind power forecasting as designed for the Xcel Energy System in the USA, the general approach is similar to many other successful wind power forecast systems (Giebel and Kariniotakis 2007; Monteiro et al. 2009). Section 2 provides more detail on the technologies and approaches used in the system, and Sect. 3 provides a brief treatment of numerical weather prediction capabilities employed in the Xcel Energy system. Section 4 discusses the statistical post processing techniques while short-term nowcasting is discussed in Sect. 5. Grid integration from a utility's point of view is discussed in Sect. 6. Areas of further research and conclusions are presented in Sect. 7.

2 A System's Approach to Forecasting

The most important step in designing any decision support system is understanding the needs and requirements of the end user. The goal of the NCAR/Xcel Energy Wind Power Forecasting System was to provide necessary information to both energy traders and the grid operators to optimize the economics of variable generation integration while providing an essential element of reliability. The grid operators require information on the time scales of minutes to hours while the traders have traditionally operated on a day-ahead schedule.[1] Longer term planning for operations and maintenance can also benefit from longer lead forecasts. Currently, no single weather forecasting technology is able to optimally perform across temporal scales that range from tens of minutes to several days. Statistical prediction and very short-term forecasting, also known as nowcasting, techniques may be most beneficial for the very short term, say less than 3 h, but NWP will best predict the weather, and thus, wind at times beyond a few hours. Spatial scales are also an issue to consider when designing a forecast system. A wind plant operator may be most interested in very local-scale phenomena, and thus, need a system tuned for a particular wind farm, or even at the turbine level. The balancing authority, on the other hand, may take the larger view of considering wind energy availability from all wind plants and utilities in their service area.

To meet all of these needs, a best practices approach to predicting the power output combines what is known about the physics and dynamics of the atmosphere with statistical methods that utilize local and regional weather observations and

---

[1] Note that some energy markets, including the U.S. Midwest System Operator (MISO) that integrates Xcel Energy's power in the Northern States Power region, now allow economic dispatch in 5 min increments and also dispatches wind energy.

Fig. 1 A system's approach to wind power forecasting that emphasizes blending various technologies at appropriate time scales. *FDDA* Four-dimensional data assimilation, *3DVar* Three-dimensional variational methods, *NWP* Numerical weather prediction

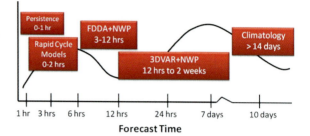

power generation data. Different approaches are preferable for differing time frames and it is important to blend the technologies in a manner to produce the best forecast for each time period and spatial scale. The overarching goal is to build a decision support system that leverages the best technologies for each spatial and temporal scale. Figure 1 illustrates where along the forecasting period various weather prediction techniques have demonstrated optimal performance. It is a useful diagram for assessing how one might consider which technologies to use for the specific time periods based on knowledge of the physics of the system, the strengths and weaknesses of each system component, and how to blend the technologies to optimize the forecast skill. At the shortest time scales, it is currently difficult to beat a persistence forecast; that is, 5 min from now, the wind speed is likely to be much like now. Thus, it is appropriate to predict no change at these very short timescales if real-time observations are available.[2] However, as we get further out in forecast lead time (beyond about an hour), one would expect changes to occur; thus implementing predictive models becomes essential. For the shorter modeled scales (1 to 3 h), it is most important to blend real-time observational data with predictive models. These "nowcasting" models assimilate local data and use known dynamics and physics relationships to intelligently extrapolate the solution forward in time according to the local weather conditions. A bit further along in time, around 3 h, high-resolution NWP becomes the prediction technology of choice. Typically NWP models require time to spin-up or come to a balanced state, but with the utilization of local data, using dynamic data assimilation methods such as the real-time four-dimensional data assimilation (RTFDDA) method described below, the spin-up period can be significantly shortened resulting in skillful predictions in the 3–12 h time range. Beyond about 6 h, more distant weather phenomena and patterns play a larger role on the local weather conditions. Therefore, many NWP modelers choose to use data assimilation methods that leverage spatial correlations over larger spatial scales so that observational data and atmospheric boundary conditions derived from global

---

[2] Note, however, that various statistical techniques are now being investigated to provide 5-min forecasts.

Wind Power Forecasting

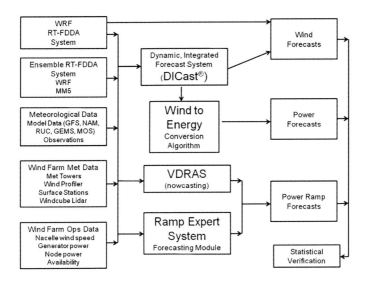

Fig. 2 Flowchart of NCAR's Xcel Energy power prediction system

weather models further away begin to influence the NWP model solution. Additionally, it is expedient to blend the results of multiple NWP models where available. Typically, this "ensemble" approach to prediction produces a better forecast than any single model. It is important, of course, to consider how best to blend the information from the various models, which is the topic of Sect. 4.

System components that comprise the NCAR/Xcel Energy Wind Power Forecasting System are shown in Fig. 2. The system ingests external, publically available weather model data and observations. In order to provide information specific to Xcel Energy's region, high-resolution NWP simulations assimilate specific local weather observations. The weather observations range from routine meteorological surface and upper air data to data from the wind farms, including wind speed data from the Nacelle anemometers. An ensemble of somewhat coarser NWP runs provides an additional best estimate of wind speed and also includes uncertainty information. Finally, to optimize estimates of short-term changes in wind power requires nowcasting technologies such as the Variational Doppler Radar Analysis System (VDRAS) and an Expert System.

The strength of the wind power forecasting system lies in blending the various components to produce a power forecast that can be used by Xcel Energy grid operators and energy traders. That blending is accomplished with the Dynamical Integrated Forecast (DICast) System. The wind speeds predicted by DICast must then be translated into power using NCAR's empirical power conversion algorithms. Each of these system components is described in more detail below.

3 Numerical Weather Prediction

NWP is based on the Navier–Stokes equations of fluid flow (Dutton 1976; Vallis 2006), includes the physics that provide the forcing functions, and in modern application, assimilates observations to initialize the flow to begin with a best initial state (Warner 2010). At a basic level, standard operational NWP models run at national centers can provide an estimate of expected changes in weather, and thus wind speed, due to large-scale forcings. Customized NWP, however, allows prediction of the wind or solar resource that is tuned to the specific application and location of the wind or solar plant and also provides an opportunity to assimilate specialized local observations in and around the power plant.

As applied to the NCAR/Xcel Energy Wind Power Forecasting System, both standard operational and customized NWP models play important roles. The integration of the standard operational forecasts from the national centers is treated in Sect. 4. Here we describe the customized application of the Weather Research and Forecasting (WRF) model. WRF has been primarily developed and maintained at NCAR and is an accepted forefront NWP model (Skamarock et al. 2005a, b, 2008) with over 10,000 users across the world. It includes many options for physics packages and uses state-of-the-science numerics and can be configured to run over small or large domains. The physics schemes chosen for use in this deterministic system include the Lin microphysics package[3], Rapid Radiative Transfer Model (RRTM) for longwave radiation, Dudhia shortwave radiation model, Grell-Devenyi ensemble cumulus parameterization, Noah land surface model, Monin–Obukhov surface layer scheme, and the Yonsei University (YSU) planetary boundary layer model.

To provide a best possible forecast requires that we know the initial state of the atmosphere, that we provide the model with up-to-date boundary conditions from the global models, and that we continually assimilate real-time observations. To that end, WRF is initialized with the U.S. National Centers for Environmental Prediction (NCEP) Global Forecasting System (GFS) initial and boundary conditions. Observations are assimilated into the model using the RTFDDA system (Liu et al. 2006, 2008a, b, 2011). This system employs a Newtonian Relaxation method to continually "nudge" the model solution toward the observations (Hoke and Anthes 1976; Stauffer and Seamon 1990, 1994). This nudging approach adds an additional forcing term to the momentum equations that, when tuned with optimal weighting factors and spread spatially and temporally, can produce a solution that is grounded in the truth of the observations yet maintains the smooth solution faithful to the fluid equations of motion. The observations that are assimilated include standard meteorological observations as well as specialized wind farm specific observations, satellite data fields, lidar data, measurements obtained from regional aircraft, local sensor data, and others. Figure 3a shows the deterministic model domains.

---

[3] The microphysics package was recently changed to the Thompson microphysics scheme.

Wind Power Forecasting

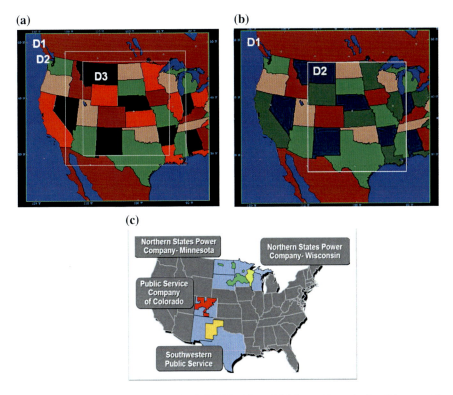

Fig. 3 Domains for the **a** deterministic WRF (30, 10, and 3.3 km grid spacing) and **b** ensemble system (30 and 10 km grid spacing). Panel **c** indicates Xcel Energy's service areas

In addition, an ensemble of NWP runs was also included in the Xcel Energy Wind Power Forecasting System. The rationale for including an ensemble is due to the atmosphere's chaotic sensitivity to initial conditions; its predictability degrades time, and theoretically, any of a number of possible realizations become likely scenarios (Buizza et al. 2009). Thus, it makes sense to define the probability density function (pdf) of likely scenarios. The mean (or median) of that pdf is typically a better prediction than any consistent single member. In addition, the standard deviation of the pdf provides information regarding the uncertainty of the forecast (Grimit and Mass 2007; Kolczynski et al. 2009, 2011; Lee et al. 2009, 2012).

To provide this probabilistic forecast, NCAR configured a 30 member ensemble system to both improve forecast accuracy and to indicate forecast uncertainty. This system includes 15 WRF members and 15 members produced by the Penn State/NCAR Mesoscale Model (MM5). These members use initial and boundary conditions from both the GFS and North American Model (NAM), include multiple boundary layer physics schemes, include differing nudging radius values, and include realistic initial condition phase perturbations in some members. That system is run on the outer two nests at 30 and 10-km resolutions (Fig. 3b).

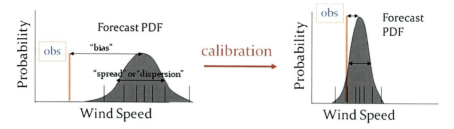

Fig. 4 Calibration improves the best estimate and the spread of the ensemble forecast

The output of the models can then be calibrated and postprocessed to improve upon the probabilistic forecasts (Fig. 4). This postprocessing relies on a substantial amount (on the order of a year or more) of historical data of matched prediction and observations pairs. First, each forecast is calibrated using an Analog Kalman Filter (AnKF) approach. This approach searches for similar (analog) situations to that of the current forecast and corrects the bias in the forecast according to the difference between the historical forecast and the actual observations of these analog events (Delle Monache et al. 2011a, b). The Kalman Filter portion of the correction weights the closest matched analog the highest. This procedure corrects the bias of the forecast.

In addition, quantile regression (QR) is used to improve the forecast reliability, meaning that the predicted probability occurs the same percentage of the time as that observed. QR fits each wind speed quantile using multi-linear regression. This regression is conditioned on ranked ensemble member ranking, ensemble mean, ensemble standard deviation, persistence, and each ensemble member (Hopson et al. 2010). In this process, the climatological quantiles are first determined. Then for each quantile, a forward step-wise cross-validation is used to select the best regression set that minimizes for that cost function quantile. The forecasts are then segregated based on ensemble spread and the models are refit for each range. The output is a "sharper" posterior pdf that is represented by the interpolated quantiles. This two-step process also calibrates the bias and the reliability of the ensemble (Hopson et al. 2010). These probabilistic approaches have been developed and are in the process of being validated and implemented.

4 Statistical Postprocessing

An important component of any advanced wind power forecasting system is a statistical learning algorithm that matches historical model data to observations to develop relationships that can subsequently be used for forecasting (Giebel and Kariniotakis 2007; Monteiro et al. 2009; Myers et al. 2011; Haupt et al. 2012; Mahoney et al. 2012). Such algorithms could be based on any of a number of artificial intelligence approaches, including Artificial Neural Networks, Random

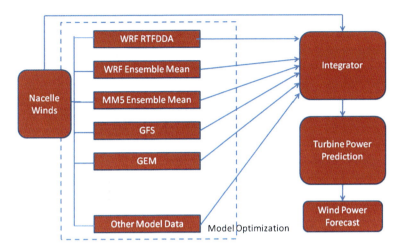

Fig. 5 Schematic of the DICast optimization system

Forests, etc. The statistical learning technology used in NCAR's wind power forecasting system is the Dynamic Integrated Forecast System (DICast). DICast is a weather forecasting system that was designed to emulate the human forecast process (Myers et al. 2011). It has been developed, refined, and implemented into various operational forecast systems by NCAR for more than a decade. It first postprocesses output of several individual NWP models separately, and then generates an intelligent consensus forecast from these optimized modules (Myers et al. 2011; Mahoney et al. 2012). A strength of DICast is that it continually "learns" how to make a better forecast based on comparisons of recent forecasts and observations.

A key to the success of this wind power forecasting system is that Xcel Energy provides NCAR with high temporal resolution observations of both hub height wind speed and power from roughly 90 % of the turbines from which it obtains wind energy. Forecasts that are tuned using statistical and artificial intelligence methods like DICast require observations as targets. The most relevant observational datasets are hub height wind speed and turbine power. The wind speeds are measured by the anemometers sited on top of the turbine Nacelles. Although one does not expect that anemometer, which is placed behind the rotating blades, to accurately represent the inflow (ambient) wind speed, it is more important that the Nacelle wind speed correlates highly with the turbine power output. DICast uses that input to produce forecasts of wind speed at hub height for each wind turbine. Figure 5 illustrates the data ingest to the integrator, which compares the input to the observed Nacelle winds to optimize the wind forecast, which is then transformed to a power forecast. The deterministic and ensemble model output plus data from various global and regional models (including NCEP's GFS and the Canadian Global Environmental Multiscale (GEM) Model) are fed into DICast.

NCAR uses a two-step approach to the forecast process. First, DICast is used to predict the wind speed at hub height. Then the advanced power conversion system converts those wind speeds to power forecasts that can be summed to provide farm and/or connection node power forecasts.

DICast "learns" how to optimize the wind speed predictions by comparing the historical model forecasts to the actual wind speeds observed. This optimization occurs in two steps. The first forecast optimization step is dynamical model output statistics (DMOS), which is a statistical postprocessing step that optimizes the raw forecast from each NWP model. This process is based on Model Output Statistics (MOS) originally designed by Glahn and Lowry (1972) and now employed by most operational NWP centers. A key difference is that DMOS has been designed to work on relatively short forecast/observational histories. By carefully choosing the inputs from the NWP model to generate the statistical relationships, reasonable DMOS can be generated with only 30 days of history, although a period of 90 days is preferred. New DMOS relations are generated weekly from the last 90-day history. The optimization uses the root mean squared error (RMSE) of the wind speed as the metric to be minimized, but additional variables can be used in the DMOS equations.

The second optimization step in DICast is the integration of the DMOS forecasts into a single consensus forecast. The DICast forecast integrator does this by objectively determining the optimal weights for blending the DMOS model forecasts. As is true for the DMOS step, the optimal combination varies depending on location, time of day, forecast generation time, lead time, and season. By design, on average the DICast integrator will outperform any ingredient forecast made from one particular model. The result is a robust system that consistently outperforms a single model-based forecast system. Figure 6 shows the errors of the DMOS-optimized forecasts compared to the DICast integrated forecast over a 4-month period. The figure indicates that while the relative performance of the NWP models varies from month to month, the DICast integrated consensus forecast substantially outperforms all the individual models' forecasts. The reduction in error is usually around 10–15 % over the best forecast model in the mix.

The end product of the wind power forecast system is a forecast of power; thus, NCAR has developed an empirical power conversion methodology that transforms the wind speed forecasts produced by DICast for each Xcel power connection node. Power production often deviates substantially from the manufacturer's power curve, leading to significant forecast power error. The left panel of Fig. 7 demonstrates that a single wind speed can correspond to many power outputs. Many of the data points corresponding to very low or very high wind speeds are anomalous and may not be a good forecast of actual power output (turbine down but not reporting nonavailability, icing conditions causing decreased power output, curtailments, high wind speed cutouts, etc.) Thus, as seen on the right panel of Fig. 7, a much more distinct power curve can be created by just considering the middle 50 % of the power values. This quality control measure allows use of an empirical power curve that predicts the power output from the wind speed more accurately than the manufacturer's power curve.

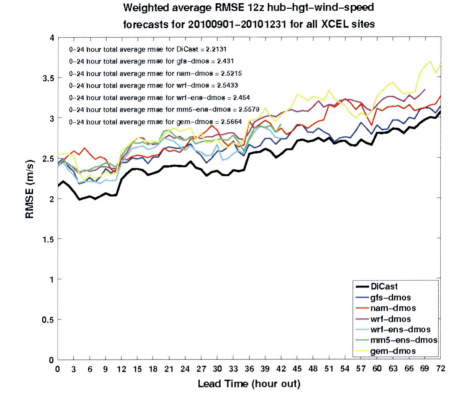

Fig. 6 Errors for DMOS forecasts compared to the DICast integrated forecast made for 3 months in 2010. The DICast integrated forecast is the *black line* with consistently lowest RMSE (Mahoney et al. 2012)

Due to these discrepancies, NCAR designed the wind power system to fit historical curves between Nacelle wind speed and power at each wind farm for each turbine type. Several artificial intelligence techniques were tested, including Random Forests, Regression Trees, K-Nearest Neighbor, etc. The regression tree, Cubist®,[4] performed well and proved easy to apply, so it was chosen for inclusion in the final system. Cubist® was trained on 15 min average observed wind/power pairs. Then during real-time operation, the equivalent data are used to compute the best historical fit given the prior behavior. If observed Nacelle wind speeds and turbine power data are missing, the system substitutes forecast wind speeds and ideal power curve estimated power. In the most recent applications, the historical data have been divided into quantiles and the extreme quantiles have been removed from the training set to avoid erroneous fitting to high-speed cutouts, curtailments,

---

[4] Cubist is a trademark of RuleQuest Research.

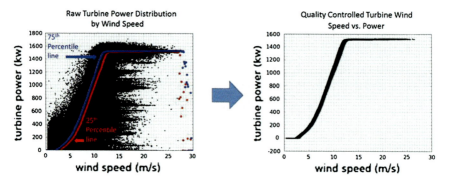

Fig. 7 *Left* Example scatter plot of empirical wind speed versus power relationship, binned into 0.1 m/s bins. *Right* Same power curve with power outside the 25th and 75th percentiles eliminated

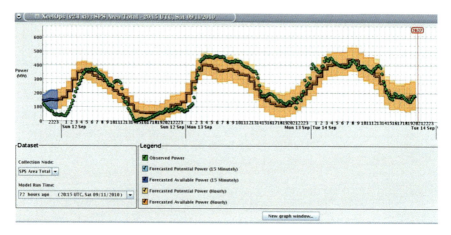

Fig. 8 Example power output GUI

and other events that may not be representative of the event being predicted. Turbine power is forecast for each turbine, then the predicted powers are summed to derive connection node and operating region power forecasts. The power predictions incorporate any available information on planned outages (such as for maintenance) in order to avoid over-forecasting for turbines that are out of service.

The system includes ongoing verification to monitor its health and performance. The metric used in this internal monitoring system is the normalized mean absolute error (NMAE),[5] which is computed at each connection node for each forecast lead time by 15 min intervals out to 3 h and then for each 60 min forecasts out to 168 h.

The results of the power forecast are displayed in a Graphical User Interface (GUI) as shown in Fig. 8. The forecast power is the solid orange line and an

---

[5] The MAE is normalized by connection node maximum capacity to produce a percent error.

estimate of the uncertainty (25 and 75 % confidence limits based on historic analysis) is the lighter orange shading. The actual power output appears as green dots. The user can select the time period, geographical region of the forecast (e.g., wind farm, connection node, or service area), and forecast initialization time to display.

5 Short-Term Forecasting

A major issue for forecasting wind energy is accurately predicting rapid changes of substantial amounts of the wind speed over short time periods, an event commonly referred to as a power ramp. Figure 9 illustrates the issue for a series of wind farms located in the same region of Colorado for a ramp event that occurred on 3 August 2009. Each color of that figure represents power output from a different wind farm. One can see that as a major weather system passed through the state, the changes in the wind power aggregated across all of the farms added to produce large changes in total wind power. Initially (between 17:10 and 20:30 local time) some small thunderstorms caused substantial power fluctuations, but these were followed by a major event, which produced a ramp up of 800 MW when a cold front passed through followed by a corresponding ramp down. This example illustrates the importance of having an accurate forecast of such events. If the utility and balancing authority have prior knowledge of such significant events, they are able to take appropriate action so that they can use the increased energy during the ramp up, yet cover the ramp down with alternative generation sources. Although cold and warm fronts and prefrontal thunderstorm systems are common causes of ramps, other causes include low-level jets, sea breezes, microbursts, eroding surface inversions, momentum (high wind) mixing from upper levels, and other meteorological phenomena.

What are the best ways to forecast such short-term ramps? Forecasting the timing, magnitude, and duration of these ramps can be challenging. NWP is often able to predict such phenomena, but because of the spin-up issues, the lag between model runs and issuing the forecast, and inherent uncertainties, there are often phase shifts and magnitude errors in those forecasts. The best strategies rely on observations near the wind or solar plants. Two such strategies that have been evaluated in the Xcel system are discussed in more detail here: the Variational Doppler Radar Analysis System (VDRAS) and an observation-based expert system.

VDRAS assimilates radar reflectivity and radial velocity data into a numerical cloud-scale model and produces high-resolution boundary layer wind fields. The VDRAS system has been used both in research and in real-time operations as a stand-alone system for homeland security applications associated with hazardous plumes and has been installed for real-time nowcasting of convective weather domestically in a number of NWS offices, in U.S. Army test ranges, for homeland security applications, and internationally for the last two summer Olympics (Sun

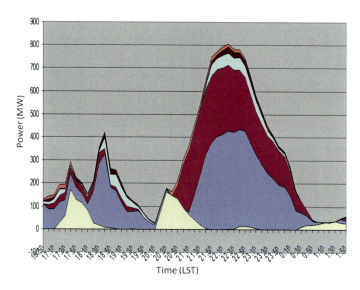

Fig. 9 Example of a wind power ramp due to a cold front passage, preceded by several small thunderstorm outflows. Each *color* represents power from a wind farm in the region

and Crook 1997, 2001; Crook and Sun 2002; Sun et al. 2010, 2011; Sun and Zhang 2008). VDRAS provides wind, thermodynamical, and microphysical analyses with a typical spatial resolution of 1–4 km and temporal update frequency of 15–20 min. Figure 10 is a flowchart of the VDRAS system. It begins with data ingest from various sources, including radar data, surface network observations, and background weather data from the WRF deterministic model with RTFDDA. The domain is chosen to be large enough to include the surrounding Doppler radar sites yet small enough to run efficiently. Figure 11 shows the three VDRAS domains that are configured for the portions of the Xcel Energy system that encompass eastern Colorado, northeast New Mexico, and the panhandle region of Texas.

VDRAS is used in this application for nowcasting the current winds. The goal is to distinguish between large-scale features such as cold fronts, thunderstorm gust fronts, low-level jets, and other weather phenomena that have strong wind gradients. Figure 12 shows a convergence feature identified by VDRAS in a case study. To estimate the future location of the features identified, several different approaches can be taken. One method is feature extrapolation, in which a feature that has been identified is extrapolated with the wind, which can have either directional shifts or speed gradients within the domain. A second approach is to run the underlying VDRAS cloud-scale model in forecast mode.

NCAR also takes an artificial intelligence approach to ramp forecasting. An expert system uses observational data available in concentric rings at specified distances from the wind farm. This expert system was developed to use publicly available observational data in eight concentric rings with 50 km spacing centered on a Colorado wind farm. The current configuration is built to predict weather

Wind Power Forecasting

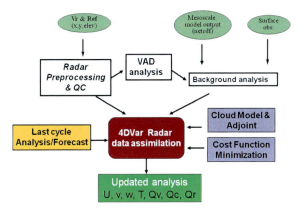

Fig. 10 Flow chart of the VDRAS Nowcasting system. Output variables include U (zonal wind speed), v (meridional wind speed), w (vertical wind speed), and Q (mixing ration for water vapor—subscript v, cloud water—subscript c, and rain water—subscript r)

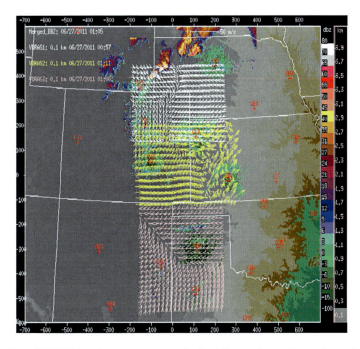

Fig. 11 Three VDRAS domains are set up over the Xcel Energy forecasting region in the central U.S. vectors indicate wind speed and direction

patterns advancing from the northwest, the predominant direction for synoptic patterns in this region. This rule-based expert system searches for wind ramp signatures in upstream observations and uses these observations to infer the time

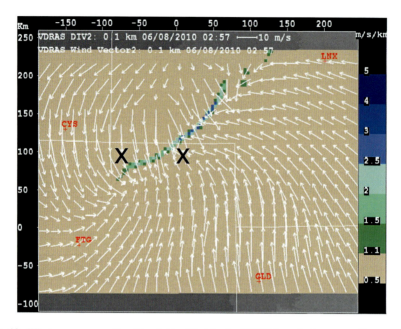

Fig. 12 Wind convergence line (*front*) identified by the VDRAS analysis system. Convergence is indicated by the *colored line* and wind speed and directions by the *arrows*

and magnitude of the wind ramp that is expected to affect the wind farm. For each site and each historical hour, a ramp metric is computed using the current hour and the previous hour's observations. The observed wind at 10 m is extrapolated to hub height (80 m) and changes in wind speed and direction are evaluated. The percentage of sites that indicate a ramp (defined here as a change of at least 25 % of capacity in an hour) is tabulated. These percentages are averaged across rings for each lead time. A ramp indicator is computed that depends on a threshold of that average percentage. The expert system is applied for up to 6 h lead time and results are displayed to advise system operators of imminent ramping events.

The results of the expert system forecast for a ramp event that occurred on 12 October 2010 appear in Table 1. As seen in the table at 0600 UTC, a ramp was forecast for 1000 UTC. By 0700 UTC, the ramping conditions by 0900 UTC were added. By 0800 UTC, a 3 h ramp event was imminently predicted, which did indeed occur. The ramp finished shortly after 1200 UTC as predicted.

6 Utility Grid Integration

The challenges of integrating wind and solar resources into the grid come in power utility operations—or the day-to-day operation and coordination of power plants, their fuel supplies, power transmission, and the interaction with the energy

Table 1 Results of expert system forecast of ramp times on 12 October 2010—predicted, X—occurred. See text for description

Valid time (UTC)	0700 - 0800	0800 - 0900	0900 - 1000	1000 - 1100	1100 - 1200	1200 - 1300	1300 - 1400	1400 - 1500	1500 - 1600	1600 - 1700
Observed Farm Ramp	X		X	X	X		X	X		
06z Forecast	01-hr	02-hr 1	03-hr 1	04-hr 1	05-hr	06-hr				
07z Forecast		01-hr	02-hr 1	03-hr 2	04-hr	05-hr	06-hr			
08z Forecast			01-hr 2	02-hr 2	03-hr	04-hr	05-hr	06-hr		
09z Forecast				01-hr 4	02-hr	03-hr	04-hr	05-hr	06-hr	
10z Forecast					01-hr 1	02-hr	03-hr	04-hr	05-hr	06-hr

markets. The uncertain and variable nature of wind and solar energy resources disrupt operational planning—if renewable resources are planned, or scheduled, and they do not produce power, then another generation resource (coal, hydro, natural gas, etc.) must make up the difference. If renewables were not scheduled, and then produce, they offset resources that were already scheduled creating inefficient operation, or worse, are curtailed until it is technically or economically feasible to bring the renewable resources online.

Uncertainty and variability are not new to utility operations. Power plants trip offline unexpectedly, but this uncertainty has been mitigated through a combination of best practices and rulemaking to ensure reliable delivery of power to end consumers. Energy demand is both uncertain and variable—but not to the same extent as renewable energy. This will be demonstrated shortly, but first it is important to understand a little about how the grid is operated. In utility operations, energy demand and resources are forecast every working day through to the next working day for operational planning purposes. Thus, for a weekend, a 3-day forecast is required. For holiday weekends, the required timeframe is even longer. Thus, the forecast is typically 18–42 h ahead, although it can be as long as 5 days in advance. Resources are scheduled in a least-cost manner to meet the forecasted energy demand. This is called day-ahead commitment, as many decisions from choosing which power plants to run and which fuel supplies to deliver are made or committed. In Fig. 13, the forecasted energy demand and wind energy are shown for 1 week for the Public Service of Colorado (PSCo) operation system. In the lower graph, the errors associated with demand and winds are shown for the same period. The wind error swamps the load error—even though the amount of wind energy is only a fraction of the total energy demand. The challenge is that comparatively small amounts of wind energy creates a disproportionate amount of uncertainty to the existing power system and rules and best practices are not fully matured to mitigate the added uncertainty.

The traditional utility operating paradigm separates power plants into three duty cycles: baseload, mid-merit, and peakers (see Fig. 14). Baseload facilities run 24/7 at near maximum output for highest utility (approaching 95 %). These are typically low variable cost facilities such as coal, large hydro, or nuclear power plants. Mid-merit facilities are the workhorses of the system, turning up and down, sometimes off/on every day—and may run anywhere from 40 to 60 % of their full capacity. Peakers are typically natural gas or diesel generators that only turn on during the peak time of the day, season, or year. These have low capital costs, but often high variable costs, but are only used 15 % of the time and often much lower.

Renewable energy disrupts this traditional paradigm. Renewable energy has incredibly low variable cost, but depends on the availability of the wind or solar energy—typically 25–40 % of its maximum capacity. Renewable energy displaces expensive peakers, mid-merit generators, or if present in large enough quantities during periods of low demand, inexpensive baseload facilities. As renewable energy makes up a larger share of the generation portfolio, the duty of plants will change. Mid-merit plants will run less and more intermittently. Baseload facilities

Wind Power Forecasting

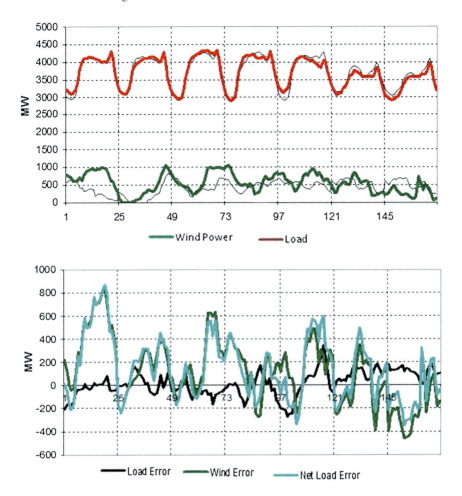

Fig. 13 Energy demand and wind energy; forecast and actual

will operate, at times, more like mid-merit facilities. Peakers will be brought on to deal with short, unexpected drops in renewable power. The *balance portfolio*, or the portion of the generation portfolio that is controllable, works in concert to mitigate the variability and uncertainty of not just energy demand, but renewable energy supply.

The cost of renewable energy uncertainty can be seen in energy markets. For example, in the Midwest Independent transmission System Operator (MISO) market, there is a day-ahead market and a real-time market. In the day-ahead market, generators offer services for a price and are granted a day-ahead award based on their price and the expected energy demand. A day-ahead award is a financially binding forward contract—if the generator does not produce the amount awarded at the time specified, the MISO market forces that participant to

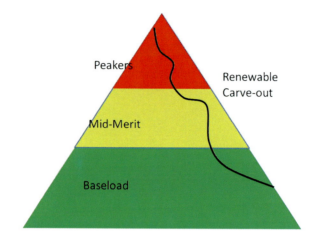

Fig. 14 Traditional utility operating paradigm and renewable carve out

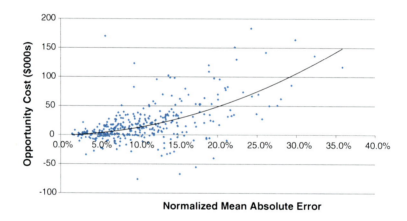

Fig. 15 Opportunity cost for Northern states power (Oct 2011–Sept 2012)

purchase the power in the real-time market, which can be much more expensive than if the day-ahead plan had been met. If the generator overproduces, then they are forced to sell in the real-time market and may not obtain the price they could have obtained if the forecast had been correct. In the case of renewable energy, a generator is awarded based on its day-ahead forecast. In real-time, the MISO market forces the renewable facility to buy or sell in real-time to make up the difference between forecast and actual—often at unfavorable rates—creating opportunity costs in the marketplace.

Opportunity cost can be significant. For example, Northern States Power, an operating company within Xcel Energy, participates in the MISO market with over 1600 MW of wind energy. In Fig. 15, every dot represents the opportunity cost associated with NSP's entire wind portfolio. As one would expect, the opportunity cost grows with increased forecast error. Additionally, the risk is asymmetric—there are far more opportunities for losing money (i.e., positive opportunity cost) than there are in making money due to forecast error (i.e., negative opportunity cost). The total opportunity cost for this period was over $6.7 M.

Therefore, improving forecast accuracy can lower the integrated opportunity cost. The NCAR/Xcel Wind Power Forecasting System showed substantial improvement in mean absolute percentage error (MAPE) compared to Xcel Energy's previous approach of using NWP forecasts based on the North American Model (NAM) and local soundings. Under that prior forecasting approach, MAPE averaged 18 % in 2009. Upon the initial implementation of the Wind Power Forecasting System in 2010 MAPE dropped to 14.3 % (a 20 % decrease in MAPE), saved Xcel Energy's ratepayers over US $6 M. As system improvements are implemented, MAPE continues to decrease, even into the single digits for some forecasting periods. Xcel Energy has calculated that using the Wind Power Forecast System has saved them $21.8 M for the 3 years 2010–2012 (Barlett 2013).

7 Concluding Remarks

NCAR has conducted research and configured a Wind Power Forecasting System for Xcel Energy that integrates high-resolution and ensemble modeling with artificial intelligence methods. This state-of-the-science forecasting system includes specific technologies for short-term detection of wind power ramps, including a Variational Doppler Radar Analysis System and an expert system. Wind power forecasting can significantly improve grid integration by improving reliability in a manner that can minimize costs. Errors in forecasts become opportunity costs in the energy market; thus, more accurate forecasts have the potential to save substantial amounts of money for the utilities and their ratepayers. Xcel Energy has shown a savings of $21.8 M in the past 3 years due to lowered forecast errors (Barlett 2013). As renewable energy expands, it becomes more important to provide high-quality forecasts so that renewable energy can carve out its place in the energy mix.

The wind forecasting approaches reported here are typical of some of the most successful approaches applied throughout the world. The state of the science has greatly increased over the past few years, but further advances are possible. For instance, Vanderwende and Lundquist (2012) have recently demonstrated the dependence of wind power output on atmospheric stability and shear and its diurnal dependence. Such findings suggest that stability dependent power curves may improve the power conversion systems. In addition, ramping forecast systems have been shown to improve when local observations are included. If one had enough historical data to establish weather regime-specific short-range forecast systems, one could expect to improve forecast skill by identifying the weather regime and making specific forecasts for the currently observed regime. There are also enhancements possible by identifying specific conditions that could degrade wind turbine performance, such as icing conditions. Research is being performed to correlate the level of icing to turbine performance and to apply established techniques to predict locale-specific icing conditions. Finally, enhancements to the probabilistic approaches are expected to be exploited by the utility industry in the near future to further enhance the economic value of the forecasts.

It has now been demonstrated in locations throughout the world that accurate forecasts of variable resources of renewable energy can greatly enhance a utility's ability to integrate that power source into their operations. Thus, improvements in such forecast systems are paramount to improving usage of such energy sources and will help in advancing the deployment of renewable energy.

Acknowledgments This chapter reports a collaboration between NCAR and Xcel Energy. In addition to the authors, quite a few people contributed to configuring, building, testing, and validating the system, including Gerry Wiener, Bill Myers, Yubao Liu, Jenny Sun, David Johnson, Luca Delle Monache, Thomas Hopson, Seth Linden, Branko Kosovic, Arnaud Dumont, Julia Pearson, Gregory Roux, William Cheng, Yuewei Liu, Frank McDonough, Becky Ruttenberg, Brice Lambi, John Exby, Doug Small, Wanli Wu, Alemu Tadesse, and Drake Bartlett. The system is now operated for Xcel Energy by Global Weather Corporation and NCAR continues to perform research to enhance wind and solar power prediction.

References

Barlett D (2013) What is the value of a variable generation forecast? Utility variable generation integration group annual forecasting workshop, Salt Lake City, 26–27 Feb. http://www.uwig.org/slcforework/BartlettSession1.pdf

Buizza R, Houtekamer PL, Toth Z, Pellerin G, Wei M, Zhu Y (2009) A comparison of the ECMWF, MSC, and NCEP global ensemble prediction systems. Mon Weather Rev 133:1176–1197

Crook A, Sun J (2002) Assimilating radar, surface and profiler data for the Sydney 2000 forecast demonstration project. J Atmos Oceanic Technol 19:888–898

Delle Monache L, Nipen T, Liu Y, Roux G, Stull R (2011a) Kalman filter and analog schemes to post-process numerical weather predictions. Mon Weather Rev 139:3554–3570

Delle Monache LA, Fournier T, Hopson Y, Liu B, Mahoney G, Roux and T Warner, (2011b) Kalman filter, analog and wavelet postprocessing in the NCAR-Xcel operational wind-energy forecasting system. 91st American Meteorological Society Annual Meeting, Seattle. http://ams.confex.com/ams/91Annual/webprogram/Paper186510.html

Dutton JA (1976) The ceaseless wind: an introduction to the theory of atmospheric motion. McGraw-Hill, New York, p 579

Ela E, Kemper J (2009) Wind plant ramping behavior. National renewable energy laboratory technical report NREL/TP-550-46938

Giebel G, Kariniotakis G (2007) Best practice in short-term forecasting—A Users Guide. European wind energy conference and exhibition, Milan, 7–10 May 2007. http://orbit.dtu.dk/en/publications/best-practice-in-shortterm-forecasting-a-users-guide%284127f7c3-b6ed-47b5-b71c-cc24dfa16917%29/export.html

Glahn HR, Lowry DA (1972) The use of model output statistics (MOS) in objective weather forecasting. J Appl Meteor 11:1203–1211

Grimit EP, Mass CF (2007) Measuring the ensemble spread-error relationship with a probabilistic approach: stochastic ensemble results. Mon Weather Rev 135:203–221

Haupt SE, Wiener G, Mahoney WP, Pearson J (2012) The need for wind power forecasting and review of current methods, 10th conference on artificial intelligence applications to environmental science, held in conjunction with the 92nd AMS Annual meeting, New Orleans, 24 Jan. https://ams.confex.com/ams/92Annual/webprogram/Paper206470.html

Hoke JE, Anthes RA (1976) The initialization of numerical models by a dynamic initialization technique. Mon Weather Rev 104:1551–1556

Hopson T, Hacker J, Liu Y, Roux G, Wu W, Knievel J, Warner T, Swerdlin S, Pace J, Halvorson S (2010) Quantile regression as a means of calibrating and verifying a mesoscale NWP ensemble, Prob Fcst Symp, American Meteorological Society, Atlanta, 17–23 Jan. https://ams.confex.com/ams/90annual/techprogram/paper_163208.htm

Kolczynski WC, Stauffer DR, Haupt SE, Altman NS, Deng A (2011) Investigation of linear variance calibration for spread-error relationship using a stochastic model. Mon Weather Rev 139:3954–3963

Kolczynski WC, Stauffer DR, Haupt SE, Deng A (2009) Ensemble variance calibration for representing meteorological uncertainty for atmospheric transport and dispersion modeling. J Appl Meteorol Climatol 48:2001–2021

Lee JA, Kolczynski WC, McCandless TC, Haupt SE (2012) Objective techniques for configuring and down-selecting an NWP ensemble for low-level wind predictions. Mon Weather Rev 140:2270–2286

Lee JA, Peltier LJ, Haupt SE, Wyngaard JC, Stauffer DR, Deng A (2009) Improving SCIPUFF dispersion forecasts with NWP ensembles. J Appl Meteorol Climatol 48:2305–2319

Liu Y, Chen F, Warner T, Basara J (2006) Verification of a mesoscale data assimilation and forecasting system for the Oklahoma City area during the joint urban 2003 field project. J Appl Meteor Climatol 45:912–929

Liu Y, Warner T, Swerdlin S, Betancourt T, Knievel J, Mahoney B, Pace J, Rostkier-Edelstein D, Jacobs NA, Childs P, Parks K (2011) NCAR ensemble RTFDDA: real-time operational forecasting applications and new data assimilation developments, Joint session for the 24th conference on weather analysis and forecasting and the 20th conference on numerical weather prediction, Seattle, 23–27 Jan. http://ams.confex.com/ams/91Annual/webprogram/Paper185473.html

Liu Y, Warner TT, Bowers JF, Carson LP, Chen F, Clough CA, Davis CA, Egeland CH, Halvorson S, Huck TW Jr, Lachapelle L, Malone RE, Rife DL, Sheu R-S, Swerdlin SP, Weingarten DS (2008a) The operational mesogamma-scale analysis and forecast system of the U.S. army test and evaluation command. Part 1: overview of the modeling system, the forecast products. J Appl Meteorol Climatol 47:1077–1092

Liu Y, Warner TT, Astling EG, Bowers JF, Davis CA, Halvorson SF, Rife DL, Sheu R-S, Swerdlin SP, Xu M, (2008b) The operational mesogamma-scale analysis and forecast system of the U.S. army test and evaluation command. Part 2: inter-range comparison of the accuracy of model analyses and forecasts. J Appl Meteorol Climatol 47:1093–1104

Liu Y, Warner T, Liu Y, Vincent C, Wu W, Mahoney B, Swerdlin S, Parks K, Boehnert J (2011b) Simultaneous nested modeling from the synoptic scale to the LES scale for wind energy applications. J Wind Eng Ind Aerodyn. doi:10.1016/j.jweia.2011.01.013

Lorenz E (1963) Deterministic nonperiodic flow. J Atmos Sci 20:130–141

Mahoney WP, Parks K, Wiener G, Liu Y, Myers B, Sun J, Delle Monache L, Johnson D, Hopson T, Haupt SE (2012) A wind power forecasting system to optimize grid integration. In: Special issue of IEEE transactions on sustainable energy on applications of wind energy to power systems 3(4):670–682

Monteiro C, Bessa R, Miranda V, Botterud A, Wang J, Conzelmann G (2009) Wind power forecasting: state-of-the-art 2009. Argonne National Laboratory, Argonne. (ANL/DIS-10-1)

Myers W, Wiener G, Linden S, Haupt SE (2011) A consensus forecasting approach for improved turbine hub height wind speed predictions. In: Proceedings of wind power 2011. http://ams.confex.com/ams/91Annual/webprogram/Paper187355.html

Parks K, Wan Y-H, Wiener G, Liu Y (2011) Wind energy forecasting—a collaboration of the national center for atmospheric research (ncar) and Xcel Energy. National Renewable Energy Laboratory, Golden

Skamarock WC, Klemp JB, Dudhia J, Gill DO, Barker DM, Duda MG, Huang X-Y, Wang W, Powers JG (2005a) A description of the advanced research WRF version 3. NCAR/TN-475 + STR

Skamarock WC, Klemp JB, Dudhia J, Gill DO, Barker DM, Wang W, Powers JG (2005b) A description of the advanced research WRF version 2. NCAR Tech Note, NCAR/TN-468 + STR, pp 88

Skamarock WC, Klemp JB, Dudhia J, Gill DO, Barker DM, Duda MG, Huang X-Y, Wang W, Powers JG (2008) A description of the advanced research WRF version 3. NCAR Tech Note NCAR/TN 475 STR, pp 125

Stauffer DR, Seaman NL (1990) Use of four-dimensional data assimilation in a limited-area mesoscale model. Part I: experiments with synoptic-scale data. Mon Weather Rev 181:1250–1277

Stauffer DR, Seaman NL (1994) Multiscale four-dimensional data assimilation. J Appl Meteor 33:416–434

Sun J, Crook NA (1997) Dynamical and microphysical retrieval from Doppler radar observations using a cloud model and its adjoint: part I model development and simulated data experiments. J Atmos Sci 54:1642–1661

Sun J, Crook NA (2001) Real-time low-level wind and temperature analysis using single WSR-88D data. Weather Forecast 16:117–132

Sun J, Chen M, Wang Y (2010) A frequent-updating analysis system based on radar, surface, and mesoscale model data for the Beijing 2008 forecast demonstration project. Weather Forecast 25:1715–1735

Sun J, Zhang Y (2008) Assimilation of multiple WSR_88D radar observations and prediction of a squall line observed during IHOP. Mon Weather Rev 136:2364–2388

Sun J, Zhang Y, Wiener G, Oien N, Mahoney W (2011) A rapid-updated wind analysis system based on mesoscale model, radar, and surface data for ramp-event wind energy forecasting. In: AMS second conference on weather, climate and the new energy economy, Seattle, 23–27 Jan. http://ams.confex.com/ams/91Annual/webprogram/Paper182972.html

Vallis GK (2006) Atmospheric and oceanic fluid dynamics: fundamentals and large-scale circulation. Cambridge University Press, Cambridge, p 745

Vanderwende B, Lundquist JK (2012) The modification of wind turbine performance by statistically distinct atmospheric regimes. Environ Res Lett 7:034035. doi:10.1088/1748-9326/7/3/034035

Warner TT (2010) Numerical Weather and Climate Prediction. Cambridge University Press, Cambridge, 526 pp.

Wyngaard JC (2010) Turbulence in the atmosphere. Cambridge University Press, Cambridge

Regional Climate Modelling for the Energy Sector

Jack Katzfey

Abstract Detailed, regionally specific information about the present and future climate is useful to the energy industry. Climate change impacts both energy demand (e.g. heating and cooling) and also generation (e.g. wind regime for turbines). New projects need to consider the current climate as well as future climate projections for the lifespan of the infrastructure. Typical global climate models give information for a 100–200 km grid box but dynamically downscaling using a regional climate model (RCM) can give more detailed information. This chapter describes the different approaches to dynamical downscaling, the issues associated with them and possible applications for the energy sector.

1 Introduction

The energy sector has a significant need for regional climate information for planning and risk management. In the past, climatic conditions were assumed to be relatively constant. With the enhanced emissions of greenhouse gases, there is now a consensus that the climate is changing, in particular becoming warmer (IPCC 2007; Garnaut 2011). It is also predicted that the frequency of extreme events may increase in some areas (IPCC 2012). There may be shifts in seasonal start dates, with more likelihood of extended warm, cold or dry periods, which will also impact on energy needs.

The energy industry is primarily concerned with providing sufficient generating capacity or supply to meet anticipated peak energy demands. During warmer summers, people use more air conditioning, which requires more electricity and more consumption of fuels used in generators, but during warmer winters, there

---

J. Katzfey (✉)
CSIRO Marine and Atmospheric Research, Aspendale, Australia
e-mail: Jack.Katzfey@csiro.au

will be less heating (Thatcher 2007) and reduced consumption of fuels such as natural gas and fuel oil. The net effect on total energy demand will depend on local climate conditions, but also on technology and the behavioural response of consumers to weather conditions (Considine 2004), as well as population growth, price and other social factors. However, realistic climate change information on the regional (local) scale for the energy resource being evaluated is of great importance to ensure that current and future energy needs are met.

Global climate models (GCMs) are our best tools for understanding the global climate system and projecting changes under future scenarios, since climate is influenced by factors on a whole-earth scale, and the GCMs are "coupled", with both atmosphere and ocean components that interact. However, GCMs are generally run at a resolution of around 200 km due to computational limitations. At this coarse resolution, they are not able to capture the effects of mountains and the proximity of large bodies of water such as oceans and other local factors which greatly influence regional climate, and in turn, energy demand. There are several techniques for producing regional climate change information from GCMs, including statistical and statistical-dynamical methods,[1] but the technique discussed here is the use of Regional Climate Models (RCMs) to dynamically downscale GCM projections to the local level, since this is the most physically consistent method. An RCM is a computer model that simulates the atmosphere and land surface for a region. In dynamical downscaling, the fields (such as pressure, temperature, moisture and momentum) of the GCM (called the host model) are used by the RCM, along with fine-scale information on features such as topography and land use that affect the physics of the atmosphere at the regional scale, to produce a more detailed, coherent picture of current and future climate.[2]

Before RCMs can be used for study of current weather events and climatic conditions or production of future climate projections, they require initialisation, or input of an atmospheric field dataset as a starting point. This can be GCM output, as discussed above, or analyses of the atmosphere. Actual observations, such as those from weather stations, tend to be available only for selected variables, are irregularly spaced and may be sparse over some areas outside of heavily populated areas and over oceans. They may also lack quality control or be strongly influenced by local considerations and not be representative of a larger area. Often they are not available for a sufficient length of time to capture true long-term climate signals as well as inter-annual variability. And of course they cannot provide information about future conditions. These factors make them unsuitable for driving an RCM without careful modification to produce a gridded global dataset.

Gridded global analyses take available observations (surface, upper-air, satellite, etc.) and assimilate them onto a regularly spaced grid. The data assimilation system typically has some atmospheric model as a basis for the first guess of the

---

[1] For a discussion of non-dynamic, statistical downscaling techniques, see Wilby et al. (2004).

[2] For a review of the numerous previous studies that have examined the capability of RCMs to downscale the climate to the regional and local scale see, for example, Rummukainen (2010).

state of the atmosphere, which is then modified by the observations, producing what is called a reanalysis dataset. There are numerous reanalysis products available from groups such as National Centers for Environmental Prediction (NCEP) in the US or the European Centre for Medium-Range Weather Forecasts (ECMWF). Typically they are provided globally at 80 km or larger resolution, which is generally too coarse to provide the detail necessary for assessing regional climate for energy needs (see also 'Current Status and Challenges in Wind Energy Assessment' by Gryning et al. and 'Wind Power Forecasting' by Haupt et al.). This reanalysis data can be used to drive an RCM in order to produce a gridded dataset that is more spatially and internally consistent and at a finer scale (down to 10 km or less). However, care must be taken in using these datasets, since the interpolation scheme or biases in the reanalysis may have an influence on the RCM results. Special care should be used in regions with few observations since the reanalyses may be less accurate there.

The large-scale analyses of the atmosphere produced by GCMs or the gridded datasets produced by the process just described can then be used to drive the regional model that produces the high-resolution data for current climate applications such as energy assessments or seasonal predictions of factors including temperature, wind speed, humidity and precipitation that influence energy use (Li and Sailor 1995). To produce high-resolution future climate projections necessary for long-term planning in the energy sector, RCMs must be fed data from GCM simulations made under greenhouse gas emissions scenarios.

For both current and future climate projections, the effect of the orography and land surface on the local climate can be resolved more realistically by the finer resolution RCMs than by GCMs or observations. This gives the potential to show a more detailed pattern of response to climate change. That is, the RCM may indicate that mean rainfall response to a warming climate is different than it is in the broader region in an area of notable topography (e.g. on the leeward side of a mountain range), or over regions of different land use (e.g. over cleared land). Achieving this detail is the goal of regional climate projections, and should be of strong interest to the energy sector.

The importance of high resolution can be seen very clearly in climate projections for Pacific Islands such as Fiji (see Fig. 1), where the main island, Viti Levu, is not resolved in the 200 km GCM grid but is only represented by an ocean grid point (not shown), while at 60 km the island is resolved as three grid boxes but with only minor elevation differences indicated and a maximum height of less than 150 m. At the finer 8 km grid possible with an RCM, the land mass and coastline of the island are more accurately represented. The mountains in the interior, a key feature, are also present, with elevations of nearly 800 m (the highest actual peak is 1,324 m). Topography affects wind, temperature and rainfall patterns, as can be seen in the simulated annual rainfall for Fiji (Fig. 2), where there is no effect from the land in the GCM simulation (left), since Fiji is not resolved. At 60 km (centre), there is some indication that there is more rain in the interior of the island. But at 8 km (right), a more detailed and complex pattern with higher precipitation amounts on the eastern (windward) side of the island is shown.

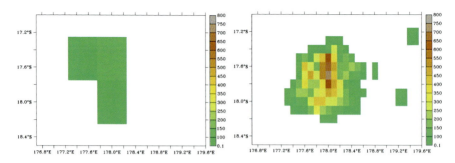

Fig. 1 Surface topography for Fiji (with heights in m) at 60 km resolution (*left*) and 8 km resolution (*right*)

Fig. 2 Simulated Fiji annual rainfall (mm day^{-1}) for 1980–2000 from GCM at 200 km resolution (*left*), RCM at 60 km resolution (*centre*) and RCM at 8 km resolution (*right*). The RCM simulations were produced using CSIRO's stretched-grid Conformal Cubic Atmospheric Model (CCAM)

In summary, the main aims of regional climate modelling are:

- To provide more detailed (and potentially more accurate) information on current climate and future climate change through higher resolution simulations with better resolved physical processes and surface inputs. The more detailed topography and land-use information input into the regional model produces more information than coarser resolution analyses and more spatially consistent information than gridded observations. In addition, the finer resolution of the regional model should more realistically represent atmospheric phenomena and dynamics.
- To validate the simulations of current climate against observations and improve modelling techniques so that accuracy continues to improve. Future climate change simulations will then be superimposed on a more realistic current climate, which in turn should give more confidence in climate change projections.

- To produce a range of downscaled climate change signals in order to capture the range of possible future climate projections, consistent with the range seen in the large-scale GCMs on which the RCMs are based, and to identify key climate change features related to higher resolution and surface information, especially terrain. This is the approach taken in the IPCC projections of climate change (IPCC 2007, 2012), where it is acknowledged that there are biases and uncertainty in model projections, so that the consideration of the output of as many models as feasible, in what is known as ensemble projections, is the most reasonable way to deal with uncertainty.

In the following sections, the various approaches to regional climate modelling through dynamical downscaling, such as use of time-slice simulations, Limited Area Models (LAMs) and stretched-grid simulations are discussed, and some of the advantages and disadvantages of each are addressed, along with examples of why this information is important for the energy sector and how it can be used effectively.

2 Methods/Techniques

As mentioned above, regional climate models are designed to produce higher resolution climate information than the host (input or driving) model or dataset. There are three main components to a regional climate modelling system: (1) the atmospheric model (the RCM), (2) the host data used as inputs to the RCM and (3) the surface data (land use, orography, etc.) used by the land surface scheme in the RCM (see Fig. 3). The host models or datasets that are typically used to drive the regional model are global coupled atmospheric-oceanic models or GCMs (for climate change) or gridded analyses of the atmosphere (for the current climate). For climate change simulations, there may also be other inputs or forcings, such as varying levels of CO_2 and other greenhouse gases, to test their effect on temperature rise projections. The modelling process may involve multiple downscaling, where a coarser resolution version of an RCM is downscaled further to the desired resolution (e.g. from GCM scale to 60 to 8 km).

The core of the downscaling system is the atmospheric model, which needs to (a) run efficiently for long simulation time periods (up to 100 years or more); (b) use the host and other input data appropriately (explained later); (c) process sophisticated mathematical equations for atmospheric physics and dynamics to produce a realistic climate response and (d) output the relevant fields for the application for which it is to be used. The model needs to be efficient since the simulations are computationally intensive and it may take weeks or months to produce a simulation for 100 years into the future in order to capture a statistically significant change signal relative to the internal and natural variability of the climate.

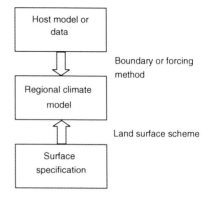

Fig. 3 The three main components of a regional climate modelling system: the input (host) data, the modelling system and the land surface specification

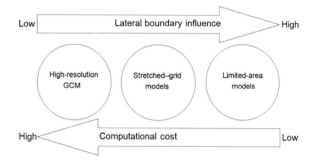

Fig. 4 Comparison of lateral boundary influence and computational expense of GCMs, stretched-grid (or variable-resolution) models and LAMs

The RCM must be able to incorporate the host data, which is typically input every 6 hours as the model runs. There are different techniques to accomplish this, dependent upon the RCM used. The surface specifications for the finer resolution RCM (topography, land use, soil properties, albedo and surface roughness) should be more detailed than in the host model.

Producing regional downscaled projections in practice will always be a balance of the scientific ideal and computation limitations. Therefore, decisions about the method to use and the approach will follow a cost–benefit analysis (Fig. 4). Ideally, one would run a global coupled model (GCM) at the resolution required in order to capture the global atmosphere at the finer resolution desired. However, this is too computationally expensive to be practical. If the information is needed for only a local region, the information outside the region of interest may be superfluous, so generating it wastes time and resources. A way to reduce computational costs is to create time-slice simulations, where only the atmospheric component of the coarse-resolution GCM is re-run at higher resolution, usually only for short time periods, typically 20–30 years. However, the gaps in time of the data can cause problems in the analyses of trends or in calculation of statistical significance. Also, running models at different resolutions requires a lengthy process of model testing and adjustment to ensure a good simulation. To avoid these problems, there are two main alternative approaches, as given below.

2.1 Limited Area Models

The traditional approach for dynamical downscaling is to run a limited area atmospheric model (LAM) over the region of interest only instead of over the whole globe. However, one needs to specify information for all atmospheric fields at the lateral boundaries of the LAM. The LAM then takes this boundary atmospheric information and develops it as it propagates through the model grid. There are several issues related to these lateral boundary conditions, due to the differing resolutions, physics and dynamics of the LAM and the host:

1. There can be differences in evolution of weather systems within the LAM. There will be situations where internally generated weather systems that are not represented in the host data will propagate to the lateral boundaries and cause inconsistencies between the internal LAM and the lateral boundary data coming from the host. These internal systems need to be damped near the boundaries so they are not reflected, influencing the internal climate of the LAM.
2. Potentially large gradients can occur in the boundary region as the internal model (LAM) develops differently than the host model (GCM). For example, if the LAM has a warm bias relative to the host, producing slightly higher temperatures, an artificial gradient of temperature will develop near the boundaries.
3. There may be undesirable effects in the LAM when the host data is interpolated to the finer LAM grid. The type of interpolation scheme used will affect the data quality being used at the lateral boundaries. The change in resolution from host to LAM may produce inconsistencies in the surface specification (specifically topography) between the host and the LAM. Thus, the vertical atmospheric profile may need to be adjusted for the different surface heights of the host data and the LAM, which may again introduce inconsistencies.
4. Large amounts of data are needed in order to specify the vertical profile of data at all lateral boundary points frequently enough (typically 6 h) to capture the atmospheric flow realistically, including possible diurnal effects. In models, the atmosphere is represented by 18 or more vertical levels extending to either the top of the atmosphere or approximately 20 km above the surface of the earth, with five variables (pressure, temperature, moisture and two wind components) specified at each grid box in the boundary region as a minimum.

A significant emphasis in LAM design has been on developing systems that minimise these problems. In particular, most models now have a boundary zone that provides a transition from the internal to the lateral boundary condition data and damps systems that impinge upon the lateral boundary to prevent reflections. Care must still be taken with:

1. The location of lateral boundaries (i.e. they should ideally not be in regions of significant topography);
2. The amount of change in resolution between host and LAM (one does not want to introduce damping or inconsistent data when the host model data needs to be interpolated in space and in time); and

3. How much large-scale variability, such as El Niño-Southern Oscillation (ENSO), in the large-scale model is transferred into the regional model.

One way to help minimise the potential for problems at the lateral boundaries between the regional model and the host data is to apply some form of large-scale filter or 'nudging' of the host data into the internal domain of the regional model (e.g. Laprise et al. 2012; Yhang and Hong 2011). With a large-scale filter, features from the host model at scales of greater than, say, 2,000 km can be introduced into the RCM, while those at smaller scales are excluded. With nudging, the grid point data of the LAM is 'relaxed' to the interpolated host data over a certain time period (e.g. Davies and Turner 1977). These techniques will help ensure that the regional model cannot drift or deviate significantly from the host model, at least for the time and space scales for which the techniques are applied. This can help solve some of the lateral boundary compatibility issues, but may also distort the results generated by the regional model if not applied appropriately. Care must also be taken to ensure that the high-resolution information from the regional model that is necessary to determine the climate change signal, which may be subtle, is not damped.

In summary, the strengths and weaknesses of limited area models are:

Strengths of LAMs

1. There is less computational cost than for GCMs, since you only compute for the area of interest (it should be noted that the computational cost is still much greater than for statistical methods).
2. Only regional high-resolution surface input datasets are required.
3. The model can be more optimally configured for a given resolution, which is not possible if there are a variety of resolutions, as for stretched-grid RCMs (see following section).
4. There are several established LAM systems that have had a lot of development and large user communities, so various technical aspects are well researched.

Weaknesses of LAMs

These are primarily due to lateral boundary problems, as discussed previously:

1. Difficulties passing the large amount of atmospheric data that is required from the host model through the lateral boundaries.
2. Scale incompatibilities between RCM and boundary data, requiring interpolation that may introduce error.
3. Differences between internal model climatology and boundary data due to differing model formulations that may cause unrealistic gradients and climate features.
4. Lack of two-way interaction of the regional domain with the larger scales.

2.2 Stretched-Grid RCMs

Another solution for generating detailed future climate information at reasonable computational cost is to use stretched-grid atmospheric models (Fox-Rabinovitz et al. 2006, 2008). Stretched-grid or variable-resolution models, hereafter referred to as stretched-grid regional climate models (SGRCMs), have come to be used more commonly in the last 10 years, since they help address the lateral boundary problems inherent in LAMs. SGRCMs are constructed with a model grid that changes resolution from higher in the region of interest to lower in regions further away.[3] For example, for simulations of a relatively small region with high degree of climate variation such as Tasmania, the grid resolution may be 10 km over the area of interest, while it is much greater than the typical GCM grid of 200 km for areas on the opposite side of the globe in the North Atlantic. Generally, SGRCMs are global, but some limited area models have been developed with variable resolutions as well (Tang et al. 2012) in an attempt to minimise some of the lateral boundary problems related to abrupt changes in resolution from host to regional model. In terms of overall costs and benefits, the advantages gained by not having to deal with lateral boundary problems in SGRCMs offset the added computational cost associated with doing computations for regions outside the area of interest. However, the model dynamics and physics must be designed for the range of horizontal resolutions of the grid used.

Although SGRCMs do not have lateral boundaries, they do still need some input of atmospheric data from the host model, especially in regions with resolution equal to or greater than the host model. Stretched-grid models can typically be run in two modes: (1) with only surface data such as sea-surface temperature and sea ice from the host model as input, so the RCM generates the atmospheric fields through its own physics and (2) with some sort of grid point nudging, or with application of a large-scale filter as discussed in the previous section (see Thatcher and McGregor 2009, for an example of use of a scale-selective digital filter in an SGRCM). Using only surface data from the host model allows more flexibility for potential adjustment of some of the biases in the host model, for example bias correction of sea-surface temperatures (see Katzfey et al. 2009, for a description of the technique).

There are two cases where applying atmospheric nudging in an SGRCM is useful: (1) when downscaling coarser scale atmospheric analyses, either for a multi-year run to test the ability of the model to capture actual, observed weather systems or for shorter term simulations where one is trying to accurately simulate an event as closely as possible or (2) for multiple downscaling, i.e. to downscale one simulation to even higher resolution (also called multiple nesting). In the latter case, the higher resolution domain typically has a much higher stretching factor (amount of change in resolution, such as from 8 km in the high-resolution domain

---

[3] Although the term used here is SGRCM, atmospheric-only models that have uniform resolution over the globe (i.e. they are not coupled atmosphere–ocean models, like GCMs) are also included.

to greater than 1,000 km in the coarse-resolution portion of the grid). If the grid is stretched even more, for example, to 1 km resolution, there may be hundreds of grid boxes over the area of interest, but only three or four grid boxes representing most of the rest of the world. In order to ensure the information influencing the high-resolution region is realistic, the coarser resolution domain must be nudged, generally by pushing it towards the analyses or the output of another model that was not run with as large a change in resolution.

One advantage of SGRCMs is the freedom to have some feedback from the high-resolution portion of the domain to other regions. This allows interactions between regions—called teleconnections. Global phenomena, such as El Niño, while primarily associated with changes in sea-surface temperatures in the central tropical Pacific, cause changes in weather and climates around the world through these teleconnections. In LAMs, this teleconnection information needs to be transmitted through the lateral boundaries. In SGRCMs, these atmospheric teleconnections to ocean temperature are developed within the model itself.

Strengths of Stretched-Grid RCMs

1. There are no lateral boundaries.
2. There is a potential to correct some of the biases in the input data through bias correction techniques.
3. There is a potential for two-way interaction between the high-resolution area and outside regions.
4. Potentially less data is required than for a LAM if only surface input data from the host model is required. Thus, input data may be less 'biased'.

Weaknesses of Stretched-Grid RCMs

1. The model needs to be configured to run correctly at a range of horizontal resolutions.
2. When run with only lower boundary conditions (such as sea-surface temperatures) from the host model, the SGRCM generates its own unique atmosphere that can differ from the host model. This means that the downscaling takes on much of the character of a new model, rather than simply adding detail to the host model simulation.
3. Global surface datasets are required to run the model.
4. Global datasets of other atmospheric variables are needed if atmospheric nudging is required. These datasets will be larger than the regional ones required for LAMs.

Point 2 can be viewed as a strength, since the SGRCM can avoid propagating biases in the host model, or a weakness, since now it must realistically simulate the full atmospheric dynamics and thermodynamics, with a greater need for conservation of atmospheric properties.

Figure 5 summarises some of the processes involved and the various grids used for regional climate modelling. Decisions on which method to use must take into consideration their different strengths and weaknesses, and their suitability for the intended purpose.

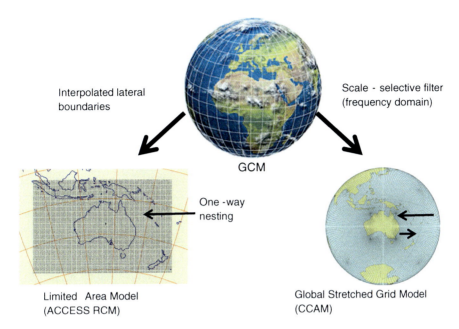

Fig. 5 Comparison of grids in a typical GCM, a LAM (example shown is ACCESS RCM) and an SGRCM (example shown is CCAM). To change from the coarse scale of the GCM to the LAM, atmospheric data must be interpolated at the lateral boundaries, and two-way interaction is difficult, unlike the Stretched-Grid Model (SGRCM), where information can be passed in both directions. Nudging or a scale-selective filter may be used in an SGRCM to introduce input from the host model

3 Application of Regional Climate Models for the Energy Industry

There are two main applications of regional climate models in the energy industry. The first application is for a more consistent assessment of the current climate. A key feature of the RCM is the consistency and completeness of its outputs, including winds, radiation, temperatures and clouds. The model can provide data that is spatially and temporally self-consistent (since it was created using one model) and at greater spatial and temporal resolution than observations. This allows assessment, for example, of wind or solar power potential in regions without sufficient fine-resolution observed data. The user can also investigate interrelationships of the various fields potentially relevant to the energy industry (e.g. Schaeffer et al. 2012).

The other application of regional climate modelling of most use to the energy industry is to investigate the potential changes in atmospheric data over time, especially for future climates and changes due to greenhouse warming. The consensus is that an enhanced greenhouse effect will cause warming of every region of the world to varying amounts, which will affect energy demand. Generally in warmer locations, there is more reliance on electricity than on direct use of natural gas, oil and other fuels, and these areas also tend to use more energy. "Climate change will likely increase electricity consumption on cooling but reduce the use of other fuels for heating" (Mansur et al. 2008). Studies such as that by Warren and Leduc (1981) have investigated the interaction of climate with the energy sector of the economy by quantifying the relationship between heating degree days and natural gas consumption for space heating, for use by utilities, public service commissions and other agencies at the local, state, regional and national levels. They found that demand responds to price and class of consumer, as well as seasonal climate, but it is still necessary to take into account known causes such as climate when undertaking this kind of economic analysis.

Because energy demand is a complex phenomenon involving technological, social and climate factors, each of which is complex and has uncertainty, care must be taken when making projections for the future. Although the science of climate modelling is constantly advancing, there are uncertainties due to internal climate variability, model uncertainty due to different internal climatologies and physics, and scenario uncertainty, since new technologies or economic changes may impact on emissions of greenhouse gases (Hawkins and Sutton 2009).

In order to run a regional model to investigate climate change, a global coupled model run, which includes both atmospheric and ocean information, is needed to provide the host data. An ensemble approach is recommended in order to address the uncertainties noted above and attempt to capture the range of possible futures. The ensemble should sample the fundamental uncertainty of future emissions by running simulations under several emission scenarios. Also, the ensemble should sample the 'model uncertainty' by using simulations from several coupled models and ideally several regional climate models. An example of this technique is the study by Kjellström et al. (2010), who used an ensemble of 16 RCMs with four emissions scenarios to produce simulations of seasonal mean temperature, precipitation and wind speed over Europe for the years 1961–2100. Boundary conditions were from 7 GCMs. They found significant temperature increases for all of Europe in the next decades, with precipitation increases in northern and decreases in southern Europe, and general wind speed decreases, except in the northern seas and parts of the Mediterranean. They traced uncertainties in current climate projections to representation of large-scale circulations by GCMs, while uncertainty related to radiative and other forcings was most important by the end of the century. For an example using regional climate model ensemble projections for hydrology applications, where output was used to drive a streamflow model, see Bennett et al. (2012).

4 Summary and Discussion

Predicting future energy use is complex, involving a mix of socio-economic and technological factors as well as consideration of changes in climate, including likely impacts of projected global warming. The limitations in spatial resolution of general circulation models (GCMs) make them less useful for regional climate impact assessments, but regional climate models (RCMs) can provide higher spatial and temporal resolution, self-consistent atmospheric datasets with detailed information on seasonal climate factors such as temperature, wind speed, humidity and precipitation that affect energy demand. These datasets can be used for a range of applications, including projections of future energy needs, as well as for shorter term planning of infrastructure and demand, such as solar and wind potential. As with any data used for energy applications, however, the end user needs to ensure that the nature and quality of the information used is sufficient for the purposes required.

All climate model simulations generate a view of the range of possible climate responses under specific emission scenarios, given our best understanding of climate processes. These future scenarios are a useful tool for planning, but they are uncertain themselves, so predictions based upon them must be seen as the best possible with the current state of knowledge. Although regional climate models can provide high-resolution datasets that look highly realistic, care must be taken when using model outputs directly, without an understanding of the modelling technique chosen and its possible limitations. Since model data inevitably have some offsets (bias) compared to a set of observations, some post-processing of model outputs is suggested to remove any significant biases before the output is used in any application.

The combined climate and non-climate induced changes in energy demand may pose significant challenges to policy and investment decisions (Matthias and Lin 2006). As noted by Schaeffer et al. (2012) in their review of energy sector vulnerability to climate change, factors with regional weather and climate components need to be identified so that atmospheric climatology or forecasts can be transformed into decision aids, especially since increases in energy costs and projected increases in the intensity of extreme weather events, affecting transportation and energy transmission in particular, have the potential for dramatic social impacts. For this reason, it is important that energy projects take weather and climate sensitivities into account in their project designs, management and mitigation and adaptation plans (Mansur et al. 2008).

Realistic regional climate simulations are necessary as part of the energy industry decision-making process (Schaeffer et al. 2012). The techniques used in regional climate modelling are constantly improving, and the use of ensemble predictions addresses many of the problems of uncertainty due to model physics and dynamics and choice of emission scenario.

References

Bennett J, Ling FLN, Post DA, Grose MR, Corney SP, Graham B, Holz GK, Katzfey JJ, Bindoff NL (2012) High-resolution projections of surface water availability for Tasmania, Australia. Hydrol Earth Syst Sci 16(5):1287–1303

Considine TJ (2004) Climate change: impact on the demand for energy. In: Cleveland, CJ (ed) Encyclopedia of energy. Elsevier, New York, pp 393–400. ISBN 9780121764807, doi: 10.1016/B0-12-176480-X/00400-9

Davies H, Turner R (1977) Updating prediction models by dynamical relaxation: an examination of the technique. Q J Am Meteorol Soc 103:225–245

Review Garnaut (2011) Australia in the global response to climate change. Cambridge University Press, Melbourne

Fox-Rabinovitz MS, Côté J, Dugas B, Déqué M, McGregor JL (2006) Variable resolution general circulation models: Stretched-Grid Model Intercomparison Project (SGMIP). J Geophys Res 111. D16104, doi: 10.1029/2005JD006520

Fox-Rabinovitz MS, Côté J, Dugas B, Déqué M, McGregor JL, Belochitski A (2008) Stretched-grid model intercomparison project: decadal regional climate simulations with enhanced variable and uniform-resolution GCMs. Meteor Atmos Phys 100:159–177

Hawkins E, Sutton R (2009) The potential to narrow uncertainty in regional climate predictions. Bull Am Meteorol Soc 90:1095–1107. doi: http://dx.doi.org/10.1175/2009BAMS2607.1

IPCC (2007) Climate change 2007: synthesis report. Contribution of Working Groups I, II and III to the Fourth Assessment Report of the Intergovernmental Panel on Climate Change.In: Pachauri RK, Reisinger A (eds). IPCC, Geneva, p 104

IPCC (2012) Summary for policymakers. In: Field CB, Barros V, Stocker TF, Qin D, Dokken DJ, Ebi KL, Mastrandrea MD, Mach KJ, Plattner G-K, Allen SK, Tignor M, Midgley PM (eds) Managing the risks of extreme events and disasters to advance climate change adaptation. A special report of Working Groups I and II of the Intergovernmental Panel on Climate Change. Cambridge University Press, Cambridge, pp 1–19

Katzfey JJ, McGregor JL, Nguyen KC, Thatcher M (2009) Dynamical downscaling techniques: impacts on regional climate change signals. In: Anderssen RS, Braddock RD and Newham LTH (eds) 18th World IMACS congress and MODSIM09 international congress on modelling and simulation. Modelling and Simulation Society of Australia and New Zealand and International Association for Mathematics and Computers in Simulation, July 2009, pp 2377–2383. ISBN: 978-0-9758400-7-8. http://www.mssanz.org.au/modsim09/I13/katzfey_I13.pdf

Kjellström E, Nikulin G, Hansson U, Strandberg G, Ullerstig, A (2010) 21st century changes in the European climate: uncertainties derived from an ensemble of regional climate model simulations. Tellus A 63(1):24–40. http://dx.doi.org/10.1111/j.1600-0870.2010.00475.x doi: 10.1111/j.1600-0870.2010.00475.x

Laprise R, Kornic D, Rapaic′ M, Šeparovic′ L, Leduc M, Nikiema O, Di Luca A, Diaconescu E, Alexandru A, Lucas-Picher P, de Elı́'a R, Caya D, Biner S (2012) Considerations of domain size and large-scale driving for nested regional climate models: impact on internal variability and ability at developing small-scale details. In: Berger A, Mesinger F, Sijacki D (eds) Climate change inferences from paleoclimate and regional aspects. ISBN: 978-3-7091-0972-4

Li X, Sailor DJ (1995) Electricity use sensitivity to climate and climate change. World Resource Review 7(3 September 1995):334–346

Mansur ET, Mendelsohn R, Morrison W (2008) Climate change adaptation: a study of fuel choice and consumption in the US energy sector. J Environ Econ Manage55(2 March 2008):175–193. ISSN 0095-0696, doi: 10.1016/j.jeem.2007.10.001

Matthias R, Lin A-C (2006) Regional energy demand and adaptations to climate change: Methodology and application to the state of maryland, USA. Energy Policy 34(17):2820–2833.10.1016/j.enpol.2005.04.016

Rummukainen M (2010) State-of-the-art with regional climate models. WIREs Clim Change, 1: 82–96. doi: 10.1002/wcc.8

Schaeffer R, Salem Szklo A, Frossard Pereira de Lucena A, Soares Moreira Cesar Borba B, Pinheiro Pupo Nogueira L, Pereira Fleming F, Troccoli A, Harrison M, Sadeck Boulahya M, (2012). Energy sector vulnerability to climate change: a review. Energy 38(11 February 2012):1–12

Tang Y, Lean HW, Bornemann J (2012) The benefits of the Met Office variable resolution NWP model for forecasting convection. Meteorol Appl 16(2):129–141

Thatcher MJ (2007) Modelling changes to electricity demand load duration curves as a consequence of predicted climate change for Australia. Energy 32(9):1647–1659

Thatcher M, McGregor JL (2009) Using a scale-selective filter for dynamical downscaling with the conformal cubic atmospheric model. Mon Wea Rev 137:1742–1752

Warren HE, Leduc SK (1981) Impact of climate on energy sector in economic analysis. Journal of Applied Meteorology 20(12):1431-1439. doi: 10.1175/1520-0450(1981)020<1431:IOCOES>2.0.CO;2

Wilby RL, Charles SP, Zorita E, Timbal B, Whetton P, Mearns LO (2004) Guidelines for use of climate scenarios developed from statistical downscaling methods. http://www.ipcc-data.org/guidelines/dgm_no2_v1_09_2004.pdf

Yhang Y-B, Hong S-Y (2011) A study on large-scale nudging effects in regional climate model simulation. Asia-Pacific J Atmos Sci 47(3): 235-243. doi: 10.1007/s13143-011-0012-0

In Search of the Best Possible Weather Forecast for the Energy Industry

Pascal Mailier, Brian Peters, Devin Kilminster and Meghan Stephens

Abstract Weather forecasts will never be perfect and hence they will always contain some degree of uncertainty. In this chapter, we argue that uncertain weather forecasts expressed in a probabilistic format can provide more value to users in the energy sector than simple, apparently confident deterministic forecasts, even when the latter show a satisfactory level of accuracy. Knowledge of the user-specific loss function is required to achieve best value.

1 Introduction

Weather forecasts cannot be perfect as they always contain intrinsic errors that grow inexorably with time. The question arises then of what constitutes the 'best possible' forecast. The answer to this question is not unique as the notion of 'goodness' is not absolute and depends strongly on forecast application.

Each end of the forecasting line involves indeed two different communities. At one end, atmospheric scientists produce weather forecast models and techniques, whereas at the receiving end, users buy weather forecast products to aid their decision-making. The available tools and methods to assess the performance of these forecasts have been developed primarily by scientists in order to monitor and improve weather and climate forecast systems with a focus on meteorological variables. For a meteorologist, the notion of a 'good' forecast typically refers to how close the forecasts are to the verifying observations. On the other hand, forecast users look at forecast performance from an operational and commercial

---

P. Mailier (✉)
Royal Meteorological Institute of Belgium, Brussels, Belgium
e-mail: pascalm@oma.be

B. Peters · D. Kilminster · M. Stephens
Meteorological Service of New Zealand Ltd. (Metra), Christchurch, New Zealand

perspective. For users in the energy sector, the economic advantage provided by weather forecasts is fundamental. The cost of forecast error becomes the dominant criterion and non-meteorological factors may be looked at, e.g. timeliness. As will be shown in this chapter, these two views are not necessarily equivalent and they can even conflict.

Since errors are inherent in forecasting, i.e. forecasts are always uncertain, another basic requirement for a 'good' weather forecast is that it should contain sufficient information on forecast uncertainty. A sensible way to do this is to present the forecast in probabilistic format. This is not only necessary to ensure coherence between the forecasters' judgements and their forecasts, but above all to enable the user to make proper use of the forecasts according to their own cost and loss functions.

In the next section, the three distinct types of forecast goodness (consistency, quality and value), as identified by Murphy (1993), are introduced and discussed. We also argue that the need for consistency leads to the necessity of expressing forecasts in a probabilistic format. In Sect. 3, we demonstrate with a practical example that optimising forecast quality without taking the user's need into account does not necessarily lead to optimal value. In Sect. 4, the superiority of probability forecasts over deterministic forecasts is exemplified. We show how the full distribution of ensemble forecasts provides useful probabilistic information on forecast uncertainty that helps the user to take optimal decisions. Section 5 sums up the results and presents some conclusions.

2 What Makes a 'Good' Forecast?

Murphy (1993) identifies three basic forecast properties that each defines a distinct type of forecast goodness.

2.1 Consistency

This property refers to the correspondence between the forecasters' judgements and their forecasts in the sense that a 'good' forecast must reflect the forecasters' degree of belief. This means that a 'good' forecaster should refrain from hedging, i.e. artificially change his level of confidence in some or all possible forecast outcomes so as to be on the 'safe' side. The requirement for consistency also entails that deterministic forecasts (simple forecasts of the most likely outcome) should be avoided since they do not convey any information on the underlying forecast uncertainty. Forecasters are rarely 100 % confident of their own forecasts, but they may be forced to select only the most likely scenario even though other outcomes are also plausible. This practice provides the user with information that is simple to understand and to handle. However, deterministic forecasts are incomplete as they do not contain all the information available. In contrast, forecasts issued in a probabilistic format are more consistent, and therefore 'better'. Despite the superiority of probability forecasts, deterministic forecasts are much more widely used owing to their simplicity and (in the case of numerical prediction) their higher resolution. Another explanation for the success of deterministic forecasts is convenience as that they transfer the onus of threshold-based 'Yes/No' decisions from the user to the forecaster. Moreover, probabilities are often misinterpreted as a reflection of the forecaster's ignorance instead of useful information that can positively feed the decision process.

2.2 Quality

This aspect of forecast goodness refers to the correspondence between the forecasts and the matching observations. Measures of quality provide the most straightforward and natural way to assess forecast goodness in the sense that forecasts that are 'closer' to the verifying observations are 'better'. A wide range of statistical methods have been developed to measure forecast quality and the reader is referred to, e.g. Jolliffe and Stephenson (2011) for a detailed survey of these techniques. Measures of forecast quality such as bias (mean error), accuracy (precision) and association (correlation), for example, belong to this category.

2.3 Value

Consistency and quality are facets of forecast goodness that are essentially focused on the content of meteorological information present in the forecasts. Hence, they are not user-specific and in particular they ignore the impact, positive or negative, this information has on the decisions the users take on the basis of these forecasts. Forecast value, however, refers to the incremental economic advantage and/or other benefits realised by decision makers through the use of the forecasts. Estimating forecast value is in general not straightforward because it requires knowledge of a particular user's cost and loss functions, which in practice can be very complex and commercially sensitive. Very simple cost/loss models have been devised to demonstrate the economic advantage of basing decisions on probability forecasts rather than on deterministic forecasts, see, e.g. Richardson (2000). In Sect. 4, a concrete example is given for a user in the energy sector.

From these three definitions, it follows that forecast goodness can have different meanings depending on which aspect (consistency, quality or value) we consider. We have also explained that in the real world, it can be very difficult—and sometimes even impossible—to quantify forecast value. For this reason, it is common practice to assume that value is positively associated with quality. In many instances, this is a reasonable assumption to make. One may expect, for example, that the more accurate temperature forecasts are, the more valuable they will be to an energy company. This assumption, however, is not always correct. In the next two sections, we will demonstrate with examples that using the most accurate forecast obtained from an ensemble of forecasts does not necessarily maximise value, and that the latter can be best achieved through a probabilistic approach.

3 The Fallacy of Accuracy

Deterministic forecasts that are optimised for maximum accuracy (minimum error) do not necessarily provide optimal value on time horizons ranging from a few hours to several days (forecast accuracy drops more quickly when dealing with volatile variables such as solar radiation, precipitation and wind speed than when predicting air temperature). It is widely known that the mean or median of an ensemble of forecasts is on average more accurate than any individual member of that ensemble. The meteorological interpretation of this behaviour is that the unpredictable weather systems are removed from the forecast through the averaging process. However, the reason for the ensemble mean's higher accuracy also has a statistical origin. For an ensemble of n members, the variance of the ensemble mean is reduced by a factor $1/n$. As a result, the error of the ensemble mean forecast is more constrained in magnitude than the error of any of the ensemble members. However, the price to pay for this greater accuracy is an increasingly unrealistic (too low) variability of the predicted meteorological variables. In situations with significant uncertainty (large ensemble spread) the ensemble mean lingers in the centre of the forecast distribution, whereas potentially important information in the tails is ignored. With temperature forecasts, this is often the case at lead times beyond 1 week. The example below illustrates this point.

During the first half of October 2011, Southern England experienced a period of unseasonably warm temperatures with daytime maxima sometimes well above 20 °C. An abrupt change to significantly colder conditions with overnight frost took place between 16 and 20 October. We show that this event could be much better anticipated using the full ensemble probabilistic information rather than the ensemble mean alone. The European Centre for Medium-range Weather Forecasts (ECMWF) produces ensemble forecasts twice a day (forecasts starting from 00 and 12 UTC) out to 15 days. Each ensemble consists of 51 global numerical simulations (50 perturbed members + 1 unperturbed) run at a horizontal resolution of ~ 32 km. More information on the ECMWF Ensemble Prediction System (EPS) can be found in Persson (2011). Figures 1, 2 and 3 show ensemble temperature forecasts above Reading (UK) at the pressure level of 850 hPa (about 1,500 m above sea level) on 10th October 2011 (runs of 00 and 12 UTC) and on 11th October 2011 (run of 00 UTC) out to 10 days. At this level, the air is no longer subject to diurnal heating and nocturnal cooling from the surface and the temperatures only fluctuate as a function of the air mass (warm/cold). A reasonable

In Search of the Best Possible Weather Forecast 339

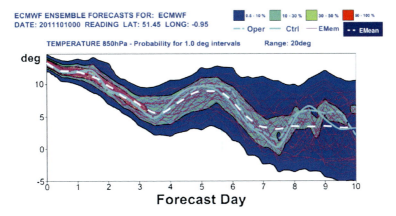

Fig. 1 Ensemble temperature forecast plume above Reading (UK) at the pressure level of 850 hPa (about 1,500 m above sea level) starting from 10th October 2011, 00 UTC and out to 10 days. The *thin purple lines* represent each of the 50 perturbed ensemble members (EMem). The continuous *blue line* represents the ensemble control (unperturbed) member (Ctrl). An additional unperturbed forecast run at twice the horizontal resolution of the ensemble control is also plotted and indicated by the *dashed blue line* (Oper). The dashed *white line* shows the ensemble mean (Emean). The ensemble probability density function estimated by means of Gaussian kernels is depicted by the shaded areas. *Source* ECMWF

estimate of the corresponding daytime surface temperatures can be obtained simply by adding 9 °C. Temperature probabilities have also been estimated by applying Gaussian kernel functions. The reader is referred to the figure captions for more explanations of the plot features.

The ensemble forecast starting from 10 October 2011, 00 UTC in Fig. 1 predicts a gradual cooling followed by a temporary warming with peak at day D + 5, then a further cooling. As expected, the ensemble mean 'consensus' forecast (dashed white line) sticks within the main body of the ensemble distribution and levels near 2.5 °C from D + 7. The fitted probability density function has a mode (peak of increased likelihood) that is also relatively central near 5 °C. Individual ensemble members, however, show large fluctuations beyond D + 7 that result in a fairly large spread (between −5 °C and 10 °C). This information rather suggests that the ensemble is 'hesitating' between cooler and warmer scenarios beyond D + 7. In fact, there is no evidence that the temperatures are likely to settle near 3 °C in that time horizon as inspection of the ensemble mean alone might imply.

Figure 2 displays the ensemble distribution produced 12 h later on 10th October 2011, starting from 12 UTC. The individual ensemble members still show substantial disagreement beyond D + 7. The ensemble mean remains steady in its central position near 2.5 °C. Nonetheless, the fitted probability density function features two modes now instead of one. This bifurcation shows that a warmer or a cooler outlook than the ensemble mean (temperature rising back to 5 °C and above, or decreasing to 0 °C or even below) is more likely than the consensus scenario in-between.

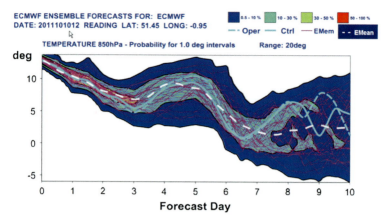

Fig. 2 Ensemble temperature forecast plume above Reading (UK) at the pressure level of 850 hPa (about 1500 m above sea level) starting from 10th October 2011, 12 UTC. Please refer to Fig. 1 for details. *Source* ECMWF

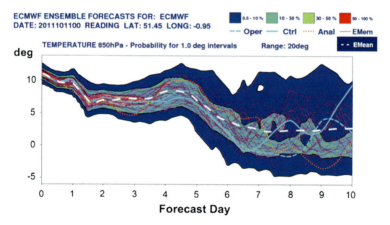

Fig. 3 Ensemble temperature forecast plume above Reading (UK) at the pressure level of 850 hPa (about 1,500 m above sea level) starting from 11th October 2011, 00 UTC. Please refer to Fig. 1 for details. The dotted orange line shows the actual temperature estimated from the available observations (Anal). *Source* ECMWF

In Fig. 3, which shows the next ensemble forecast starting from 11th October 2012 at 00 UTC, a larger proportion of members have rallied in support of the cooler scenario with the cooler mode becoming clearly dominant. The confidence in a cooler outcome has therefore increased significantly. In contrast, the ensemble mean still lingers around more conservative consensus values. The verification provided 10 days later and represented on the graph by the dotted line confirms a cooling to nearly −4 °C at D + 9. On the one hand, this cooling could not be well anticipated 1 week ahead by the ensemble mean alone despite its higher accuracy on average compared with individual ensemble members. On the other hand, this cooling could be spotted using probabilistic information contained in the ensemble forecast distribution, more in particular by inspection of its modes.

4 The Value of a Probability Forecast

No one would argue much with the statement that most weather forecast users in the energy industry would prefer a perfect forecast to a probabilistic forecast. Meteorology will continue to work towards better forecasts, but despite all the progress in atmospheric science uncertainties remain that are unlikely to be completely resolved in the future due to imperfect model representation and the chaotic nature of the atmospheric system. The best way to quantify and deal with these uncertainties is through ensemble forecasting and a probabilistic treatment of ensemble output. By recognising that the forecast is unlikely to be perfect, a large amount of new and potentially valuable information becomes available.

Ensemble forecasts provide a distribution of predicted values. This distribution may or may not be representative of what the actual possible distribution of outcomes is. If the user assumes that each ensemble member has an equal chance of occurring and then compares the ensemble spread with historical performance it is possible to 'calibrate' the ensemble forecast such that the spread (dispersion) is representative of the historic spread of actual outcomes. The calibrated ensemble then provides an estimate of a probability density function (pdf). This is the most common means of creating probabilistic forecasts. If an ensemble is not available, one can create a pdf using older forecasts.

4.1 The Probability Density Function

There are many ways to generate a probabilistic forecast and many formats for display of the information. Ultimately, they should all derive from a probability density function (Fig. 4), either empirical or estimated. The probability density function represents the distribution of possible outcomes for a given situation along with the probability that a value in any given range will actually occur. For the example in Fig. 4, we show the distribution of possible temperatures for a given location and time. This distribution is called 'Normal' because it has a Gaussian shape. Not all probability density functions are Gaussian. There are a few features of the probability density function which require a simple explanation for understanding as we proceed.

First and foremost, the probability of an outcome in a specific range, say between 21 and 22°, is the area under the curve between the range boundaries. The Confidence Level at a particular point indicates the probability that the outcome will be below that value. For example, the 5 % Confidence Level in Fig. 4

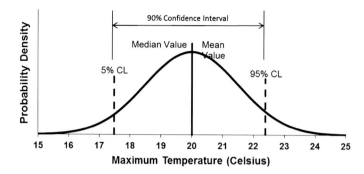

Fig. 4 A Normal (Gaussian) probability density function. The median and mean values are both at the peak. The 90 % Confidence Interval is the range between the 5 and 95 % Confidence Levels

indicates that there is a 5 % chance the actual temperature will be below 17.4°. The 95 % Confidence Level indicates that there is a 95 % chance that the actual temperature will be below 22.3°. Taken together these values represent a 90 % Confidence Interval, meaning there is a 90 % chance the actual outcome will be between 17.4 and 22.3°.

There are two additional points on the curve that are important for further discussion. The first is the distribution Median value. The Median value is also the 50 % Confidence Level and represents the point where there is equal probability of an outcome above or below. The second point of interest is the distribution Mean value. This is also referred to in mathematics as the Expected value. It is calculated by taking a weighted average of all the possible outcomes. The weights correspond to the probability of that value occurring. For a symmetric distribution, the Median and Mean values are the same. Unfortunately, not all probability density functions we encounter will be symmetric.

4.2 Choosing the 'Best' Forecast Value

So now we have a bit of a quandary: most people want a forecast of temperature, or wind speed, etc. to be a single (deterministic) number, a number which represents the best estimate of what the conditions will be. But the probability density function provides a continuum of numbers which could be chosen. How do we determine which value represents this 'best' forecast? Considering the forecast situation represented by the probability density function shown in Fig. 5, there is a strong likelihood that a temperature value between 17 and 22 will be the outcome. There is also a chance that the temperature will be closer to 28. So what would the 'best' deterministic forecast value be? The forecast pdf alone does not provide the answer, because the cost if the selected forecast value is wrong must also be considered.

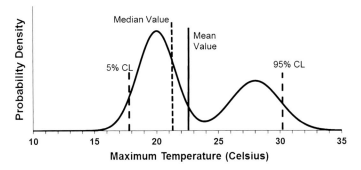

Fig. 5 A bi-modal probability density function. Note that the median value is offset from the peak and the mean value is actually close to a temperature range that is relatively unlikely to occur

4.3 What Will it Cost?

The primary question the user of a probabilistic forecast must answer is 'How much will it cost if the forecast is wrong?' Once this question has been answered, there is an entire branch of mathematics dedicated to providing the optimal value to be used from the probability density function. The focus in determining the 'best' forecast value is around balancing cost with probability of outcome, generally over a large number of forecasts, and minimising the cost in the long term. Unfortunately, this question often does not have a simple answer. We can make some simplifications which will bring the problem into clearer focus even if the situations are somewhat idealised. If the reader is interested in the details of these cases or general Decision Theory, further information can be found in Berger (1985) or Jaynes (2003).

4.3.1 Case 1: Symmetric Linear Costs

Consider the situation where it makes no difference to the forecast user whether the forecast is too warm or too cold in terms of cost (symmetric). If the forecast is out by 1° it costs the user $1, and if the forecast is wrong by 5° it costs the user $5 (linear). In this situation the user wants to minimise the average day-to-day error. In terms of verification measures this corresponds to minimising the Mean Absolute Error (MAE). The mathematics of the situation dictates that in the long run the best value to use from the probability density function is the Median value (Fig. 6).

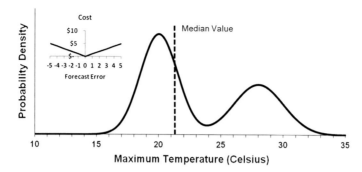

Fig. 6 If the cost function is symmetric and linear, the optimal value to use for a forecast to minimise the long-term costs is the median value

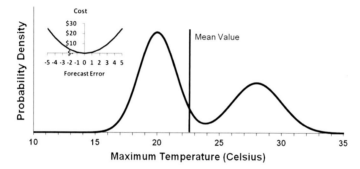

Fig. 7 When the cost function is symmetric but quadratic, with significantly larger costs associated with large errors, the optimal value to use to minimise long-term costs is the mean value

4.3.2 Case 2: Symmetric Quadratic Costs—Large Error Averse

Consider the situation where it makes no difference to the forecast user whether the forecast is too warm or too cold in terms of cost (symmetric), but if the forecast is out by 1° it costs the user $1, and if the forecast is wrong by 5° it costs the user $25. In this situation the user wants to minimise the impact of large errors. In terms of verification measures this corresponds to minimising the Root Mean Squared Error (RMSE). The mathematics of the situation dictates that in this case the best value to use from the probability density function is the Mean value (Fig. 7).

4.3.3 Case 3: Asymmetric Linear Costs

Now let us look at the case where if the forecast is too warm by 1° it costs the user $1 but if the forecast is too cold by 1° it costs the user $3. In this situation the user wants to balance the costs with the probability of the forecast being too warm or too cold. A 3:1 probability ratio will balance the costs, and this ratio corresponds to the 75 % Confidence Level value from the probability density function (Fig. 8).

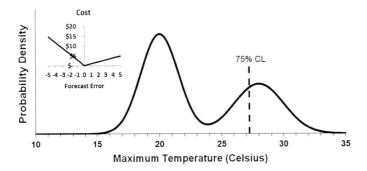

Fig. 8 When the cost function is asymmetric the optimal forecast value to use will be a confidence level that balances the cost with the probability of outcome. In this case where the cost of forecasting too cold is 3 times the cost of forecasting too warm, the optimal confidence level is 75 %

4.4 Applied Example—Probabilistic Forecasting for a CCGT

A Combined Cycle Gas Turbine (CCGT) is an electricity generation plant built around a large gas turbine, similar to a jet engine. The electricity production efficiency of a CCGT is directly related to the air density which is a function of atmospheric pressure, air temperature and relative humidity. The most significant portion of the efficiency variation is due to fluctuations in air temperature. This efficiency can be expressed mathematically in Mega Watts (MW) by a formula similar to the equation below (KEMA 2009):

$$\text{Efficiency} = \left(1.0503 - 0.0018 * T - 0.00007 * T^2 - 0.0000002 * T^3\right) * \text{CCGT\_Rating},$$

where T is the air temperature in Celsius.

When estimating how much power the CCGT will be able to produce during the day, the forecast temperature is used to calculate the expected output. Then when the generation time comes around for a particular trading period the actual possible generation will depend on the actual temperature.

In wholesale electricity markets such as the UK, if the actual generation possible is less than the forecast generation, the operator needs to buy electricity on the spot market (cost) to make up the difference. If the actual generation possible is higher than forecast, they have extra power that could have been sold (lost revenue counted as a cost) but will not be accepted on the grid.

Peak period is generally defined as being between 7 am and 10 pm. During this period electricity prices are significantly higher than during the overnight period (Base period) due to much higher demand (Fig. 9). During both periods the buy price is higher than the sell price. Consequently, we have an asymmetric cost function with respect to forecast temperature error.

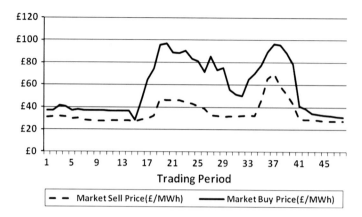

Fig. 9 An example of the daily profile of the UK Wholesale Electricity Market buy and sell prices. The day is divided into 48 half-hour trading periods. The cost to buy electricity on the 'spot market' is higher than the return on selling data, especially during the peak between trading periods 15 and 42 in this case

Now let us say that we hypothetically built a 450 MW CCGT at Heathrow Airport, in August 2010. If we then take 1 year of hourly probabilistic (probability density function) forecasts (September 2010 to August 2011), along with 1 year of actual hourly data, using the equation above, we can calculate the difference between the forecast generation and the hypothetical actual generation. To do this we assume forecasts are issued once per day at 4:30 am and that the forecast is used from 5 am to 4 am the next morning. We then combine these results with actual wholesale electricity prices at the time (https://www.elexonportal.co.uk/) to determine the cost of forecast temperature inaccuracies. If the forecast was too cold, efficiency would be lower than expected and the CCGT would need to buy on the spot market to make up the difference. If the temperature forecast was too warm, efficiency would be higher than expected and there would be extra generation available which could have been sold but was not, this being counted as an operating cost as well.

The majority of CCGT operators use the forecast with the lowest MAE. As mentioned above, this forecast value corresponds with the probability density function median value. Using the median value (50 % Confidence Level) as the deterministic forecast in the above procedure yields a total cost of temperature inaccuracies over the course of the year on the order of £552,896.

We then repeat the same procedure as above, but using Confidence Levels ranging from 55 to 80 % as forecast values from the identical probability density function forecasts. During the spring and summer, when demand is generally lower and consequently there is less spread between the buy and sell price, the optimal Confidence Level to use was near 55 %. During the autumn and winter when the price spread is generally higher, the optimal Confidence Level to use was closer to 65 %. Simply by using these Confidence Levels from the probability density function forecast would have saved the CCGT operator £15,475 over the course of the year. Admittedly the savings is small in the context of the total cost of inaccuracies, but it shows how even slight changes using probabilistic forecasts can lead to cost savings.

Importantly, the forecasts derived from the 55 and 65 % Confidence Levels would have a higher Mean Absolute Error than the forecast derived from the 50 % (median) level, even though they provided better value to the user.

5 Best Versus Most Useful

Although ensemble and probabilistic forecasting have been used within the meteorological community for almost two decades, they are still new to the energy community. The energy community and the systems it has built to exploit weather information typically uses deterministic forecasts. The cost of modifying complex software to accommodate a probability density function or 50+ ensemble member forecasts would be considerable. Is the gain worth the expense even in light of the arguments we presented which indicate that a probabilistic forecast may be better than a single forecast?

First, let us look at the example above with the CCGT. If the weather forecast provider was able to provide the 65 % Confidence Level forecast to the CCGT as a simple deterministic forecast, there would be little or no cost to implement and value to be gained. Hence we would deem it to be 'useful'.

The usefulness of a probabilistic or ensemble forecast will depend not only on the cost of implementation, but also on the type of cost function the decision maker is working with. If the company's cost function is symmetric, the standard deterministic forecast would be eminently useful as most weather providers aim to reduce the MAE or RMSE in their forecast. Any information beyond this would be extraneous and not very useful.

Probabilistic- or ensemble-based information would be most useful in assessing the asymmetric, high cost/return scenarios. It is often very hard to quantify the cost function, but the decision maker will have a general feel for it and know whether or not there are scenarios where, if the deterministic forecast is incorrect by a conceivable amount, the cost to the company will be very high. Provision by the weather service provider of values derived from a probabilistic forecasting approach, such as confidence intervals or the probability of exceeding certain thresholds, will assist in assessing risk associated with these critical situations. The decision maker will be able to make adjustments or decisions which will reduce the risk of loss based on the likelihood of the costly scenario occurring. The usefulness of obtaining and displaying this additional information for reducing high cost episodes due to forecast uncertainty will depend somewhat on the decision maker and their understanding of forecast uncertainty.

Many energy companies see value in employing their own meteorologist to assist in interpreting weather data to suit their in-house requirements, particularly for trading. Those that do will find that one of the first forecast datasets the meteorologist requests is an ensemble forecast. The meteorologists understand that traders are looking for the high cost/return scenarios and they use the ensemble data to reduce the time required to assess the likelihood of these scenarios. There is no doubt about the usefulness of the approach in these situations.

But what about further down the calculation chain? It is not a mathematically simple process to propagate uncertainty and probability through the various mathematical models used to quantify demand and plant efficiency. It can be done, but if it is not done correctly it can lead to unreliable results and then a deterministic forecast would certainly be more useful. Indeed it might be argued that the applied example of the CCGT used earlier is overly simplified given the many factors other than temperature which also determine plant efficiency. Combining uncertainty information for all of the weather factors makes rigorous calculation of the probability of the various outcomes for plant efficiency almost intractable. The CCGT example is an all-else-being-equal scenario. Introduction of probability information for other parameters would eliminate the 'all-else-being-equal' part of the example.

This does, however, lead us to the next principle in probabilistic forecasting. To get the most reliable and the most useful probabilistic results you need to forecast what you want to know. In the example of the CCGT, we would create a probabilistic forecast of plant efficiency. In terms of electricity demand forecasting, rather than trying to feed probabilistic weather forecasts into a demand model, we would create a probabilistic demand forecast model. It is important to move the probabilistic treatment as close to the final decision point as possible to maximise its usefulness.

6 Conclusions

There is no unique definition of what constitutes a good forecast. In the first section of this chapter, we introduced the three pillars that define forecast goodness: consistency (agreement with the forecaster's belief), quality (agreement with the meteorological observations) and value (agreement with the user's economic objective). Ideally, the performance of weather forecasts should be assessed taking these three facets into account, but in practice it is gauged by quality measures only such as accuracy. We have argued above that in order to satisfy the requirements of consistency and value, weather forecasts should, if possible, be expressed in a probabilistic format.

Simple deterministic forecasts (single values, no uncertainty but generally higher resolution) are cheap and easy to use, which makes them very popular. They are typically calibrated so as to optimise accuracy. However, because information on uncertainty and possible alternative scenarios has been stripped from them, deterministic forecasts are not necessarily optimised for value even when they have been optimised for quality (e.g. accuracy). In Sect. 3, we have shown with a concrete example how a deterministic forecast based on the mean of an ensemble of forecasts so as to deliver maximum accuracy fails to predict a temperature downturn, whereas the use of the ensemble forecast probability density function allows a more useful prediction. In Sect. 4, another example demonstrates again that a probabilistic approach is more adequate when seeking to achieve better value than the systematic recourse to the most accurate forecast. Although this may sound counter-intuitive, focusing solely on forecast accuracy is a strategy that does not naturally lead to best value and can even conflict with it. The example also shows that in order to optimise forecast value, users of probabilistic forecasts must be able to quantify their specific loss function, which is not always straightforward. For simplicity, a linear asymmetric loss function was assumed, but in reality these functions are non-linear.

Weather forecasts will never be perfect and will therefore always contain some degree of uncertainty. Because they do not acknowledge this inherent uncertainty and are simple to use, deterministic forecasts may give the false impression of being more useful than probabilistic forecasts. But quantified uncertainty is useful information, and removing it from the forecasts has a negative effect on their potential value. In other words: proper use of probabilistic forecasts can save the user money. However, for optimal use of probabilistic forecasts, it is important to appreciate the user's cost and loss functions. One should also move away from the traditional quality control methodologies that are exclusively based on agreement with weather observations without taking account of the users' specific needs. A synergy between the meteorologists who make and provide the forecasts and the users of these forecasts in the energy sector is vital to guaranteeing that weather forecasts are truly fit for purpose.

References

Berger JO (1985) Statistical decision theory and Bayesian analysis. Springer Series in Statistics, Springer, New York, p 627
Jaynes ET (2003) Probability theory: the logic of science. Cambridge University Press, New York, pp 417–450
Jolliffe IT, Stephenson DB (eds) (2011) Forecast verification: a practitioner's guide in atmospheric science. John Wiley and Sons, Chichester, p 274
KEMA (2009) LRMC of CCGT generation in Singapore for technical parameters used for setting the Vesting Price for the period 1 January 2009 to 31 December 2010, Energy Market Authority of Singapore, 9
Murphy AH (1993) What is a good forecast? An essay on the nature of goodness in weather forecasting. Weather Forecast 8:281–293
Persson A (2011) User guide to ECMWF forecast products. European Centre for Medium-Range Weather Forecasts, Reading, p 127. Accessible online at http://www.ecmwf.int/products/forecasts/guide/
Richardson DS (2000) Applications of cost–loss models.In: Proceedings of the seventh ECMWF workshop on meteorological operational systems. November 15–19, 1999, Reading, England, pp 209–213

Part IV
How is the Energy Industry Applying State-of-the-Science Meteorology?

A Probabilistic View of Weather, Climate, and the Energy Industry

John A. Dutton, Richard P. James and Jeremy D. Ross

Abstract Probability methods provide quantitative insight into the implications of atmospheric variability for the energy industry. Contemporary computer probability forecasts of climate anomalies for the weeks, months, or seasons ahead offer new precision in managing both risk and opportunity. The forecasts of two major international centers and a multi-model constructed from them by the World Climate Service demonstrate that the contemporary probability forecasts have sufficient skill and reliability to provide advantageous guidance for energy decisions. An analytical model of choices available in response to predicted anomalies illustrates how and when to act on forecasts. Atmospheric informatics is introduced as a system for creating, transferring, and applying atmospheric information in important endeavors. The aim is to show energy and other industry decision-makers what they need to know—now.

1 Introduction

The ever-changing weather and seasonal climate are a major source of uncertainty for key components of the energy industry, affecting both performance and profit. As shown by frequent comments in annual reports of utilities and other energy companies, weather and climate events often pose risk or offer opportunity.

The impacts of weather and seasonal variability range widely across the energy industry and include severe-weather damage to facilities and infrastructure, wide swings in demands and loads, fluctuations in availability of hydropower, and wind

---

J. A. Dutton (✉) · R. P. James · J. D. Ross
Prescient Weather Ltd, State College, PA, USA
e-mail: john.dutton@prescientweather.com

J. A. Dutton
The Pennsylvania State University, University Park, PA, USA

Table 1 World energy demand and cost. Demand and gross domestic product (GDP) estimates are from Exxon-Mobil (2012) and U.S. energy cost from U.S. Energy Information Agency (2012)

	2000	2010	2025
Energy demand (Quadrillion BTUs)			
World	415	525	633
US	96	94	96
US/World	0.231	0.179	0.152
Cost of energy ($ billion)			
World		6,724	*13,405*
U.S.		1,204	2,033
Gross domestic product ($ Billion)			
World		51	81
U.S.		13	19
Energy cost/GDP (percent)			
World		13	17
U.S.		9	11

The estimate of world energy cost is inferred from World Cost = U.S. Cost (World Demand/U.S. Demand)

and solar power, and unexpected costs through indirect effects. Long-term planning to anticipate demand, generation, and delivery facilities must consider the potential direct and indirect effects of possible climate change as well as evolving public attitudes about the environment. The energy industry is servant to civilization but often captive to the weather and the climate.

In the United States, the total annual expenditure for energy of $1.2 trillion is nearly 10 % of the gross domestic product (Table 1), and if average world energy prices were about equal to U.S. prices, then the annual global cost of energy would be $7 trillion and 13 % of global GDP (in 2010 data). Fractionally small perturbations of the energy cash flow owing to weather and climate may seem quite large relative to other parts of the economy.

The energy industry, like many other components of the economy, is operating within continuously decreasing margins as it becomes more sophisticated and economically efficient, and it thus becomes increasingly sensitive to weather and climate. Continued progress in managing the energy industry requires greater skill in coping with the uncertainties owing to weather and climate variability. Thus this chapter will explore methods for describing and managing the continually evolving energy uncertainties associated with weather and climate events.

The multifaceted mission of the energy industry includes several components:

- Serve the energy needs of customers and communities;
- Meet the expectations of investors;
- Ensure safety and security for customers, employees, and communities;
- Contribute to economic vitality and environmental quality and sustainability.

Several sources of opportunity, challenge, and risk attend efforts to achieve the mission. They include:

- Technological change and advance
- Economic pressures and competition
- Evolution of societal expectations
- Weather, climate, and other environmental phenomena—from hours to decades.

These sources of opportunity and risk lead to a set of questions:

- What is the range of possible events in each of these domains of opportunity and risk?
- Which of the possible events is most likely?
- Which extreme events would be of most significance?
- What is the likelihood they might occur?
- What would their impact be?

If we attempt to think through these issues about sources of risk and opportunity as they apply to each component of the mission, we will soon find ourselves constructing a framework—perhaps formal, perhaps informal—that separates likely events from the unlikely ones. We will discover sequences of events that together imply success or failure. We might turn to the concept of *probability of mission success* as it appears in analysis of military operations or space flight and seek to ascertain how expectations about opportunity and risk could be transformed into expectations about the performance or profit of energy operations.

We might then resonate with the client plea: *Don't tell me about the weather, tell me about the money.*

Converting an outlook about the weather into an outlook about the money requires converting one set of probabilities into another.

2 Probability Methods

Probabilities that reliably estimate the likelihood of future events help us to make effective decisions despite the inevitable uncertainties inherent in our business and physical environments. The study of probability began in the late sixteenth century as gamblers engaged mathematicians to look for rigorous ways to analyze games of chance. Laplace (1812) applied probability to a broader set of applications and called it "a science...[that is] the most important object of human knowledge."

The foundations for modern mathematical theory of probability were proposed by Kolmogorov (1933), commenting that he was offering "...an axiomatic foundation for probability...putting in their natural place, among the notions of mathematics, the basic concepts of probability—concepts which until recently were considered to be quite peculiar."

Fig. 1 A histogram and an empirical frequency distribution constructed from 100 simulated temperature observations. As the number of observations increase, the histogram and the distribution will converge to the Gaussian forms

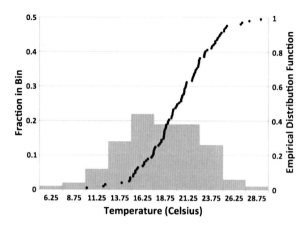

Probability methods describe and predict the likelihood of outcomes in a set of events. For the purposes here, we will use *frequencies* and *statistics* to refer to quantities observed in the past and present and *probabilities* to refer to expectations about the future.

Suppose we have N observations of a quantity of interest, say temperatures near the surface. We sort and renumber the observations so that we have a non-decreasing sequence from T_1 to T_N. Now we have two choices. We might define a frequency distribution $F(T_n) = n/(N+1)$ or we might divide the real line into K intervals, with T_1 in the first and T_N in the last, and then count the number of values that fall in each interval to obtain a histogram, as illustrated in Fig. 1. Summing the bin counts over the intervals and dividing by N produces a curve that converges to $F(T_n)$. Frequency distributions and histograms like those in Fig. 1 often occur as approximations to observed data and are known as *normal* or *Gaussian*. The analytical forms for the normal probability distributions and densities are

$$F(x) = \frac{1}{\sqrt{2\pi}\sigma} \int_{-\infty}^{x} e^{-\frac{1}{2}(\frac{\xi-\mu}{\sigma})^2} d\xi, \quad f(x) = \frac{1}{\sqrt{2\pi}\sigma} e^{-\frac{1}{2}(\frac{x-\mu}{\sigma})^2} \quad (1)$$

in which the functions are centered at the arithmetic average μ and the width is proportional to the standard deviation σ. We often assume that future observations will be distributed like those we have already observed and in that case the two quantities in (1) would give the probabilities

$$\text{Prob}[x \leq X] = F(X), \quad \text{Prob}[x \geq X] = 1 - F(X) \quad (2)$$

for normal variates. More generally, whenever a probability distribution can be represented by some sufficiently smooth analytical form $P(X)$, we always have

$$\text{Prob}[x \leq X] = P(X), \quad p(x) = dP(x)/dx \quad (3)$$

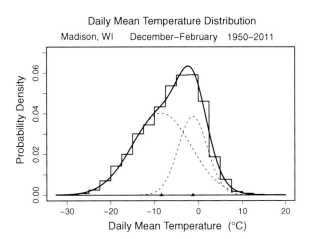

Fig. 2 The frequency distribution of average daily winter (DJF) temperatures at Madison WI, 1950–2011. The histogram is modeled with two Gaussian distributions with means and standard deviations {(−8.44, 6.68 °C), {−1.27, 3.35 °C)} which combine with weights (0.67, 0.33) to form the empirical frequency function

While the normal distribution sometimes provides a useful approximation to the statistics of meteorological observations, the situation is often more complex. For example, Fig. 2 shows how the winter temperature distribution at Madison, WI, USA, can be approximated as a mixture of two Gaussian distributions, representing the characteristics of the two air masses that alternate in the Madison midcontinental winter.

Precipitation and wind speed cannot possibly be normally distributed because negative values are impossible. The distributions of these variables can be difficult to model because there can be many small or zero observations and a few very large observations. The gamma distribution is often used for precipitation and the Weibull distribution for wind speed (e.g., Dutton 2002).

The variables used in studying or managing energy operations are sometimes derived from atmospheric variables by nonlinear operations that complicate statistical modeling. For example, the commonly used degree days involve the distributions of daily average temperature values above and below a reference of 18 °C for cooling and heating. The available wind power is proportional to V^3 for wind speed V, with cutoffs at small and large wind speeds that depend on the specific installation.

For all such variables, we need reliable historical statistics and we need forecasts to help with decisions that depend on how they will evolve in the days, weeks, months, and seasons ahead.

3 Probability Forecasts of Atmospheric Events

Forecasts of events in the atmosphere and ocean for lead times ranging from days to seasons are generally created with computer programs that analyze historical data (van den Dool 2007) or that solve initial- and boundary-value problems based on differential equations that describe the evolution of momentum and energy (Kalnay 2003; Troccoli 2010a, b; Palmer and Hagedorn 2006).

Meteorologists today take advantage of accelerating supercomputer capability to create ensembles of many forecasts computed simultaneously. The members of the ensembles can be arranged, as discussed in Sect. 2, to create probability distributions describing the likelihood of a range of future events. A further strategy of constructing multi-ensembles by combining solutions from independent computer models has proved to be advantageous, as shown by the results of two joint European projects: DEMETER as described by Hagedorn et al. (2006) and ENSEMBLES reported at www.ensembles-eu.org. A similar but still developing collection of U.S. and Canadian models is known as the National Multi-Model Ensemble (NMME).

The contemporary weather and climate prediction process has been profoundly shaped by the discovery (Lorenz 1963) that small differences in initial conditions can lead to large variations in the solutions of the dynamical equations—a mathematical phenomenon known as chaos. Forecasts of specific events remain skillful for lead times of a week or more and can be enhanced by statistical methods (see Ruth et al. 2009; Glahn et al. 2009, for a recent assessment). Seasonal forecasts portraying monthly average conditions a month or more in advance are also somewhat skillful because of the slow evolution of boundary conditions at the sea and land surface. The period of 2–4 weeks in between has been a continuing challenge, as pointed out by Dubus 2013. Most of the discussion here will concern seasonal forecasts with averaging and lead times of months or more, but we report new results on weekly forecasts. An important strategy for prediction on the range of weeks to seasons is to forecast average conditions over periods comparable to the lead—weeks for weeks, months for months, seasons for seasons, with the hope that capabilities will improve to meet the needs in the energy industry for finer temporal and spatial resolution.

Computer probability forecasts of seasonal climate variability must be calibrated by comparing them with the corresponding observations over as long a history as possible in order to remove bias by centering the forecasts correctly and to adjust the ensemble spread by scaling it with the spread of the observations. The corrections determined from the history are then applied to new forecasts.

We focus here on three seasonal temperature forecasts: the Climate Forecast System, version 2, (CFSv2) of the U S. National Weather Service (NWS) (see Saha et al. 2012), the Seasonal Forecast System, version 4, (SFSv4) of the European Centre for Medium-Range Weather Forecasts (ECMWF) (see Anderson et al. 2007; ECMWF 2011), and a World Climate Service[1] multi-model ensemble (WCS MME) created as a Bayesian mixture of the NWS and ECMWF models.

---

[1] The World Climate Service is a joint venture of Prescient Weather Ltd. in the U.S. and MeteoGroup in Europe and the U.S. For more information see www.worldclimateservice.com.

A Probabilistic View of Weather, Climate, and the Energy Industry

Table 2 Contingency table for assessing forecast performance

		Observations		
		Above normal	Below normal	Sum
Forecasts	Above normal	$a = A/N$	$b = B/N$	n_a
	Below normal	$c = C/N$	$d = D/N$	n_b
	Sum	f_a	f_b	1
Forecasts	Above normal	$n_a F_a$	$(1 - F_a) n_a$	n_a
	Below normal	$(1 - F_b) n_b$	$n_b F_b$	n_b
	Sum	f_a	f_b	1

The distribution of N forecasts into the four possibilities is described by the numbers A, B, C, and D; $n_a = a + b$ is the fraction of forecasts that were for above normal, and $f_a = a + c$ the fraction of observations that were above normal. The fractions F correct are defined in (4)

3.1 Answering the Question: How Good Are the Forecasts?

Seasonal probability forecasts are often defined relative to historical normal values. Thus a binary forecast may be for above normal or below normal conditions and a ternary forecast for above normal, nearly normal, or below normal conditions. The boundaries of the categories are usually defined from observations so that the frequencies of occurrence are usually equal for each category: one-half for binary forecasts, one-third for ternary.

Contingency tables provide a convenient way to assess the quality of forecasts and thus answer the first question a potential user will ask: *How good are the forecasts?* Table 2 provides an example for binary forecasts. The numbers A, B, C, D are the numbers of N forecasts that occur in each of the four possibilities for forecasts and observations with $a = A/N$, $b = B/N$, $c = C/N$, $d = D/N$ the corresponding fractions. These quantities can be used to form the ratio S of forecasts that correctly predict the event and the fraction F of correct forecasts

$$S_a = A/(A+C) = a/f_a, \quad F_a = A/(A+B) = a/n_a \\ S_b = D/(B+D) = d/f_b, \quad F_b = D/(C+D) = d/n_b \quad (4)$$

in which $f_a + f_b = n_a + n_b = 1$. The statistics S and F are identical when the matrix is symmetric about the main diagonal. Table 3 summarizes the performance of the WCS MME binary forecasts for the period 2000–2009; in each case the 18 years preceding the forecast year were used to create the training set for calibration and the climatological normal for verification. Table 4 provides similar information for WCS MME ternary forecasts. Table 5 illustrates how the ratios improve as the threshold probability level required to issue a forecast for above, near, or below normal increases.

It is evident from the tables presented that the performance of these seasonal prediction models is not symmetric with respect to the forecast categories. One reason seems to be that the decadal and longer term trends are different in the

Table 3 Percent of correct binary forecasts for 2 m temperature from the World Climate Service multi-model ensemble, 2000–2009

Season	October -> DJF			April -> JJA		
Forecast	Below-normal	Above-normal	All forecasts	Below-normal	Above-normal	All forecasts
Globe	61	69	67	46	68	62
NA	60	65	63	60	60	60
EU	53	60	58	46	65	64
AU	54	71	65	64	69	68
TP	95	70	77	82	69	72

DJF stands for December, January, and February; JJA for June, July, August. The forecasts were initialized in April and October. NA stands for North America, EU for Europe, AU for Austral-Asia, and TP for the tropical Pacific Ocean. The AU JJA statistics are with DJF and AU DJF is with JJA in this and the following tables. The fractions correct would be expected to be ½ for binary forecasts constructed from random numbers

Table 4 Percent of correct ternary forecasts for 2 m temperature from the World Climate Service multi-model ensemble, 2000–2009

Season	October -> DJF				April -> JJA			
Forecast	Below normal	Near-normal	Above normal	All terciles	Below normal	Near-normal	Above normal	All terciles
Globe	49	41	54	50	32	42	52	46
NA	46	42	47	46	41	44	43	43
EU	39	41	43	42	34	41	48	46
AU	38	45	56	48	59	40	52	50
TP	83	41	59	56	68	45	56	55

The fractions correct would be expected to be 1/3 for ternary forecasts constructed from random numbers

observations and historical forecasts, perhaps because of the inability of the models to accurately anticipate climate change effects.

Returns achieved in trading hypothetical weather derivatives provide a metric for assessing the economic value of the forecasts. Considering the binary case first, we assume an option on one of the binary outcomes costs P and that it pays $2P$ if the binary event occurs; for the ternary case a successful option costing P will pay $3P$. Then the average rate of return R in trading these options on the basis of a set of forecasts with fraction F correct will be.

$$R_2 = (2F - 1)P/P = 2F - 1, \quad R_3 = 3F - 1 \quad (5)$$

The average of the fractions correct for all forecasts in Tables 3 is 66 % and the average in Table 4 is 48 %, and so the expected returns in trading the hypothetical derivatives are 32 % for the binary forecasts and 44 % for the three-category forecasts.

Table 5 Percent of correct ternary forecasts for 2 m temperature (all three categories combined) from the World Climate Service multi-model ensemble for all forecasts with probabilities exceeding the three probability levels

October -> DJF			
	Forecast criterion (percent)		
	33	50	67
Globe	50	57	66
NA	46	54	66
EU	42	53	84
AU	48	52	58
TP	56	59	67
April -> JJA			
	33	50	67
Globe	46	52	60
NA	43	49	58
EU	46	48	56
AU	50	56	65
TP	55	57	62

3.2 Reliability of Seasonal Forecasts

Probability forecasts are considered reliable when the events being predicted occur with a frequency equal to the predicted probability. To illustrate, a binary forecast for rain or no rain is reliable if it rains on one-third of the days for which we predicted a probability of one-third for rain. A reliability assessment compares the predicted probabilities with the frequencies of occurrence over the entire range of predicted probabilities; the curve relating observed frequencies to predicted probabilities would lie along the diagonal for perfectly reliable forecasts. The reliability diagrams shown in Figs. 3 and 4 demonstrate that the WCS MME seasonal forecasts are reasonably reliable over much of the probability range.

3.3 One- to Four-Week Forecasts

The World Climate Service recently applied the methods used to calibrate seasonal forecasts to computer probability forecasts for the range of 1–4 weeks ahead with surprisingly successful results. A component of the NWS CFSv2 has been focused on this range, producing detailed information out to 45 days.

The WCS process was designed to test CFSv2 ensemble forecasts for the 1–4 week range constructed from sequences of model runs ending on the 1st, 8th, 15th, and 22nd day of each month. The significant challenge was to create a climatology with averages over one-week periods corresponding to each grid-point and each target forecast week in order to calibrate the forecasts. The training period was

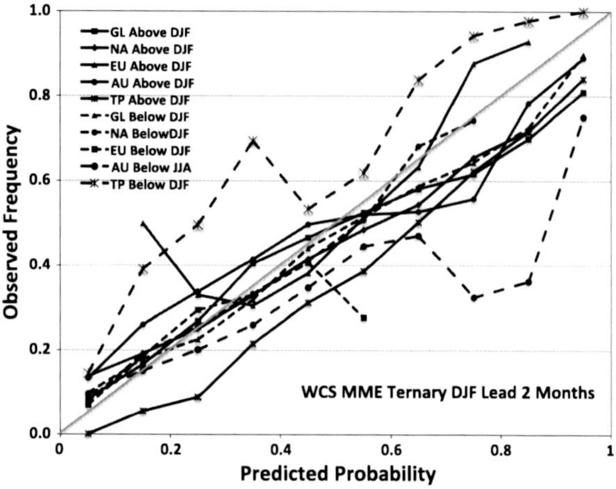

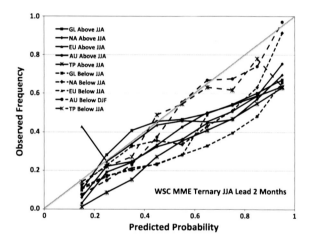

Fig. 3 Reliability diagrams for the World Climate Service multi-model ensemble forecasts for surface temperature, October forecasts for December, January, February, 2000–2009. The abbreviations are GL for Global, NA for North America, EU for Europe, AU for Austral-Asia, and TP for Tropical Pacific. The AU forecasts for the austral winter are included in this set

Fig. 4 Reliability diagrams for the World Climate Service multi-model ensemble forecasts for surface temperature, April forecasts for June, July, August, 2000–2009. The AU forecasts for the austral summer are included in this set

2000–2006 and the forecasts were calibrated and verified for the 4 years 2007–2010 using the new CFS Reanalysis as the verification data.

The fractions correct for the forecasts for weeks 2 and 4 ahead are shown in Table 6 and demonstrate skill similar to that of the seasonal forecasts and better balance between categories, as in the findings of Dubus (2013) for monthly forecasts for France. The success ratios were comparable and are not shown here. The average fractions correct for the week-2 and week-4 forecasts of 53 and 46 % imply returns of 59 and 38 % in trading the hypothetical weather derivatives. The reliability curves in Fig. 5 demonstrate that these forecasts are remarkably reliable and slightly overconfident at larger values of the probability. Because the reliability curves are linear and rather tightly grouped, a correction could easily be applied as subsequent forecasts are issued to rotate the predicted probabilities onto the diagonal. Additional skill statistics are available in Dutton et al. (2013).

Table 6 Percent of correct CFSv2 weekly 2 m temperature tercile probability forecasts for weeks two and four ahead, initialized in DJF and JJA, 2007–2010

Season	Winter—DJF				Summer—JJA			
Forecast	Below-normal	Near-normal	Above-normal	All forecasts	Below-normal	Near-normal	Above-normal	All forecasts
Forecasts for week 2								
Globe	61	48	55	55	57	46	54	52
NA	62	50	55	56	62	45	53	53
EU	60	49	58	56	53	48	48	49
Forecasts for week 4								
Globe	53	45	48	49	49	43	47	46
NA	49	48	43	47	53	41	43	45
EU	45	48	48	47	42	44	37	41

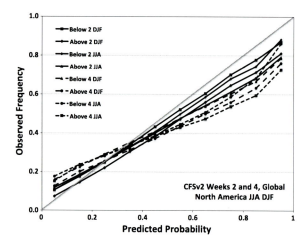

Fig. 5 Reliability diagrams for the World Climate Service forecasts for week 2 and week 4 surface temperature prepared using the CFSv2 ensemble forecasts. This set is for North America and for the winter and summer seasons, DJF and JJA

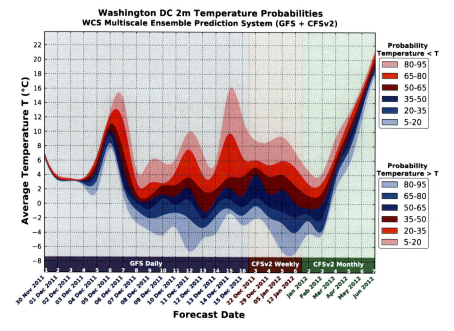

Fig. 6 A World Climate Service multiscale ensemble forecast for daily average surface temperature for Washington, D.C., constructed with members of the Global Forecast System and the Climate Forecast System (v2) ensembles. The daily forecasts span the period 30 Nov 2011 to 15 Dec, the weekly forecasts 22 Dec to 12 Jan 2012, and the monthly forecasts Jan through June 2012

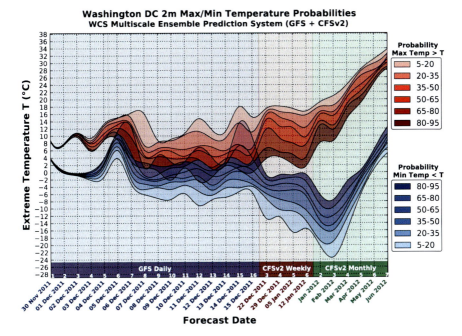

Fig. 7 A World Climate Service multiscale ensemble forecast of maximum and minimum surface temperature for Washington, D.C., constructed with members of the Global Forecast System and the Climate Forecast System (v2) ensembles. The daily forecasts span the period 30 Nov 2011 to 15 Dec, the weekly forecasts 22 Dec to 12 Jan 2012, and the monthly forecasts Jan through June 2012

3.4 Multi-Scale Ensemble Forecasts

The apparent success of the 2- and 4-week forecasts allows us to construct a comprehensive and integrated probabilistic forecast simultaneously spanning periods of days, weeks, and months. Figure 6 shows such a multi-scale forecast for surface temperature and Fig. 7 depicts probabilities of maximum and minimum temperature.

3.5 Degree-Day Forecasts

The interest in climate sensitive industries usually centers on industry-relevant variables rather than the atmospheric variables in the computer forecasts. Thus energy interests often focus on degree days computed by summing the difference between the daily average temperature and a base value, usually 18 °C. A day with an average temperature of 10 °C contributes 8 heating degree days; a day with 28 °C contributes 10 cooling degree days.

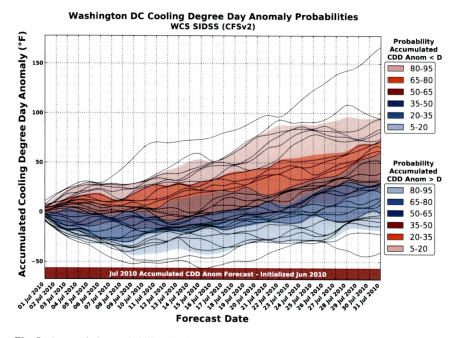

Fig. 8 Accumulating probability distributions for cooling degree-day anomalies for Washington D.C. over the month of July 2010, computed from the CFSv2 ensembles

It is relatively straightforward to obtain degree days from calibrated temperature forecasts numerically. An example of an evolving probability distribution of accumulated cooling degree days is shown in Fig. 8.

3.6 Modeling Forecast Performance

In issuing a forecast, the category with the larger probability is usually chosen as the expected event and so a probability exceeding 50 % would determine the binary forecast. But we might improve the utility of the forecasts in business decisions by choosing to act only if the probability of the event with which we are concerned exceeded, say 60 %. In the ternary case, we could encounter probability values of 34, 33, and 33 % for the three categories, but again we could decide to act only if the probability of the event of interest exceeded, say, 50 %.

In order to explore this idea and take advantage of the increased skill with increased probability thresholds shown in Table 5, it is convenient to create a synthetic, symmetric model of the contingency statistics presented above. We observe that statistics for binary forecasts demonstrate a useful symmetry that arises because an incorrect forecast for above normal corresponds to a correct forecast for below normal. This property can be seen in the portion of Table 7

Table 7 Percent of correct binary forecasts for 2 m temperature for North America DJF from the World Climate Service Multi-model ensemble, 2000–2009, in probability bins with midpoints shown

Midpoint probability	North America October -> DJF				Symmetric model			
	Below correct	Sharpness below	Above correct	Sharpness above	Below correct model	Sharpness model	Fraction correct average	Fraction correct model
0.05	0.4	2.4	0.0	0.6	0.1	0.7	14	14
0.15	2.2	8.8	0.3	1.2	1.1	4.9	25	22
0.25	6.5	17.3	1.4	4.1	3.2	10.5	37	30
0.35	9.6	21.9	3.3	8.4	5.9	15.5	43	38
0.45	10.1	21.0	6.4	14.3	8.5	18.4	47	46
0.55	7.9	14.3	11.0	21.0	9.9	18.4	53	54
0.65	5.0	8.4	12.3	21.9	9.6	15.5	57	62
0.75	2.8	4.1	10.8	17.3	7.4	10.5	63	70
0.85	0.9	1.2	6.6	8.8	3.8	4.9	75	78
0.95	0.6	0.6	2.0	2.4	0.6	0.7	86	86

giving the verification statistics for the WCS multi-model forecasts for the North American winter months.

The fraction $s(f)$ of forecasts in a bin with midpoint f is known as the sharpness function or diagram and we will represent the fraction correct as $r(f)$. Using the tables above as a guide we create a model of the observed results in Table 7 as.

$$s(f) = 30f^2(1-f)^2 \quad r(f) = 0.1 + 0.8f \quad v_b(f) = r(f)s(f) \qquad (6)$$

in which $v_b(f)$ is the normalized number of correct below normal forecasts. We integrate over a range of below normal forecasts to obtain.

$$C(p) = \int_p^1 v_b(\eta)\,d\eta, \quad S(p) = \int_p^1 s(\eta)\,d\eta \qquad (7)$$

The complement to these integrals over the domain $p \leq f \leq 1$ for the forecasts of below normal is the integrals of above normal statistics over the domain $1 - p \leq f \leq 1$, thus covering the entire range $0 \leq f \leq 1$. Taking advantage of the symmetries of the contingency table and (7), we find that integrated probabilities can be summarized as functions of p in Table 8.

The numerical values for the synthetic symmetric case derived by summing in Table 7 and evaluating the quantities in Table 8 are available in Table 9. They will be used in the next section to illustrate how an enterprise can take advantage of the forecasts to optimize the performance of the climate-sensitive business.

Table 8 Contingency table for all binary forecasts with probabilities greater than p for the integrated sharpness $S(p)$ and normalized number $C(p)$ of correct forecasts from (6) and (7)

Forecasts	Verification		Sum
	Above	Below	
Above	$1 - C(0) - (S(p) - C(p))$	$C(0) - C(p)$	$1 - S(p)$
Below	$S(p) - C(p)$	$C(p)$	$S(p)$
Sum	$1 - C(0)$	$C(0)$	

Table 9 Numerical version of the cumulative functions in Table 8 for the binary forecasts for 2 m temperature for North America DJF from the World Climate Service multi-model ensemble, 2000–2009, and for the symmetric model of forecast performance

Midpoint probability	North America October -> DJF				Symmetric Model			
	Below correct	Sharpness below	Above correct	Sharpness above	Below correct model	Sharpness model	Fraction correct average	Fraction correct model
0.05	46.1	100.0	53.9	100.0	50.0	100.0	50	50
0.15	45.6	97.6	53.9	99.4	49.9	99.3	51	50
0.25	43.4	88.8	53.7	98.2	48.8	94.5	52	52
0.35	36.9	71.6	52.3	94.1	45.7	83.9	54	54
0.45	27.3	49.6	48.9	85.7	39.8	68.4	56	58
0.55	17.2	28.6	42.6	71.4	31.3	50.0	60	63
0.65	9.3	14.3	31.6	50.4	21.4	31.6	63	68
0.75	4.3	5.9	19.3	28.4	11.8	16.1	69	73
0.85	1.5	1.8	8.6	11.2	4.4	5.6	78	79
0.95	0.6	0.6	2.0	2.4	0.6	0.7	86	86

4 Modeling Probabilities of Business Results

Seasonal forecasts offer potential improvement in the performance or profitability of endeavors exposed to weather and climate risk that may be realized with decision strategies that link the statistical characteristics of the forecasts and the enterprise in ways that lead to optimum results. The decision strategies must answer the question: *When and how should I act on the forecast?*

4.1 Climatology and Hedges

Some of the advantages and disadvantages of hedges against adverse conditions can be illustrated by using them to reduce the impact of adverse events at the occurrence rate predicted by climatology. We let X represent the gain when conditions are favorable, $-Y$ the loss in unfavorable conditions, and we assume that we could purchase a hedge that will pay H if unfavorable conditions prevail;

A Probabilistic View of Weather, Climate, and the Energy Industry

Table 10 A business model and climate statistics for computing the cost and consequences of hedging below normal conditions

	Above normal	Below normal
Business model	$X - cH$	$-Y + H(1 - c)$
Climate frequencies	$f_a = f$	$f_b = 1 - f_a = 1 - f$

the preseason cost of the hedge will be assumed to be cH. Table 10 shows a business model describing the cost and effect of the options.

The expected mean and variance of the revenue for climate without the hedge are

$$R_c = f_a X - f_b Y = fX - (1-f)Y$$
$$V_c = f_a X^2 + f_b Y^2 - R_c^2 = f(1-f)(X+Y)^2 \quad (8)$$

and with the hedge are

$$R_h = f_a X - f_b Y + f_b H - cH = R_c + f_b H - cH$$
$$V_h = f_a (X - cH)^2 + f_b (H - Y - cH)^2 - R_h^2 \quad (9)$$
$$= f(1-f)(X + Y - H)^2$$

As noted by Dutton (2002), the cash flow M of the hedge contract to the issuing counterparty is $M = cH - f_b H$ and thus the market will require a premium so that $c > f_b$; as a consequence the revenue with the hedge is less than the revenue that would obtain without it.

But hedges may provide significant reduction of the variance of earnings. If we continue with this case for $X = Y$ and $f_a = f_b = 1/2$ so that $R_c = 0$ and $R_h = H(\frac{1}{2} - c)$, we find that

$$\frac{V_h}{V_c} = \frac{V_h}{X^2} = \frac{1}{4}(\frac{H}{X} - 2)^2 = \frac{1}{4}(h-2)^2 \quad (10)$$

for $h = H/X$. The variance with the hedge reaches zero at $h = 2$ and is less than the variance with climate for all $h < 4$, but the net revenue decreases concomitantly. The relation between revenue and variance in this model is illustrated with a parametric graph in Fig. 9 for two values of the premium c. There is obviously no advantage to be gained in this case with hedges $h > 2$.

4.2 Modeling the Performance of a Weather- and Climate-Dependent Enterprise

The sensitivity of an enterprise to weather or seasonal variability presumably can be described and modeled quantitatively to some adequate degree of accuracy. For example, a parabolic function often approximates the load on a utility, with the

Fig. 9 Relation between size of hedge and variance for various values of the hedge parameter h in (10) and two values of the cost of the hedge, $c = 1/2 + 1/8$ and $c = 1/2 + 1/4$

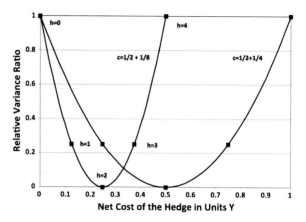

Fig. 10 A model of business response to favorable and adverse conditions relative to normal. The continuous curve showing the loss and gain is approximated by two values, a loss $-Y$ in adverse conditions, a gain X in favorable conditions

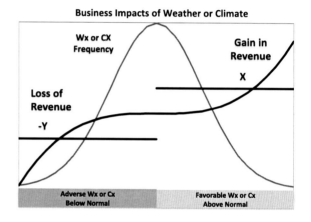

load increasing for daily average temperatures above and below about 18 °C (for an example, see Dutton 2010). Deviations from the loads expected on the basis of climatology can either increase costs or increase profit, depending on whether the utility is prepared for them. For example, warm summer temperatures may increase profit while cool conditions could lead to less.

A model of the financial results of a weather- or climate-sensitive business simplified to match the binary forecasts is shown in Fig. 10, which might be the model of dependence on degree day variation around some average or reference value, or of the level of a reservoir below and above a normal level as a function of seasonal precipitation, or of the net profit in a greenhouse operation as a function of heating or cooling loads imposed by the external temperature.

As indicated in Fig. 10, the business expects an increase X of revenue when above normal or favorable conditions prevail and a relative loss $-Y$ otherwise. Taking advantage of the weather and climate risk market, management can purchase a hedge that will pay H if the adverse conditions prevail; the cost is assumed to be a preseason premium of cH.

A Probabilistic View of Weather, Climate, and the Energy Industry 371

Table 11 A model for linking business operations and options with binary forecasts

Forecasts	Verification	
	Above	Below
Above	$X(1 + g_X(1 - \gamma_X))$	$-Y - g_X\gamma_X X$
Below	$X - \gamma_Y g_Y Y - cH$	$-Y(1 - g_Y(1 - \gamma_Y)) + H(1 - c)$

We also assume that management has the option of increasing the gain in favorable conditions by an amount $g_X X$ at a preseason cost of $\gamma_X g_X X$ and of reducing the loss in unfavorable conditions by an amount $g_Y Y$ at a preseason cost of $\gamma_Y g_Y Y$. Table 11 describes the options, benefits, and costs relative to forecasts in the same form as the forecast performance matrix W defined from Table 8 as

$$W = \begin{Bmatrix} 1 - C(0) - (S(p) - C(p)) & C(0) - C(p) \\ S(p) - C(p) & C(p) \end{Bmatrix} \quad (11)$$

Although significant asymmetries may exist in response to conditions above or below normal, it will be convenient to assume here that $X = Y$, $h = H/Y$, and that $\gamma_X = \gamma_Y$. Then the matrix describing the business performance relative to $X = Y$ becomes

$$B = \begin{Bmatrix} 1 + g_X(1 - \gamma) & -1 - g_X\gamma \\ 1 - \gamma g_Y - ch & -1 + (1 - \gamma)g_Y + h(1 - c) \end{Bmatrix} \quad (12)$$

We define a matrix product operator $*$ as a term-by-term sum in the form

$$P * Q = \sum_{i=1}^{i=2} \sum_{j=1}^{j=2} P_{i,j} Q_{i,j} \quad (13)$$

and then the dimensionless expected revenue and variance are

$$R(p, \Gamma) = B * W$$
$$V(p, \Gamma) = B^2 * W - (B * W)^2 \quad (14)$$

in which the elements of B^2 are the squared elements of B and the vector $\Gamma = \{g_x, g_y, h, c, \gamma\}$ represents the model parameters. Some examples of the revenue and variance for various combinations of the parameters are shown in Fig. 11.

This wide range of possible performance demonstrates that choices must be made relative to some definition of optimum response or economic utility. To illustrate, it is convenient to define a utility function

$$U(p, \Gamma) = \alpha R(p, \Gamma) + (1 - \alpha) \frac{1}{V(p, \Gamma)} \quad (15)$$

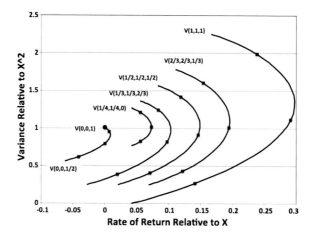

Fig. 11 Parametric curves showing relative return and variance for a variety of configurations of the business model as a function of the probability threshold p above which action is taken. The curves all begin at $p = 0$ at the lower end and proceed to $p = 1$ at the upper end with the points $p = 0.25$, 0.5 and 0.75 shown on each curve with a square marker. The parameters are $V(g_x, g_y, h)$ with $c = 5/8$ and $\gamma = 1/3$. The solution $R(0, 0, 0) = 0$, $V(0, 0, 0) = 1$ is shown with a dot at that point

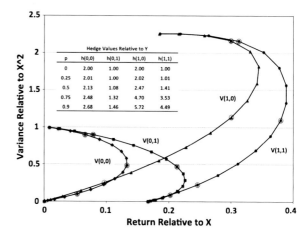

Fig. 12 Parametric curves showing relative return and variance for solutions maximizing the utility (14) as a function of the probability threshold p above which action is taken. The curves all begin at $p = 0$ at the lower end and proceed to $p = 0.9$ at the upper end. The points with $p = 0.25$, 0.5 and 0.75 are shown on each curve with a square marker. The values of the optimum hedge are shown in the *inset*

that increases as the dimensionless revenue increases or as the dimensionless variance decreases. The weight α would enable a preference between the two quantities but for convenience we will use equal weighting.

We suppose a charitable counterparty will offer a hedge at a cost $c = 1/2$. We set $\gamma = 1/3$ and proceed to use numerical methods to find hedge values $h(p, g_x, g_y)$ that maximize the utility function for four sets of the g's. The results are shown in Fig. 12.

The implication is that managers can respond to forecasts at various levels of probability with the expectation of achieving specific results when averaged over a sufficiently large number of events—provided, of course, that the statistics of both the events and the forecast performance are reasonably stationary.

The World Climate Service is examining a three-component version of this model that would be applicable with ternary forecasts. A more complex, time-dependent approach will be outlined in the next section.

5 Atmospheric Informatics

The increasing volume, diversity, and quality of information about atmospheric variability create an intensifying challenge and opportunity to improve the performance and profitability of weather and climate sensitive enterprises. The previous section demonstrated how atmospheric and business information might be linked with a quantitative model to provide probabilistic information about the returns and variability to be expected by taking action in response to forecasts. Regarding this as a first step, we attempt to foresee how present trends may evolve in the years to come.

The atmospheric sciences have advanced remarkably in the past few decades by combining improving observations, enhanced theoretical understanding, and rapidly accelerating computer capability. The grand challenges of atmospheric science as they relate to energy can be summarized[2] as

- Extend range, accuracy, and utility of weather forecasts;
- Improve understanding and prediction of severe and extreme weather and climate events;
- Organize the climate record more fully and effectively;
- Improve understanding and prediction of seasonal-, decadal-, and century-scale climate variation on global, regional, and local scales;
- Develop improved understanding of nonlinearity and atmospheric tipping points.

Success in meeting these challenges will lead to ever-increasing flows of information that offer potential improvement in the management of weather and climate risk and opportunity. Indeed, the U.S. National Research Council (2008) foresaw the creation of an entirely new capability:

The Virtual Earth System (VES) will run in an Internet cloud of petascale computers, assimilating data from satellites in space and from observation sites all over the world.

VES will maintain a continuous, dynamically consistent portrait of the atmosphere, oceans, and land—a digital mirror reflecting events all over the planet.

---

[2] Adapted for the present purposes from National Research Council (2008).

It will be the foundation for a prediction system...

From this, Dutton (2010) imagined that

> A Future Earth System (FES) will also run in a cloud of computers, maintaining a continuous simulation of Earth System events expected in the days to decades ahead.
>
> These simulations will drive numerical models of energy enterprises and their interaction with society, providing a continuous outlook on opportunity and risk—from weather, climate variability, and from new realities created by an evolving climate—a digital telescope peering into the future.

As the flows of atmospheric information available to enable effective decisions become more complex, we must try to understand the

- Desired outcomes in energy and other industries;
- Realities that constrain decision and actions;
- Critical complexities and uncertainties;

and then collaborate to create the mechanisms that will ensure more effective decisions.

It is essential to recognize that information flows exploding in variety and volume may degrade both understanding and the quality of decisions. We must develop a strategy to filter the information and focus only on the critical issues. An engineering team designing an avionics system for a new U.S. Air Force fighter airplane faced the same challenge of filtering and focusing an otherwise overwhelming flow of information. The team arrived at an effective guiding design principle: *Show the pilot what he needs to know, NOW.*

It will be advantageous to define *atmospheric informatics* as a process and system for formalizing methods and mechanisms that create, transfer, and apply atmospheric information in important endeavors. Atmospheric informatics will concentrate on assembling information about past and future atmospheric states and reshaping it into forms that prove useful and advantageous in the management of weather and climate risk and opportunity. It will develop as an emerging combination of atmospheric and information science, computer and communications technology, and the psychology of human–computer interaction that has the goal of improving key decisions in weather and climate sensitive endeavors. The aim and aspiration of atmospheric informatics will be: *Show the decision-makers what they need to know, NOW.*

Thus atmospheric informatics will shape information flows to be relevant to the decision context. The key tasks are to

- Identify and understand the *critical decisions* in energy and other industries on the full range of temporal and spatial scales.
- Create *effective processes* for transferring information between prediction systems and decision systems.
- Collaborate in the design and implementation of *decision systems* that predict the probability of success of alternative actions.

It is essential to transform probabilities about future atmospheric and other environmental events into probabilities about enterprise performance. And then we must develop the capability for the decision-makers—or perhaps their computers—to make the decisions and take the actions that will increase the probabilities of favorable and profitable outcomes. The business model of Sect. 4 is a simple and static approach based on a deterministic view of business response. But like the atmosphere, business responses are uncertain because of the cumulative impact of a wide variety of forcing effects.

In order to consider a more comprehensive and realistic model as a framework for decisions, as a mechanism for transforming probabilities, we consider the vectors

$$\mathbf{Y} = \{y_1, y_2, \cdots, y_N\} \text{ Performance variables (results)}$$
$$\mathbf{C} = \{c_1, c_2, \cdots, c_M\} \text{ Management variables (controls)}$$
$$\mathbf{B} = \{b_1, b_2, \cdots, b_I\} \text{ Business environment variables} \quad (16)$$
$$\text{(realities, constraints)}$$
$$\mathbf{W} = \{w_1, w_2, \cdots w_J\} \text{ Weather and climate variables}$$

and then presume that equations governing the evolution of the business exist in the form

$$\frac{d\mathbf{Y}}{dt} = \mathbf{F}(\mathbf{Y}, \mathbf{C}, \mathbf{B}, \mathbf{W}, t) \quad \text{with solution } \mathbf{Y}(t) = \mathbf{G}(\mathbf{C}, \mathbf{B}, \mathbf{W}, t) \quad (17)$$

and will describe the probabilities of performance

$$\text{Prob}(\mathbf{Y}, t) = \mathbf{H}(C, B, \text{Prob}(\mathbf{W}, t), t) \quad (18)$$

Since the model (17) will undoubtedly require numerical rather than analytic solutions, the same strategy used in numerical weather and climate prediction can be used: The solutions (17) can be forced by each member of the forecast ensemble and then combined into a probability distribution (18) of the business performance as explained in Sect. 2. We will then know the probability of mission success relative to the predicted conditions. Moreover, with a forecast performance history we will be able to anticipate performance statistics expected for acting on specific predicted probabilities of environmental events.

Figure 13 envisions how the energy information and decision systems of a number of enterprises might interact with an atmospheric informatics system that organizes the information available from the national and international observation and prediction systems. As a less grand and more immediate attempt to provide decision support to the energy and other industries, the World Climate Service is developing a new Internet-based workspace that assembles historical information, probability forecasts, and information from some trading environments. Figure 14 shows the system diagram and Fig. 15 shows an interactive chart displaying expected degree-days. It is an attempt to tell the users what they need to know, now.

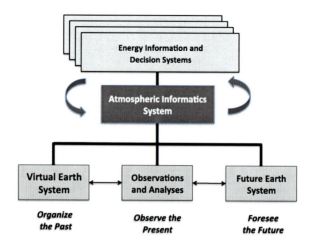

Fig. 13 A configuration of observation and prediction systems, an atmospheric informatics system, and energy information decision systems that may evolve over the years ahead with the aid of the collaboration stimulated by ICEM and similar efforts

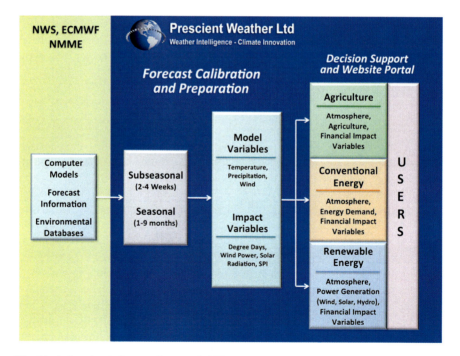

Fig. 14 A functional diagram for a probability prediction and decision support system being developed by Prescient Weather and the World Climate Service to serve the agricultural and energy industries, designed in part to tell decision-makers about the money and about what they need to know, now

A Probabilistic View of Weather, Climate, and the Energy Industry 377

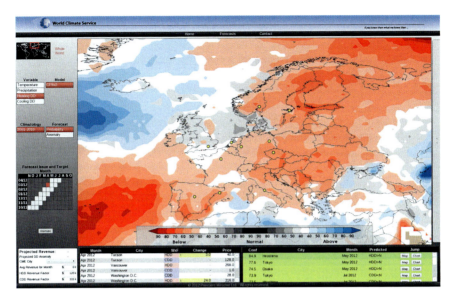

Fig. 15 A chart of degree-day probabilities in the short-term climate information and decision support system being developed by the World Climate Service to augment its present Internet site at www.worldclimateservice.com. The data boxes at the bottom of the screen scroll information about exchange-traded weather derivatives

6 Conclusion

The maturing capability for linking the computational systems of atmospheric prediction facilities and weather and climate sensitive enterprises is creating a new era in the management of environmental risk and opportunity. The energy and other industries will be able to estimate probabilities of success associated with forecast probabilities and manage for results more precisely. We will see an increasingly sophisticated and quantitative version of the strategy:

Expect the mean,
Hedge the extreme;
Use the forecast,
To go in between.

References

Anderson D, Stockdale T, Balmaseda M, Ferranti L, Vitart F, Molteni F, Doblas-Reyes F, Mogenson K, Vidard A (2007)Development of the ECMWF seasonal forecast system 3. ECMWF Tech Memo 503:56

Dubus L (2013) Weather and climate and the power sector: needs, recent developments, and challenges. Chap. 18, this volume

Dutton JA (2002) The weather in weather risk. In: Dischel RS (ed) Climate risk and the weather market., Risk Books, Risk Waters Group Ltd., London, pp 185–211

Dutton JA (2010) Weather, climate, and the energy industry: a story of sunlight—some old, some new. In: Troccoli A (ed) Management of weather and climate risk in the energy industry. Springer, Netherlands, pp 3–19

Dutton JA, James RP, Ross JD (2013) Calibration and combination of dynamical seasonal forecasts to enhance the value of predicted probabilities for managing risk. Clim Dyn 40:3089–3105

European Centre for Medium-Range Weather Forecasts (2011) Seasonal forecast user guide (system 4). http://www.ecmwf.int/products/forecasts/seasonal/documentation/system4/index.html

ExxonMobil (2012) The outlook for energy: a view to 2040. ExxonMobil. www.exxonmobil.com/energyoutlook

Glahn B, Peroutka M, Wiedenfeld J, Wagner J, Zylstra G, Schuknecht B, Jackson B (2009) MOS uncertainty estimates in an ensemble framework. Mon Weather Rev 137:246–268

Hagedorn R, Doblas-Reyes FJ, Palmer TN (2006) DEMETER and the application of seasonal forecasts. In: Palmer T, Hagedorn R (eds) Predictability of weather and climate. Cambridge University Press, Cambridge, pp 674–692

Kalnay E (2003) Atmospheric modeling, data assimilation, and predictability. Cambridge University Press, Cambridge 341 pp

Kolmogorov A (1933) Foundations of the theory of probability. Chelsea (translated and published), New York, 1956, 84 pp

LaPlace PS (1812) Théorie Analytique des Probabilités. Mme. Ve, Courcier, Paris

Lorenz EN (1963) Deterministic non-periodic flow. J Atmos Sci 20:130–141

National Research Council (2008) The potential impact of high-end computing on four illustrative fields of science and engineering. National Academy Press, Washington, 142 pp

Palmer T, Hagedorn R (2006) Predictability of weather and climate. Cambridge University Press, Cambridge, pp 674–692

Ruth DP, Glahn B, Dagostaro V, Gilbert K (2009) The performance of MOS in the digital age. Weather Forecast 24:504–519

Saha S et al. (2012) The NCEP climate forecast system version 2. J Clim (to be submitted)

Troccoli, Alberto (2010a) Weather and climate predictions for the energy sector. In: Troccoli A (ed) Management of weather and climate risk in the energy industry. Springer, Netherlands, pp 25–37

Troccoli, Alberto (2010b) Seasonal climate forecasting: a review. Meteorol Appl 17:251–268

U.S. Energy Information Agency (2012) Annual energy outlook, early release overview. http://www.eia.gov/forecasts/aeo/er/

Van den Dool H (2007) Empirical methods in short-term climate prediction. Oxford University Press, Oxford, 215 pp

Weather and Climate and the Power Sector: Needs, Recent Developments and Challenges

Laurent Dubus

Abstract Weather and climate information is essential to the energy sector. The power sector in particular has been using both observations and forecasts of many meteorological and hydrological parameters for several decades. In the last 10 years, a clear upward trend has been observed in the number, complexity, and value of data provided by National Meteorological and Hydrological Services (NMHSs) or produced by the energy sector itself. Much progress has been made, especially in the medium-term and longer time ranges; the development of reliable probabilistic forecasting systems has allowed many improvements in demand and production forecasts, although there is still a lot to do because of the difficulty in integrating probabilistic weather forecasts in management tools. In addition, the rise of renewable energy (RE) production systems, in particular wind and solar energy, has emphasized new needs for more accurate and reliable short-term forecasts, from real-time to a few days ahead. Rapid fluctuations in wind and solar radiation at local scale certainly raise a serious problem for the management of power grids. Significant and swift improvements in local forecasts, at hourly or even sub-hourly time step, become increasingly important and will be among the drivers for the large-scale development of RE systems. In this paper, we present some important results concerning monthly ensemble forecasts of temperature and river streamflows in France. We then point to the principal needs in weather forecasting associated with the development of RE. We also discuss the importance of collaboration and relationships between providers and users of weather, water, and climate information.

L. Dubus (✉)
R&D/Atmospheric Environment and Applied Meteorology,
6 Quai Watier, 78400 Chatou, France
e-mail: laurent.dubus@edf.fr

Fig. 1 Aerial view of Migouélou dam and lake, © EDF, Gilles De Fayet

1 Introduction: The Power Sector is Increasingly Weather Dependent

The power sector is constantly evolving and this has, in particular, been the case in the last 15 years in France, because of the liberalization of the energy market. In addition to physical constraints on the systems, financial factors have become ever more important, bringing even greater complexity to an already complex optimization problem.

Most utility operations are influenced by climatic variables: demand of course depends on temperature, either for heating in winter or cooling in summer; RE production depends on the respective source (wind for wind energy, solar radiation for solar energy, precipitation and river discharge for hydropower, etc.) (Fig. 1)

The importance of weather and climate for economic activity has been the subject of many studies (Marteau et al. 2004; Teisberg et al. 2005; Dubus 2007; Lazo 2007; Rogers et al. 2007; Dutton 2010; Frei 2010, etc.). French energy companies gave figures explicitly on the climatic impact on their activity for the first time in 2010, in their corporate results communication. Electricité De France (EDF), the French leading power company, in particular, evaluated the impact of

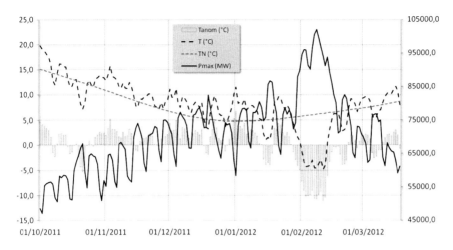

Fig. 2 Temperature and peak demand in France. Data from Météo-France and RTE (www.rte-france.com): normal temperature (*grey dotted line*), daily temperature (*black dotted line*), temperature anomaly (*bars*), and daily maximum demand (*solid black line*)

weather on the variation in sales and EBITDA[1] (EDF 2011). In 2010, more power was sold, in particular due to cold conditions in winter, and this resulted as a positive impact on both indicators (+€337 million on sales and +€215 million on EBITDA, respectively). Both demand and production in fact depend on weather.

Power demand depends foremost on air temperature, as shown in Fig. 2 which represents, for October 2011 to March 2012, the time evolution of the average temperature over France and of power demand (together with the climate normal and the anomaly with respect to this normal). There is a clear correlation between both variables: when temperature decreases, power demand increases and vice versa. This relationship is commonly defined as the "demand gradient". In France, the winter gradient is 2,300 MW/ °C[2] at around 7:00 PM (the time of peak demand in winter). This means that for an extra anomaly of −1 °C (or respectively +1 °C), the demand (and hence, the production required to meet it) increases (respectively decreases) by 2,300 MW, which corresponds to twice the electricity consumption of a large city such as Marseille (∼850,000 inhabitants). The value of this gradient depends on both the time of day and the day of the year. In summer, the maximum value is 500 MW/ °C and is reached at around 1:00 PM.

Clearly, power generation also depends on climate variables. Temperature and river flow determine the cooling capacities of (standard and nuclear) thermal power plants which are located along rivers. Summer heat waves and/or low river

---

[1] Earnings Before Interest, Taxes, Depreciation, and Amortization.
[2] The power of a production unit is expressed in megawatts (MW). A nuclear plant has a production capacity of 900–1,600 MW, depending on the technology; the production capacity of a typical windmill is around 1–5 MW.

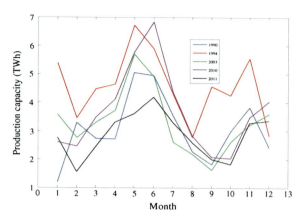

Fig. 3 From EDF's Financial communication, 2011/02/15. Monthly evolution of hydropower generation potential (TWh), for five different years

water levels, as in 2003, can reduce cooling capacity and thus entail a reduction in production capacity (Dubus and Parey 2009).

The influence of weather parameters is also crucial for RE sources (hydro, wind, and solar). Figure 3 shows the hydropower generation potential for four different years. In addition to a strong seasonal cycle, the production capacity is also marked by a strong interannual variability: in the last 25 years, the difference between highest and lowest annual generation potential was 23 TWh, for a theoretical maximum generation of 44.4 TWh.[3] In 1994, for instance, the autumn was characterized by many perturbations affecting most watersheds, and then strong precipitations that explain the high level of production capacity for this particular year (redline on Fig. 3).

In order to meet both the major challenges facing the power sector (IEA 2011) and political objectives, it is necessary to vastly develop wind and solar power production over the next 30 years. Due to their fluctuating nature, however, wind and solar energy cannot be scheduled in the same way as conventional power plants. This can lead to security problems for the networks and hence to power disruption for customers. Improving the quality of production forecasts is therefore crucial, to enable the development of solar and wind energy suited to the challenges of climate, energy demand and fossil fuel prices in the decades ahead (see also the chapters by George and Hindsberger, Love et al., Renne, Gryning and Haupt in this book).

Demand and production forecasts are thus crucial to the management of power systems, at all timescales. The new market organization over the last 15 years has even emphasized the need for longer term forecasts, in order to optimize the use of the different production means, in particular hydropower reservoirs. This paper is organized as follows: parts 2 and 3 respectively present some recent results from monthly forecasts of temperature and river streamflows in France and show their

---

[3] 1 TWh (Terawatt. hours) = 10^{12} W-h, is a measure of energy, the product of power capacity and the time during which it runs (maximum 8,760 h per year).

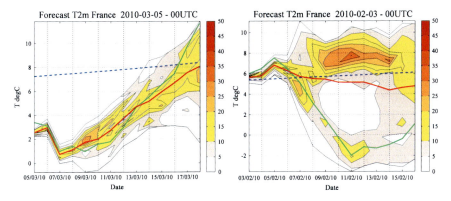

Fig. 4 14-day probabilistic forecasts of temperature over France, from ECMWF VarEPS system

improvement in quality, with respect to current reference forecasts. Part 4 discusses some important challenges in the coming years, to improve the use of probabilistic forecasts and the quality of short-term forecasts for RE. The conclusion summarizes the results and gives some important points about collaboration and partnerships between providers and users of weather and climate information.

2 Probabilistic Temperature Forecasts of a Few Days to One Month

As seen above, power demand in France depends on air temperature, the winter peak time gradient being on average 2,300 MW/ °C, and around 500 MW/ °C in summer. Temperature forecasts are therefore crucial to the supply/demand balance optimization problem. Deterministic forecasts from Météo-France and the European Center for Medium-range Weather Forecasts (ECMWF) are used routinely for short-term forecasts (Dubus 2010). For more than 10 years now, EDF has been using ECMWF EPS 14-days temperature forecasts and it seemed natural to test the benefits of using longer lead-time forecasts. Figure 4 illustrates the advantages of probabilistic versus deterministic forecasts. The plots represent two different forecasts, up to 14 days, of temperature averaged over France. The color corresponds to the density of the 51 runs of the ensemble predicting the corresponding temperature. For March 5th 2010 (left panel), the forecast dispersion is small, indicating a rather predictable situation, and the ensemble mean (red line) is very close to the a posteriori observed temperature (green line; the blue dotted line is the climate normal for that period): the difference between observation and forecast, up to day 9, is less than 1 °C. In this case, using the ensemble mean as a single deterministic forecast seems quite reasonable and would not lead to large errors, at least up to day 10. On the other hand, the forecast made on February 3rd

2010 (right panel) shows a marked bimodal distribution: most of the ensemble members being around the normal (blue dotted line), with a few indicating much lower temperatures (8 °C lower than the climate normal on February 11th). As shown by the observed values (green curve), the ensemble mean in red is, in this case, far from the observation: on the 12th, the error made using this crude deterministic forecast is 7.1 °C, equivalent to some 16,300 MW at demand peak time or 16.5 % of France's total installed capacity. Taking into account the whole probability distribution would therefore lead decision makers to act differently in the management of the system, with an evident reduction in risk. This clearly illustrates the superiority of probabilistic forecasts, even if the information is much more difficult to deal with and to integrate in existing power system management tools (see also the chapters by Mailier and Dutton in this book).

Although monthly weather forecasting was being studied as early as 1980 (Nap et al. 1981), numerical weather predictions with this lead time only improved significantly about 10 years ago. Since the early 2000s, ECMWF has been developing a monthly forecasting system which is now fully integrated in the VarEPS-Monthly system. It consists of a twice-weekly extension to 32 days of the EPS runs, an ensemble of 51 members at the global scale. The horizontal resolution is around 30 km up to day 10 and from then around 50 km up to day 32 (Vitart 2004; Vitart et al. 2008). In 2004/2005, a subjective evaluation of the forecasts was conducted on the basis of the graphical charts displayed on the ECMWF website, involving end users in the system optimization branch of EDF. Positive feedbacks allowed to study more deeply the potential benefits of such forecasts and to make a quantitative evaluation. A rather extensive study was undertaken, of which only the key results are given here. The evaluation was carried out on forecasts from October 2004, date of the operational release of the monthly forecasting system, up to April 2012 (395 forecasts). The variable of interest is air temperature, averaged over France (the figure is a weighted average of 26 stations in France, with the different weights corresponding to the proportion of total energy demand allocated to the 26 areas). Deterministic and probabilistic scores were calculated and compared to those of 2 reference forecasts: (1) from a historical dataset of 120 years of observed daily data, taken as a reference climatology (this 120-member ensemble always gives the same forecast for a given period) and (2) from a ∼15,000-year time series dataset, obtained with a statistical model, which has the same statistical characteristics as the 120-year dataset.[4] These references will henceforth be called REF1 and REF2.

Classical deterministic and probabilistic scores and skill scores (Jolliffe and Stephenson 2011[5]) have been calculated: bias, MAE, RMSE, ACC, rank diagrams, ROC scores, Brier Scores, and reliability diagrams. For the probabilistic scores,

---

[4] This 15,000-scenario dataset was established to deal with probability distribution tails (e.g., 1 % quantile), which cannot be estimated accurately with only 120 years of data.

[5] See also the web site maintained by Beth Ebert at http://www.cawcr.gov.au/projects/verification.

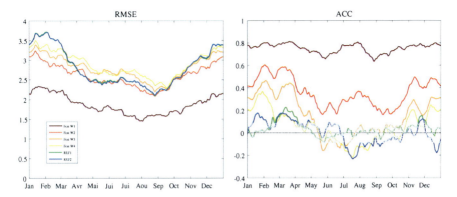

Fig. 5 RMSE (*left*) and ACC (*right*) of monthly temperature forecasts over France. Weeks 1–4 of the forecasts are in *brown, red, orange,* and *yellow,* respectively. *Green* and *blue lines* are reference forecasts (see text for details)

different events were considered (Temperature anomaly <−4 °C, <−2 °C, <0 °C, >+2 °C, and >+4 °C; Temperature anomaly <20 % percentile of the climatological distribution and >80 % percentile of the climatological distribution). The verification of each set of forecasts is made against a posteriori observation of temperature, weight-averaged over the 26 reference stations according to the aforementioned procedure.

Figure 5 shows the yearly evolution of two deterministic scores (Root Mean Squared Error and Anomaly Correlation Coefficient) for weeks 1–4 of the forecasts, together with those of the forecasts REF1 and REF2. Monthly forecasts display better scores up to week 2 than REF1 and REF2, throughout the year. The scores continue to be better in weeks 3 and 4 during winter (Dec–Jan–Feb). The Mean Error (not shown here) is of the same order of magnitude for the forecasts and REFs, with a yearly average value approaching 0. Evidently, these forecasts should, due to their nature, be evaluated instead in terms of probabilistic scores, which is presented below.

Only ROC skill scores for temperature forecasts falling below the 20th percentile or above the 80th percentile of the observed distribution are shown here. Figure 6 shows the time evolution of these scores depending on lead time (1–32 days), averaged over all forecasts.

The ROC skill scores (ROCSS) are always positive for both events, hence the forecasts are better than the climatology throughout the period. When compared to forecasts REF1 and REF2, the monthly system is better up to day 20 for both events and for the other thresholds considered (not shown here), although, the higher the amplitude of the anomaly considered (either positive or negative), the better the monthly forecasts.

Figure 7 shows the evolution of the same ROCSS throughout the year, for each week of the forecast. The plots show, first, that there is strong variability, denoted by the high-frequency oscillations, even if the scores were calculated using a

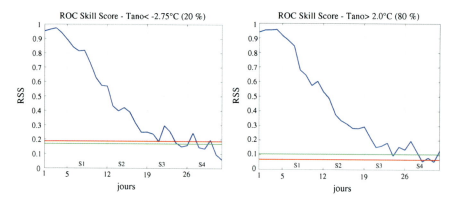

Fig. 6 ROC skill scores of monthly temperature forecasts over France (*blue line*), for the events: temperature anomaly within the <20 % (>80 %) percentile of the observations. The *red* and *green lines* are the ROCSS of the two reference datasets

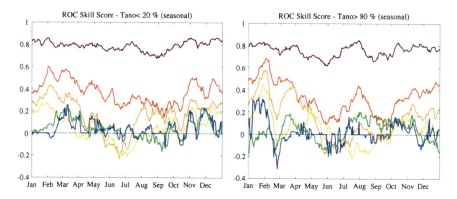

Fig. 7 ROC skill scores for each individual week and REF1 and REF2 forecasts (same *colors* as Fig. 4 and events as Fig. 5)

smoothing procedure. Monthly forecasts are better than the reference forecasts for weeks 1 and 2, throughout the year and for both events. Secondly, the skill of the forecasts varies through the year, with a maximum ROCSS during winter months (from November to March). In weeks 3 and 4 the conclusions must be moderated, but there is accuracy up to week 3 and even week 4 in December, January, and February, as well as in summer. This is, however, less evident in the intermediate seasons (spring and fall). With the exceptions of June, July, and August for week 4 and June–July for week 3, the ROCSS of the monthly forecasts is always positive and, for the majority of the time, higher than those of the reference forecasts.

The different plots and computed scores all confirm that monthly forecasts provide better information, at least up to week 3 in winter and week 2 (corresponding to days 11–18, that is to say 4 days more than the EPS) over most of the year.

The recent implementation of a second run of the system on Mondays has reinforced the value of these forecasts, which have now been used in operations for more than 3 years. The limiting factor to their use, at present, lies in the lack of integration between the forecasts and the existing tools: the forecasts are not used formally within the operational tools, but instead used as extra information which aids managers in taking their decisions on the management of the power system. A quantitative estimation of the economic benefit of such forecasts is rather difficult to produce, because they are not yet explicitly taken into account in optimization models. However, it is clear that these forecasts can be very useful to decision makers, in particular to anticipate cold spells in winter and heat waves in summer. Certain limitations have been identified and ways to progress will be discussed in Part 4 of this chapter.

3 Improvement in Monthly River Flow Forecasts

Hydropower represents 20.6 % of EDF's installed capacity in France, EDF being ranked number 5 in Europe for total installed renewable capacity, at 25 GW in late 2010 (EDF 2012). Hydropower production is very important in the French power system, as it provides a relatively partitionable energy stock, due to the presence of high capacity reservoirs. It therefore provides very attractive flexibility during peaks in demand. The difficulty, however, is that it is essential to manage the storage capacities and therefore to accurately forecast the annual water cycle inflow. At a given time, managers of the system are faced with making the optimal choice between using the water to produce energy in response to a peak in demand, or choosing alternative solutions as e.g. buying energy on the European market and keeping the water available in the reservoirs, should some forecasts show that the water will have a greater value in the days/weeks/months ahead. The problem is not only a question of financial optimization, but perhaps more importantly a physical problem, because rivers have to be managed in coordination with other users (agriculture, tourism, etc.)

Operational forecasts of river flow and water stocks are therefore crucial for the managers of the system. At present, they are generated everyday for the next 7 days, using deterministic and probabilistic forecasts from Météo-France and ECMWF, through an analog method (Zorita and von Storch 1999; Obled et al. 2002; Paquet 2004; Andréassian et al. 2006). Following from the studies made on temperature and reported in part 2, it was decided to evaluate the usefulness of ECMWF monthly forecasts for river flows. These were generated using the same analog method. For each of the 32 days and 50 members of a forecast, the method uses the geopotential fields forecasts at 700 and 1,000 hPa (Z700 and Z1000 respectively) over North Atlantic/Europe and search for analogs in the NCEP reanalysis. Fifty analog dates are kept for each member, so that it produces 2,500 analog weather patterns to the current forecast. The assumption, then, is to consider that for a given large-scale circulation pattern, the local precipitation and temperature at the given site will be the same. Then, referring to EDF's

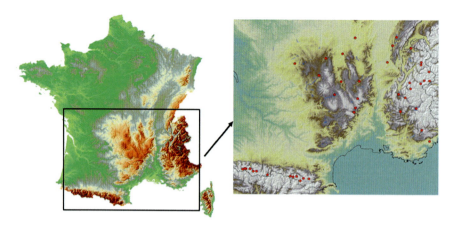

Fig. 8 Locations of the 43 basins considered for the monthly forecasts of precipitation and stream flows

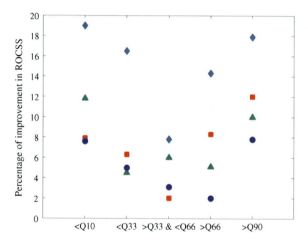

Fig. 9 Improvement in precipitation forecast ROCSS for analog versus ECMWF raw forecasts

high-quality precipitation and temperature database from 1953 to 2010, one obtains 2,500 precipitation and temperature forecasts for each time step and each station point considered. The study presented here focused on 43 basins, presented in Fig. 8, and forecasts from October 2004 to April 2010 (291 forecast dates).

A preliminary comparison of direct ECMWF model precipitation forecasts and analog forecasts showed that the analog method improves the local forecasts of precipitation on average over all basins, and over the course of the year. Similar results are observed for 2 m temperature (not shown here). Figure 9 shows the relative gain in ROC skill score taking analog forecasts of precipitation, with respect to the nearest ECMWF grid point forecast, for different events, averaged over each week (1–4) of the 291 forecast start dates. The improvement varies between 2 % for the central tercile in week 4 and about 18–20 % for week 1 and

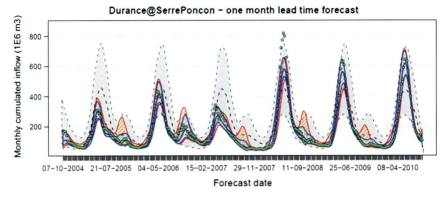

Fig. 10 Monthly cumulated inflow forecasts for CLIM (*grey*), REF (*orange*) and ANA (*blue*) methods, for the river Durance at Serre-Ponçon. Observed values are in *green dots*

for more extreme events (in the lowest 10 % and highest 90 % of the climatological distribution). Naturally, some local and seasonal discrepancies exist, but the improvements with the analog method are obvious.

The precipitation and temperature forecasts thus obtained are then used in the MORDOR hydrological model (Paquet 2004), to forecast river flows. The model is initialized with observed conditions (water stocks, observed inflows, snow stocks, etc.); the acquaintance with these initial conditions allows the model to make rather good forecasts in mountainous areas in spring, where flow is determined by the melting of the winter snow stock when temperature begins to rise. Of course, the quality of the model is not as good in plains and during the other seasons, because the flow is then less determined by initial conditions, but rather by direct precipitation. During the integration, the hydrological model requires temperature and precipitation forecasts. The current method, for lead times longer than 7 days, consists in using historical time series (1953–2010) in an ensemble climatological approach. It will henceforth be referred to as REF. The alternative method, tested here, is to use the monthly forecasts obtained using the analog method with ECMWF forecasts (referred to as ANA below). A third method can be used, which consists simply in using the streamflow climatology as a forecast (this CLIM is obtained from the 1953–2010 streamflow database).

Figure 10 shows forecasts of the monthly cumulated streamflow obtained with the three methods (CLIM, REF, and ANA) described above, and the observed values (in green) for the river Durance at Serre-Ponçon (French Alps), for the 291 start dates. This plot is a typical one, and summarizes the overall results: first, both REF and ANA methods give better results than the CLIM method, because they are based on the hydrological model, which takes advantage of the acquaintance with initial conditions and the physics of the water cycle. Considering only these two versions of the MORDOR model, monthly forecasts coupled with the analog method allow a better simulation of the inflows: in particular, they provide a narrower dispersion of the forecasts with respect to the observed time series (REF method). This dispersion is nonetheless sometimes too narrow and there are some outliers, but these generally correspond to extremely high inflows due to specific floods, which are very difficult to forecast more than a few days in advance. In the autumn of 2008, for example, the observed inflows were outside of the climatological distribution. There are some examples in which, even if ANA does not forecast high enough inflows, larger values are given than with the REF method. Overall, the most noteworthy point is that the ANA method was much more accurate during the last five autumns, which were characterized by very low water levels: the REF method considers the last 58 years of temperature and precipitation, whereas the ANA method only incorporates the most similar examples with respect to the current large-scale atmospheric pattern, thus excluding not relevant situations from the past.

This study shows, therefore, that even if the raw precipitation forecasts from ECMWF are not very accurate beyond days 10–15, post-processing, via an analog method applied to geopotential fields, can significantly increase the skill of precipitation forecasts and subsequently of water level forecasts. Another important consideration is that better results are obtained when forecasting monthly cumulated inflows, rather than daily time series, in accordance with the general result that long lead-time forecasts have better scores when one looks at integrated measures (Troccoli 2010).

The method used here has already been extensively tested in its 7-day operational configuration, with different predictors, distance criteria (to define analogs), and other key parameters. Although further refinements could be possible, it already gives positive results and has now been released operational. In addition to the better management of hydropower on a monthly timescale, the forecasts can also be used to schedule maintenance operations on dams and production units. An economic assessment of the benefits of such a revised model is planned, even if a difficult exercise.

4 Some Challenging Problems

NWP models have significantly improved in the last 10–15 years, in particular at lead times between 10 days and 1 month. In parallel, many efforts were made to better assess the quality and benefits of weather and climate forecasts in conjunction with the sector's economic needs (Dutton 2010; Lemaître 2010; Buontempo et al. 2010). If National Meteorological and Hydrological Services(NMHSs) are pursuing the development of forecasting systems, their customers play (or should play) an important role in defining the priorities to be addressed, in order for their needs to be answered. The paragraphs that follow emphasize some key considerations for the energy sector.

4.1 Further Use of Probabilistic Information

Ensemble forecasting is now routinely processed in several NWP centers and used in many sectors: energy, insurance, tourism, etc. (Dutton, in this book). Associated with increasing computational power, it has allowed the limit of predictability to be pushed beyond the 2 weeks suggested by Lorentz in 1963 (Buontempo et al. 2010), as was demonstrated for example in parts 2 and 3 of this chapter. However, one has to deal with several problems when using ensemble forecasts in operational applications.

First, existing tools are often complex systems, with a long history of development and evolution, as is the case with supply/demand optimization models in the energy sector (Dereu and Grellier 2009; Hechme-Doukopoulos et al. 2010; Charousset-Brignol et al. 2011). The integration of weather ensemble forecasts, for example those from the ECMWF VarEPS-Monthly system, is a difficult task because users' systems were not initially built to use such information. In addition, probabilistic information from ensemble forecasting systems is not simple to understand and manipulate for end users, who often have to deal with much information, from many different sources, in real-time decision-making processes.

A second limitation in the use of ensemble forecasts comes from the restricted number of members (typically, 51 at ECMWF). Although this is considered to be sufficient from a meteorological point of view, probabilistic forecasts are notably used to assess extremes, but calculating for instance the 1 % percentile of temperature distribution from 51 members is not straightforward. Current methods generally make the assumption that the temperature is normally distributed and use the mean and standard deviation of the 51 members to then estimate the necessary quantiles. This method gives accurate results as long as the temperature anomaly is not too significant, but can lead to suboptimal decisions when the deviation from normal is significant or when the forecast distribution is bimodal and hence very different from a Gaussian distribution, as is the case in Fig. 4. An internal study has shown that extreme quantiles of temperature distributions can be better estimated using a kernel density estimation and bootstrap resampling from ECMWF EPS ensembles. Further work and research is therefore needed to improve the estimation of forecast distributions from a finite number of members, in particular for distribution tails. As this will have to deal with extreme forecasts and risk optimization, it is a sensitive point which may bring extra value to probabilistic weather forecasts. In addition, "jumpiness" in successive forecasts is very often equated with "bad" forecasting by end users. As stated in Persson and Riddaway (2011), this is a natural characteristic of NWP models, but ways should be found to avoid conveying it to end users, in order to prevent confusion and misunderstanding. A third important point is linked to the fact that optimization models in the power sector generally need the same type of information, whatever the lead time; in particular, temperature information is used at a 3 h time step, for lead times of 1 day to 1 year. If weather forecasts are used up to days 12–14, historical time series (observations) are used in annual optimization models. In the same vein as the seamless forecasting concept developed in NWP (Vitart 2004; Rodwell and Doblas-Reyes 2006; Buontempo et al. 2010), research is under way to find solutions to achieve consistency between medium-range and annual forecasts. The initial idea, unsurprisingly, is to use medium-term forecasts at the beginning of the annual ones, rather than running independent simulations, but this raises the question of how to combine 14 days of 51 members' forecasts with (e.g.,) 100-year-long daily (observed) time series.

Long-term investment strategy and planning are important for the energy sector, with the scope between 10 and 50–60 years ahead. For the longest ranges, climate projections are used. For instance, EDF uses IPCC and CMIP scenarios and complex statistical methods to estimate future extreme temperatures in France and in the UK, in the context of climate change (Parey et al. 2007). Projections are also very important to aid decision-making processes for the next 10–30 years. Renewable energies investment or the adaptation/reinforcement of current facilities and networks require information about the probable climate for the next couple of decades. Decadal predictions for the next 10 years have been used by the UK Met Office to help the energy sector in the UK (Buontempo et al. 2010). Météo-France has developed a method which consists in extrapolating observed trends of the last 30 years to the next decade and then creating a new climatology, centered on the extrapolated mean with the past variability. The homoscedasticity assumption seems fair for extrapolation one decade ahead, but it would need deeper investigation for longer term projections. This method has the advantage of not using decadal climate predictions, which are not yet mature and about which many questions still remain. However, this emerging field of research seems promising and many efforts are currently under way to develop climate services applicable to economic activity, such as the EU FP7 EUPORIAS[6] project for instance.

4.2 Local Short-Term Forecasts for Renewables: Wind and Solar PV Power

Although the global use of energy is critical to contemporary human society, the power involved is quite small compared to that in the Earth's environment (Dutton 2010). However, extracting this "natural potential" energy is far from trivial due to its unequal distribution over the Earth, technical challenges, and the characteristics of the different sources. The projected growth of renewables in the decades ahead (IEA 2011) will, moreover, make energy systems increasingly dependent on weather and climate, which calls for a rapid improvement in production forecasting. In particular, the most mature technologies, wind and solar

---

[6] "European Provision of Regional Impact Assessment on a Seasonal-to-decadal timescale", www.euporias.eu.

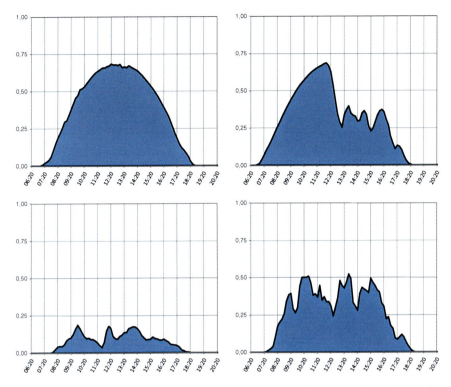

Fig. 11 Typical daily profiles of PV power production at a site on the island of Réunion (10 mins' data)

energy, are largely dependent on weather conditions. Therefore, research and development is essential to assist energy companies in developing these production means, by improving the reliability of integrating these variable resources and improving economic feasibility (Mahoney et al. 2012). Due to the characteristics of wind and solar radiation, the problem is complex and multi-faceted: both parameters vary quickly in time and space with non-linear impacts on the corresponding power generation.[7] Figure 11, for example, shows 4 typical daily profiles of photovoltaic power production at a single site on the Réunion Island (Indian Ocean). For reviews on wind and solar energy forecasting, one can refer to Lei et al. (2009), Lorenz et al. (2009), Heinemann et al. (2006), and the chapters by George and Hindsberger, Renné, Lorenz, Gryning, Haupt and Coppin in this book.

With the correct weather information, it is generally possible to make rather good power generation forecasts, even though the weather/power relationship is non-linear. Figure 12 shows 1 year of daily photovoltaic (PV) power production at one site on the Réunion island, estimated with two different statistical models

---

[7] Wind power, for instance, varies with the cube of wind speed.

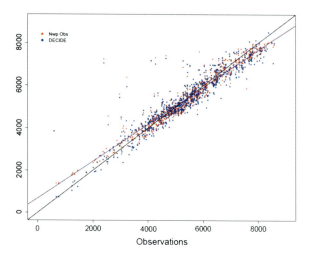

Fig. 12 Observation (*x*-axis) versus forecasts (*y*-axis) of PV power at one site on the Réunion island. *Red dots* multi-linear regression model; *blue dots* Multivariate Adaptive Regression Splines model. Both models are forced with observed solar radiation data from the nearest weather station

(a multi-linear regression model, and a Multivariate Adaptive Regression Splines based model), and with forcing provided by observed solar radiation from the nearest Météo-France station. Although neither model is perfect, the power production can be modeled with good accuracy as long as the input weather variable is "good." However, when switching from observations to forecasts from NWP, the results are different and model errors grow very markedly. Over the Réunion island, an internal study showed that RMSE can reach 50 % of the average power production at day +1. When considering production at a number of sites dispersed over a large area (a country for instance), the spread allows a significant reduction in the errors, by a factor of ~3 for example over France or Germany, in comparison with a single site. However, in small areas like islands, this averaging effect does not exist; hence local forecasts suffer, in particular, from the lack of resolution in NWP models. Model deficiencies and weather characteristics combine to make the forecasting problem very difficult. Evidently, predictions for locations such as the Réunion island, which are characterized by a very sharp orography and complex convective systems, are even more difficult to make.

However, recent studies showed that significant improvements are possible and involve the integration of multiple technologies. Mahoney et al. (2012) in particular (see also the chapter by Haupt in this book) have developed a complex system which takes advantage of the respective prediction capacities of its different components across the different forecast horizons. Such a system is not commonly used by energy companies at present, because it requires substantial computational resources and many sources of information (about both weather and power production) which are not universally available.

Very short-term prediction is also of great importance, because grid operators need to know, in real-time, how the whole power system will behave in the coming minutes to hours: if some production from a site is to decrease (or increase), the system manager has to adjust other production means, in order to ensure the equilibrium of the system. If the manager does not do so, there are risks to the stability of the grid and a subsequent risk of black-out. For these very short lead times, NWP is useless because models run, in general, from every 6 h (for global models) to every hour (for high-resolution limited area models). For this reason, other forecasting methods are generally used, based on real-time weather observations, both in situ measurements and satellite data and images (Gauchet et al. 2012; Lorenz et al. 2012) on the one hand and recent power production on the other. The latter type of data, in particular, seems promising because it only requires the real-time management of a utility operation's own data. Gomez Berdugo et al. (2012) showed that using only past production measurements allows forecasts up to 3 h with better accuracy than, for example, persistence. This type of method appears particularly interesting when collaboration between neighboring sites is used in the model, which requires centralized or distributed communication architecture. Of course, combining production data and weather data should further increase forecast skill, and further efforts are needed to develop such methods.

Studies (e.g., Mahoney et al. 2012) have shown that improvements in power production forecasting would provide significant financial benefits which would facilitate the faster development of renewables. In order to accommodate a deeper penetration of RE sources into power networks, many challenges still have to be addressed: first, it is essential that weather forecasting centers should provide better forecasts of wind speed at wind turbine height, and of solar radiation. "Better," in this case, means of higher resolution, both in time and space. Naturally, these forecasts should be delivered in a timely manner, so that the lead time of the forecasts is sufficient for them to be taken into account by system management operators.[8] A particularly important point is the prediction of ramps (very rapid fluctuations in power production due to snap changes in wind or solar radiation), which can have serious consequences for grid stability or even cause physical damage in the case of wind turbines. In addition to improving wind and solar radiation forecasts, RE development requires high-quality observations in the dimensioning phase of the projects, in order to evaluate the potential resources. The development of offshore wind energy, in particular, demands offshore wind observations at 100 m height, or, even better, vertical profiles from the surface to 200 m. These are only some examples, and there is no doubt that new data and forecast variables will become essential to the power sector in the future.

---

[8] For example, a D + 1 forecast at Réunion should be available for the grid operator no later than 16:00 local time on day D, and provide information up to D + 1 at 20:00 local time, in order to be useful. This means that the forecast should be issued at 10:00 UTC up to H + 30, considering a running delivery time of 2 h and the 4-h time lag at Réunion. At the moment, forecasts from Météo-France are issued at 00:00 UTC and 12:00 UTC, for H to H + 30 with the AROME model. In the first case, the forecast does not completely cover D + 1; in the second case, the D + 1 forecast is complete, but arrives too late to be taken into consideration in the planning of the operators.

5 Conclusion: Importance of Collaboration Between Users and Providers

The energy industry is exposed to weather and climate variability in the whole range of its activities. The impacts concern all time and space scales. The sector is one of the most important users of weather and climate information and forecasts, and its rapid evolution constantly creates new needs. Long-range forecasts (seasonal to annual and even decadal) become ever more important to the—physical and financial—optimization of the systems, especially for temperature and precipitation, which drive demand and hydropower production. Notwithstanding this, wind and solar radiation observations and short-term forecasts have also become invaluable, and their quality will certainly be among the drivers for the development of RE in years to come.

Scientific progress on its own is not sufficient to increase the value of weather forecasts. There are, in fact, three ways to increase this value (Lazo 2007; Rogers et al. 2007): by increasing forecast quality, by improving communication between providers and users, or by improving the decision-making processes. Each of these three components may be improved separately, but the whole process is undeniably more efficient if the whole chain is improved. This can only be achieved if a close collaboration is set up between the parties. Although state-of-the-art scientific knowledge may put some limitations to possible developments, it remains that the users' needs should be taken into account upstream, and then considered in an iterative process. Only this kind of collaboration can ensure an improvement of operational decision-making processes.

Further communication, collaboration, and partnerships between NMHSs and energy companies are then essential. These synergies will allow to develop better answers to operational needs, but also to add extra value to services provided by weather agencies. Finally, it will be beneficial to the entire society.

References

Andréassian V, Bergström S, Chahinian N, Duan Q, Littlewood I, Mathevet T, Michel C, Montanari A, Moretti G, Moussa R, Nasonova O, O'Connor K, Paquet E, Perrin C, Rousseau A, Schaake J, Wagener T, Xie Z (2006) Catalogue of the models used in MOPEX 2004/2005. In Andréassian V, Hall A, Chahinian N, Schaake J (eds) Large sample basin experiments for hydrological model parameterization. IAHS (Red Book Series N°307), Wallingford, pp 41–93

Buontempo C, Brookshaw A, Arribas A, Mylne K (2010) Multi-scale projections of weather and climate at the UK Met Office. In: Troccoli A (ed) Management of weather and climate risk for the energy sector, NATO Science Series, Springer Academic Publisher. ISBN 978-90-481-3691-9

Charousset-Brignol S, Doukopoulos G, Lemaréchal C, Malick J, Quenu J (2011) Optimization of electricity production. In: Lery T, Primicerio M,. Esteban MJ, Fontes M, Maday Y, Mehrmann V, Quadros G, Schilders W, Schuppert A, Tewkesbury H (eds) European success stories in industrial mathematics Springer, 1st edn. vol XII, 136 p. ISBN 978-3-642-23847-5, 2011.

Dereu G, Grellier V (2009) Latest improvements of EDF mid-term power generation management. Handbook of Power Systems (Energy Systems Series). Springer, Berlin and GmbH & Co. K, Heidelberg, 900 p. ISBN-10: 3642024920, ISBN-13: 978-3642024924

Dubus L (2007) Weather, water and climate information and the energy sector. In: Rose Tudor (ed) Elements for life, a WMO publication for the Madrid Conference. WMO, Geneva, Switzerland

Dubus L (2010) Practises, needs and impediments in the use of weather/climate information in the electricity sector. In: Troccoli A (ed) Management of weather and climate risk for the energy sector. NATO Science Series, Springer Academic Publisher. ISBN 978-90-481-3691-9

Dubus L, Parey S (2009) EDF's perspectives on adaptation to climate change and variability. In: Rose T (ed) Climate sense, a WMO—WCC3 publication. ISBN 978-92-63-11043-5

Dutton JA (2010) Weather, climate, and the energy industry. In: Troccoli A (ed) Management of weather and climate risk for the energy sector. NATO Science Series, Springer Academic Publisher, ISBN 978-90-481-3691-9

EDF (2011) Group results 2010. http://shareholders-and-investors.edf.com/fichiers/fckeditor/Commun/Finance/Publications/Annee/2011/2010EDFGroupResultats_3_va.pdf. Accessed 15 Feb 2011

EDF (2012) Facts and figures. http://shareholders-and-investors.edf.com/fichiers/fckeditor/Commun/Finance/Publications/Annee/2012/EDF2011_Fact-Figures_20120223_va.pdf

Frei T (2010) Economic and social benefits of meteorology and climatology in Switzerland. Meteorol Appl 17:39–44

Gauchet C, Blanc P, Espinar B (2012) Surface solar irradiance estimation with low-cost fish-eye camera. In: COST WIRE workshop on "Remote Sensing Measurements for Renewable Energy", DTU Risoe

Gomez Berdugo V, Chaussin C, Dubus L, Hebrail G, Leboucher V (2012) Collaborative methods for very short term PV predictions. Submitted to Solar Energy

Hechme-Doukopoulos G, Brignol-Charousset S, Malick J, Lemaréchal C (2010) The short-term electricity production management problem at EDF. OPTIMA Math Optim Soc Newsl 84:2–6

Heinemann D, Lorenz E, Girodo M (2006) Solar energy resource management for electricity generation from local level to global scale, chapter Forecasting of Solar Radiation. Nova Science Publishers, Inc, New York

IEA (International Energy Agency) (2011) World energy outlook 2011. IEA, Paris

Jolliffe IT, Stephenson DB (eds) (2011) Forecast verification: a practitioner's guide in atmospheric science. Wiley, New York, 274 pp

Lazo JK (2007) Economics of weather impacts and weather forecasts. In: Rose T (ed) Elements for life, a WMO publication for the Madrid Conference. ISBN: 92-63-11021-2

Lei M, Shiyan L, Chuanwen J, Hongling L, Yan Z (2009) A review on the forecasting of wind speed and generated power. Renew Sustain Energy Rev 13:915–920

Lemaître O (2010) Meteorology, climate and energy. In: Troccoli A (ed) Management of weather and climate risk for the energy sector. NATO Science Series, Springer Academic Publisher. ISBN 978-90-481-3691-9

Lorenz E et al (2009) Benchmarking of different approaches to forecast solar irradiance. In: 24th European photovoltaic solar energy conference, Hamburg, Germany.

Lorenz E, Kühnert J, Heinemann D, (2012) Short term forecasting of solar irradiance by combining satellite data and numerical weather predictions. In: Proceedings of 27th EUPVSEC, Frankfurt, Germany, 25–27 Sep 2012, pp 4401–4405

Mahoney WP, Parks K, Wiener G, Liu Y, Myers B, Sun J, DelleMonache L, Hopson T, Johnson D, Haupt SE (2012) A wind power forecasting system to optimize grid integration. IEEE Trans

Marteau D, Carle J, Fourneaux S, Holz R, Moreno M (2004) La Gestion du Risque Climatique, Economica, Paris, France, 211 pp

Nap JL, Van Den Dool HM, Oerlemans J (1981) A verification of monthly weather forecasts in the seventies. Mon Weather Rev 109:306–312. doi: http://dx.doi.org/10.1175/1520-0493(1981)109<0306:AVOMWF>2.0.CO;2

Obled C, Bontron G, Garçon R (2002) Quantitative precipitation forecasts: a statistical adaptation of model outputs through an analogues sorting approach. Atmos Res 63(3–4):303–324

Paquet E (2004) A new version of the hydrological model MORDOR: snowpack model at different elevations. Houille Blanche-Revue international de l'eau (2):75–82

Parey S, Malek F, Laurent C, Dacunha-Xastelle D (2007) Trends and climate evolution: statistical approach for very high temperatures in France. Clim Change 81(3–4):331–352

Persson A, Riddaway B (2011) Increasing trust in medium-range weather forecasts. ECMWF Newsl 129:8–12

Rodwell M, Doblas-Reyes FJ (2006) Predictability and prediction of European monthly to seasonal climate anomalies. J Clim 19:6025–6046

Rogers D, Clark S, Connor SJ, Dexter P, Dubus L, Guddal J, Korshunov AI, Lazo JK, Smetanina MI, Stewart B, Tang Xu, Tsirkunov VV, Ulatov SI, Whung P-Y, Wilhite DA (2007) Deriving societal and economic benefits from meteorological and hydrological services. WMO Bull 56(1):15–22

Teisberg TJ, Weiher RF, Khotanzad A (2005) The economic value of temperature forecasts in electricity generation. Bull Am Meterol Soc 86:1765–1771

Troccoli A (2010) Weather and climate predictions for the energy sector. In: Troccoli A (ed) Management of weather and climate risk for the energy sector. NATO Science Series, Springer Academic Publisher. ISBN 978-90-481-3691-9

Vitart F (2004) Monthly forecasting at ECMWF. Mon Weather Rev 132:2761–2779. doi: http://dx.doi.org/10.1175/MWR2826.1

Vitart F, Buizza R, Balmaseda MA, Balsamo G, Bidlot JR, Bonet A, Fuentes M, Hofstadler A, Molteni F, Palmer T (2008) The new VarEPS-monthly forecasting system: a first step towards seamless prediction. Q J R Meteorol Soc 134:1789–1799. doi:10.1002/qj.322

Zorita E, von Storch H (1999) The analog method as a simple statistical downscaling technique: comparison with more complicated methods. J Clim 12(8):2474–2489

Unlocking the Potential of Renewable Energy with Storage

Peter Coppin, John Wood, Chris Price, Andreas Ernst and Lan Lam

Abstract The role of storage in managing the variability in wind, solar and wave energy generation is well understood. Shifting energy from periods of high generation to low generation is seen as an ideal role for storage. However, the rapid fluctuations in wind and solar generation with periods less than 1 h can lead to very significant problems on the grid, reducing carrying capacity of lines and increasing the amount of spinning reserve and regulation services required to unachievable levels. A number of electrical storage technologies which are aimed at removing these rapid fluctuations are now being successfully demonstrated at the MW-scale. The various storage technologies are outlined and an example of wind smoothing system is examined in detail.

1 Introduction

The issue of how to cope with the inherent variability in renewable energy generation from weather-driven sources such as wind, solar and wave is well known. Shifting energy from windy or sunny days to those with less wind or sun is seen as an ideal role for storage utilising technologies such as pumped hydro and compressed air. The current power generation, transmission and market systems are, in fact able to cope with these longer term trends over several hours or days while variable renewables remain at modest penetration levels. The more pressing issue is fluctuations in wind speeds with periods of 1 h or less which occur during highly convective or stormy weather conditions. Similarly, intermittent cloud can produce very sharp changes in PV solar power generation. These conditions can lead to

P. Coppin (✉) · C. Price · A. Ernst · L. Lam
CSIRO Energy Transformed Flagship, Newcastle, Australia
e-mail: peter.coppin@csiro.au

J. Wood
Ecoult Pty. Ltd., Sydney, Australia

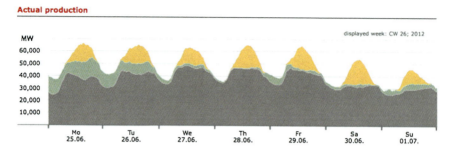

Fig. 1 PV and Wind contribution to power generation in Germany, June 25th to July 1st 2012 (*Yellow*—solar, *green*—wind, *grey*—conventional). Burger (2012)

very significant problems on the grid, reducing carrying capacity of power lines and increasing the amount of spinning reserve and regulation services required to unachievable levels. The only alternative is to curtail the renewable generation which is already being done in several markets.

A number of electrical storage technologies are being developed to both remove these rapid fluctuations and provide support to grid systems with large amounts of solar and wind power. Most of these are now being demonstrated at the MW-scale.

Whilst the variability of renewable energy sources such as wind and solar is well known, what is less understood is that the variability occurs over a wide range of timescales, including quite short timescales. The longer timescale variability (longer than 1 h) is associated with the passage of weather systems and daily cycles and the shorter timescale variability (shorter than 1 h) is associated with the turbulence or gustiness of the wind. The longer timescale variability in wind power can be mitigated by moving energy from periods of high generation to times of lower generation. The shorter timescale variability in wind power can cause instabilities in the power system.

Figure 1 shows the consequences of longer timescale variability in an examination of generation sources in the German grid from a week in Summer 2012 (Burger 2012). Clearly, the contribution of wind and solar power to satisfying the load is continuously varying. Longer-term storage, such as pumped hydro can play a significant role at these timescales.

Figure 2 shows short-term variability in wind power and the predicted consequences. It shows some modelling results, calculated in 2004 by NEMMCO, predicting flows in the main interconnector between the states of South Australia and Victoria, based on simulations of wind power production from wind speed records. With 400 MW of wind power installed in South Australia (the approximate installed capacity at the time) there is little difference from the no-wind power case, but with a projected 1,000 MW installed the wind power causes short-term violation of the connector flow limits. Fast storage systems could smooth this excess variability. The alternative solutions are to curtail the wind farm output or reduce the interconnector flow. As of October 2011, South Australia has 1,200 MW of installed wind power.

Unlocking the Potential of Renewable Energy with Storage 401

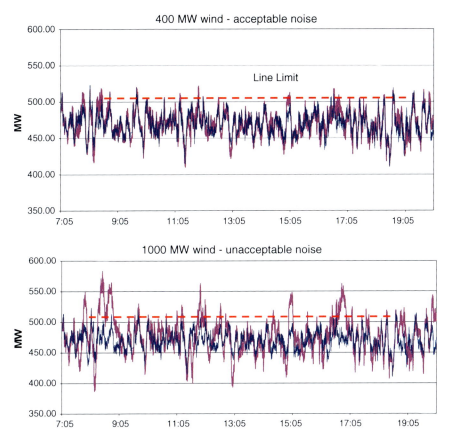

Fig. 2 Effect of increasing wind power in South Australia on Heywood interconnector flow—model results (12 h sample—*blue trace* is background flow, *pink trace* is with indicated installed wind power)

Rapid fluctuations in generation can cause an increased demand in regulation power, which is used to compensate between predicted demand and actual demand. This is shown in Fig. 3, which shows the difference between the predicted demand (the smooth blue curve) and the actual demand. Normally, the generation is scheduled against the predicted demand and the difference (the red curve) is supplied by separate regulation services. Wind power is generally treated as negative demand and most systems with significant contributions from wind power have wind forecasting systems, which calculate the likely contribution from wind. However, these forecasting systems do not predict the rapid fluctuation component which remains as an error, which needs to be corrected through regulation services.

Studies by the New York Independent System Operator (ISO), predict a significant rise in the regulation power requirement with increasing wind power on

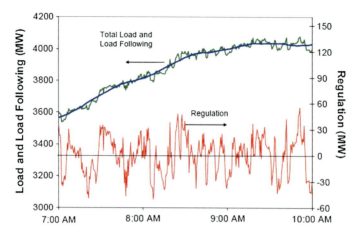

Fig. 3 Definition of regulation services (Kirby 2004), *blue trace* is predicted demand, *green trace* is actual demand and the *red trace* is the difference of the two—the regulation power required

their system (Fig. 4). The Californian ISO also projected an increase in regulation power requirement with increasing intermittent renewable energy on their system. They predicted a rise from a frequency regulation requirement of 1 % of peak load dispatch (approx. 350 MW) to 2 % regulation as renewables rose to 20 % contribution in 2010 and to 4 % regulation required (1,400 MW) as the contribution rises to 33 % by 2020. Regulation services are mainly supplied by peak generating plants with high emission levels or by modulating the output of base load generating plant. This is an ideal application for fast storage systems and we will see later that several technologies are already being trialled in this role.

In a report on the role of energy storage with renewable electricity generation, NREL (Denholm et al. 2010) concluded that high penetration of variable generation increases the need for all flexibility options including storage. They also concluded that it creates market opportunities for these technologies; however, storage has been difficult to sell into the market because of the challenges it has in quantifying the value of its services. Indeed the role which storage systems can play in the future power grids is very diverse. A recent study by Sandia Labs (Eyer and Corey 2010) lists 17 applications in five major categories such as Electric Supply, Ancillary (regulation) Services, Grid System, End User/Utility Customer and Renewables Integration (Table ES 1 from Sandia report) shows the potential value of the application, the likely market in the USA in a 10-year period and the timescale of the application. There is a very wide range in each case.

Figure 5 shows a similar view from a report by EPRI (Rastler 2010) which orders the storage applications by value and indicates the size of the market. For some high value applications, Transmission and Distribution (T&D) system support and area frequency regulation (part of regulation services), some technologies are already economic (See Table ES-4 and Figure ES-14 in the EPRI report). One

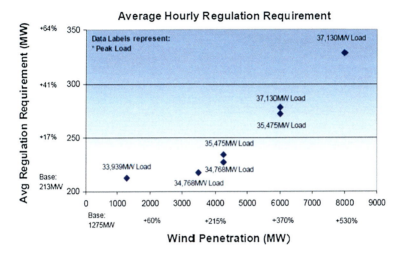

Fig. 4 Regulation requirement versus installed wind power, New York ISO

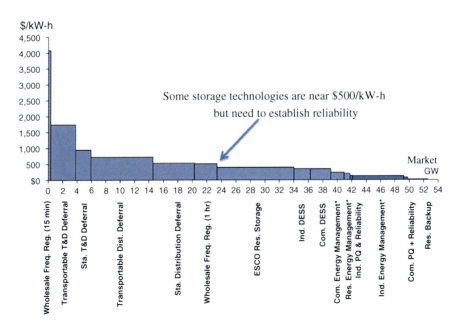

Fig. 5 Estimated target market size and target value analysis (Rastler 2010)

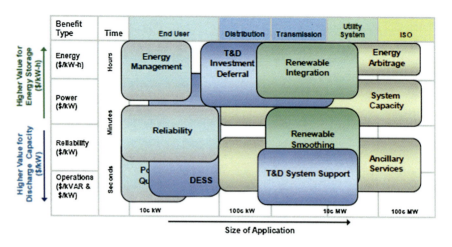

Fig. 6 EPRI analysis of storage value (Rastler 2010)

problem identified is that many technologies are still only at the large-scale demonstration phase and need more experience to earn the trust of Utilities and other customers.

The EPRI report also shows the range of storage applications in a diagrammatic form (Fig. 6). This representation is intended to show that there is often an overlap between applications in when size and timescale are considered. In fact, appropriately designed storage systems could provide multiple roles in a single system—achieving so-called "benefit stacking" and multiplying the possible monetary value which could be achieved. This may be essential to the initial viability of storage solutions.

However, it is acknowledged that currently storage value is difficult to extract when its operation may be simultaneously spread over a number of service areas. While it may be possible to calculate the benefits in areas where there are established markets or where there established markets—e.g. retail energy trading in time-shifted wind and solar energy (arbitrage) and regulation services, it is more difficult to access the value where there is strong regulation such as grid infrastructure services and upgrade deferral. Indeed it becomes even more complex when devices may fill both roles simultaneously.

Regulatory incentives are starting to appear. The US Federal Energy Regulatory Commission (FERC) issued a ruling on pay for performance in October 2011, which gives extra payment for faster ramping services, benefitting fast storage-based services. The California State Energy Storage Bill AB2514, March 2010 now requires electrical corporations and locally owned utilities to create energy storage systems in their distribution networks to either reduce emissions of greenhouse gases, reduce demand for peak electrical generation, or improve the reliable operation of the electrical transmission or distribution grid. This law mandates storage equal to 2.25 % of daytime peak power by 2014 and 5 % of daytime peak power by 2020.

Fig. 7 Taum sauk 450 MW custom pumped hydro storage system in Missouri

2 The Technologies

Rastler (2010) (Table ES-4) lists a number of technology options and costs (in today's terms) when applied to applications in bulk energy storage, fast frequency regulation, renewables integration and grid support. The list of technologies which are deemed to be commercial is quite short with many more in the demonstration phase of development. Some examples are listed below.

For bulk energy storage (several hours duration) pumped hydro, underground compressed air, sodium sulphur and advanced lead-acid is deemed to be commercial. More innovative pumped storage systems include custom built units such as at Taum Sauk in Missouri—a 450 MW system first installed in 1963 (Fig. 7) and the Okinawa Yanbaru Seawater Pumped Storage System. Constructed in 1999, this system utilises a cliff-top reservoir adjacent to the ocean to give a system capable of a throughput of 31 MW with approximately 400 MWhr of storage.

Compressed air storage systems use reversible turbines to compress or expand air which is stored in depleted underground hydrocarbon reservoirs, which can provide many hours of storage. There have been two long established systems at Huntdorf, Germany (290 MW peak power, established 1978) and McIntosh, Alabama (110 MW peak power, established 1991). An example of commercial electrical storage is the Sodium-sulphur (NaS) system from NGK from Japan which is rated at 1.2 MW/7.2 MWhr (6 h of storage), ideally suited to substation upgrade deferral.

Of the faster response systems, A123 has demonstrated a modular Li-ion system rated at 2 MW/0.5 MWhr (15 min storage). This multi-purpose system can address a range of applications such as regulation services. Beacon Power has constructed a 20 MW/5 MWhr (15-min storage) flywheel demonstration in New York State designed for regulation services (Fig. 8).

Fig. 8 Beacon power 20 MW/5 MWhr (15 min storage) flywheel demonstration in New York State

Fig. 9 Prosperity solar energy storage project, New Mexico featuring both voltage smoothing and peak shifting

East Penn Manufacturing Co. through its subsidiary Ecoult has constructed a 3 MW/3 MWh (1 h storage) system utilising UltraBatteries (modified lead-acid) for regulation services. This technology will be discussed further below. Ecoult has also supplied storage systems to the Prosperity Solar Energy Storage Project,

Fig. 10 1 MW UltraBattery storage system at Hampton Wind Farm

New Mexico (Fig. 9), simultaneously providing voltage smoothing and peak shifting of power from the 500 kW PV plant. The 500 kW/500 kWh (1 h) smoothing system uses the East Penn UltraBattery and the 250 kW/1,000 kWh (4 h) shifting system uses a more conventional high-performance lead-acid battery (East Penn Unigy II).

3 A Case Study: Wind Farm Smoothing

This case study describes the energy storage trial implemented at Hampton Wind Farm in NSW, Australia. The objective of the trial at Hampton is to smooth the ramp rate of the wind farm before presenting it to the grid. In turn the impact objective is to achieve higher penetration of wind and renewable energy in grid systems. While the Hampton system smooths the energy produced "at the source" on the wind farm, it is an objective of the work that the system and learning are transferable wherever the benefit of reducing renewable energy variability exists, for example at grid nodes (or substations) or via the provision of ancillary services generally.

The implementation a wind smoothing system followed a path of progression from laboratory trials, through the attachment of larger scale systems to a Vestas V47-660 kW wind turbine (Fig. 10). Stage 1, commissioned in mid 2010, consisted of a custom built system rated at 144 kW/240 kWh, and which featured four battery banks of different lead-acid battery types, including prototype UltraBatteries from Furukawa. Stage 2, commissioned in mid-2011, consisted of commercial modular building blocks rated at 1 MW (capped 660 kW)/500 kWh utilising East Penn UltraBatteries (Fig. 11).

Fig. 11 UltaBattery cell bank at Hampton Wind Farm

The UltraBattery is a hybrid energy storage device that integrates a supercapacitor with a lead-acid battery in one unit cell, without the need for extra electronic control (Fig. 12). This unique design, harnessing the best of both technologies, produces a battery which can provide high power discharge and charge with a long, low-cost life (Lam and Louey 2006).

Developed in Australia by the CSIRO Energy Transformed Flagship research program, the UltraBattery already serves applications for use in hybrid electric vehicles (HEVs) with further variants aimed at resolving issues of intermittency in capturing energy produced from renewable sources. It is manufactured in various forms by Furukawa in Japan and East Penn in the USA and is available in production quantities. Testing by Sandia National Laboratories Single Cell Testing under a regulation services profile has shown it to have several times the life of conventional lead-acid batteries (Hund et al. 2012).

Initial results from the first stage of the system show that with a simple proportional-integral (PI), fixed-parameter algorithm, significant reductions in rates of change of power output (ramp rates) can be achieved. Figure 13 shows results from 1 day with a variety of wind conditions. The lower traces show the raw turbine input and the smoothed output when combined with the storage system. The upper traces show the reduction in 5-min ramp rate which averages a factor of 7. The 1-min ramp rate reduction achieved by the system is a factor of 10.

Unlocking the Potential of Renewable Energy with Storage   409

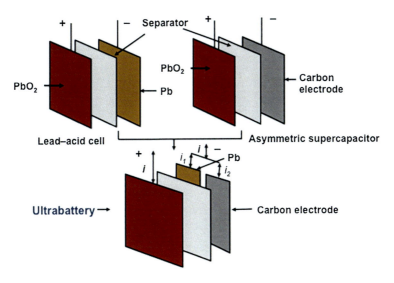

Fig. 12 The principle of UltraBattery technology

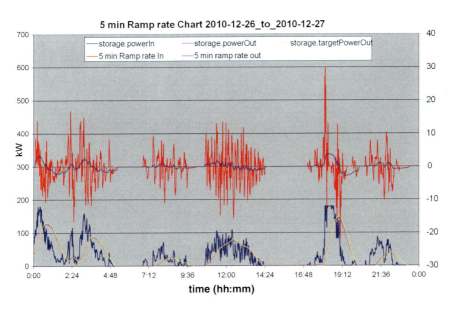

Fig. 13 Smoothing of Wind output and ramp rate reduction with fixed-parameter controller algorithm

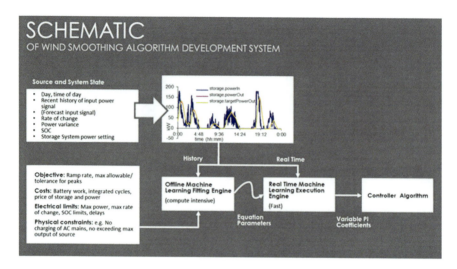

Fig. 14 Schematic of Wind smoothing algorithm development system

4 A More Advanced Algorithm

A more advanced algorithm system is now being developed. The system, shown in Fig. 14, works as an adaptive scheme which allows the smoothing parameters to be continuously changed. An offline optimising scheme is used to design the functions used in real-time. The optimisation takes into account a number of objectives (goals) and costs while being aware of system electrical and physical constraints. The system can be re-optimised for each installation.

The offline learning mechanism uses a quantum particle swarm optimisation algorithm. This is a meta-heuristic search method that copes well with highly non-linear objective functions and lends itself to parallelisation. This allows the optimiser to use of all 32 cores on a CSIRO multicore-computing server. A 21-day training data set composed of a combination of challenging periods is used to evaluate the quality of adaptive parameter settings.

The learning engine is used to create functions for the execution engine. This engine generates PI variables in real-time based on current conditions and derived from the learning that are in turn provided to the storage controller and used to generate charge and discharge commands to the batteries.

5 Advanced Algorithm Results

Initial trials show a significant improvement is possible using this approach. Figure 15 shows the result of a simulated comparison where the standard fixed parameter PI algorithm is run against the same wind data sample as the new

Unlocking the Potential of Renewable Energy with Storage

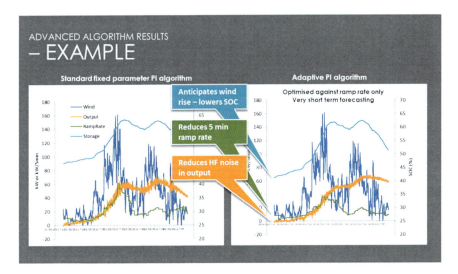

Fig. 15 Comparison of standard fixed parameter PI algorithm and adaptive PI algorithm

adaptive algorithm. In this case, the adaptive mechanism has achieved a result where it has achieved a superior reduction in the 5-min ramp rate over the standard fixed parameter algorithm and significantly reduced the high frequency noise in the signal. As a result of the training, the adaptive system recognised that wind conditions had been comparatively low for a period and anticipated the storm front that moved through by moving the state of charge of the energy store lower to where it had headroom to react more efficiently, reducing sudden output changes.

The architecture approach with the adaptive mechanism allows for many variables to be considered during the learning and for performance to be optimised against a number of targets. For example, minimisation of battery use has now been combined into the parameters optimised by the Hampton algorithm and CSIRO testing has shown that using the methodology the amount of energy passed through the energy store to achieve the ramp rate objectives can be significantly reduced while maintaining system performance. Further developments will test simultaneous energy shifting and smoothing by utilising the continuously variable control parameters. There is also potential to translate this simultaneous approach to solar PV.

6 Conclusions

Proven and emerging storage technologies have a wide variety of roles to play on the electricity grid, particularly in integrating renewable energy. The large range of storage timescales and capacities required give opportunities for all technologies, including performing multiple functions, simultaneously. The ability to extract value is currently limited and further regulatory changes are needed before the full potential of storage can be realised.

Acknowledgments We would like to acknowledge contributions to the wind farm smoothing study from the Australian Government Department of Resources, Energy and Tourism (DRET); and the NSW Government Department of Environment, Climate Change and Water (DECCW).

References

Burger B (2012) Electricity production from solar and wind in Germany in 2012. Fraunhofer Institute for Solar Energy System. www.ise.fraunhofer.de/en/renewable-energy-data. Accessed June 2013

Denholm P, Ela E, Kirby B, Milligan M (2010) The role of energy storage with renewable electricity generation. Technical Report NREL/TP-6A2-47187, Jan 2010

Eyer J, Corey G (2010) Energy storage for the electricity grid: benefits and market potential assessment guide. Sandia Report SAND2010-0815, Feb 2010

Hund T, Clark N, Baca W (2012) UltraBattery test results for utility cycling applications. In: 18th international seminar on double layer capacitors and hybrid energy storage devices, 12/08

Kirby B (2004) Frequency regulation basics and trends. Oak Ridge National Laboratory, ORNL/TM 2004/291, Dec 2004

Lam LT, Louey R (2006) Development of ultra-battery for hybrid-electric vehicle applications. J Power Sources 158:1140–1148

Rastler D (2010) Electricity energy storage technology options—a white paper primer on applications, costs and benefits. EPRI Electricity Power Research Institute, Final Report, 1020676, Dec 2010

Improving NWP Forecasts for the Wind Energy Sector

Merlinde Kay and Iain MacGill

Abstract Weather forecasting has traditionally been primarily used in the energy industry to estimate the impact of weather, particularly temperature, on future electrical demand. As a growing proportion of electricity generation comes from intermittent renewable sources such as wind, weather forecasting techniques need to be extended to this highly variable and site-specific resource. We demonstrate that wind speed forecasts from Numerical Weather Prediction (NWP) models can be significantly improved by implementing a bias correction methodology. For the study presented here, we used the Australian Bureau of Meteorology (BoM) MesoLAPS 5 km limited domain NWP model, focused over the Victoria/Tasmania region of Australia. The site for this study is the Woolnorth wind farm, situated in north-west Tasmania. We present a comparison of the accuracy of uncorrected hourly NWP forecasts and bias-corrected forecasts over the period March 2005 to May 2006. This comparison includes both the wind speed regimes of importance for typical daily wind farm operation, as well as infrequent but highly important weather risk scenarios that require turbine shutdown. In addition to the improved accuracy that can be obtained with a basic bias correction method, we show that further improvement can be gained from an additional correction that makes use of real-time wind turbine data and a smoothing function to correct for timing-related issues that result from use of the basic correction alone. With full correction applied, we obtain a reduction in the magnitude of the wind speed error by as much as 50 % for 'hour ahead' forecasts specific to the wind farm site.

M. Kay (✉)
School of Photovoltaic and Renewable Energy Engineering,
University of New South Wales, Sydney, Australia
e-mail: m.kay@unsw.edu.au

I. MacGill
Centre for Energy and Environmental Markets, School of Electrical Engineering,
University of New South Wales, Sydney, Australia

1 Introduction

The past decade has seen substantial growth in the wind energy sector, and this form of energy generation is now emerging as an important contributor to electricity generation around the world (Archer and Jacobson 2005; Sanchez 2006; Wiser and Bolinger 2011). Weather forecasting plays a vital operational role within the electricity industry (Feinberg and Genethliou 2005) but this has traditionally been primarily on estimating future, temperature-dependent demand. In an Australian context, 1-hour ahead forecasts are useful because the onus is on the grid operator to match electricity supply and demand. In other parts of the world, the onus is more on the wind farm operator to predict output further ahead, e.g. a day in advance. Intermittent renewable energies such as wind present new challenges with regard to the potential impacts of weather on future generation availability. From a meteorological perspective, a particular challenge is one of scale; at the wind farm level what is required are accurate, near real-time wind speed and direction forecasts for a specific height over a small area, often for remote locations subject to highly variable conditions (e.g. coastal sites). This is not what traditional NWP models are designed to provide.

There are currently three approaches to the challenge of wind farm specific forecasting, with the choice-dependent upon the forecast time horizon. The first approach uses purely statistical techniques to forecast minutes to a few hours ahead (Milligan et al. 2003; Torres et al. 2005; Potter and Negnevitsky 2006). The second is to design new meso- and micro-scale NWPs devoted to providing local wind forecasts. Usually, this approach involves modifying an already existing NWP with some meso-scale development (Vincent et al. 2008). The third approach is to devise methods that combine existing NWP outputs with real-time local data and advanced statistical analyses to produce accurate site-specific predictions (Giebel et al. 2003; Landberg et al. 2003; Focken and Lange 2008; Jørgensen and Möhrlen 2008).

In this work, we concentrate on the third approach and show that by taking NWP outputs and applying a bias correction methodology we can obtain improved predictions of wind speeds of importance to wind farm performance. The bias correction methodology allows us to minimise systematic errors in future wind speed forecasts by using a knowledge of the bias in errors from past forecasts. This technique extends beyond 'fair weather' conditions, and we show its successful use in predicting an extreme weather event that required shutdown of the wind farm. The wind farm site and forecast model used in this study are described in Sect. 2, with results obtained for the seasons spanning 2005–2006. As discussed in Sect. 4, we have successfully managed to reduce the forecast error by 50 % by application of the bias methodology described in Sect. 3. Some key conclusions from this study are then considered in Sect. 5.

Improving NWP Forecasts for the Wind Energy Sector   415

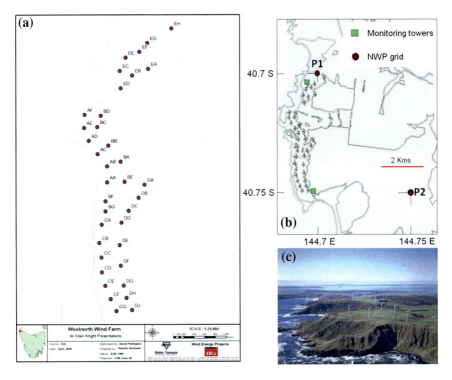

Fig. 1 (**a**) Geographical layout of the wind turbine groups relative to the topography; (**b**) Schematic layout of the wind turbines at Bluff Point, along with the original meteorological monitoring towers and the two nearest NWP grid points (P1 and P2); (**c**) The bottom inset is an actual image of the turbines along the coastline (image courtesy of the Clean Energy Council — http://www.cleanenergycouncil.org.au/cec/resourcecentre/casestudies/Wind/Woolnorth1.html)

2 Wind Farm Data and the Forecast Model Used in This Study

The site used for this study is the Woolnorth wind farm, which is situated on the west-facing coastline of the north-western tip of Tasmania, Australia at a latitude of 40.7° S. The Woolnorth wind farm has a maximum generation capacity of 140 MW arising from two 'stages' of development at the site. The first stage, known as Bluff Point, consists of 37 Vestas V66 1.75 MW wind turbines in two rows along a 6 km stretch of coastline, as shown in Fig. 1. The second stage, known as Studland Bay, consists of 25 Vestas V90 3 MW turbines and was completed in 2007, 6 years after the Bluff Point site (Roaring 40s 2011). This study focuses on the Bluff point site alone as the available NWP data and Bluff Point wind speed and direction, and power output data cover the same time period from 2005 to 2006, a period before Studland Bay became operational. Figure 1a, b contains the relevant geographical information regarding the Bluff Point site, with the locations of the 37 turbines shown relative to the topography in Fig. 1a, and the two meteorological monitoring towers (green squares) and two nearest NWP grid points (red circles) in Fig. 1b. A photograph of part of the site is shown in Fig. 1c highlighting the cliff top siting of the turbines.

Measurements of wind speed and direction are obtained at 10 min intervals from a sonic anemometer mounted on the nacelle behind the blades of each of the 37 turbines. These anemometers are located 60 m above ground level. Measurements from the meteorological towers at the northern and southern ends of the site are not used in this study as their data are inferior to the turbine anemometer data, as discussed in previous work (Kay et al. 2009). We used a 24 h time series from 12 UTC forecasts of the limited domain Vic/Tas 5 km NWP model developed by the Australian BoM (Puri et al. 1998) for this study. To obtain better correspondence between this forecast data and the measured turbine anemometer observations, two averaging processes are performed. The first is averaging of the 10 min observations to produce hourly observations for each turbine and the second is an averaging of these hourly observations over the set of 37 turbines to produce a wind farm-wide hourly observation to compare to the hourly forecast for the chosen NWP grid point. Although there are two NWP grid points in close proximity to the wind farm (Fig. 1b) we focus on grid point P2 over P1. This may seem counterintuitive at first, since P1 is at the northern tip of the site and, whereas P2 is 5 km inland from the southern tip of the site. However, we have shown in previous work (Kay et al. 2009) that P2 more capably predicts the conditions of the wind farm, a fact which we attribute to topographical aspects of this particular site. More details on the layout and specifics of measurements for Woolnorth wind farm can be found in (Kay et al. 2009).

Regarding height-related aspects of the data, the NWP forecasts were obtained at sigma levels of 0.998 and 0.9943. The 0.9943 sigma level converts to a forecast height of 45 m above the surface, which is the closest to the turbine anemometer height. Our methodology was applied to this forecast height alone for this study.

2.1 Wind Speed Regime

Not all wind speeds are of equal importance in wind energy forecasting. The expected electrical power output from a wind turbine at different wind speeds is depicted by a power curve. Figure 2a shows the manufacturer's power curve for the Vestas V80 turbines at Bluff Point. A power curve can be divided into four regions of interest, which are identified and labelled in Fig. 2a. The first region labelled (i) is the cut-in wind speed. This is the wind speed that is sufficient for the wind turbine to commence operation and occurs at around 3 m/s for this model of turbine. We refer to region (ii) as the cubic region and it extends from cut-in to the rated output point for the turbine, which is the lowest wind speed where maximum power output is achieved. This region is known as the cubic region by reference to the wind power equation, $P = 1/2\ \rho A v^3$, where ρ is the air density, A is the area swept by the wind turbine blades, and v is the wind speed. The output power rises

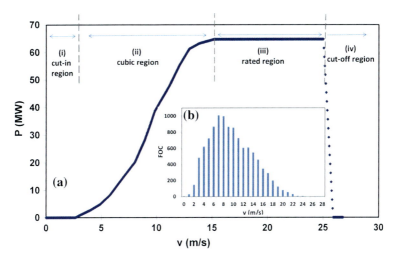

Fig. 2 (**a**) The manufacturer's power curve for the turbines on site at Bluff Point. The power curve has been scaled to the total power output for the wind farm which is rated at 64.75 MW. Four regions of interest are identified and labelled; (**b**) The inset is a frequency histogram of wind speed for Bluff Point over the period March 2005–May 2006

approximately as the cube of the wind speed over much of region (ii), which spans the range 5–15 m/s, although note that the increase in output tapers off as the wind approaches the rated output speed in order to ensure a smooth transition to region (iii). Region (iii) is referred to as the rated output region and spans the approximate range 15–25 m/s. In this region the power output remains constant despite further increases in wind speed. This curtailing of output is done deliberately to prevent structural damage to the turbine and typically involves both blade design and active control of blade pitch to 'spill' the excess wind energy (Manwell et al. 2006). Region (iv) is known as the cut-off region, and the boundary between (iii) and (iv) is called the cut-off wind speed, which occurs at around 25 m/s for this model of turbine. For wind speeds exceeding the cut-off speed, the turbine is shutdown to prevent structural/mechanical damage to the turbine.

In terms of forecasting, the most critical considerations are region (ii) and the boundary between (iii) and (iv), as these are where changes in wind speed have their greatest impact on turbine output. Of these two, region (ii) is most important, simply because, as Fig. 2b shows, the measured wind speed sits in this region for 77 % of the time that the turbine is operating over a single year for this site. Although Fig. 2b shows that wind speeds above the cut-off level are infrequent, they still need consideration due to the sharp change in turbine power output that they cause. Such large and rapid power out transitions can pose particular operational challenges for power system operation (Culter et al. 2011). In Sect. 4, we address forecast improvements for both cases, focussing first on region (ii) due to its greater frequency of occurrence. However, first we turn to the methodology used to obtain those improvements in forecast accuracy.

3 Bias Correction Methodology

The quality of a forecast can be quantified by a forecast verification process that involves determining how well the forecast corresponds to the observations. There are different measures that allow us to focus on one or many aspects of the forecast quality (Murphy 1993, 1995), with bias being one of the measures. Bias in a forecast means that it systematically over- or under-predicts against observations. It is detected statistically as a difference between the means of the historical forecast and historical observation distributions. As pointed out by Woodcock and Engel (2005), bias is one of the key errors found in NWP forecasts.

Bias is a systematic error, which means that it can be determined relatively accurately from an analysis of past forecasts and observations, and then corrected for in the process of producing new forecasts. While the concept is straightforward, the most appropriate implementation is more challenging, as there are a number of factors to consider in developing a bias correction methodology. These include, for example, the best statistical measures from which to quantify the bias and how to apply them, the historical data sample size and issues of forecast scale and timing. There are numerous statistical measures to choose from to quantify the bias, ranging from simple to complex. Initial investigations were carried out using simple linear and multiple regression. However, both these measures were not able to take into account sudden changes in wind speed or extreme wind speed events. For wind speed correction, a statistical measure needs to be robust and sensitive to extreme values as wind speed can vary quite rapidly across a wind farm site. We therefore found the most appropriate parameter to use was the trimean or Best Easy Systematic (BES) estimator:

$$BES = (Q_1 + 2Q_2 + Q_3)/4 \quad (1)$$

where the quartiles Q_1, Q_2 and Q_3 statistically divide the distribution over the errors in the sample size into the lowest 25 % of the data, the median and the highest 25 %, respectively (Turkey 1977; Wonnacott and Wonnacott 1977). The errors are calculated as the observation minus the forecast. The BES is essentially a weighted average of the error distribution's quartiles and median, with a heavier weighting given to the median. This approach is a compromise that allows information about the extreme values within the distribution to not be lost but, at the same time, does not disproportionally contribute to the applied correction.

A negative forecast error indicates an over-predicted forecast and a positive forecast error indicates an under-predicted forecast. The calculated BES can also be positive or negative. Therefore, before a corrected prediction (CP) can be calculated, conditions are put in place that identify the sign of the forecast error and the sign of the BES to ensure that the new forecast is correctly adjusted. After sign considerations, the corrected forecast is determined by adding or subtracting the BES from the future hour's NWP prediction to produce a bias CP.

Improving NWP Forecasts for the Wind Energy Sector

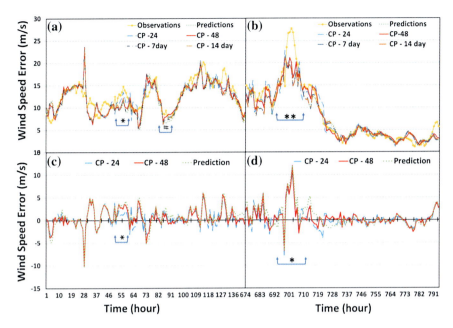

Fig. 3 (**a**) A 140 h time period during 2005–2006 illustrating the effects of bias width for windows of length: 24 h, 48 h, 7 days and 14 days for the corrected prediction methodology; (**b**) A 120 h time period during 2005–2006 illustrating the effects of bias width for windows of length: 24 h, 48 h, 7 days and 14 days for the observations, prediction and corrected prediction methodology; (**c**) Forecast error for the same period as in Fig. 3a; (**d**) Forecast error for the same period as in Fig. 3b

The historical sample data size, also known as the bias window, can significantly affect the level of improvement that is achieved using this approach (Woodcock and Engel 2005). The bias window length is a sequence of time series data that comprises a moving window of a specified length, i.e. the number of data points that is used for calculating the BES. We have investigated the effect of bias window length by considering four different window lengths (24 h, 48 h, 7 days and 14 days) as shown in Fig. 3. Figure 3a, b shows two time periods within 2005–2006 of 140 and 120 h respectively to investigate the differences between four chosen bias window lengths. At first glance, over a majority of both time periods, there is only a minimal difference between the results obtained using the different bias windows. There are, however, periods indicated by [*] and [#] in Fig. 3a, where the 24 and 48 h bias windows outperform the 7 and 14 day window length. This is more evident in Fig. 3b, where for at least half of the time period the 24 and 48 h bias window outperforms the 7 and 14 day windows, indicated by [**]. Figure 3c, d shows the forecast error for the uncorrected prediction, 24 h and 48 h bias windows for the time periods shown in Fig. 3a, b, respectively. There are times when the 24 h window is a better performer in Fig. 3c as indicated by [#], and other times when the 48 h prediction has a smaller error. This same trend is evident in Fig. 3d as indicated by [*], but overall the 48 h bias window gives slightly better performance than the 24 h bias window. We find that increasing the bias window beyond 48 h resulted in a minimal decrease to the resulting forecast accuracy, and as a result, we completed the remainder of our study with a 48 h bias window.

The bias correction methodology described above, when implemented alone, enhances the forecast accuracy, as we will show. However, further improvement can be obtained by applying an additional correction to account for spatio-temporal factors. For example, NWP models often produce hourly forecasts, which are not adequate for capturing sudden changes in wind speed or direction (e.g. the arrival of a rapidly moving frontal system). The resolution of NWP models is also often too coarse to predict the fine scale winds that a wind farm may experience (Archer and Jacobson 2005), and the NWP forecast grid point is typically not at the exact location of the wind farm either. These factors are not taken into account in the simple bias correction methodology above, and although the resulting CP is an improvement on an uncorrected forecast, it can still be far from ideal. Therefore, after an initial bias correction is applied via the methodology above, we apply an extra correction (referred to in the text below as a Double Corrected Prediction—DCP) to the bias CP to improve not just the accuracy, but also some of the timing and scale errors, associated with the forecast.

The DCP is implemented by taking the weighted sum of the CP and an observation O_t made some time t earlier:

$$DCP = \alpha O_t + (1 - \alpha)CP \qquad (2)$$

where the smoothing factor $0 < \alpha < 1$. Because the observations are obtained at 10 min intervals we take O_t as the observation at 10 min to the hour at which CP is made. For the work we have reported here we used $\alpha = 0.5$, which we have found provides the best results for both cases where the forecast errors are large and small. A detailed study of how t and α influence the accuracy of the DCP will be the subject of future work.

4 Results

We begin by presenting the results from a study of the performance of the CP and DCP forecasts for everyday operation of the wind farm. This involved comparing the performance of each of the prediction methods over the 72 h period 27–29 July 2005; the results are shown in Fig. 4a, b. In Fig. 4a, we present the NWP prediction and the CP and DCP results obtained using a 48 h bias window. For most of the period shown, the deviation in wind speed between all three forecast methods and the observations is relatively small, but there are two regions highlighted [*] and [#] that demonstrate the improvement that can be obtained through bias correction, and its potential importance. The first, labelled [*] occurs late in the period and here there is a large deviation between forecast and observation,

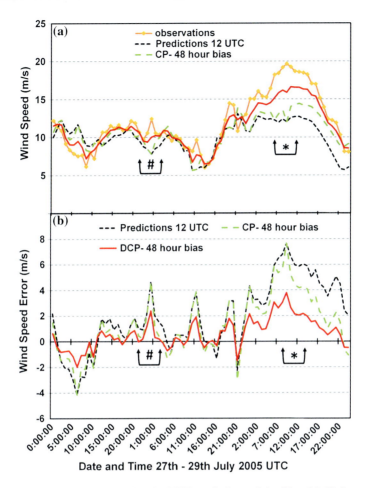

Fig. 4 (a) Observations for Bluff Point, the NWP prediction and the CP and DCP forecasts with a 48 h bias window for the corrected predictions for the period 27th–29th July 2005; (b) The magnitude wind speed error for all predictions (original and corrected) over the 27th–29th July 2005 to give a closer look at the difference in error statistics between all prediction methods for this time period

particularly for the uncorrected and CP forecasts. At only 5–10 m/s, these wind speed errors may seem insignificant, but it is important to bear in mind that they occur in regions (ii) and (iii) of the power curve where fluctuations in wind speed have their greatest effect on real or predicted output. The observed wind speed occurs in region (iii) of the power curve that corresponds to rated power (operating at full capacity); however, the uncorrected forecast is below this region and would therefore predict that the wind farm would be operating below full capacity. Predictions of future wind output can play a significant role in appropriately committing (starting) and dispatching other generation to meet expected demand. Although the CP forecast is an improvement on the uncorrected forecast, the DCP forecast fares much better, highlighting the importance of taking the scale and timing issues discussed earlier into consideration in the correction methodology. The second region, labelled [#], features a smaller magnitude of discrepancy between forecast and observation but highlights a benefit of having delayed observation data as a component of the correction, as in the DCP implementation. The benefit is that the trend is more correct, as both the uncorrected and CP forecasts show a minima for [#], whereas the DCP shows a maxima as the observation does. A similar event occurs approximately 10 h later. Note that when the observation exceeds 10 m/s in Fig. 4a the NWP forecast is an under-prediction. This is especially notable on the last day of the dataset in Fig. 4a.

To more clearly demonstrate the effectiveness of the DCP over CP, in Fig. 4b, we plot the magnitude wind speed error for the uncorrected, CP (48 h) and DCP (48 h) forecasts over the same time period in Fig. 4a. There is a clear skew towards positive errors that represents a tendency for under-prediction within the framework used for calculating errors in this chapter. However, the most notable aspect is that the DCP error is generally $\sim 50\ \%$ smaller than the other two methods throughout the 72 h dataset. This improvement holds potential significance for wind farm management and the scheduling of controllable (dispatchable) generation within the power system to meet future demand. Improving wind speed forecasts will facilitate the successful integration of wind generation within the electricity industry.

We now extend the time period studied from 3 days to 3 months to demonstrate that the enhanced effectiveness of DCP holds more generally. In doing so, we change the data presentation such that in Fig. 5a, b, c we plot the prediction error versus the predicted wind speed from that particular correction for each hour where the observed speed is between 5 and 15 m/s. Figure 2 shows why this wind speed range was chosen—it is within the cubic region of the power curve where small changes in wind speed can lead to large changes in power, and also because 77 % of measured wind speeds fall within that range. This gives a total of 1,522 data points in each panel. We look at outliers (i.e. predicted speed less than 5 or greater than 15 m/s) in Fig. 6.

To understand this data, we begin by explaining a few features that occur in Fig. 5. First, we address the data point immediately above the upward-pointing vertical arrow near the centre of the graph. Here, the predicted wind speed was 10 m/s and the prediction error was minus 5 m/s, which means that the observed wind speed was 5 m/s in this instance. In the case, where predictions were perfect, we would expect all of these points to cluster along the horizontal dotted line at an error of zero; scatter away from this horizontal line represents imperfection in the forecast. There is also a significant diagonal banding in the scatter of the data in Fig. 5a, which is emphasised by the two diagonal dashed lines. If we return to the point indicated by the upward vertical arrow, the nature of this banding becomes clear. This point corresponds to an observation of 5 m/s, which is at the edge of the observed wind speed range of 5–15 m/s. Immediately below this point the error is larger than 5 m/s, the observed speed is less than 5 m/s and the data point is

Improving NWP Forecasts for the Wind Energy Sector

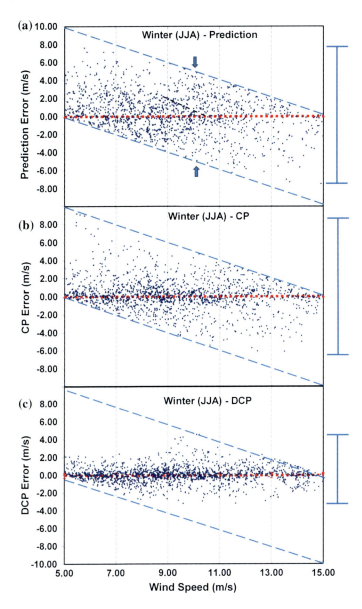

Fig. 5 Error statistics for Winter (JJA) 2005 where the observation lies between 5–15 m/s for (**a**) the NWP predictions; (**b**) the CP methodology and (**c**) the DCP methodology

excluded. This argument, with appropriate sign reversal, holds for the upper diagonal line, and also explains the slopes of these lines. For example, at a predicted speed of 5 m/s, any error less than zero excludes a data point. Thus, the diagonal dashed lines indicate a boundary for the scatter in these plots.

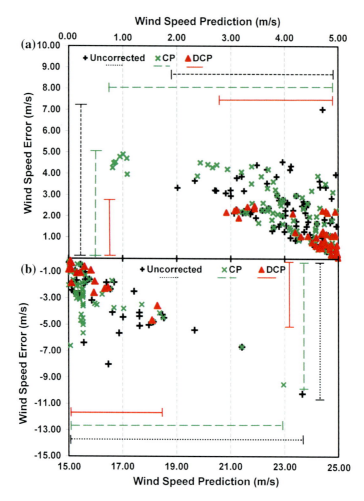

Fig. 6 Error statistics for the same conditions as Fig. 5a, b, c for the outliers when the wind speed predictions fall outside the 5–15 m/s range for (**a**) less than 5 m/s and (**b**) greater than 15 m/s respectively from for the same conditions

If we now compare the data points for CP and DCP in Fig. 5b, c respectively, it is clear that the points move closer to the ideal (i.e. the horizontal dotted line) with each applied correction. This is also evident if we look at the error bars displayed to the right-hand side of each prediction figure. The overall spread in error reduces with the application of each correction, once again confirming the efficacy of the DCP methodology for everyday wind speed forecasting for the site.

Figure 6a, b shows the outliers from Fig. 5 for predicted wind speeds that were less than 5 m/s and greater than 15 m/s, respectively for all prediction methods. Using the same reasoning as Fig. 5a, if we have a prediction of 2 m/s and an error of 3 m/s, this indicates an under-prediction by 3 m/s. Figure 6a, b also shows the range of errors in both the *x* and *y* direction on each graph, as depicted by the error bars. Generally, we see a reduction in the error bars along each axis, however, for the case of under-prediction the CP does not perform as well as the uncorrected prediction. We can see that the use of real wind farm data has corrected this problem in the DCP output. Figure 6b shows the wind speed error for each prediction methodology for the higher wind speed ranges. It is evident that the magnitude error is greater in this case than the lower wind speed ranges shown in Fig. 6a. Over-predicting the wind speed is more problematic than under-prediction at the lower wind speed range. If a wind farm operator was relying on the wind speed forecast, it would erroneously indicate that the wind farm would be operating at rated power at that future point in time. Once again implementation of the DCP methodology has reduced the wind speed error compared to the uncorrected prediction, with better improvement observed in the higher wind speed ranges.

The encouraging improvement in wind speed forecast accuracy led to the next investigation of forecast performance—how well do the correction methodologies improve the forecasting of extreme wind events? Extreme wind events are particularly important to forecast for the wind energy industry, as sudden changes in wind speed or direction can change the output of a wind farm from full power to zero output in a matter of minutes. Figure 7a shows an extreme wind event that occurred for a 4 h period within the period 30–31 August 2005. The observed wind speed, the NWP prediction and corresponding power output of the Bluff point wind farm are shown in Fig. 7a. The important features illustrated in Fig. 7a are the sudden and complete drop in power output around 00:00 UTC on the 31st August, which occurred because the wind speed exceeded 25 m/s for a significant period of time. The meteorological features identified as the cause of this extreme wind event were a deepening low pressure system, accompanied by a strong front and trough following closely behind (Kay et al. 2009). This in turn kept wind speeds high for a prolonged period. The other important feature of Fig. 7a is the apparent inability of the NWP model to predict this wind speed event. The NWP model under-predicted the wind speed for a majority of the time over the 48 h period. The model also failed to capture the timing and duration of this wind speed event. Figure 7b shows the wind speed observations, original uncorrected prediction, CP and DCP. It is clear that the DCP is an improvement over the uncorrected prediction for both the magnitude and timing of the wind speed event, as illustrated by the region denoted [#]. To quantify the error, Fig. 7c shows the magnitude error in wind speed for the bias-corrected forecasts, CP and DCP along with the original uncorrected 12UTC predictions. The CP forecast has improved some of the magnitude issues over most of the 48 h period, and the error has been reduced by almost half, as evident towards the end of the time period. However, the most noticeable improvement in accuracy is in the DCP. Another event that the DCP has improved the forecast for is indicated by the [*] in Fig. 7b. The uncorrected prediction and CP incorrectly predicted the magnitude and direction of the short timescale minima ('dip') in wind speed. The DCP is an improvement over

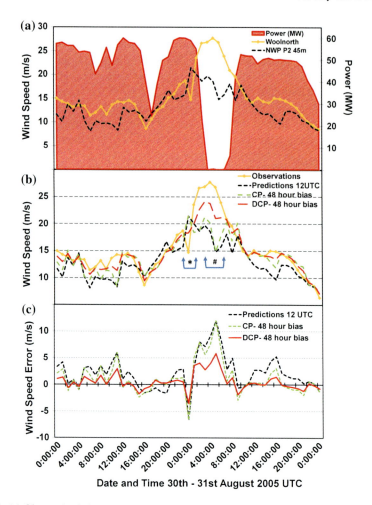

Fig. 7 (a) Observed wind speed and corresponding power output for Bluff Point over a duration of 48 h for an extreme wind event in August 2005. The NWP prediction for that duration is also shown; (b) the wind speed observations, original prediction, and CP and DCP for a 48 h bias window applied to Figure 6a and (c) Magnitude wind speed error for Bluff Point, predictions and bias corrected forecasts at 12 UTC for the 30–31 August 2005

the uncorrected prediction and CP, but this method was still not able to fully show that there would be a sudden dip in the wind speed. The high speed cut-out was still not fully captured, and further improvement is still necessary, nevertheless, for this extreme wind event the DCP has successfully reduced the forecast error by at least 50 %.

5 Conclusion

Weather forecasting within the electricity sector to date has largely concentrated on demand side prediction—how the weather, particularly temperature, affects consumers' electricity usage and hence overall system demand. However, as intermittent renewable generation achieves higher levels of penetration within power systems globally, it becomes increasingly important to provide useful forecasts for this generation. This study investigated the potential application for a correction methodology for wind energy forecasting, with the aim to improve the accuracy and timing of wind speed forecasts from a NWP model. The wind farm chosen for the case study is notable for experiencing highly variable and extreme wind events. Results covered the period March 2005 to May 2006, and focused- on the wind speed ranges of greatest importance to wind power output variability and an extreme wind speed event. Forecast biases can often be location- dependent, and for this location, a systematic under-prediction occurred for a majority of wind speed events.

Numerous bias windows were tested, with 48 h proving to be optimum. The bias methodologies improved the prediction for the two specific wind events that were investigated, with the wind speed error more than halved by the DCP methodology for the extreme wind event and also an improvement in the timing of the event. This halving of forecast error was also evident during general wind speed conditions.

For the more extensive time series, the frequency of under-prediction was reduced by as much as 10–19 % across all seasons by the DCP method, and by 2–6 % for over prediction. Overall smaller magnitudes of error were evident by the CP and DCP methods for under and over prediction. The use of the bias correction methodology has been shown to greatly enhance the timing and accuracy of forecasts from the NWP, with the use of actual wind farm data providing an added benefit to scaling the forecast to better represent the wind farm.

Acknowledgements The authors wish to thank Hydro Tasmania and Roaring 40s for access to data, and A. Micolich for useful discussions. This project was funded partly by the Australian Greenhouse Office, as part of their Australian Wind Energy Forecasting Capability (WEFC) initiative and partly by the Australian Government through the Australian Solar Institute (ASI), part of the Clean Energy Initiative.

References

Archer CL, Jacobson MZ (2005) Evaluation of global wind power. J Geophys Res 11:D1211. doi:10.1029/2004JD005462

Culter N, Outhred HR, MacGill IF (2011) Final report on UNSW project for AEMO to develop a prototype wind power forecasting tool for potential large rapid changes in wind power. http://www.aemo.com.au

Feinberg EA, Genethliou D (2005) Load forecasting In: Chow JH, Wu FF, Momoh JJ (eds) Applied mathematics for restructured electric power systems: optimization, control, and computational intelligence. Springer, New York, pp 269–285

Focken U, Lange M (2008) Final report—wind power forecasting pilot project in Alberta, Canada. In: Wind power forecasting pilot project. Energy and Meteo Systems, Oldenburg, p 21 http://www.aeso.ca/gridoperations/13825.html

Giebel G, Landberg L, Kariniotakis G, Brownsword R (2003) State-of-the-art on methods and software tools for short-term prediction of wind energy production. EWEC, Madrid, pp 16–19

Jørgensen J, Möhrlen C (2008) AESO wind power forecasting PILOT project: final project report. In: Wind power forecasting PILOT project. Weprog Ebberup, Denmark, p 102. http://www.aeso.ca/gridoperations/13825.html

Kay MJ, Cutler N, Micolich A, MacGill I, Outhred H (2009) Emerging challenges in wind energy forecasting for Australia. Aust Meteor Ocean J 58(2):99–106

Landberg L, Giebel G, Nielsen HA, Nielsen T, Madsen H (2003) Short-term prediction: an overview. Wind Energy 6:273–280

Manwell JF, McGowan JG, Rogers AL (2006) Wind energy explained, theory, design and application. John Wiley and Sons Ltd, West Sussex

Milligan M, Schwartz M, Wan Y (2003) Statistical wind power forecasting models: results for U.S. wind farms. NREL/CP-500-33956, p 17

Murphy AH (1993) What is a good forecast? An essay on the nature of goodness in weather forecasting. Wea Forecast 8:281–293

Murphy AH (1995) The coefficients of correlation and determination as measures of performance in forecast verification. Weather Forecast 10:681–688

Potter CW, Negnevitsky M (2006) Very short-term wind forecasting for Tasmanian power generation. IEEE Trans Power Syst 21(2):965–972

Puri K, Dietachmayer G, Mills G, Davidson N, Bowen R, Logan L (1998) The new BMRC limited area prediction system. LAPS Aust Meteor Mag 47(3):203–223

Roaring 40s (2011). http://www.roaring40s.com.au/article.php?Doo=Redirect&id=296

Sanchez I (2006) Short-term prediction of wind energy production. Int J Forecast 22:43–65

Torres JL, Garcia A, de Blas M, Francisco A (2005) Forecast of hourly averages wind speed with ARMA models in Navarre. Sol Energy 70(1):65–77

Tukey JW (1977) Exploratory data analysis. Addison-Wesley, Reading, p 688

Vincent C, Bourke W, Kepert JD, Chattopadhyay M, Ma Y, Steinle PJ, Tingwell CIW (2008) Verification of a high-resolution mesoscale NWP system. Aust Met Mag 57:213–233

Wiser R, Bolinger M (2011) 2010 Wind energy technologies market report, Lawrence Berkeley Laboratories, US Department of Energy, June 2011. http://eetd.lbl.gov/ea/ems/reports/lbnl-4820e.pdf

Woodcock F, Engel C (2005) Operational consensus forecasts. Wea Forecast 20:101–111

Wonnacott TH, Wonnacott RJ (1977) Introductory statistics. Wiley, New York, p 650

Overview of Irradiance and Photovoltaic Power Prediction

Elke Lorenz, Jan Kühnert and Detlev Heinemann

Abstract Power generation from solar and wind energy systems is highly variable due to its dependence on meteorological conditions. With the constantly increasing contribution of photovoltaic (PV) power to the electricity mix, reliable predictions of the expected PV power production are getting more and more important as a basis for management and operation strategies. We give an overview of different approaches for solar irradiance and PV power prediction, including numerical weather predictions for forecast horizons of several days, very short-term forecasts based on the detection of cloud motion in satellite or ground-based sky images, and statistical methods to optimize and combine different data sources as well as methods for PV simulation and upscaling to regional PV power predictions. Evaluation results for selected irradiance and power prediction schemes show the benefit of different approaches for different timescales.

1 Introduction

Solar and wind power will contribute a major share of the future global energy supply, introducing new challenges to the electricity system's operation. Both, solar and wind energy systems show fundamentally different generation characteristics than conventional energy sources. While the power supply by the latter can be directly adapted to the electricity demand, the availability of solar and wind power is largely determined by the prevailing weather conditions and therefore highly variable. Any adaptation within the electric power system to this new supply structure—may it be intelligent power plant scheduling, demand side

---

E. Lorenz (✉) · J. Kühnert · D. Heinemann
Energy Meteorology Unit, Energy and Semiconductor Research Laboratory,
Institute of Physics, University of Oldenburg, Carl von Ossietzky Strasse 9-11,
26129 Oldenburg, Germany
e-mail: elke.lorenz@uni-oldenburg.de

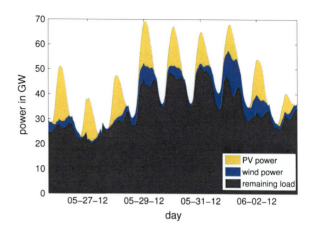

Fig. 1 Contribution of solar and wind energy to the total power supply in Germany for 26.5–3.6.2012. *Remaining load* contribution from conventional power plants ≥ 100 MW; *data source* (Data from Energy Exchange Leipzig EEX, http://www.transparency.eex.com/)

management, or on the long-term restructuring of the grid topology and the introduction of storage capacities—will require detailed information on the expected power production from these sources on various temporal and spatial scales. Reliable forecasts of solar power production therefore are an important factor for an efficient integration of large amounts of solar power into the electric supply system.

Already today, solar power prediction systems are an essential part of the grid and system control in countries with substantial solar power generation. For example, in Germany with an installed photovoltaic (PV) power capacity of 32 GW_{peak} at the end of 2012, a share of 5 % of the total electricity in 2012 was covered by PV power. On sunny summer days, PV power can contribute up to 42 % of the electricity demand during peak load hours at noon (Fig. 1). Grid-integration in Germany is regulated through the Renewable Energy Sources act (RES). According to the RES, PV system operators may feed in electricity with priority to conventional plants for a fixed price. The transmission system operators (TSOs) are in charge of marketing and balancing the overall fluctuating PV power feed-in, making regional forecasts necessary. PV power first is offered on the day-ahead auction at the European Power Exchange (EPEX), requiring day-ahead forecasts of PV power. Additionally, the EPEX offers the possibility of intraday trading, where electricity may be traded until 45 min before delivery begins based on forecasts for several hours ahead. Remaining deviations between scheduled and needed power are adjusted using balancing power, which has much higher costs than regular trading. Therefore, a high accuracy of PV power forecasts is very important for a cost-efficient grid-integration.

Following the new and rapidly evolving situation on the energy market with a strong need for accurate solar power predictions increasing effort has been spent on the development of irradiance and PV power prediction models during the last years. In this article, we intend to give an overview of basic concepts of irradiance and PV power prediction by referring to selected examples rather than giving an extensive review of existing models, that can be found, e.g., in Lorenz and

Heinemann (2012). First, the basic modeling steps in PV power prediction are briefly presented, followed by an introduction to different data sources and models for irradiance forecasting. An example evaluation illustrates concepts of accuracy assessment and gives typical accuracies for some of the forecasting approach introduced before. Next, models to derive PV power forecasts are shortly described and evaluated for an example system.

2 Typical Outline of PV Power Prediction Systems

Several systems for PV power prediction have been introduced recently (e.g., Remund et al. 2008; Bacher et al. 2009; Lorenz et al. 2010; Pelland et al. 2013), most of them consisting of some or all of the basic elements illustrated in Fig. 2. All given modeling steps may involve physical or statistical models or a combination of both.

Forecasting surface solar irradiance is the first and most essential step in most PV power prediction systems. Depending on the application and the corresponding requirements with respect to forecast horizon and temporal and spatial resolution, different models and data sources are used. Numerical weather prediction (NWP) models are applied to derive forecasts of several days ahead. For very short-term horizons, irradiance forecasts may be obtained by detection and extrapolation of cloud motion, based on satellite images for forecasts of several hours ahead and on ground-based sky imagers for sub-hourly forecasts with a very high spatial and temporal resolution. Measured irradiance data, forming the basic input to time series models, are another valuable data source for very short-term forecasting in the range of minutes to hours. Furthermore, measured data are required for any statistical post-processing procedure, applied to optimize forecasts derived with a physical model for a given location.

To derive PV power forecasts from the predicted global horizontal irradiance different approaches may be applied. Explicit physical modeling involves conversion of the irradiance from the horizontal to the angle of tilt of the module plane, followed by the application of a PV simulation model. Here, characteristics of the PV system configuration are required in addition to the meteorological input data, implying information on nominal power, tilt and orientation of a PV system as well as a characterization of the module efficiency in dependence of irradiance and temperature. Alternatively, the relation between PV power output and irradiance forecasts and other input variables may be established on the basis of historical datasets of measured PV power with statistical or learning approaches. In practice, often both approaches are combined and statistical post-processing using measured PV power data is applied to improve predictions with a physical model.

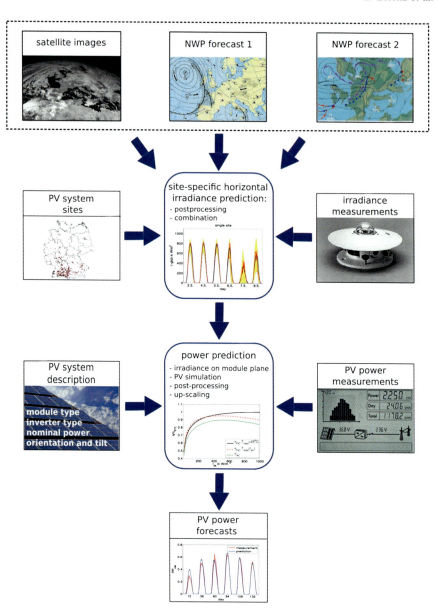

Fig. 2 Overview of basic modeling steps in PV power prediction

PV power prediction for utility applications usually requires forecasts of the cumulative PV power generation for a specified area, i.e., a regional forecast, instead of single site forecasts. These regional predictions are typically obtained by up-scaling the predicted PV power production from a representative set of single PV systems.

3 Irradiance Forecasting

Solar power production is essentially determined by the incoming solar irradiance, and therefore the quality of PV power forecasts also strongly depends on the quality of the underlying irradiance forecasts. Forecasting of irradiance is still a comparatively new research field with strong need for further development. Consequently, major research efforts for reliable PV power forecasting focus on accurate irradiance prediction. Basic features and characteristics of the different possible approaches to predict surface solar irradiance listed in the previous section are briefly described below.

3.1 Irradiance Forecasting with Numerical Weather Prediction Models

Irradiance and PV power predictions for the day-ahead power market, at present probably one of the most important applications, are based on numerical weather prediction models. NWP models are routinely operated by weather services to forecast the state of the atmosphere for several days ahead. The atmospheric dynamics, i.e., the temporal changes of the atmospheric conditions, are modeled by numerically solving the basic differential equations describing the physical laws governing the weather (prognostic equations). This physical modeling is essential for any forecast more than several hours ahead.

In the first step, the future state of the atmosphere is calculated with a global NWP model covering the complete Earth. To initialize the forecasts, data assimilation tools are applied to make efficient use of worldwide meteorological observations and measurements, including, e.g., measurement data from weather stations and buoys and satellite observations. The prognostic equations are then solved on a grid using numerical methods that involve spatial and temporal discretization, where the grid resolution is determined by the computational costs. Nowadays, the spatial resolution of global NWP models is in the range of 16–50 km; the temporal resolution is 1, 3, or 6 h. Global NWP models are currently run by about 15 weather services, such as the Global Forecast System (GFS) by the US National Oceanic and Atmospheric Administration (NOAA), and the Integrated Forecast System (IFS) operated at the European Centre for Medium-Range Weather Forecasts (ECMWF).

In a second step, the spatial and temporal resolution of the weather forecasts may be increased by the application of a mesoscale model covering only part of the Earth and taking initial and lateral boundary conditions from a global NWP model. Weather services typically operate mesoscale models with a spatial resolution in the range of 3–20 km and provide hourly forecasts, but also higher resolutions are feasible. Solving the governing equations on a finer grid makes mesoscale models capable of resolving smaller atmospheric phenomena such as land-sea breezes, thunderstorms, and topographically forced wind fields. In particular, the spatial and temporal dynamics of surface solar irradiance are strongly influenced by small-scale features like broken cloud fields and heterogeneous surface characteristics, which makes the application of mesoscale models for solar irradiance forecasting potentially advantageous. An example of a mesoscale model with a wide application in energy meteorology is the Weather Research and Forecasting (WRF, Skamarock et al. 2008) model, developed as a series of open source models in a collaborative effort of several institutes lead by the National Center for Atmospheric Research (NCAR) in the US. WRF is a nonhydrostatic model with multiple nesting capabilities that offers various parameterizations for the different physical processes, which allows it to be configured for specific conditions for the region of interest.

Both global and mesoscale models require parameterizations to describe processes on spatial scales smaller than the scale explicitly resolved by the grid of an NWP model. For irradiance forecasting, parameterizations of radiation transfer and clouds are of special importance. Radiation transfer modeling requires the characterization of all interactive processes in the atmosphere, i.e., scattering and absorption, in a variety of spectral domains. These processes are related to the occurrence of clouds in the atmosphere, which are highly variable in space. The representation of their optical and microphysical properties in atmospheric models with coarse resolution is one of the major challenges in atmospheric modeling. Solar energy applications of the mesoscale model WRF mostly use the radiation scheme proposed in Dudhia et al. (1989). With respect to global NWP modeling, the radiation scheme currently implemented at the ECMWF (Morcrette et al. 2008) is viewed as today's state of the art for these processes.

Most NWP models nowadays offer surface solar irradiance as direct model output (DMO), and therefore in principle these forecasts can be directly utilized without involving any additional measurement data. In practice however, post-processing methods using historic or recent measured data are often applied to improve the accuracy of the irradiance forecasts (see also Sect. 3.4).

3.2 Irradiance Forecasting with Cloud Motion Vectors

For forecast horizons up to several hours ahead, the temporal change of cloud structures is mainly governed by horizontal advection while the shape of clouds in the relevant spatial scale often remains rather stable. Any technique which is able to detect this horizontal cloud motion in sufficient detail should therefore provide valuable information for irradiance forecasting in the corresponding timescales. Only when local cloud formation and dissipation processes, such as strong thermal convection, are dominant, will the forecast capability using this approach be reduced.

A forecasting scheme based on cloud motion vectors comprises the following basic steps:

- Images with cloud information ("cloud images") are derived from the raw—satellite or ground-based sky images.
- Cloud motion vectors (CMV) are determined by identifying matching cloud structures in consecutive "cloud images."
- Future "cloud images" are forecasted by extrapolation of cloud motion, i.e., by applying the calculated motion vectors to the latest image. Here the basic assumption is applied, that cloud speed and velocity are persistent as well as the structure and optical properties of the moving clouds.
- Forecasts of site-specific global horizontal are derived from the predicted "cloud images."

Forecasts of several hours ahead require observations of cloud fields in large areas (approx. 2,000 km × 2,000 km for an assumed maximum cloud velocity of 160 km/h for 6 h ahead). Satellite data with their broad coverage are an appropriate data source for these horizons. Forecasts in the sub-hourly range with a much higher temporal and spatial resolution may be inferred from ground-based sky images.

3.2.1 Satellite-Based Forecasts

As an example of an operational satellite-based irradiance forecasting scheme, our approach (Lorenz et al. 2004, Kühnert et al. 2013) based on images of the geostationary Meteosat second-generation (MSG) satellites positioned at (0°N/0°E) is briefly described here. The forecasts use the broadband high resolution visible (hrv, 600–900 nm) channel with a spatial resolution of 1 km × 1 km at the sub-satellite point and a temporal resolution of 15 min.

Cloud information and surface solar irradiance are derived from the satellite images using the Heliosat method (Hammer et al. 2003), which is widely used for solar energy applications. A characteristic feature of the method is the dimensionless cloud index, giving information on the cloudiness of the image pixels. A basically linear relationship is assumed to describe the influence of the cloud index on cloud transmissivity. Surface solar irradiance is calculated by combining the information on cloud transmissivity with a clear sky model.

To derive cloud motion vectors, corresponding areas in two consecutive cloud index images are identified. First, rectangular regions (approx. 90 km × 90 km) are defined, large enough to contain information on temporally stable cloud structures and small enough so that the same vector describes the motion for the whole region. Next, mean square pixel differences between rectangular regions in consecutive images are calculated for displacements in all directions, the maximum possible displacement given by maximum wind speeds at typical cloud heights. Finally, the motion vectors are determined by the displacements that yield the minimum mean square pixel differences.

Besides irradiance forecasting, atmospheric cloud motion vectors are commonly used in operational weather forecasting to describe wind fields at upper levels in the atmosphere.

3.2.2 Forecasts Using Ground-Based Sky Imagers

Short-term forecasting based on ground-based sky imagers is a very new research field; a first approach is reported in Chow et al. (2011). With their very high spatial and temporal resolution sky imagers have the potential for capturing sudden changes in irradiance—often referred to as ramps—on a temporal scale of less than a minute. Cloud fields may be resolved in high detail allowing to model and forecast partial cloud cover on large PV installations. The maximum possible forecast horizon strongly depends on the cloud height and velocity. It is limited by the time the monitored cloud scene has passed the location or area of interest, ranging from 5 to 25 min for the examples investigated in Chow et al. (2011).

Cloud detection according to Chow et al. (2011) is performed by evaluating the red-to-blue ratio of the sky images in comparison to a clear sky library resulting in binary cloud decision maps. Cloud motion vectors are generated by maximizing the cross-correlation between shifted areas in two consecutive images. To infer irradiance from the forecasted cloud maps, cloud shadows at the surface were estimated and combined with a simple model describing irradiances for clear sky and cloud shadows.

3.3 Time Series Modeling

Solar irradiance forecasting with times series models uses recent measurements of irradiance as a basic input, possibly complemented by related variables. The functional dependence between previous values (predictors) and forecast values (pedictands) is learned with statistical algorithms in a training phase on historic data, assuming that patterns in the historical datasets are repeated in the future and thus may be exploited for forecasting.

Time series models take advantage of the high autocorrelation for short time lags in time series of solar irradiance and cloud cover. Figure 3 shows the autocorrelation coefficients of the clear sky index k^*, which is defined as the ratio of measured global irradiance I_{meas} to irradiance for clear sky conditions I_{clear}. The clear sky index can be considered an inverse measure of cloudiness. The autocorrelation is high for time lags of a few hours, but decreases rapidly with increasing time lags, limiting the effectiveness of the time series approach for horizons beyond several hours ahead. For these horizons it is essential to account for dynamic phenomena like motion and formation or dissolution of clouds, which make physical modeling necessary and cannot be determined on the basis of local measurements. However, any model has an inherent uncertainty regardless of the forecast horizon. This is caused, for example, by limits in spatial and temporal

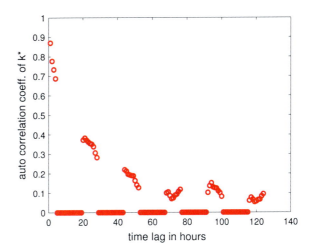

Fig. 3 Evolution of the autocorrelation of the clear sky index k^* over several days. The correlation coefficient of an NWP model forecast for the first forecast day was found to be 0.54

resolution, uncertainty in input parameters, and simplifying assumptions within the model. Therefore, for the very short-term scale of typically up to 1 or 2 h ahead, forecasts based on accurate on-site measurements are advantageous.

Time series approaches include classical regression methods, such as simple autoregressive (AR) models, autoregressive-integrated moving average (ARIMA) models, and artificial intelligence methods such as Artificial Neural Networks (ANN), k Nearest Neighbors (kNN), or support vector regression (SVR). Detailed descriptions and comparisons of different time series approaches are given, e.g., in Reikard (2009) or Pedro and Coimbra (2012). For any time series model, the selection and availability of appropriate input variables as well as an optimized pre-processing of these data is of critical importance for good forecast accuracies.

Statistical models may include not only on-site measured data as predictors but also additional exogenous data, such as NWP model or satellite based forecasts. This approach of combining statistical and physical forecast models allows for an extension of the forecast horizon from some hours to some days and can significantly increase forecast accuracy (see e.g., (Bacher et al. 2009)). Depending on the perspective, different terminology is used for this combined forecasting technique: The community of statistical modeling and artificial intelligence refers to these models as "statistical models with exogenous input," while meteorologists commonly use the terms "post-processing" or, more specifically, "model output statistics" (MOS), which is the terminology adopted here.

3.4 Post-processing

Post-processing to improve forecasts of a physical model is a standard approach in weather prediction. In particular, the output of NWP models is frequently refined to account for detailed local weather features that are generally not resolved by these models. Furthermore, also cloud motion vector forecasts may benefit from post-processing.

Post-processing methods may be used to

- reduce systematic forecast errors,
- account for local effects (e.g., topography, surface),
- account for the influence of selected variables in more detail (e.g., aerosols),
- derive parameters that are not directly provided by the NWP models (e.g., direct and diffuse surface solar irradiance, which usually are not standard output parameters of NWP models),
- combine the output of different models.

Different types of post-processing procedures are briefly presented here, with statistical post-processing being the most common technique applied in this context.

3.4.1 Model Output Statistics

The method of Model Output Statistics (MOS) is widely used to refine the output of NWP models, mainly to account for local variations in surface weather (Glahn and Lowry 1972) with surface observations and/or climatology for specific locations as basic input. In the case of solar irradiance, satellite-derived values may also be used instead of ground measurements. Measurements of the forecast variable, the predictand, are related to model forecast variables, the predictors, with a statistical approach. The set of predictors may be extended by including any relevant information, for example, prior observations and climatological values.

Traditionally, the term MOS is associated with the use of regression equations. A wider concept of MOS includes any statistical approach that relates observed variables to NWP output. Examples are a weather-dependent bias correction applied to ECMWF irradiance forecasts (Lorenz et al. 2009), an approach using Kalman-filtering to improve irradiance forecasts of the Canadian Global Environmental Multiscale (GEM) model (Pelland et al. 2013) or artificial intelligence methods as the application of an ANN adopting predicted variables from US National Weather Service's (NWS) forecasting data base (Marquez and Coimbra 2011).

3.4.2 Spatial Averaging

A possibility to improve forecast accuracy without requiring additional measurement data is the application of smoothing filters or spatial averaging. This approach reduces fluctuations of forecast values in variable cloud situations where this is favorable because the correlation between forecasts and measurements is low, but preserves the quality of the original forecasts in homogeneous clear sky and overcast situations. In Lorenz et al. (2004), the benefit of horizon-dependent smoothing filters for CMV forecasts is shown. With respect to NWP forecasts the impact of spatial averaging is largest for models with a high temporal and spatial resolution. A detailed evaluation of irradiance forecasts from the Canadian GEM model reveals a reduction of forecast errors by 10–15 % when averaged over several hundred kilometers (Pelland et al. 2013).

3.4.3 Temporal Interpolation

Global model forecasts are often provided with a temporal resolution of 3–6 h. Many applications in the energy sector, however, need forecasts of solar power at least on an hourly basis, which may be inferred with temporal interpolation techniques. In Lorenz et al. (2009), the combination of irradiance forecast data with a clear sky model to account for the typical diurnal course of irradiance is proposed.

3.4.4 Physical Post-processing Approaches

Physical post-processing using radiation transfer calculations may be applied for a better consideration of parameters that are not handled in detail in NWP models, such as aerosols. They also may be used to calculate parameters that are not provided as DMO, such as direct normal irradiance, which is relevant for all concentrating solar power plants. A forecasting system for global and direct irradiance forecasting in Southern Spain based on WRF model output and satellite retrievals is proposed and evaluated in Lara-Fanego et al. (2012).

3.4.5 Human Interpretation of NWP Output

A traditional method to obtain improved local forecasts from NWP model output is to involve forecast experts' human knowledge. In Traunmüller and Steinmaurer (2010), a clear sky model is combined with cloud cover forecasts, derived from meteorologists by analyzing and comparing the output of different global and local NWP models. Especially, in difficult forecast situations, such as fog, this method offers potential for improvements of the irradiance forecasts.

3.4.6 Combination of Different Models

Finally, combing the output of different models may also increase the forecast accuracy considerably when compared to single model forecasts. Simple averaging is beneficial for models with similar accuracy, e.g., several NWP models, exploiting the fact that forecast errors of different models are usually not perfectly correlated. Furthermore, combination methods using statistical tools may account for strengths and weaknesses of the different models for certain situations, e.g., by adapting the contribution of each model depending on the weather situation. In particular, they may also be applied to establish an irradiance forecasting tool covering horizons from several minutes to several days ahead by integrating measurements, CMV and NWP forecasts with an optimized weighting depending on the forecast horizon (see e.g., (Lorenz et al. 2012)).

4 Evaluation of Irradiance Forecasts

The irradiance forecasting scheme of Lorenz et al. (2012) also used in an operational PV power prediction system (Lorenz et al. 2011) is evaluated here as an example forecasting system, including satellite and NWP-based irradiance predictions as well as a combination of the different models. A focus will be on the accuracy of regional predictions relevant for utility applications and on the comparison of different models in dependence of the forecast horizon.

4.1 Measurement and Forecast Data

The irradiance forecasts are evaluated in comparison to hourly pyranometer measurements of 290 weather stations in Germany (see Fig. 4) for the period for January–September 2012, unless specified otherwise.

The irradiance predictions covering intraday and day-ahead forecast horizons include three models: (1) The global model irradiance forecasts (Morcrette et al. 2008) of the ECMWF Integrated Forecast System have been the basis of the PV power prediction service since starting in 2009. The forecasts were available as 3-h values on a grid of $0.25° \times 0.25°$ then, which is still the resolution used today, though higher resolved IFS forecast are available now. (2) Irradiance forecasts of the COSMO-EU model, operated by the German weather service DWD are used as a second NWP model. The COSMO-EU forecasts are provided as hourly values with a spatial resolution of $0.0625° \times 0.0625°$. The evaluations shown here are based on the 12:00 UTC run of the ECMWF model and on the 18:00 UTC COSMO-EU run, which are used for the PV power predictions delivered every day at 7:00 local time to the transmission system operators. Day-ahead predictions are based on a combination of these two NWP models. (3) For intraday forecast horizons satellite-based CMV forecasts (see Sect. 3.2.1), which are updated every 15 min, are additionally combined with the NWP predictions.

In the first step, the post-processing scheme applied to enhance and combine the forecasts implies a bias correction in dependence on the cloud situation and the solar elevation (Lorenz et al. 2009) for the single model forecasts. In the second step, linear regression is applied to determine each model's contribution to the combined predictions for each hour of the day separately (Lorenz et al. 2012). The

Fig. 4 Locations of the meteorological stations with irradiance measurements

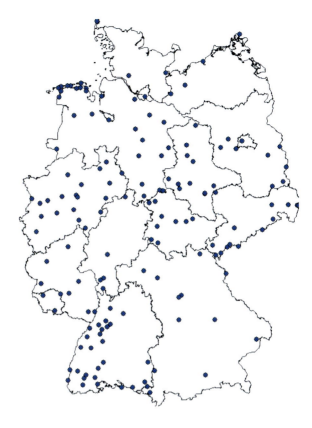

model coefficients are updated daily, based on training with measured irradiance values of the last 30 days.

4.2 Accuracy Measures

To evaluate the performance of a forecasting model many different aspects have to be taken into account. In particular for model development, a detailed analysis of the forecast accuracy is very important as a basis for further research. From the forecast users' point of view a basic set of accuracy measures should highlight the aspects of forecast performance most relevant for the application.

For the evaluation of solar and wind power predictions it is common practice to use the root mean square error as the primary score for assessing forecast accuracy:

$$\text{RMSE} = \sqrt{\frac{1}{N} \sum_{i=1}^{N} (I_{\text{pred},i} - I_{\text{meas},i})^2}. \quad (1)$$

Here, $I_{\text{pred},i}$ denotes predicted values, $I_{\text{meas},i}$ measured values and N the number of evaluated data pairs. Using the RMSE, large deviations between predicted and measured values are weighted more strongly than small deviations. This is suitable for the application addressed here, because large forecast errors have a disproportionately high impact for grid management issues. Other statistical errors measures often used are the mean bias error (MBE), the mean absolute error (MAE) or the correlation coefficient (CC) (see e.g., (Lorenz and Heinemann 2012)).

The evaluations are performed for single site predictions and average irradiance values for Germany to assess the quality of regional forecasts. Only daytime values are considered for the evaluation. With respect to the evaluation of the CMV forecasts, there is the additional constraint that only forecasts derived for solar elevations larger than 10° give reliable results when using the visible range of the satellite measuring reflected solar irradiance for cloud detection. The RMSE values for each forecast horizon are determined using a consistent dataset of all forecasts available for this horizon, which is limited by the CMV forecasts (e.g., one hour ahead forecasts are already available in the morning, while 6 h ahead forecasts are available earliest around midday, with an additional variation due to seasonal changes).

In addition to statistical error measures, the comparison to trivial reference models is a frequently applied check to assess forecast quality. The most common reference model for short-term forecasts is persistence, i.e., the assumption that the current situation does not change and therefore actual or recently measured values can be taken as forecast values. For solar irradiance and power forecasting, the deterministic component of solar irradiance due to the geometrically determined path of the sun should be considered as an additional constraint. A simple approach of persistence reproducing the daily course of irradiance is to consider the measured value of the previous day at the same time as a forecast value, which does not require any model for the daily irradiance pattern. A more advanced concept uses persistence of the dimensionless clear sky index k^*_{meas} in combination with a clear sky model instead of persistence of irradiance values (Lorenz and Heinemann 2012). For forecast horizons of several hours (Δt) ahead, persistence $I_{\text{per},\Delta t}$ for the time t is then defined as:

$$I_{\text{per},\Delta t}(t) = k^*_{\text{meas}}(t - \Delta t) I_{\text{clear}}(t) = \frac{I_{\text{meas}}(t - \Delta t)}{I_{\text{clear}}(t - \Delta t)} I_{\text{clear}}(t). \qquad (2)$$

For more than about 6 h ahead, persistence of the average cloud situation of the previous day is a suitable reference model.

Another common reference is the use of climatological mean values. An investigation for German weather stations (Lorenz and Heinemann 2012) showed a superior performance when using climatological mean values of k^* in comparison to persistence from 2 days onwards. This might be different in different climates; at any case it is worthwhile to investigate the use of climatological mean values as a second reference model besides persistence.

Overview of Irradiance and Photovoltaic Power Prediction

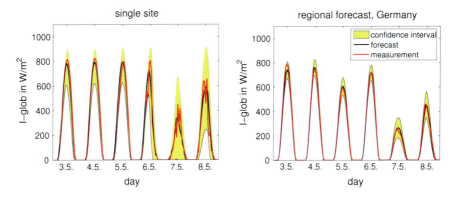

Fig. 5 Forecast of global irradiance I_{glob} with confidence intervals of an uncertainty level of 95 % compared with measured irradiance for 6 days in May 2007 for a single site (*left*) and for the average of 290 measurement stations in Germany (*right*). *Forecast data* ECMWF-based forecasts with post processing (Lorenz et al. 2009)

The performance of the investigated forecast model in comparison to a reference model is evaluated with skill scores, defined as the difference between the forecast and the reference forecast normalized by the difference between a perfect and the reference forecast. For example the RMSE skill score ss_{RMSE} is given as:

$$ss_{RMSE} = \frac{RMSE_{ref} - RMSE_{model}}{RMSE_{ref}}, \quad (3)$$

where $RMSE_{ref}$ refers to the reference model and $RMSE_{model}$ to the investigated forecasting algorithm. Its value thus ranges between 1 (perfect forecast) and 0 (reference forecast). A negative value indicates performance worse than the reference. Skill scores may not only be applied for comparison with a simple reference model but also for inter-comparison of different forecasting approaches (improvement scores).

4.3 Uncertainty Information

Specifying the expected uncertainty of solar irradiance or power predictions gives valuable additional information for forecast users. Confidence or prediction intervals indicate the range in which the actual value is expected to appear with a quantified probability (see Fig. 5). Moreover, probability density functions of the expected future values give even more detailed information of the expected uncertainty.

Two essentially different approaches can be applied to estimate forecast uncertainty:

- In ensemble prediction systems (e.g., (Molteni et al. 1996)), several runs of the same model are computed at the same time starting from slightly different initial conditions or with changes in the physical parameterizations or using different lateral boundary conditions in the case of mesoscale models. Due to the high complexity and nonlinearity of atmospheric processes, even small changes of the initial conditions can lead to a significant spread in forecast results. The ensemble of forecasts is interpreted in terms of a probability distribution, giving a direct measure of uncertainty.
- Information on the distribution of forecast errors (or, seen from a different perspective, on the distribution of possible measurement values for a given forecast value) may also be obtained by analyzing historical time series of forecasted and measured values. With quantile regression this distribution can be directly determined (Bacher et al. 2009). Alternatively, the distribution can be described with a suitable distribution function (e.g., Gaussian or beta distribution) fitting parameters on the basis of forecasted and corresponding measured values. Here, again the deterministic daily course of the sun should be considered, because it determines maximum possible irradiance values and therefore also has a strong impact on the magnitude of forecast errors. In addition, the dependence of the expected uncertainty on the weather conditions can also be modeled (e.g., (Lorenz et al. 2009), see also Fig. 5, left).

4.4 Regional Forecasting with Spatial Averaging Effects

Regional forecasts derived as average values of a set of distributed stations show higher accuracy than single site forecasts. Due to spatial averaging effects, the course of average irradiance forecasts and measurements is smoother than for single sites, which show in particular a high variability for partly cloudy conditions (Fig. 5). The deviations between forecast and measurement are smaller for the regional average values, because forecast errors of the single stations partly compensate each other. The reduction of errors when considering a set of stations instead of a single station is determined by the cross-correlation of forecast errors between the sites, depending on their distance from each other (Fig. 6, left).

Modeling this dependence using an exponential fit and applying basic statistics (e.g., (Beyer 1993)) the forecast errors $RMSE_{region}$ may be estimated for arbitrary of sets of stations (Lorenz et al. 2009). The $RMSE_{region}$ gets smaller with increasing region size, which leads to smaller error reduction factors $f = RMSE_{region}/RMSE_{single}$ (Fig. 6, right). For example, the mean of sites, which are distributed rather uniformly (Fig. 4), over a $3° \times 3°$ region, shows an $RMSE_{region}$ about half $RMSE_{single}$. For sites distributed over Germany $RMSE_{single}$ is reduced to less than 40 %. A good agreement is observed between modeled error reduction factors and error reduction factors directly determined from the data, demonstrating that the proposed model can be applied to estimate the expected RMSE for arbitrary scenarios with a good accuracy.

Overview of Irradiance and Photovoltaic Power Prediction

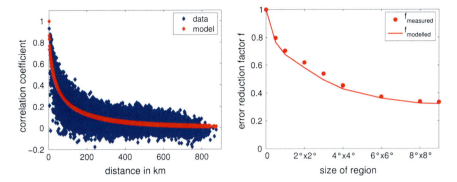

Fig. 6 *Left* Correlation coefficient of forecast errors of two stations depending on the distance between both. *Blue dots* observations, *red dots* exponential fit function. *Right* Error reduction factor $f = \text{RMSE}_{\text{region}}/\text{RMSE}_{\text{single}}$ for regions with increasing size. *Forecast data* ECMWF-based forecasts with postprocessing (Lorenz et al. 2009), *measurement data* 290 stations in Germany (see Fig. 4), *period* January–October 2007

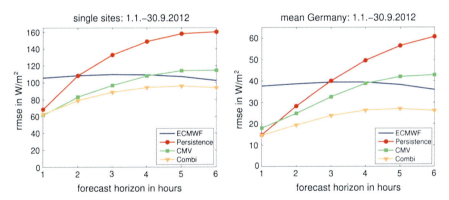

Fig. 7 RMSE of ECMWF-based, CMV and combined forecasts in comparison to persistence (Eq. 2) in dependence on the forecast horizon. For the ECMWF-based forecast, the 12:00 UTC run of the previous day is evaluated independent of the forecast horizon and variations of the RMSE with the forecast horizon are due to the horizon-dependent datasets (see Sect. 4.2) *Left* single sites. *Right* German mean

4.5 Results: Accuracy of Different Approaches for Irradiance Forecasting

A comparison of the performance of the different approaches (Sect. 4.1) in dependence on the forecast horizon (Fig. 7) shows the benefits of the investigated models for various intraday forecast lead times:

- For forecast horizons of 1 h ahead, persistence of ground-measured clear sky index values performs similar to the satellite-based approach. Forecasts in the

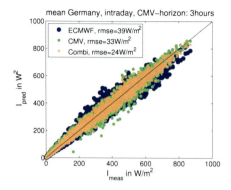 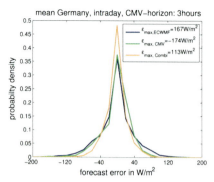

Fig. 8 Scatterplot of predicted versus measured irradiances (*left*) and frequency distributions of forecasts errors (*right*) for German average values of ECMWF-based, CMV, and combined forecasts

sub-hourly range (not shown here) are likely to benefit from any use of ground-based irradiance measurements or sky imagers.
- For about 1–4 h ahead the satellite-based approach outperforms both ground-based persistence and the ECMWF-based forecasts. Satellite data are a valuable data source for irradiance forecasting in these time horizons.
- Beyond 5 h ahead ECMWF-based forecasts are the best choice when referring to single model forecasts. Any forecasting system aiming at these horizons should include NWP-based forecasts.
- The forecasts combining all three models show a significant improvement compared to both ECMWF and CMV single model forecasts. This demonstrates the strong potential for improving the forecast performance by combining different models.
- For all forecasting methods forecast errors of the German mean values are considerably smaller than for single site forecasts (see also previous section). However, it is notable that the impact of spatial averaging is not the same for all methods, e.g., the improvement by the combined approach is much more pronounced for regional forecasts than for single site forecasts.

A more detailed view on the performance of the different forecasting approaches is given in Fig. 8 for the example of 3 h ahead regional forecasts. The scatterplot (Fig. 8, left) shows many situations where CMV forecasts perform better than the ECMWF-based forecasts, but there are also some outliers with poor performance of the CMV forecasts. These CMV outliers correspond to situations with large-scale formation or dissolution of cloud fields, not modeled with the CMV approach. Largest NWP forecast errors occur for fog situations, hardly predictable with current state-of-the-art NWP models.

The combined approach effectively reduces the occurrence of large forecast errors. This is clearly notable also, when comparing the frequency distributions of the forecast errors of the different methods (Fig. 8, right). Particularly, the maximum occurring forecast error of the combined approach is much smaller than the maximum forecast errors of both single model forecasts. Avoiding these large forecast errors is of special importance in the context of grid-integration of solar power, because the amount of reserve power that must be maintained is determined by maximum possible forecast errors.

With respect to day-ahead forecasts based on the two NWP models, a considerably better performance of the ECMWF-based irradiance forecasts ($RMSE_{ECMWF,mean} = 41$ W/m^2, $RMSE_{ECMWF,single} = 104$ W/m^2) than the DWD irradiance forecasts ($RMSW_{DWD,mean} = 49$ W/m^2, $RMSE_{DWD,single} = 112$ W/m^2) is found. Still, the combined forecasts ($RMSE_{Combi,mean} = 37$ W/m^2, $RMSE_{Combi,single} = 100$ W/m^2) are better than both single model forecasts. The forecast errors of the two models with a correlation coefficient of 0.64 for the average values and of 0.77 for the single site forecasts partly compensate each other. The improvement (see Eq. 3) compared to the originally used ECMWF forecasts is 10 % for the German average and 4 % for the single site forecasts.

5 PV Power Forecasting

Most PV power forecasting schemes use global irradiance forecasts as basic input. In addition, it is important to include temperature forecasts, because of the impact of the module temperature on the efficiency of PV power generation.

Generally, approaches for predicting PV power output may be grouped into the following categories:

- Physical approaches explicitly model the processes determining the conversion of solar irradiance to electricity. This includes conversion of irradiance from the horizontal to the plane of the array (giving consideration to its tilt and the axis of tracking, if any) and then simulating PV output in dependence on the irradiance on the array plane and the temperature.
- Statistical or learning algorithms do not model the physical processes directly but try to establish the relation between PV power output and irradiance forecasts and/or other input variables by training on historic data.
- Combined or hybrid approaches first apply physical models and then enhance the results with a statistical model.

A short overview of the basic steps in PV simulation and examples of statistical approaches for PV power prediction are given below. This is complemented by a section on up-scaling to regional power, which is usually the final step for utility applications.

5.1 PV Simulation

Existing models for PV simulation and tilted irradiance conversion, many of them developed and applied in the context of planning of PV systems and for yield estimations, may also be directly used for PV power forecasting. Application of these models requires knowledge of the main characteristics of a PV system, including the nominal power, system orientation, information on part load behavior, temperature coefficients, and mounting type. Here, we exemplarily introduce the tilted irradiance and PV simulation models applied in our operational PV power prediction system (Lorenz et al. 2011).

In a first step, the forecasted horizontal global irradiance is converted to the titled irradiance according to the orientation and declination of the modules. The tilted irradiance model (Klucher 1979) requires information on the direct beam, diffuse sky, and ground-reflected radiation, which are converted separately to the tilted plane. For the decomposition of global horizontal irradiance into horizontal beam and diffuse irradiance an empirical model is applied, based on the clearness index (i.e., ratio of global irradiance to extraterrestrial irradiance) and solar elevation. The conversion of the direct irradiance component is straightforward and subject to geometric considerations only. The conversion of the diffuse irradiance component implies a model for the directional distribution of radiance over the sky, describing anisotropic effects like horizon brightening and circumsolar irradiance.

Next, simulation of direct current (DC) PV system output is performed with the model proposed in Beyer et al. (2004). The efficiency of the PV generator operating at maximum power point (MPP) is estimated with a two-stage approach with solar irradiance on the module plane and module temperature as input. First, the basic influence of the irradiance for a module temperature of 25 °C on the efficiency is described with a three parameter model, where the model parameters may be derived either from data sheet information or fitted to historic data. Second, the performance at module operating temperatures other than 25 °C is modeled by a standard approach using a single temperature coefficient. The module temperature may be approximated from the ambient temperature and irradiance on the module plane taking also the mounting type of the module (e.g., free standing or roof top) into account. Conversion of direct current to alternating current (AC) is done with an inverter model, describing the inverter efficiency as a function of DC input.

Finally, miscellaneous other losses, including reflection and spectral losses or mismatch between modules, are taken into account by empirical factors.

5.2 Statistical Methods

Statistical methods may be applied to forecast PV power directly from NWP output or previous PV power measurements as well as to improve PV power forecasts derived with explicit physical modeling. The methods used in this

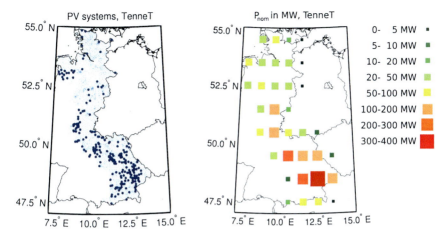

Fig. 9 *Left* Example of a representative set of PV systems (*large dark blue dots*) as a basis for upscaling to the power production of all systems (*small light blue dots*) in the area of the German transmission system TenneT. The position of the systems in the overall dataset is determined by the postal codes, which leads to a limited spatial resolution. *Right* Distribution of nominal power in the control area of TenneT with a spatial resolution of 1°, based on "RES data" from December 2008 (Lorenz and Heinemann 2012)

context are basically the same as for irradiance forecasts (Sect. 3.3) ranging from simple regression techniques to advanced artificial intelligence techniques.

A fully statistical algorithm has been proposed in Bacher et al. (2009). An autoregressive model with exogenous input is used to predict PV power on the basis of measured power data and NWP global irradiance predictions. The weight for both input data sources is adjusted depending on the forecast horizon. The model coefficients are obtained by recursive least-squares fitting with forgetting using online measured data in order to cope with temporal changes that affect PV power output, such as snow covered modules or temporal shading effects.

An example for statistical post-processing of PV power predictions is given in Lorenz et al. (2012b). The proposed empirical approach aims at improving PV power predictions during periods of snow cover, where the original forecasts often show a strong overestimation. The approach integrates online measured PV power output and additional meteorological forecast parameters.

5.3 Up-Scaling to Regional Power Prediction

Renewable power prediction for utility applications usually requires forecasts of the cumulative power production in an area. For both wind and solar power forecasts, this is routinely done by up-scaling from a representative set of single sites (Fig. 9, left), reducing computational and data handling efforts to a practicable amount. Another reason for upscaling is that basic information necessary for predicting PV power output (system parameters or measurement data) is generally not available for all PV systems. For example, in Germany according to the RES, all PV systems receiving the feed-in tariff have to be registered with address and nominal power ("RES data"), but more detailed information on module types or system orientations is not recorded.

Because of the high correlation of the power output of nearby PV systems, there is almost no loss in accuracy by the upscaling approach, given that the representative set approximates the basic properties of the total dataset with good accuracy. A correct representation of the spatial distribution of the nominal power is most important in this respect, which for Germany may be inferred from the "RES data" (Fig. 9, right).

6 Evaluation of PV Power Forecasts

Typical accuracies of local and regional PV power forecasts are shown, again using our operational PV power prediction system (Lorenz et al. 2011) as an example.

6.1 Measurement and Forecast Data

Intraday and day-ahead hourly, regional PV power forecasts are evaluated for the control area of the German transmission system operator "50Hertz" for the period October 2009–September 2010. Because PV power is not measured for the majority of PV systems, the regional power production has to be estimated from available measurements by upscaling from a representative dataset, as done also for the forecasts. Measurements of more than 500 PV systems monitored in this control area by Meteocontrol GmbH are used as a basis to infer the regional power production with a detailed upscaling approach (Lorenz et al. 2011).

Two versions of the PV power prediction system are compared here, both based on ECMWF predictions (12:00 UTC run) only. The comparisons here do not include the additional models (DWD and CMV forecasts) that are currently also part of the service. The basic approach for PV power prediction is based on a simple upscaling procedure from 77 PV systems. No correction for snow-covered modules is applied. A more advanced approach includes a more detailed upscaling with better modeling of the spatial distribution of the systems (Lorenz et al. 2011) using the same representative dataset and an algorithm for snow detection (Lorenz et al. 2012b).

Table 1 Relative RMSE with respect to the installed power of different forecasting approaches for the control area of "50Hertz"; 1.10.2009–30.9.2010 (24 h of the day)

	Regional		Single sites	
	Intraday (%)	Day-ahead (%)	Intraday (%)	Day-ahead (%)
Basic	4.8	5.3	8.1	8.6
New	3.9	4.6	8.0	8.5
Persistence	8.2	10.2	12.4	14.0

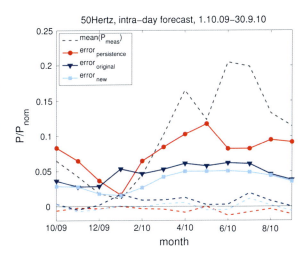

Fig. 10 Monthly mean values (*dashed gray line*) of measured PV power and error measures for persistence (*red line*), the original PV power forecasts (*dark blue line*), and the new PV power forecasts (*light blue line*) for the control area of 50Hertz. Solid lines with markers correspond to RMSE values and dashed-colored lines represent the Bias

6.2 Results

A comparison of the RMSE values for the different forecast versions and persistence for intraday and day-ahead forecast horizons is given in Table 1. All RMSE values are normalized to the nominal power as this is common practice for utility applications. As expected, significantly smaller errors are found for regional than for single site predictions (compare also Fig. 6). Besides the size of the region, the spatial distribution of the installed power causes an additional affect on the error reduction factor. Smallest RMSE values are to be expected for uniformly distributed systems.

For single site forecasts snow detection results in a minor reduction of the RMSE values. But for the regional forecasts, there is a considerable improvement with the new approach. In particular, during winter months forecast errors are strongly reduced by applying the snow detection algorithm (see Fig. 10). The improvement during the rest of the year is due to the more detailed upscaling approach. This emphasizes the importance of a correct modeling of the spatial distribution of the installed PV power for upscaling.

7 Summary and Outlook

Solar power forecasting will be an essential component of any future energy supply system that uses large amounts of fluctuating solar power. Already today PV power forecasting systems contribute to successful grid-integration of considerable amounts of solar power, as shown for the example of Germany. As a consequence of the need for precise and detailed forecast data in the energy sector, increasing research is underway on forecasting of solar irradiance and solar electricity generation.

Different approaches of irradiance forecasting have been introduced. Their benefit for different forecast horizons has been discussed and evaluated for typical examples. Techniques based on NWPs with forecast horizons up to several days ahead are of high importance for most grid-integration issues. For shorter timescales of several hours ahead, forecasts based on cloud motion vectors from satellite images show a superior performance to NWP-based forecasts. For less than 1 h ahead, ground-based observations of cloud motion from sky imagers provide valuable information.

Whatever physical model is used for forecasting, partly stochastic and partly systematic errors will remain. With statistical methods these errors may be reduced based on training of historic datasets of forecasted and measured irradiance and PV power data. In particular, there is a high potential for improving forecast accuracy by combination of different models with statistical models, which has been demonstrated for an example approach including NWP and satellite-based irradiance forecasts.

A proper accuracy assessment provides valuable information for all users that rely on the forecasts as a basis for decision-making. Moreover, a detailed evaluation is a key to model testing and further development, revealing weak points of the applied models.

Current research in irradiance and PV power forecasting covers all the different approaches presented and the complete PV power prediction chain. Improvements in NWP-based irradiance forecasting may be expected as these models will develop with respect to resolution, data assimilation, and parameterizations of clouds and radiation. In particular, the development and application of rapid update cycle models has a high potential to improve intraday forecasting. A promising approach is also the use of ensemble prediction systems, providing inherent uncertainty information. Forecasting techniques based on cloud motion will benefit from enhancements in cloud detection approaches. With respect to statistical methods, apart from model development, the availability of high-quality and up-to-date measurement data of solar irradiance and solar power will be of critical importance. Finally, an optimized combination of different physical and statistical models will be of advantage for any PV power prediction system.

References

Bacher P, Madsen H, Nielsen HA (2009) Online short-term solar power forecasting. Sol Energy 83:1772–1783

Beyer HG, Luther J, Steinberger-Willms R (1993) Power fluctuations in spatially dispersed wind turbine systems. Solar Energy 50:297–305

Beyer HG, Betcke J, Drews A et al (2004) Identification of a general model for the MPP performance of PV modules for the application in a procedure for the performance check of grid connected systems. In: Proceedings of the 19th European Photovoltaic Solar Energy Conference and Exhibition, pp 3073–3076

Chow CW, Urquhart B, Lave M et al (2011) Intra-hour forecasting with a total sky imager at the UC3 San Diego solar energy testbed. Sol Energy 85(11):2881–2893. doi:10.1016/j.solener.2011.08.025

Dudhia J (1989) Numerical study of convection observed during the winter monsoon experiment using a mesoscale two-dimensional model. J Atmospheric Sci 46:3077–3107. doi:10.1175/1520-0469(1989)0462.0.CO;2

Glahn HR, Lowry DA (1972) The use of model output statistics (MOS) in objective weather forecasting. J Appl Meteorol 11:1203–1211

Hammer A, Heinemann D, Hoyer C et al (2003) Solar energy assessment using remote sensing technologies. Remote Sens Environ 86:423–432

Klucher TM (1979) Evaluation of models to predict insolation on tilted surfaces. Sol Energy 23:111–114. doi:10.1016/S0038-092X(87)80031-2

Kühnert J, Lorenz E, Heinemann D (2013) Satellite-based irradiance and power forecasting for the german energy market' in solar energy forecasting and resource assessment, Editor: Jan Kleissl (Elsevier 2013).

Lara-Fanego V, Ruiz-Arias JA, Pozo-Vázquez D, Santos-Alamillos FJ, Tovar-Pescador J (2012) Evaluation of the WRF model solar irradiance forecasts in Andalusia (Southern Spain). Sol Energy 86(8):2200–2217. doi:10.1016/j.solener.2011.02.014

Lorenz E, Heinemann D (2012) Prediction of solar irradiance and photovoltaic power. Compr Renew Energy 1:239–292. doi:10.1016/B978-0-08-087872-0.00114-1

Lorenz E, Heinemann D, Hammer A (2004) Short-term forecasting of solar radiation based on satellite data. In: Proceedings of EuroSun 2004. Freiburg, Germany, pp 841–848

Lorenz E, Hurka J, Heinemann D, Beyer HG (2009) Irradiance forecasting for the power prediction of grid-connected photovoltaic systems. IEEE J Special Topics Earth Observ Remote Sens 2:2–10

Lorenz E, Scheidsteger T, Hurka J et al (2011) Regional PV power prediction for improved grid integration. Progr Photovoltaics Res Appl 19: 757–771. doi:10.1002/pip.1033

Lorenz E, Kühnert J, Heinemann D (2012) Short term forecasting of solar irradiance by combining satellite data and numerical weather predictions. In: Proceedings of 27th European Photovoltaic Solar Energy Conference, Valencia, Spain, pp 4401–440

Lorenz E, Heinemann D, Kurz C (2012b) Local and regional photovoltaic power prediction for large scale grid integration: assessment of a new algorithm for snow detection. Progr Photovoltaics Res Appl 20:760–769. doi:10.1002/pip.1224

Marquez R, Coimbra CFM (2011) Forecasting of global and direct solar irradiance using stochastic learning methods, ground experiments and the NWS database. Sol Energy 85:746–756

Molteni F, Buizza R, Palmer TN, Petroliagis T (1996) The ECMWF ensemble prediction system: methodology and validation. Quart J Royal Meteorol Soc 122:73–119

Morcrette JJ, Barker HW, Cole JNS et al (2008) Impact of a new radiation package, McRad, in the ECMWF integrated forecasting system. Mon Weather Rev 136:4773–4798

Pedro HTC, Coimbra CFM (2012) Assessment of forecasting techniques for solar power output with no exogenous variables. Sol Energy 86:2017–2028

Pelland S, Gallanis G, Kallos G (2013) Solar and photovoltaic forecasting through post-processing of the global environmental multiscale numerical weather prediction model. Progr Photovoltaics Res Appl 21: 284–296. doi:10.1002/pip.1180

Reikard G (2009) Predicting solar radiation at high resolutions: a comparison of time series forecasts. Sol Energy 83(3):342–349. doi:10.1016/j.solener.2008.08.007

Remund J, Schilter C, Dierer S et al (2008) Operational forecast of PV production. In: Proceedings of 23rd European Photovoltaic Solar Energy Conference, Valencia, Spain, pp 3138–3140

Skamarock WC, Klemp JB, Dudhia J et al (2008) A description of the advanced research WRF version 3. TechnicalNote NCAR/TN-475+STR. Mesoscale and Microscale Meteorology Division, National Center for Atmospheric Research, Boulder

Traunmüller W, Steinmaurer G (2010) Solar irradiance forecasting, benchmarking of different techniques and applications to energy meteorology. In: Proceedings of EuroSun 2010. Graz, Austria

Spatial and Temporal Variability in the UK Wind Resource: Scales, Controlling Factors and Implications for Wind Power Output

Steve Dorling, Nick Earl and Chris Steele

Abstract The UK makes for a fascinating case study regarding variability in renewable energy resource, impacting on the safe degree of penetration of renewable energy into the electricity grid. In this chapter, we highlight temporal and spatial scale wind speed variability which impacts upon UK wind farm operation both onshore and offshore. We argue how natural variations in wind climate at the monthly, seasonal and decadal scale, linked to the changing frequency of largescale weather patterns, need to be built into assessments of renewable energy revenue streams and consideration of associated insurance products.

1 Introduction

Colloquially, it is often said it is possible to experience all four seasons in a single day in the United Kingdom, such is the variability of UK weather. Key underlying drivers for this characterisation, taking the UK to mean both the land and its neighbouring waters, include (a) being at the boundary of one of the earth's major continents, Eurasia, and oceans, the North Atlantic; (b) acting as a meeting place and area of interaction for contrasting air mass types which have, themselves, been modified en route; (c) proximity to the end of a major mid-latitude extra-tropical storm track (Dacre and Gray 2009) yet also a preferred region for anticyclonic blocking (Barriopedro et al. 2006); (d) having an exceptionally long coastline, subject to seasonal land–sea temperature contrasts and resulting thermally driven circulation patterns (Simpson 1994); (e) hosting very spatially variable terrain. These factors, and others, lead to spatial and temporal complexity with regard to onshore and offshore wind climate and challenges with respect to both regional

S. Dorling (✉) · N. Earl · C. Steele
School of Environmental Sciences, University of East Anglia, Norwich, UK
e-mail: s.dorling@uea.ac.uk

weather forecasting and climate prediction (Woollings 2010). With particular respect to the focus of this book, the UK also makes for a fascinating case study regarding variability in renewable energy resource, impacting on the safe degree of penetration of renewable energy into the electricity grid.

By combining analysis of land station measurements with high-resolution mesoscale modelling, we focus in this chapter on two particular aspects of the UK wind climate which have implications for wind energy. First, the variability observed in 10 m wind measurements since 1980 over a land measurement network which extends across the UK. Second, the seasonal influence of sea breeze circulations on the offshore environment where major wind farm development is underway.

2 Tools

Any assessment of long-term variability in climate, over a particular period, must always be considered as a snapshot in a context of broader scale change. It is now widely accepted that decadal variability is a strong feature of our climate, whether we are talking about historical rainfall in the Sahel, climate scenarios of the future or variations in renewable energy resource (Wang et al. 2009; Woollings 2010). Decadal variability is both a naturally and anthropogenically forced feature of the climate system. Drawing conclusions from analyses of relatively short records is therefore a dangerous and inappropriate practice. At the same time, changes which do occur in measurement networks lead to a compromise needing to be struck between the aspiration for a long period of record and the requirement for a stable station network providing reasonable spatial coverage. The compromise we adopt here is to analyse a 1980–2010 record of 10 m wind speed measurements from a UK Met Office network of 40 land-surface stations distributed across the UK. We thereby focus on the most recent 30-year climate normal period during which modern-day wind energy has emerged in terms of its importance for renewable energy generation. The stations all adhere to rigorous requirements relating to exposure and have excellent data completeness (Hewston and Dorling 2011; Earl et al. 2013). Nevertheless, we show how local topography remains influential.

The constraints imposed by the availability and quality of measurements can be partially addressed using modelling approaches. The practical challenges of making long-term, reliable measurements offshore lead to computer models being particularly important in the marine environment. This is especially true given the tendency for met-masts, installed in the planning and construction phases of offshore wind farms, to then be removed during the operational phase. With such a current focus on offshore wind energy development for the achievement of renewable energy generation targets, utilising models to gain a greater understanding of offshore wind variability is a natural approach. An interesting component of the marine (and coastal) renewable energy environment is the sea/land breeze system. As mesoscale features, sea breezes require mesoscale models of sufficient spatial resolution to resolve their important features. We present some results of mesoscale modelling, using the Weather Research and Forecasting (WRF) model (Skamarock and Klemp 2008), choosing to highlight a spatial and temporal scale which otherwise tends to be over shadowed by attention to synoptic (Wang et al. 2009; Brayshaw et al. 2011) and wake effect scales (Barthelmie et al. 2009). Our model sensitivity experiments demonstrate how results are dependent upon precise model setup.

3 Analysis of 10 m Station Measurements

The network of stations which formed the basis of our selection is presented and discussed in Hewston and Dorling (2011) and in Earl et al. (2013). As a measure of spatial variability in the UK wind climate, for five stations located in Figs. 1, 2 presents 1980–2010 average wind roses broadly representing the geographical extent of the UK. Of course these contrasting wind climates give a broad indication of the more favourable UK areas for onshore wind energy generation.

Earl et al. (2013) discuss in detail the temporal variability in mean wind speed which is seen over the full 1980–2010 period. They highlight sharp inter-annual variations in mean wind speed and potential power output in adjacent years (e.g. between 1986 and 1987) and longer term variations indicative of decadal variability (ending on a low point in the most recent decade 2001–2010). Here we focus on one particular example, namely December 2010 relative to the average of all Decembers in the analysed period. Figure 3 shows how unusual December 2010 was (strong anticyclonic blocking and North Atlantic Oscillation Index of − 4.62), recalling that winter months are on average particularly important with respect to high wind turbine capacity factors (Sinden 2007). The low wind energy density of December 2010 is an indication of the role which insurance policies can play in maintaining 'bankability' in a wind energy generation context.

The Weibull distribution is commonly used as a fit to observed wind data (Takle and Brown 1978; Seguro and Lambert 2000; Weisser 2003; Celik 2004). Earl et al. (2013) show how the Weibull parameters vary between the stations in our network, highlighting how inappropriate the assumption of a Rayleigh distribution, for the purposes of wind turbine performance assessment, can be (Petersen et al. 1998). Not only do the Weibull parameter values vary between sites, they also vary at individual monitoring stations as a function of wind direction, as demonstrated by way of example in Fig. 4 by the shape parameter, k, at the coastal Aberporth (west Wales) station relative to London Heathrow Airport (Fig. 1). For site-specific wind energy resource assessment, monthly and inter-annual variations in the frequency of winds from different quadrants should be accounted for since this clearly affects both the temporal mean wind speed itself and the wind speed distribution (through k), with important implications for power output when this variability is combined with a turbine power curve.

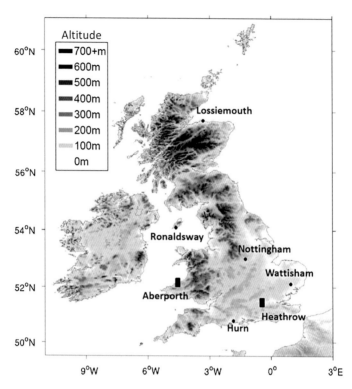

Fig. 1 Wind rose station locations (*circles*) and Weibull distribution station locations (*rectangles*)

4 Sea Breeze modelling

Sea breeze systems have been studied extensively with respect to their impacts at coastlines and in the immediate hinterland, especially with respect to air quality issues, but only rarely in terms of their offshore components (Crosman and Horel 2010; Miller et al. 2003; Simpson 1994; Lapworth 2005). Yet the most significant wind energy development, in the UK and increasingly beyond, is now strongly focused in this offshore environment and we need to fully understand the role these wind systems play in enhancing or negating the overall offshore wind resource.

Thermal contrasts between land and sea, resulting from differences in their respective heat capacities, lead to horizontal pressure gradients which may facilitate the establishment of sea breeze cells. These critical land–sea temperature differences are at a maximum in the mid-late spring and early summer when the land is heating up quickly but the sea temperature is still lagging behind. Becker and Pauly (1996) show that southern North Sea temperatures, in the months May–July, generally increase from 9–16 °C, while mean maximum temperatures over

Spatial and Temporal Variability in the UK Wind Resource 459

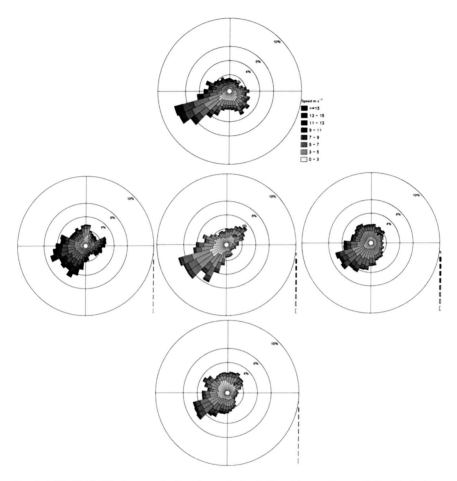

Fig. 2 1980–2010 Wind roses for Lossiemouth (*top*), Ronaldsway (*centre left*), Nottingham (*centre*), Wattisham (*centre right*) and Hurn (*bottom*)

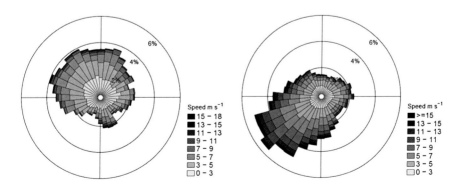

Fig. 3 Network-average wind roses for December 2010 (*left*), all Decembers (*right*)

Fig. 4 Percentage origin of wind, Weibull shape parameter (k) and mean wind speed (ū ms^{-1}) as a function of direction quadrant at a Aberporth and b London Heathrow

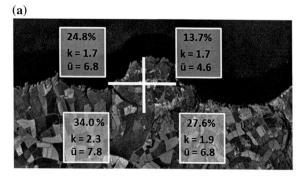

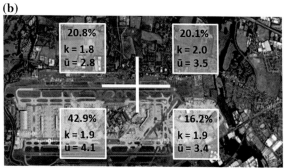

land, for example in SE England, would typically increase from 16–22 °C (and be significantly higher on some particular days) over this same period of the year.

Using idealised model experiments, Steele et al. (2012) demonstrate that the influence of sea breeze circulations (on-shore flow depth ~700 m) can be felt to distances offshore which comfortably extend from coast to coast between say East Anglia in the UK and the Dutch coast, an area of extensive current offshore wind farm development. They also show how the synoptic scale gradient wind speed and direction, relative to the orientation of a coastline, determines the type of sea breeze which may be established: *pure, corkscrew* or *backdoor*. The different sea breeze types have contrasting effects on the scale of offshore calm zones which may, as a result, be established and the strength of the gradient flow in which a sea breeze can become established also varies between sea breeze types.

WRF model wind fields at a model level equivalent to ~90 m in height were generated to analyse the impact of model horizontal resolution on offshore wind turbine capacity factor. A height of 90 m was chosen as this corresponds to the hub-height of a typical offshore wind turbine, with the capacity factor based on the power curve of a 3.6 GW Siemens offshore turbine. Figure 5 presents results of these model runs for June 2006, a month exhibiting significant sea breeze activity (-ve values indicating a higher capacity factor at higher resolution). Here, the largest changes in capacity factor, off the south and east coasts of England, are largely a function of the more realistic coastline detail in the higher resolution simulation. Meanwhile Fig. 5b shows the change in capacity factor resulting from

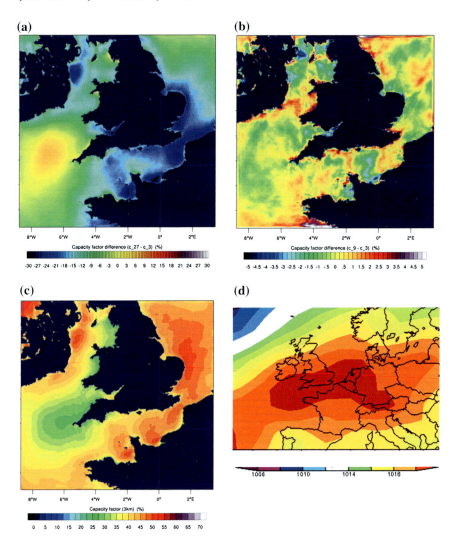

Fig. 5 Offshore capacity factor differences (%) for the month of June 2006 as a function of model resolution for **a** 27 km *minus* 3 km [*top left*]; **b** 9 km *minus* 3 km (*top right*). **c** Capacity factors (%) for June 2006 based on 3 km model resolution output (*bottom left*). **d** June 2006 mean sea-level pressure (mb) [courtesy NOAA/ESRL Physical Sciences Division based on NCEP Reanalysis data]

an increase in model resolution from 9 to 3 km, both resolutions being capable of resolving at least some of the detailed features of sea breeze cells. Figure 5b highlights how the impact of sea breezes is a function of both coastline orientation relative to prevailing gradient windflow and also of sea breeze interactions between neighbouring coastlines. Figure 5c shows 3 km model resolution results, highlighting much higher capacity factors in June 2006 off the east coast of

Table 1 Theoretical sea breeze types identified in June 2006 for the east coast of East Anglia (England) using the classification system shown in Fig. 6, with associated synoptic scale weather types as per Jones et al. (1993), gradient wind strengths and land–sea temperature differences

Date	Weather types	Predicted sea breeze type	850 hPa gradient wind strength (ms^{-1})	Land–sea temperature difference (K)
1 June 2006	AA;AA;AA;AA	None	–	–
2 June 2006	AA;AA;AA;AA	Backdoor	7.2	8.4
3 June 2006	AA;AA;AA;AA	Backdoor	9.3	9.4
4 June 2006	AA;AA;AA;AA	Backdoor	9.6	8.5
5 June 2006	AA;AA;AA;AA	Backdoor	7.7	5.3
6 June 2006	AA;AA;AA;AA	Backdoor	2.7	5.6
7 June 2006	AA;AA;AA;AA	Pure	8.7	10.8
8 June 2006	AA;AA;AA;AA	Corkscrew	2.8	11.2
9 June 2006	AA;AS;AS;ASE	None	–	–
10 June 2006	SE;S;S;S	Detached	10.9	12.7
11 June 2006	S;S;AS;AS	Corkscrew	8.4	16.6
12 June 2006	S;S;CSW;SW	None	–	–
13 June 2006	SW;W;AW;AA	Pure	6.6	6.6
14 June 2006	AA;AA;AA;AA	None	–	–
15 June 2006	AA;AA;AA;AA	None	–	–
16 June 2006	AA;AA;AA;AA	Pure	6.5	8.8
17 June 2006	AA;ASW;SW;SW	Corkscrew	3.9	11.6
18 June 2006	W;CW;CC;CC	None	–	–
19 June 2006	CC;CC;CNW;NW	None	–	–
20 June 2006	W;SW;CSW;CC	None	–	–
21 June 2006	CC;CC;CC;CC	None	–	–
22 June 2006	CC;CNW;CNW;NW	None	–	–
23 June 2006	NW;W;W;W	Pure	5.5	7.1
24 June 2006	CW;CC;CC;CC	None	–	–
25 June 2006	u;u;ANE;AA	None	–	–
26 June 2006	AA;AA;AA;AA	None	–	–
27 June 2006	AA;AA;AA;AA	None	–	–
28 June 2006	AA;AA;ASS;S	Pure	3.2	5.9
29 June 2006	SW;ASW;AS;S	Corkscrew	3.9	6.2
30 June 2006	S;SW;SW;ASW	Corkscrew	3.4	9.4

[AA = anticyclonic, SE = south-easterly, S = southerly, AS = anticyclonic southerly, CC = cyclonic, CNW = cyclonic north-westerly, etc.]

England relative to the south-west coast. In a month where the whole of southern England is the focus of relatively high sea-level pressure (Fig. 5d), we believe this capacity factor distribution to partly result from stronger land–sea temperature contrast in the east (higher land temperature and cooler sea temperatures) when compared with the south-west (which is a more maritime climate strongly influenced by the North Atlantic drift). This stronger temperature contrast encourages more frequent sea breeze development in the east (see Table 1 later).

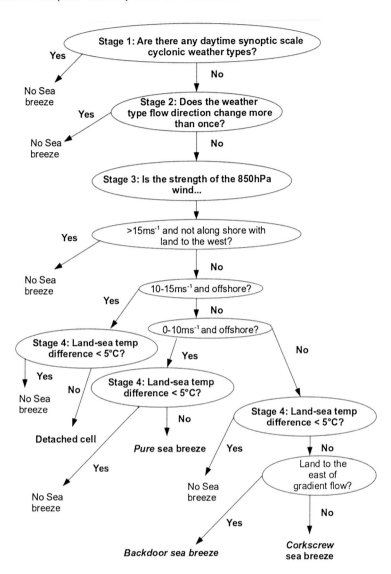

Fig. 6 Decision Tree for theoretical identification of conditions which would be expected to support the establishment of a sea breeze

Since sea breeze systems clearly impact on the offshore wind resource during a period of the year when capacity factors are generally lower than the annual average, it is instructive to assess how common such systems are. To do this, an automated classification system is established (Azorin-Molina et al. 2011) which assesses the presence of supportive conditions for sea breeze establishment, based on synoptic scale weather type (Jones et al. 1993). Land–sea temperature gradient is also checked, as is the strength of the gradient windflow. The gradient windflow direction also determines the sea breeze type if all other conditions for sea breeze establishment are supportive. Figure 6 shows the associated decision tree and Table 1 presents results based on its application to data from June 2006 for the east coast (north–south orientated) of East Anglia, England. The approach is being used to establish a 10-year climatology of sea breeze (type) occurrence and is enabling the identification of events and periods to model in order to produce a more comprehensive assessment of sea breeze (type) impact on the offshore wind resource, incorporating verification against offshore wind farm measurements. This work will assess the sensitivity of our results to, especially, the chosen model planetary boundary layer scheme.

5 Conclusions

We have shown examples, and pointed to other supporting literature, highlighting temporal and spatial scale wind speed variability which impacts upon UK wind farm operation both onshore and offshore. Offshore and close to the coast, sea breeze systems in the late spring and summer have the potential to either reduce or enhance the overall wind resource depending on sea breeze type and on the precise location of a wind farm. Onshore, wind speed distributions are a function of upwind surface roughness, and therefore also of variations in frequency of airflow from different directions, meaning that use of a constant shape parameter value, k, will lead to errors in resource assessment. Onshore and offshore, natural variations in wind climate at the monthly, seasonal and decadal scale, linked to the changing frequency of large-scale weather patterns, need to be built into assessments of renewable energy revenue streams and consideration of associated insurance products.

References

Azorin-Molina C, Tijm S, Chen D (2011) Development of selection algorithms and databases for sea breeze studies. Theor Appl Climatol 106:531–546

Barriopedro D, Garcia-Herrera R, Lupo AR, Hernández E (2006) A climatology of Northern Hemisphere blocking. J Clim 19:1042–1063

Barthelmie RJ, Hansen K, Frandsen ST, Rathmann O, Schepers JG, Schlez W, Phillips J, Rados K, Zervos A, Politis ES, Chaviaropoulos PK (2009) Modelling and measuring flow and wind turbine wakes in large wind farms offshore. Wind Energy 12(5):431–444

Becker GA, Pauly M (1996) Sea surface temperature changes in the North Sea and their causes. ICES J Mar Sci 53:887–898

Brayshaw DJ, Troccoli A, Fordham R, Methven J (2011) The impact of large scale atmospheric circulation patterns on wind power generation and its potential predictability: A case study over the UK. Renew Energy 36:2087–2096

Celik AN (2004) A statistical analysis of wind power density based on the Weibull and Rayleigh models at the southern region of Turkey. Renew Energy 29:593–604

Crosman ET, Horel JD (2010) Sea and lake breezes: a review of numerical studies. Bound-Layer Meteorol 137:1–29

Dacre HF, Gray SL (2009) The spatial distribution and evolution characteristics of North Atlantic cyclones. Mon Weather Rev 137:99–115

Earl N, Dorling SR, Hewston R, von Glasow R (2013) 1980–2010 variability in UK surface wind climate and associated implications for the insurance and wind energy sectors. J Clim 26:1172–1191

Hewston R, Dorling SR (2011) An analysis of observed daily maximum wind gusts in the UK. J Wind Eng Ind Aerodyn 99:845–856

Jones PD, Hulme M, Briffa KR (1993) A comparison of Lamb circulation types with an objective classification scheme. Int J Climatol 13:655–663

Lapworth A (2005) The diurnal variation of the marine surface wind in an offshore flow. Q J Roy Meteorol Soc 131:2367–2387

Miller STK, Keim BD, Talbot RW, Mao H (2003) Sea breeze: structure, forecasting and impacts. Rev Geophys 41:1011

Petersen EL, Mortensen NJ, Landberg L et al (1998) Wind power meteorology, part I: climate and turbulence. Wind Energy 1(S1):25–45

Seguro JV, Lambert TW (2000) Modern estimation of the parameters of the Weibull windspeed distribution for wind energy analysis. J Wind Eng Ind Aerodyn 85(1):75–84

Simpson JE (1994) Sea breeze and local winds. Cambridge University Press, Cambridge

Sinden G (2007) Characteristics of the UK wind resource: long-term patterns and relationship to electricity demand. Energy Policy 35:112–127

Skamarock WC, Klemp JB (2008) A time-split non-hydrostatic atmospheric model for weather research and forecasting applications. J Comput Phys 227:3465–3485

Steele CJ, Dorling SR, von Glasow R, Bacon JD (2012) Idealized WRF model sensitivity simulations of sea breeze types and their effects on offshore windfields. Atmos Chem Phys Discuss 12:15837–15881

Takle ES, Brown JM (1978) Note on the use of Weibull statistics to characterize wind-speed data. J Appl Meteorol 17(4):556–559

Wang XI, Zwiers FW, Swail VR, Feng Y (2009) Trends and variability of storminess in the Northeast Atlantic region, 1874–2007. Clim Dyn 33:1179–1195

Weisser D (2003) A wind energy analysis of Grenada: an estimation using the 'Weibull' density function. Renew Energy 28:1803–1812

Woollings T (2010) Dynamical influences on European climate; an uncertain future. Phil Trans R Soc A 368:3733–3756

Reducing the Energy Consumption of Existing Residential Buildings, for Climate Change and Scarce Resource Scenarios in 2050

John J. Shiel, Behdad Moghtaderi, Richard Aynsley and Adrian Page

Abstract A pilot study of energy efficiency measures, or retrofit actions, was carried out for a single-story detached (single-family) house. It used climate downscaling to project the climatic conditions of a region, and building simulation techniques with two thermal comfort approaches for scenarios of "Climate Change" and "Scarce Resources" in the year 2050. This study was the first stage of a research program to find cost-effective retrofit actions to lower greenhouse gas (GHG) emissions for existing Australian houses in a temperate climate. The pilot study ranked retrofit actions that were cost-effective in reducing the heating and cooling energy usage of a house. These actions included removing carpet from a concrete floor for added thermal mass, and adding external shading with deciduous trees to lower summer radiation from the northern windows (in the southern hemisphere). Also, the alternative thermal comfort approach showed that occupants had more control to lower their energy usage than the standard Australian approach.

1 Introduction

The current phase of climate change is due to the unprecedented rapid rise of temperature (IPCC 2008). This is ironically called global warming and is mostly due to man-made GHG emissions (IPCC 2008). Buildings contribute around 33 % of the global GHG emissions (Urge-Vorsatz et al. 2007).

J. J. Shiel (✉) · B. Moghtaderi · A. Page
Priority Research Centre for Energy, School of Engineering,
Faculty of Engineering and Built Environment, University of Newcastle,
University Drive, Callaghan, NSW 2308, Australia
e-mail: jshiel@westnet.com.au

J. J. Shiel
EnviroSustain, PO Box 265, Jesmond, NSW 2299, Australia

R. Aynsley
Building Energetics, PO Box 2153, Noosaville, BC QLD 4566, Australia

The prices of oil and other resources related to construction are increasing due to production shortages and increased demand (Frei 2012; Aleklett 2010; Heinberg 2007; Campbell 2002; Hirsch et al. 2010). Calls are being made to use materials more efficiently, including for construction (Bol 2011; McKinsey & Company 2011; Ginley and Cahen 2011). Conventional cheap oil supplies have reached a plateau of production for the last 5 years (IEA and OECD 2011; Aleklett 2010), and prices are now, on average, above AUD$100 (USD$103) per barrel for TAPIS crude.

The prices of oil and other scarce materials could continue to rise with global population growth and rising affluence due to increased resource consumption. So, more financial pressure could be placed on activities such as construction that rely on transport and new materials. This would lead to increases in the cost of retrofitting houses for energy reduction, affecting householders trying to lower their energy costs or carbon intensity.

A pilot study is described here of energy efficient retrofit actions for a single-story detached (single-family) house for two key scenarios in 2050: (i) Climate Change and (ii) Scarce Resources. These actions are compared with those of the baseline 1990 climate.

The key drivers of the research project are:

- most of the Australian residential stock is in a poor energy-performance state;
- Australia is particularly vulnerable to climate change with its large coastal population, threats to its water supply in food-growing regions, flood and bushfire exposure, and species extinction; and
- the rising costs of energy supplies and other scarce resources.

There has been little research into low-cost retrofit actions to suit Australia's stock of existing houses. However, some research has been done for 20 year futures out to 2100, and for very low energy consumption resulting from behavioral change and improved design of appliances (Robert and Kummert 2012; Ren et al. 2011; Lee 2009). While some researchers consider existing houses, they do not model house construction indicative of the existing Australian stock, e.g., with timber floors.

Following this introduction, Sect. 2 describes the methodology of the overall research program and the pilot study. It also explains how the climate scenarios for 2050 were projected, and how the retrofit actions were simulated to predict the house thermal performance for those scenarios. In particular, Sect. 2.6 describes in detail the thermal comfort approaches.

Section 3 presents the results that show the impacts of the retrofit actions on the house temperature and on the heating and cooling energy needed for comfort. This is followed by Sects. 4 and 5 which discuss the findings, and the overall conclusion.

Fig. 1 Pilot study method: the pilot study compared the impact of energy-reducing retrofit actions on room temperatures and the energy for comfort with those of an unmodified house in Adelaide. There were simulations for 1990 and two scenarios in 2050, and for two thermal comfort approaches. The retrofit actions were ranked by cost-effectiveness

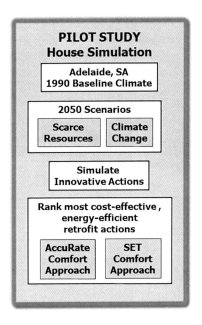

2 Method

2.1 Research Program

The pilot study is the first stage of an overall research program aimed at improving the poor thermal performance of the Australian building fabric for detached existing houses in a temperate climate.

The research program will establish cost-effective GHG-reducing retrofit actions for the house envelope, for the most common types of Australian house construction. It will compare a 1990 baseline climate with two climate scenarios of 2050:

- Climate Change: the high-range A1FI scenario (IPCC 2008) of the Intergovernmental Panel on Climate Change (IPCC); and
- Scarce Resources: where oil and other resource shortages reduce consumption.

2.2 Pilot Study

The goal of the pilot study was to find initial energy savings and cost-effectiveness of retrofit actions for one typical Australian house construction for the two climate scenarios of 2050.

The pilot study (see Fig. 1) was informed by:

- a literature review (Shiel et al. 2009) of the best practice retrofits:
  - that were recognized by professional building associations;
  - from regions with more advanced building practices than Australia, particularly Germany, Austria, Canada, and California;
- experimentation and simulation of retrofit actions on a house in Newcastle, NSW (Shiel et al. 2010) to test some innovative ideas; and
- advice from research Supervisors (Page et al. 2011; Aynsley 2012; Lehmann 2010), colleagues and the Ph.D. network on the feasibility of retrofit actions, and in techniques for calculating thermal comfort and for experimenting.

The pilot study established the infrastructure of:

- AccuRate v1.1.4.1, a house simulation package for temperature and energy, which is described further in Sect. 2.4;
- AccuBatch v2.0.0.0, which is a CSIRO batch program to run many AccuRate cases;
- JMP (v9.0.0), SAS Institute Inc's program for graphical statistical analysis; and
- ASHRAE's Thermal Comfort Tool v1.07 software program (Fountain and Hulzenga 1995) to calculate the standard effective temperature (SET), an alternative thermal comfort index to one used in AccuRate.

Figure 1 shows the pilot study method which is described in detail in the following sections. The pilot study simulations were carried out for the house temperatures and energy required to maintain comfort for the unmodified house and for each retrofit action.

The retrofit actions were simulated for the 1990 baseline climate for the two climate scenarios of 2050, and for two comfort approaches. The pilot study ranked the actions by the most cost-effective in reducing the energy demand, and by having the largest internal temperature change, compared to the unchanged house.

2.3 Scenarios for 2050

2.3.1 Climate Change

The Climate Change scenario chosen was the IPCC's highest emission scenario (IPCC 2008), also known as A1 Fossil Fuel Intensive (A1FI). This was because its projected emissions are showing signs of being exceeded (Rahmstorf et al. 2007; Cleugh et al. 2011). The eminent Australian economist Professor Ross Garnaut observed:

> Between 2000 and 2008, the annual increase in fossil fuel emissions grew ...well above the IPCC scenario with the highest emissions through to 2100... (Garnaut 2011, 3).

2.3.2 Scarce Resources

Future patterns of scarce resources concern scientists (Ginley and Cahen 2011; Heinberg 2007; Campbell 2002; Aleklett 2010; Bol 2011; Hirsch et al. 2010) as well as economists (McKinsey & Company 2011; Garnaut 2011; Curtis 2009; Pauli 2010).

The scarce resource problem, which is due to global increases in population and affluence, causes particular issues for the construction sector in regard to certain raw materials and energy supply. Global reserves are small for lead, silver, tin, zinc, copper, and nickel (Bol 2011) and this is evident in price increases. Some materials used to produce energy resources are also in scarce supply (Ginley and Cahen 2011), and there are issues with conventional oil.

Conventional oil supplies have reached a plateau level of production over the last 5 years, and prices are expected to remain around AUD$100 or more per barrel for TAPIS crude. As conventional oil supplies drop, a massive global investment is needed to meet the global demand for oil, of around AUD$2 trillion per year for 20 years (IEA and OECD 2011). This could create a greater reliance on the more expensive and carbon intensive unconventional oil sources.

So, if the available conventional oil supply is reduced and large investments are needed, these oil prices will become volatile and inelastic (Martin 2012) and supply chains will shorten. This will cause a decline in globalization, which is the global trade of goods and services, and could lead to a "peak globalization" (Curtis 2009). This could slow coal and gas exports (Vivoda 2011) which rely on oil products for transportation, further increasing energy prices by disrupting the supply of fossil fuels. This could cause GHG emissions to decrease, helping to stabilize cumulative emissions.

Rising energy prices would lower economic growth until demand is restored again by the rising consumption of population increase and affluence, when energy prices could escalate again. This pattern has been recognized in recent times (McKinsey & Company 2011; Hall and Klitgaard 2011; Aleklett 2010).

To simplify calculations, the pilot study assumed that the temperature increase of the Scarce Resources scenario would be half that of the Climate Change scenario temperature increase from the 1990 baseline climate. Future research may be able to consider the new scenario approach of the IPCC (Moss et al. 2010).

2.4 House Temperature and Energy Simulations

2.4.1 Details of the House

Figure 2 shows the plan of the brick veneer house that was modeled in the Adelaide, SA location. It had a concrete slab-on-ground floor with an area of 120 m^2. The external walls had 600 mm eaves and were constructed of brick veneer with:

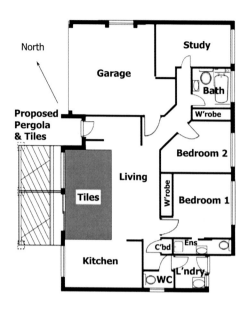

Fig. 2 House plan with pergola: the house is poorly oriented for passive solar design with a large, north-western exposed window. The Pergola simulations reported here were conducted on an enclosed conservatory with a metal roof instead of the angled roof beams shown

- an outer skin of 110 mm single brick; and
- an inner load-bearing timber frame (stud wall), lined with reflective insulation and plasterboard.

The house had single-glazed aluminium-framed windows and a metal roof with reflective insulation underneath.

The internal walls were timber stud and plasterboard, and there was a plasterboard ceiling with no insulation. The living room/kitchen external wall had 70 % glass and was oriented 50 ° west of north, which is a very poor orientation for a house in the southern hemisphere. The house also had some shading from neighboring properties and trees.

2.4.2 Retrofit Actions

After experimenting and simulating many possible retrofits, including temporary double glazing, thermally lined curtains, and a Pergola with deep angled roof beams to manage seasonal window shading (see Fig. 2), the actions investigated in detail in the pilot study were:

- adding a high value of ceiling insulation with a thermal resistance of R4 K·m^2/W (R23 h·ft^2· °F/Btu);
- growing deciduous trees beside the northwest wall;
- having a window size reduction in the living room/kitchen of 50 %;
- removing the carpet and adding black tiles to the concrete floor for added thermal mass;

- adding temporary double glazing to the bedrooms, study and living room/kitchen;
- adding a metal roof conservatory, which is a form of closed-in Pergola, to the northwestern living room—effectively to "re-orientate" the house and catch more winter sun;
- adding four large 1.4 m diameter ceiling fans; and
- weather-stripping the house to eliminate draughts, including sealing architraves, cornices, and external doors.

Thermal mass is the ability of a material to absorb heat energy. It can act like a thermal battery and absorb and release heat to a room to smooth out large daily temperature fluctuations.

The retrofit actions were modeled in isolation for the pilot study, and their combined impact will be researched further.

2.4.3 NatHERS House Rating Scheme

The Australian Nationwide House Energy Rating Scheme (NatHERS) awards star ratings from 0 to a maximum of 10 for the energy performance of new houses (DCCEE 2011).

The star rating of 10 is for dwellings that need negligible space heating and cooling energy to keep occupants thermally comfortable all year. This annual house energy required for comfort, known as the "required energy," is not to be confused with the energy consumption of the appliances themselves for space heating and cooling. The latter energy is lower and depends on the efficiency of the appliances which can be as high as a factor of 5.

NatHERS software applications calculate the annual required energy from room temperatures and its thermal comfort approach. The annual required energy is divided by the air-conditioned floor area, and adjusted according to the house area for surface effects. The resulting areal intensity value is compared with the star rating energy bands for the appropriate house climate region, and a star rating is assigned.

Currently, the Australian new house minimum NatHERS star rating is six stars.

2.4.4 AccuRate

The pilot study used AccuRate v1.1.4.1 (Hearne 2011; Delsante 2005; Chen 2008), one of the NatHERS-accredited house energy rating packages. It was used to simulate the internal room temperatures, and to calculate the house energy intensity and star rating.

AccuRate was used in a design manner where parameters can be adjusted to obtain a better result, rather than to produce a fixed star rating from a known house specification and plan. This helped to optimize the retrofit actions.

Like other NatHERS-accredited packages, AccuRate applies rules for natural ventilation, ceiling fans and then air-conditioning, according to heating and cooling criteria. It uses a frequency response building thermal model and a multiroom network ventilation model to calculate residential annual hourly room temperatures (Ren et al. 2011).

2.4.5 Energy and Temperature Simulation with AccuRate

AccuRate has standard loadings and assumptions related to the implementation of the NatHERS protocol (see Sect. 2.6.1).

In addition, the AccuRate assessor inputs are:

- a building location which selects a local reference meteorological year (RMY) weather profile;
- the geometry of
  - each construction component such as wall, floor ,and roof as well as thermal properties such as insulation resistance;
  - the rooms and their relationships to each other; and
- shading and orientation.

The AccuRate outputs are:

- the room temperatures for each hour of the year;
- the total house required energy divided by the house area and so is adjusted for house size; and
- the house star rating (from 0, bad, to 10, excellent relative to the climate).

The base case house can be simulated for these results, and a retrofit action can be applied to see if any improvement has been made in terms of energy savings or of temperature control.

2.5 Present and Future Climates

2.5.1 Present Climate

The 1990 climate was provided by AccuRate's default reference meteorological year (RMY), previously known as the typical meteorological year (TMY (Lee and Snow 2008). The RMY is annual hourly data of a range of important weather parameters for 12 of the most typical meteorological months from 41 years centered on 1987 (Lee and Snow 2008). This RMY data was assumed to be close enough to the climate centered on 1990 as used in coupled atmospheric oceanic general circulation models (AO-GCMs) (Clarke and Ricketts 2010). The GCM is also known as a global climate model.

2.5.2 Climate Projections for 2050

Monthly Climate Data Changes

A single AO-GCM climate model approach was used to forecast the climate in 2050 since it is inappropriate to combine multimodel results for a detailed risk assessment (Clarke and Ricketts 2010). Furthermore, the INM-CM3.0 climate model usually was used as the most likely downscaling model from 18 AO-GCMs for Adelaide, but mid-range summer temperature changes were sometimes needed from other models (Clarke and Ricketts 2010).

The parameters of the weather data changes were averaged over a $5°$ latitude–longitude grid-square for Adelaide for the 20 years centered on 2050 (Clarke and Ricketts 2010). They were:

- the seasonal changes from 1990 to 2050 of temperature, relative humidity, solar radiation, and wind speed, which were all converted to monthly changes; and
- the monthly changes in minimum and maximum temperatures from 1990 to 2050.

Future Hourly Weather Data

Belcher's time series adjustment approach ("morphing"), was used to further downscale the AO-GCM model climate forecast, being the most appropriate method for building design (Belcher et al. 2005). The morphing approach provided annual hourly weather data for four of the 2050 AccuRate weather parameters. It adjusted the 1990 RMY annual hourly Adelaide weather parameters by adding the most likely AO-GCM monthly climate data changes from 1990 to 2050.

This approach was considered more suited to thermal building simulation than either dynamical downscaling that uses detailed, computationally expensive regional climate models; stochastic weather generation where large data sets are used to generate weather time series; or interpolation in space and time from coarse AO-GCMs. This is because of (Belcher et al. 2005):

- the reliability of present-day weather forecasts;
- the meteorological consistency of the resulting weather sequence; and
- the present-day observations being made from a real location.

The main AccuRate input parameters of hourly ambient temperature, relative humidity, solar irradiance, and wind speed were projected for 2050 using eqs. 1–5 provided in Annex 1 (Belcher et al. 2005).

2.6 Thermal Comfort

The concept of thermal comfort inside buildings has developed over many years, and is now accepted as "that condition of mind that expresses satisfaction with the thermal environment and is assessed by subjective evaluation" (ASHRAE-55 2010, 9).

The pilot study investigated two sets of thermal comfort approaches for house occupants to determine their impact during more extreme climate conditions:

- the standard Australian NatHERS thermal comfort approach that uses the "new Effective Temperature" (ET*); and
- an alternative thermal comfort approach using ASHRAE's more recent Standard Effective Temperature (SET) thermal comfort index (ASHRAE-55 2010).

2.6.1 NatHERS Comfort Approach

The NatHERS protocol contains rules about:

- room occupancy conditions;
- internal heat loads;
- occupant behavior for operating doors, windows, curtains, and fans; and
- heating and cooling thermostat settings related to climate, to room type, and to time of day.

It adopted the most appropriate measure of comfort at that time, the new effective temperature" (ET*) thermal comfort index (Delsante 2005; Auliciems and Szokolay 1997; Aynsley and Szokolay 1998; Chen 2012).

ET* is based on the parameters: environmental temperature, which depends on the indoor radiant temperature and dry-bulb air temperature; humidity; air speed from natural ventilation or ceiling fans; and clothing levels for winter and summer.

For its adaptive comfort limit, the NatHERS protocol uses Auliciems' neutral temperature—where you are neither hot nor cold—for 90 % of participants from de Dear's survey data (de Dear 1997). This takes into account acclimatization effects over the seasons. If ET* is outside this band, heating or cooling is invoked.

The ET* cooling effects are shown in Fig. 3, and are calculated from Szokolay's approximate formula (Auliciems and Szokolay 1997).

2.6.2 An Alternative Comfort Approach with SET

The pilot study used an alternative comfort approach based on ASHRAE's standard effective temperature (SET) thermal comfort index (ASHRAE-55 2010; Fountain and Hulzenga 1995; Gagge et al. 1986), and a wider neutrality condition, the "acceptable" range from Parsons (Parsons 2003).

There were three SET combinations that were checked for the pilot study:

1. light clothes and maximum ventilation of the ceiling fan, for maximum cooling before air-conditioning is required;
2. light clothes and no ceiling fan for minimum energy while staying cool; and
3. heavy clothes and no ceiling fan for maximum personal warmth before additional heating is required.

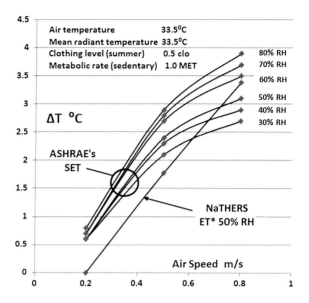

Fig. 3 The cooling effect on the human body of the SET and ET* thermal comfort indices, for air speed and relative humidity, for an air temperature of 33.5 °C. This shows the additional cooling in summer that SET provides with air speed, especially in humid conditions, compared to the NatHERS ET* values for its relative humidity (RH) of 50 %. *Source* (Based on Aynsley 2012)

The SET thermal comfort index is based on six parameters: air dry-bulb temperature; radiant temperature; humidity; air speed; the clothing level (clo); and metabolic rate (MET).

For the pilot study, the room air temperature was obtained from an AccuRate analysis without air-conditioning, and the radiant temperature was assumed equal to the air temperature. Also, the internal humidity ratio was assumed to be equal to the AccuRate external humidity ratio, and the metabolic rate was assumed to be 1.0 MET, which is that of a person in a sedentary state in the metabolic rate scale.

The other two parameters, air speed and clothing level, were varied, and this allowed the occupant to adapt his or her behavior according to the actual temperature and humidity.

ASHRAE's Thermal Comfort Tool software program (Fountain and Hulzenga 1995) was used in batch mode to calculate the SET temperatures for each retrofit action, for each scenario, for each hour of the year, and for the above three SET parameter cases. The hottest and coldest days were filtered in an Excel spreadsheet.

The SET cooling effects on the human body are shown in Fig. 3 for 33.5 °C, for an air speed of 0.15 m/s. The values were calculated in accordance with the method in Appendix F of the ASHRAE standard (ASHRAE-55 2010).

2.7 Deciding on the Methodology

The justifications for the main decisions of the methodology were:

- the retrofit embodied energy was ignored—since it was usually much smaller than the operational energy for existing houses;
- choosing the year 2050, not 2100, for analysis—to match the longevity of the large number of houses that will still be existing in 2050 that originated from the Australian housing boom after the second world war;
- using Belcher's "morphing" approach for downscaling— as most suitable to thermal building simulation (see Sect. 2.5.2);
- the INM-CM3.0 climate model as the most suitable— for Australia, at the time the calculations were carried out;
- using a predictive method based on simulation rather than an historical auditing approach—so that the retrofit impacts of energy savings and thermal comfort can be estimated during design. The historical auditing approach can only measure the impact of a retrofit after it has been implemented.
- using the Australian home energy rating program, AccuRate, rather than using other software—since it is accredited by NatHERS and was accessible;
- including a "Scarce Resources" scenario—which is a lower emissions scenario than the "Climate Change" scenario, due to possible long-term high commodity prices slowing GHG emissions;
- choosing Adelaide as the house site—since it is a temperate climate, and for its long periods of consecutive hot days; and
- using an alternative comfort approach with SET, as well as the one from NatHERS to evaluate the retrofit actions—because SET has superseded ET* in the ASHRAE standard and it promises energy savings.

3 Results

3.1 Thermal Comfort

Figure 3 compares the cooling effect on the human body for each comfort index: SET and ET*. It shows the additional one degree of cooling with SET above ET* at high values of humidity, and is discussed in Sect. 4.6.2.

ΔT is the equivalent number of degrees that the dry-bulb air temperature would need to be lowered from 33.5 °C, to achieve the same cooling effect on the human body as there is for increases in air speed at various levels of relative humidity (RH).

ΔT is calculated for the SET and ET* thermal comfort indexes, where the air and mean radiant temperatures are assumed to be 33.5 °C, with a summer clothing level of 0.5 clo and a sedentary metabolic rate of 1 MET. Air speed is varied to

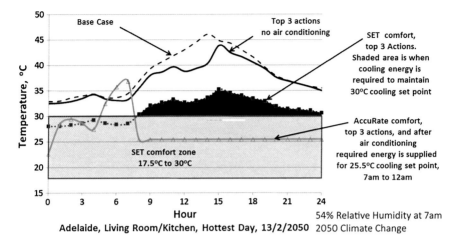

Fig. 4 Hottest day living room temperatures in 2050, with 54 % RH. The living room temperature reaches 46 °C without any modifications (base case), and after the best three actions have been applied, it still rises to 43 °C, in the 2050 Climate Change Scenario. The AccuRate comfort approach with NatHERS protocol requires that the room is cooled to 25.5 °C all waking hours, from 7 a.m. to 12 midnight. The alternative comfort approach with SET also requires cooling, but only to 30 °C with light clothing and high fan. So, the energy required to maintain occupant comfort is less than the AccuRate approach

0.8 m/s which is the maximum recommended by ASHRAE for sedentary office purposes.

The temperature of 33.5 °C is chosen because it is skin temperature, and the 30–80 % relative humidity is chosen to keep within the human comfort range.

3.2 House Base Case

The base case house, i.e., without modification, achieved a star rating of around 1.5 stars for the 1990 baseline climate.

It needs a required energy of 310 MJ/m^2/a to remain comfortable, with the NatHERS comfort approach for the 2050 Climate Change scenario.

The base case living room without air-conditioning reaches a temperature greater than 45 °C for the hottest day of the Climate Change scenario as shown in Fig. 4. It also performs poorly in the winter as Fig. 5 shows, with a temperature of around 8 °C for this scenario, without some form of heating.

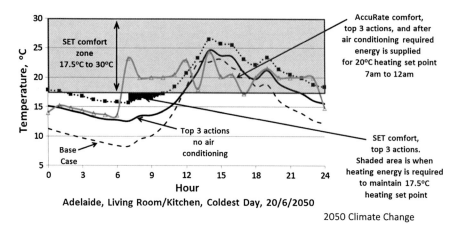

Fig. 5 Coldest day living room temperatures in 2050. The living room temperature drops to 8 °C without any modifications (base case), and after the best three actions have been applied, it still drops to 13 °C, for the 2050 Climate Change Scenario. AccuRate's NatHERS protocol requires that the living room in Adelaide is warmed to 20 °C from 7 a.m. to 12 midnight. The SET approach also requires warming, but only to 17.5 °C with heavy clothing and no fan. So the required energy of the SET approach is smaller than the Accurate approach

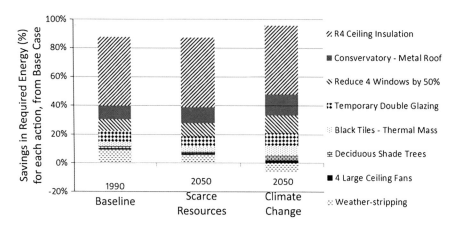

Fig. 6 Energy savings of each retrofit action for the baseline climate in 1990, and the 2050 scenarios of scarce resources and climate change. These are the savings that each of the retrofit actions provide, as a percentage of the energy required to maintain comfort (for space heating and cooling) of the unmodified house (Base Case). It uses the NatHERS comfort approach with ET*

3.3 Energy Savings: AccuRate Comfort

Figure 6 shows for each of the retrofit actions, the energy savings to be gained as a percentage of the required energy needed for the unmodified house to remain thermally comfortable. It shows these savings for the 1990 baseline climate and two scenarios of 2050.

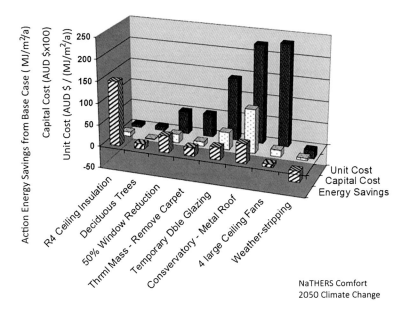

Fig. 7 Cost-effectiveness of energy-performance retrofit actions for 2050 (the unit cost). Also shown are their capital costs and their Energy Savings from the required energy required by the unmodified house (base case). Ceiling insulation is the most cost-effective action (with the lowest unit cost), followed by using deciduous trees to block the window summer sun externally. The next most cost-effective action is the 50 % single-glazed window reduction, followed by an increase in thermal mass achieved by removing the carpet, and so on

These retrofit actions are detailed in Sect. 2.4.2, and the required energy of the unmodified house (the Base Case) is based on the NatHERS comfort approach with ET*.

3.4 Cost-effective Actions for Energy Savings

The energy savings ($MJ/m^2/a$) of each retrofit action are shown in Fig. 7 for the Climate Change scenario. These savings are calculated based on the required energy of 310 $MJ/m^2/a$ to maintain comfort with the NatHERS comfort approach for the unmodified house (the base case). For example Fig. 7 shows that the savings in the required energy to maintain comfort from the base case for ceiling insulation is about 130 $MJ/m^2/a$, and lesser amounts for the other actions.

Figure 7 also shows the capital costs of the actions in AUD$ × 100, and unit cost of each action in AUD$/($MJ/m^2/a$) which is the cost per required energy savings.

3.5 Temperature Control

Figures 4 and 5 show the AccuRate predicted living room temperatures for the Climate Change scenario in 2050, for the hottest and coldest days, respectively. These figures show temperatures with and without air-conditioning. Each figure shows the temperatures for:

- the unmodified building base case (with no air-conditioning),
- the impact (with no air-conditioning) after a retrofit of the top three actions of ceiling insulation, carpet removal for added thermal mass, and shading with deciduous trees,
- the best SET cases with different clothing levels and varying ceiling fan air speeds, and its comfort approach (with air-conditioning required for the dark shaded area), and
- the AccuRate air-conditioned temperature with its stricter NaTHERS comfort approach (with the air-conditioning already invoked).

The light-shaded bands are the temperature comfort range limits for SET—17.5–30 °C for "acceptable" conditions (Parsons 2003)—and the black areas show the extent of SET time intervals requiring conditioning.

Figure 4 shows SET temperatures on the hottest day with occupants wearing a summer clothing level of 0.5 clo, which is a measure of a person's degree of clothing, and a high fan setting giving an air movement of 0.77 m/sec. The AccuRate air-conditioned temperature shows its cooling set point is lower than SET's set point.

Figure 5 shows the SET temperatures on the coldest day with occupants wearing winter clothing to a level of 1.0 clo, with no fan. AccuRate's set point is now higher than the one for SET.

4 Discussion

4.1 Base Case

The house with no modifications (the Base Case) had an energy rating of 1.5 stars, which is close to the average existing Australian house rating of around 2 stars. This is largely due to the extent of detached housing; poor insulation of the envelope, especially for ceilings and for timber floors that are not enclosed; poor orientation; high porosity; and single-glazed windows, which characterize many Australian homes.

The base case house needs a required energy of 310 MJ/m^2/a to remain comfortable, with the NatHERS comfort approach for the 2050 Climate Change scenario.

Figure 4 shows the base case living room without air-conditioning reaching a temperature greater than 45 °C for the hottest day of the Climate Change scenario. This presents a serious health concern since around 50,000 mostly elderly people died from heat stress and fires, after a week of above 40 °C temperatures during the European summer of 2003 (Brown 2012). Adelaide is renowned for its heatwaves, and will be researched further.

Figure 5 shows that additional heating is still needed for this 2050 Climate Change scenario for the base case living room in winter as the temperature reaches 8 °C overnight.

4.2 Scenarios in 2050

4.2.1 Scarce Resources

As shown in Fig. 6 with both scenarios, the retrofit actions in the Scarce Resources scenario have a lower impact than in the Climate Change scenario, from an energy savings point of view, if weather-stripping is ignored (explained further in Sect. 4.3). This is because the retrofit actions have a greater impact with larger temperature increases.

Additionally, the price volatility in the Scarce Resource scenario would be a major concern, and so it would be urgent to complete all these retrofits while scarce material and energy prices are more affordable.

4.2.2 Climate Change

Figure 6 also shows the improved energy savings of certain retrofit actions from the 1990 baseline to the Scarce Resource and even more to the Climate Change scenario. The largest improvements for the Climate Change scenario are the conservatory, reducing the area of four windows by 50 %, temporary double glazing and black tiles for added thermal mass.

Figure 4 shows the temperature effect of the top three actions which reduces the maximum living room temperature by around 3 °C on the hottest day. The same retrofit raises the lowest temperature in the living room temperature by around 5 °C on the coldest day.

4.3 Energy Analysis

As indicated in Fig. 6, the ceiling insulation of R4 K·m^2/W (R23 h·ft^2· °F/Btu) is the most effective action, accounting for almost 50 % of required energy savings from the unmodified house base case. This level of benefit agrees with other simulation research (Willrath 1998; Kordjamshidi 2011; Ren et al. 2011), as well as experimental monitoring of house temperatures for ceiling insulation (Shiel et al, 2010; Page et al. 2011).

In addition, other low-cost retrofit actions provided a combined 20 % of required energy savings. These actions were: removing the carpet and using black tiles on the concrete floor for added thermal mass; adding temporary double glazing; growing deciduous trees on the northwest wall; and adding four large 1.4 m diameter ceiling fans. Again, experimental results have agreed with the effectiveness of the carpet retrofit actions (Page et al. 2011).

Weather-stripping is very affordable, but has mixed results across the scenarios. In the 1990 baseline climate, weather-stripping reduced the required energy by 9 %. In 2050 there was only a 6 % improvement in the Scarce Resources scenario, but the extreme Climate Change scenario showed that the action needed more energy than it saved. This is because more heat is being trapped with less air escaping, especially with the large windows in a warming climate, so that more cooling is needed.

If the conservatory was added and the window size reduced, another 25 % of energy savings would be gained, although at higher cost (see Fig. 7).

4.3.1 Most Cost-effective Actions

The required energy savings of each action are shown in Fig. 7 from the total required energy needed by the unmodified house (base case).

As well as being the most effective energy-saving action, ceiling insulation is also the most cost-effective retrofit action since it has the lowest unit cost.

The next most cost-effective retrofit action is deciduous trees to block externally the window summer sun, and then the 50 % window size reduction, which lowers the heat escaping during winter, followed by removing the carpet for added thermal mass.

4.4 Temperature Analysis

On the hottest day in the Climate Change scenario (see Fig. 4), the unmodified house base case exceeds 45 °C in the living room/kitchen after the room had reached a maximum of 54 % relative humidity at 7 a.m. There is a small benefit with a lowering of temperature by the top three low-cost retrofit actions.

When the occupants adapt to conditions by wearing light clothes and operating a ceiling fan at the highest speed, the SET temperature (see Fig. 4) shows the effective benefit gained, but the 30 °C threshold is still breached and the room requires cooling with air-conditioning. This is in spite of the less stringent "acceptable" level of comfort, rather than the stricter NatHERS comfort criteria where 90 % of occupants are comfortable.

On the coldest day of the Climate Change scenario (see Fig. 5), heating will be required for SET conditions despite occupant adaptations of a winter clothing level of 1.0 clo and no fan. However, the SET conditions only need heating from 7 to 10 a.m. when the temperature is below 17.5 °C. For the NatHERS comfort approach used by AccuRate, the room needs heating above 20 °C from 7 a.m. to 12 midnight.

On each of these extreme temperature days, heating and cooling appliances will be needed for this particular house for this level of refurbishment, for both NatHERS and SET comfort approaches in 2050 for Adelaide. The additional energy used by these house appliances will produce more GHG emissions, creating a positive feedback loop and leading to more global warming.

This shows the importance of as many houses being retrofitted with as many effective actions as possible, and encouraging occupants to take an active role in adapting to the warmer climate.

4.5 Most Effective Actions

The top three actions for the temperature analysis for the Climate Change scenario as shown in Figs. 4 and 5, differed from those of the energy analysis (see Fig. 6). For temperature control they were:

- a high value of ceiling insulation of R4 K·m^2/W (R23 h·ft^2· °F/Btu);
- carpet removal for added thermal mass; and
- shading with deciduous trees.

For energy savings they were:

- a high value of ceiling insulation of R4 K·m^2/W (R23 h·ft^2· °F/Btu);
- the conservatory; and
- temporary double glazing.

The energy required to maintain comfort is different for the energy and temperature analysis because the pilot study analysis for the control of temperatures for a single room requires different actions than for the control of the whole house energy intensity.

4.6 Thermal Comfort

There are interesting issues in implementing thermal comfort approaches.

Thermal comfort is difficult to estimate because it depends partly on the internal humidity and internal air speed parameters, which both rely on the number of air changes per hour (ACH). The ACH depends on the degree of weather-stripping and the air-flow rate which changes with natural or mechanical ventilation, e.g., with window, door, or ceiling fan operation. These operations can be spatially nonuniform and highly dependent on behavior.

Also, while the clothing value (clo) appears simple, it needs to be averaged across all occupants. There could be building signs to warn occupants about appropriate dress codes and seasonal set-points, and suitable clothing could be made available, especially for commercial buildings.

The air speed and radiant temperature are assumed to be the same for occupants of the same room.

4.6.1 NATHERS Comfort Approach

AccuRate's standard input uses small envelope leakage rates compared to the larger values measured in older existing houses (MEFL 2010). This will affect the estimates of humidity and natural ventilation, and will need special modeling assumptions for older houses.

Also, ASHRAE states that the chart method used by the NatHERS ET* is not appropriate for relative humidity above 50 %, and that a computer model approach with SET should be used (ASHRAE-55 2010).

The NatHERS approach also allows the relative humidity to rise to 95 % (Delsante 2005), but there can be moisture and health problems once the humidity is above 80 %. So designers need to consider poor indoor air quality issues that can result from low values of ACH that allow higher values of humidity.

4.6.2 Alternative SET Comfort Approach

For the alternative SET comfort approach the metabolic rate would also need to be averaged across all occupants.

Also, by using Parsons' wider acceptability range, occupants would need to adapt to its lower acceptability criteria than the 90 % level in NatHERS. For example, this means that they would need to be comfortable with higher cooling set-points by changing their behavior. This could include using ceiling fans; moving to a more comfortable radiant heat location such as away from a window; or wearing different clothing, before using the air-conditioning.

For the alternative comfort approach, less required energy was needed by varying the clothing level and air speed, and by using Parsons' wider acceptability range.

4.6.3 Relative Cooling Effects

Figure 3 shows the additional cooling in summer that SET provides with air speed, especially in humid conditions, over ET*. The maximum cooling occurs at a moderate air speed of 0.5 m/s for values of relative humidity of 60 % or more, when the thermal comfort will appear around 1 °C cooler with SET than by using ET* which is used in NatHERS and AccuRate.

4.7 Thermal Modeling Complexity

Some of the innovative retrofit actions are difficult to model properly using AccuRate because:

- the thermal curtains had only a small effect in summer (AccuRate does not draw the curtains during the day to help prevent the window heat conduction);
- the weather-stripping is difficult to model since existing houses are not as airtight as newer ones, and AccuRate's standard input values are low; and
- the degree of shading is difficult to estimate for different deciduous tree species and for various foliage covers.

These difficulties will be addressed by introducing additional components into the house structure to provide a similar effect, e.g., a window covering that will screen the window during the hottest part of a summer's day, or vents that will increase air circulation.

The thermal properties that will require more research are:

- different types of thermal curtains;
- temporary double glazing plastic films; and
- the degree of tree shading.

The calculation of SET is a lengthy procedure. However, SET has been adopted by ASHRAE as the standard and there are computer programs to assist.

4.8 Reliability of Approach

4.8.1 AccuRate Reliability

The goal of NatHERS-accredited applications is to rate houses based on standardized weather data, heat load assumptions, and occupancy patterns, and these are not likely to match those of any actual house at any point in time.

There have been many studies conducted of monitoring house temperatures and their energy consumption compared to the AccuRate simulations. Some have used actual weather data, realistic house component modeling, actual internal heat loadings and occupancy patterns to better suit reality, and converted the required energy into energy consumed.

Of the comparisons that have been conducted, the thermal modeling experience of the assessor has proved to be of paramount importance. This is because there are many sources of error on assumptions for the numerous inputs required to model the house correctly (Szokolay 2004; Saman et al. 2008; Williamson et al. 2010).

Errors may be in the actual weather data; in the assumptions about wing walls and shading; in the determination of the air-tightness; in selecting occupancy levels; in setting thermostat set points; and in the thermal properties of the building, e.g., timber walls have a framing factor where the timber studs affect the thermal properties.

For weather data, AccuRate's standard RMY data is often from the nearest airport, which can under-estimate the urban temperature which can be higher due to urban heat island effects. It may also over-estimate the wind effects in the urban location. For actual weather data, hourly solar radiation is important for building simulation but not readily available, and it is also difficult to obtain precise humidity data.

Also, any error in the AccuRate implementation is unknown since the software source code is proprietary, and the software code is not available for independent analysis.

However, AccuRate has been validated using the BESTEST protocol, and the AccuRate output of internal temperature data and house energy across many houses are approximately correct (Delsante 2005; Miller and Buys 2010; Copper and Sproule 2011; Williamson et al. 2010). It should be noted that BESTEST tests the simulation of one building in one location.

4.8.2 Reliability of the Climate Projection

The Belcher morphing approach used here was verified using heating degree days calculated from a UK weather series "morphed" to a future climate using a UK climate model. The results agreed well with those directly based on output from a climate model (Belcher et al. 2005). It has also been used in other climate change building simulation studies (Ren et al. 2011; Chen et al. 2012).

The future climate parameters for the Adelaide pilot study projection were supplied by the INM-CM3.0 climate model. Since then, better climate models have been found for the Pacific (Irving et al. 2011), and a more appropriate model will be considered for future research.

In projecting the future climate, the main four weather parameters of temperature, humidity, radiation, and wind speed were used in Belcher's "morphing" approach (Belcher et al. 2005), and the time intervals were their monthly mean values. Using all these parameters and the monthly time interval provide a reliable estimate of the future climate when estimating a building's heating and cooling energy requirements (Chen et al. 2012).

4.9 Key Findings

The key findings are that:

- innovative, low-cost retrofit actions could reduce the house required energy for comfort, e.g., external shading with deciduous trees to lower summer radiation from northern windows (southern windows in the northern hemisphere), as well as carpet removal from a concrete floor for added thermal mass;

- an alternative comfort approach that uses ASHRAE's SET reduces the required energy in 2050 compared to the standard NatHERS comfort approach; and
- in the 2050 Scarce Resources scenario, more retrofit actions are needed than in the more extreme Climate Change scenario for the same impact, and they are more urgent due to the rising costs of scarce resources.

5 Conclusion

Innovative, low-cost retrofit actions can be found that lower housing envelope energy consumption in 2050 for a temperate climate, if they are suited to the house construction and site. The significant actions include adding ceiling insulation; removing the carpet for added thermal mass; shading externally with deciduous trees; adding a conservatory to reorient the living space toward the sun; and adding temporary double glazing.

Furthermore, an alternative comfort approach with ASHRAE's SET comfort index and a different acceptability condition can give occupants more control to lower their energy usage than with the current NatHERS approach.

In the more extreme Climate Change scenario the climatic conditions projected for 2050 will require much adaptation. However, if the Scarce Resources scenario eventuates, less abatement is required but the urgency of residential retrofits is higher due to rising costs.

Learnings from the pilot study are that (1) a larger range of low-cost retrofit actions should be investigated; (2) the updated AccuRate sustainability version should be used; (3) a more appropriate climate model is needed than INM-CM3.0 for Adelaide; (4) savings in operating cost as well as capital costs need to be included; (5) a smaller SET temperature band for comfort should be investigated; (6) more detailed modeling and better computer techniques are needed for complex thermal modeling issues; and (7) the more recent scenario approach of the IPCC will be considered if possible.

Acknowledgments The authors are grateful for the thoughtful advice from the anonymous reviewers.

They would also like to acknowledge the assistance of Trevor Moffiett and Dariusz Alterman from the University of Newcastle; EnviroSustain; Dong Chen, John Clarke, Leanne Webb and Jack Katzfey from the Commonwealth Scientific and Industrial Research Organization (CSIRO); and Karl Braganza, Alex Evans and Perry Wiles from the Bureau of Meteorology (BoM).

Annex 1: Future Climate Morphing Calculation

The main weather parameters of temperature, humidity, radiation, and wind speed are projected for Adelaide for 2050 using Belcher's morphing approach (Belcher et al. 2005). It creates an RMY set of annual hourly weather parameters for 2050 based on an Adelaide 1990 RMY set of parameters, and the monthly changes in parameters from 1990 to 2050 of the INM-CM3.0 climate model (Clarke and Ricketts 2010).

$$T = T_0 + \Delta T_m + \alpha_{Tm}(T_0 - \bar{T}_{0m}) \quad (1)$$

$$\text{where } \alpha_{Tm} = \frac{\Delta T_{MAXm} - \Delta T_{MINm}}{\bar{T}_{0MAXm} - \bar{T}_{0MINm}} \quad (2)$$

$$RH = RH_0(1 + \alpha_{RHm}) \quad (3)$$

$$I = I_0(1 + \alpha_{Im}) \quad (4)$$

$$WS = WS_0(1 + \alpha_{WSm}) \quad (5)$$

where

- T and T_0 are the future and present hourly ambient dry-bulb temperatures, respectively,
- $\bar{T}_{0m}, \bar{T}_{0MAXm}$ and $\bar{T}_{0MINm}$ are the monthly mean values of the ambient dry-bulb temperature, the daily maximum temperature and the daily minimum temperature, respectively, for hourly values calculated over all the averaging years to make up the baseline climate. In this case, the RMY set of one year hourly temperatures represents that base 1990 climate,
- $\Delta T_m, \Delta T_{MAXm}$ and ΔT_{MINm} are the changes projected for each month by the AO-GCMs for the mean temperature, daily maximum temperature and the daily minimum temperature of the dry-bulb temperature, respectively,
- RH and RH_o are the future and the present-day values of the relative humidity, respectively and α_{RHm} is the AO-GCM projected fractional change in the monthly mean relative humidity,
- I and I_o denote the future and the present-day solar irradiance and α_{Im} represents the AO-GCM projected fractional change in the monthly mean solar irradiance,
- WS and WS_o are the future and the present-day wind speeds respectively and α_{WSm} is the AO-GCM projected fractional change in the monthly mean wind speed.

Reference

Aleklett K (2010) Peak oil—an end to economic growth? Canberra A.C.T., Australia. http://www.aspo-australia.org.au/References/Aleklett/20090609Canberra1.pdf Accessed 30 Apr 2012

ASHRAE-55 (2010) ANSI/ASHRAE Standard 55-2010—Thermal Environmental Conditions for Human Occupancy. ASHRAE, Atlanta, GA, USA

Aynsley R (2012) Benefit of SET at high temperatures and humidity. Private Correspondence

Aynsley RM, Szokolay SV (1998) Options for Assessment of Thermal Comfort/Discomfort for Aggregation into NatHERS Star Ratings. James Cook University, Townsville, QLD, Australia

Auliciems A, Szokolay SV (1997) Thermal comfort. Design tools and techniques; note 3. PLEA notes. PLEA in association with Department of Architecture, University of Queensland, Brisbane, QLD, Australia

Belcher S, Hacker J, Powell D (2005) Constructing design weather data for future climates. Building Serv Eng Res Technol 26(1):49–61

Bol D (2011) Material Scarcity and its effects on Energy Solutions. ASPO 9 Conference presentation. Brussels, Belgium

Brown LR (2012) Plan B Updates–56: Setting the Record Straight–More than 52,000 Europeans Died from Heat in Summer 2003. http://www.earth-policy.org/plan_b_updates/2006/update56 . Accessed 30 Apr 2012

Campbell C (2002) Peak oil: an outlook on crude oil depletion. http://www.greatchange.org/ov-campbell.outlook.html Accessed 24 Sep 2010

Chen Z (2008) The Latest in Software Innovation—Update on AccuRate Development presented at the ABSA National Conference, Melbourne, Australia

Chen D (2012) Parameters and Settings in AccuRate. Personal Correspondence

Chen D, Xiaoming W, Zhengen R (2012) Selection of climatic variables and time scales for future weather preparation in building heating and cooling energy predictions. Energy and Buildings 51:223–233

Clarke J, Ricketts J (2010) Future Climate Data for Building Regulation Energy Impact Assessment. CSIRO, Aspendale, VIC, Australia

Cleugh H, Stafford Smith M, Battaglia M, Graham P (eds) (2011) Climate change—science and solutions for Australia. CSIRO Publishing, Collingwood, VIC, Australia

Copper JK, Sproule AB (2011) Simulated and Measured Performance of an 8 star Rated House in Sydney. In: Proceedings of 49th Annual Conference of the Australian Solar Energy Society (AuSES), available from http://auses.org.au/, Sydney NSW

Curtis F (2009) Peak globalization: Climate change, oil depletion and global trade. Ecol Econ 69(2):427–434

DCCEE (2011) Nationwide House Energy Rating Scheme (NatHERS)—Administrative and Governance Arrangements. Deptartment of Climate Change and Energy Efficiency, Canberra, A.C.T., Australia

de Dear R (1997) Developing an adaptive model for predicting thermal comfort and preference. ASHRAE sponsored research report RP 884, Macquarie Research Limited, Macquarie University, Sydney, NSW, Australia

Delsante A (2005) Is the New Generation of Building Energy Rating Software Up to the Task? A Review of AccuRate. In: ABCB Conference Building Australia's Future, Surfers Paradise, QLD, Australia

Fountain M, Hulzenga C (1995) A thermal sensation prediction model for use by the engineering profession. ASHRAE, Peidmont, CA, USA

Frei C (2012) Global and regional issues: The energy challenges for the future. World Energy Insight, 2012, World Energy Council, Istanbul, Turkey

Gagge AP, Fobelets AP, Berglund LG (1986) A standard predictive index of human response to the thermal environment. ASHRAE Trans 92(2B):709–731

Garnaut R (2011) The Garnaut review 2011: Australia in the global response to climate change. Commonwealth of Australia, Creative Commons. Cambridge University Press, New York, NY, USA

Ginley DS, Cahen D (2011) Fundamentals of Materials for Energy and Environmental Sustainability. Cambridge University Press, Cambridge, UK

Hall CAS, Klitgaard KA (2011) Energy and the wealth of nations: understanding the biophysical economy. Springer, New York, NY, USA

Hearne (2011) AccuRate—Hearne scientific software. http://www.hearne.com.au/products/accurate/ Accessed 21 Sep 2011

Heinberg R (2007) Peak everything: waking up to the century of declines. New Society Publishers, Canada

Hirsch RL, Bezdek RH, Wendling RM (2010) The impending world energy mess. Apogee Prime, Ontario, Canada

IEA and OECD (2011) World energy outlook 2011. International energy agency and organisation for economic co-operation and development, Paris, France

Irving D, Perkins S, Brown J, Sen Gupta A, Moise A, Murphy B, Muir L, et al (2011) Evaluating global climate models for the pacific island region. Clim Res 49(3):169–187

IPCC (2008) Climate Change 2007: Synthesis Report. In: Bernstein L, Pachauri RK, Reisinger A (eds) Intergovernmental Panel on Climate Change. Geneva, Switzerland. http://www.ipcc.ch/pdf/assessment-report/ar4/syr/ar4_syr.pdf. Accessed 21 Apr 2013

Kordjamshidi M (2011) House rating schemes from energy to comfort base. Springer, Berlin, Germany

Lee T (2009) Simulating the impact of changing climate - creation and application of ersatz future weather data files. In: Proceedings of the 47th ANZSES Annual Conference, ANZSES, Townsville, QLD, Australia

Lee T, Snow M (2008) "Australasian climate data bank projects. In: Proceedings of the ISES-AP—3rd international solar energy society conference – Asia Pacific Region, International Solar Energy Society, Sydney, Australia

Lehmann S (2010) The principles of green urbanism : transforming the city for sustainability. Earthscan, Washington, DC, USA

Martin W (2012) Peak oil and peak everything. Cornell University Lecture, Ithaca, NY, USA. http://www.youtube.com/watch?v=0HeEHKJxSA8 Accessed 21 Dec 2012

McKinsey & Company (2011) Resource revolution: meeting the world's energy, materials, food, and water needs. McKinsey Global Institute, Seoul, South Korea

MEFL (2010) On-ground assessment of the energy efficiency potential of victorian home. Moreland Energy Foundation, Sustainability Victoria, Melbourne, Australia

Miller W, Buys L (2010) Post-occupancy analysis of a sub-tropical Zero Energy Home (ZEH). In: Proceedings of 48th Annual Conference of the Australian Solar Energy Society (AuSES), available from http://auses.org.au/, Canberra, A.C.T., Australia

Moss RH, Edmonds JA, Hibbard KA, Manning MR, Rose SK, van Vuuren DP, Carter TR, et al (2010) The next generation of scenarios for climate change research and assessment. Nature 463(7282):747–756. http://www.nature.com/nature/journal/v463/n7282/full/nature08823.html Accessed 21 Apr 2013

Page AW, Moghtaderi B, Alterman D, Hands S (2011) A study of the thermal performance of Australian housing. Priority Research Centre for Energy, University of Newcastle, Callaghan, N.S.W., Australia. http://www.thinkbrick.com.au/download.php?link=assets/Uploads/TB-PHASEI-REPORT-FINAL-web.pdf&name=TB-PHASEI-REPORT-FINAL-web.pdf. Accessed 21 Apr 2013

Parsons KC (2003) Human thermal environments: the effects of hot, moderate, and cold environments on human health, comfort, and performance, Taylor & Fransis, London, UK

Pauli G (2010) Blue economy-10 years, 100 innovations, 100 million jobs. Paradigm Publications, Taos, NM, USA

Rahmstorf S, Cazenave A, Church JA, Hansen JE, Keeling RF, Parker DE, Somerville RCJ (2007) Recent climate observations compared to projections. Science 316:709

Ren Z, Chen Z, Wang X (2011) Climate change adaptation pathways for Australian residential buildings. Build Environ 46:2398–2412

Robert A, Kummert M (2012) Designing net-zero energy buildings for the future climate, not for the past. Build Environ 55:150–158. doi:10.1016/j.buildenv.2011.12.014

Saman W, Oliphant M, Mudge L, Halawa E (2008) Study of the effect of temperature settings on AccuRate cooling energy requirements and comparison with monitored data. Residential Building Sustainability, Department of Environment, Heritage and the Arts, Commonwealth of Australia, Australia

Shiel J, Lehmann S, Mackee J (2009) Reducing Greenhouse Gases in Existing Tropical Cities. In: CD Proceedings of iNTA-SEGA 2009, Kasetsart University, Bangkok, Thailand

Shiel J, Lehmann S, Mackee J (2010) A method for practical zero carbon refurbishments: a residential case study. In: Proceedings of 48th Annual Conference of the Australian Solar Energy Society (AuSES), Canberra, A.C.T., Australia

Urge-Vorsatz D, Danny Harvey L, Mirasgedis S, Levine M (2007) Mitigating CO2 Emissions from Energy Use in the World's Buildings. Build Res Inf 35(4):379–398

Szokolay SV (2004) Final report on the evaluation of the computer program AccuRate—the version of NatHERS updated to include natural ventilation in warm climates. Office of Energy and Environmental Protection Agency, Qld Government, Queensland, Australia

Vivoda V (2011) Evaluating energy security in the Asia-Pacific region: a novel methodological approach. Research seminar. Gold Coast, Queensland, Australia

Williamson T, Soebarto V, Radford A (2010) Comfort and energy use in five australian award-winning houses: regulated, measured and perceived. Build Res Inf 38(5):509–529

Willrath H (1998) The thermal performance of houses in australian climates. Dissertation, Available from the University of Queensland, Queensland

Part V
Concluding Chapter

Energy and Meteorology: Partnership for the Future

Don Gunasekera, Alberto Troccoli and Mohammed S. Boulahya

Abstract This concluding chapter draws on the main aspects covered in this book, such as the discussions on the increasing reliance of the energy sector on meteorological information. We then describe current and potential funding models of National Meteorological and Hydrological Services. These are the main, though not the only, providers of meteorological information for energy and all the other sectors affected by meteorological phenomena. It emerges that public sector funding for such Services are dwindling. This is in spite of the recognised impacts that meteorology has on the energy industry, and on other sectors. Some lessons from the important interaction between aviation and meteorology are discussed with a view to drawing some parallels with energy. We then discuss possible options for strengthening the relationship between energy and meteorology in order for society to be better prepared for the increasing vulnerability of the energy sector to the vagaries of weather and climate.

1 Introduction

There is increasing recognition that energy sector issues need to be viewed in the context of water, food and climate-related imperatives (Lior 2012; PMSEIC 2010). This view is underpinned by the ongoing impacts of growths in population and income on the energy sector and the sector's vulnerability to climate change (Schaeffer et al. 2012, "A New Era for Energy and Meteorology" by Ronalds et al.). The broader socio-economic-environmental context within which energy sector developments need to be observed are illustrated in Table 1.

D. Gunasekera (✉) · A. Troccoli
Commonwealth Scientific and Industrial Research Organisation, Canberra, ACT, Australia
e-mail: Don.Gunasekera@csiro.au

M. S. Boulahya
Climate for Development in Africa, Tunis, Tunisia

Table 1 Some key data during the period 2006–2010 (After Lior 2012)

Item	Global amount
Total primary energy use (2010)	502 EJ (476 × 10^{15} Btu)
Industry	30 %
Transportation	29 %
Residential	22 %
Commercial	19 %
Electricity	40 %
Electric power installed (2008)	4.4 TWe
Electricity generated per year (2010)	21.3 PWh
People without electricity (2009)	1.44 billion
Global temperature change—industrial period	+0.76 °C
Global temperature change—2006–2010 average	−0.04 °C
Water shortages	884 million people lack safe drinking water, 2.5 billion people have inadequate access to water for sanitation and waste disposal, ground water depletion harms agriculture
Food shortages	925 million undernourished people (1 in 7)

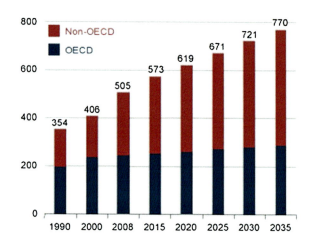

Fig. 1 World energy consumption, 1990–2035 (in 10^{15} Btu). *Source* USEIA (2011)

According to the US Energy Information Administration (2011) world energy consumption is estimated to grow by 53 % from 2008 to 2035. Total world energy use is projected to increase from 505 × 10^{15} British thermal units (Btu) in 2008 to 619 × 10^{15} Btu in 2020 and 770 × 10^{15} Btu in 2035 (Fig. 1). Much of the growth in energy consumption is expected to take place in non-OECD nations, where demand is driven by long-term population and economic growth (see also "A New Era for Energy and Meteorology" by Ronalds et al (2014)).

The supply side of the energy sector over the coming decades is likely to be affected by a range of factors, including near constant reserve-to-production ratios for oil (0.4), gas (0.6) and coal (1.2), high availability of tar sands and oil shales, growth in renewables, and continuing R&D into carbon capture and storage technologies (Lior 2012). It is important to recognise that fossil fuel generated energy is still the major energy provider, especially for electricity, and will continue to be so in the foreseeable future. By comparison, energy production through renewables is relatively small in output and in share but is gradually being expanded across many countries.

The linkages between weather, climate and energy systems are diverse. Energy producers, and in particular those in the electricity sector, are increasingly using weather and climate information. This applies to both energy generation and distribution ("Improving Resilience Challenges and Linkages of the Energy Industry in Changing Climate" by Majithia (2014), "Weather and climate impacts on Australia's National Electricity Market (NEM)" by George et al. (2014), "Weather and climate and the power sector: Needs, recent developments and challenges" by Dubus (2014)). Hence, on the supply side, for example, changes in the frequency and intensity of disruptive weather events could have important impacts on energy systems. Furthermore, use of forecasts of winds and solar radiation plays an important role in the management of power generation.

Weather and climate information is also sought to assist in energy demand management (Fischer 2010, "Reducing the energy consumption of existing, residential buildings, for climate change and scarce resource scenarios in 2050" by Shiel et al. (2014)). Hence, on the demand side, the accessibility of tailored forecasts of temperature, humidity, precipitation, wind speeds and cloud cover could influence energy demand management for heating, cooling, and lighting, and the timing of particular industrial and manufacturing processes that depend on certain weather conditions (Marquis 2011).

Across many countries different models exist for the supply (generation, transmission, and distribution) of energy to customers. Thus, energy suppliers can be publicly owned, corporatized or privately owned. Similarly, meteorological information including weather and climate information generally comes from government funded and operated National Meteorological and Hydrological Services[1] (NMHSs) or privatised or semi-privatised service providers.

The linkages between weather, climate and energy systems highlight the importance of harnessing the use of meteorological information and services in energy systems, both on the supply side and demand side. It is important to recognise that many energy utilities, particularly in developed countries, are currently working closely with weather and climate service providers (both public and private, e.g. in France EDF works closely with Météo France).

The notion of effective use of meteorological information and services in energy systems raises a range of issues. Given the dependency of weather- and climate-sensitive aspects of the energy sector on information provided by meteorological service providers, are there opportunities for greater cooperation between these two groupings? Will partnerships between meteorological and

---

[1] Sometimes two related denominations—National Meteorological Service (NMS) or National Hydrological Service (NHS)—are also used, even if NMHSs should normally be preferred.

energy sectors lead to increased efficiency and effectiveness in the latter sector, with higher overall economic and commercial benefits? Can a win-win situation be created that fulfils the meteorological sector's responsibilities in meeting the needs of the energy sector? Although benefits for the energy sector may be obvious, what would be the benefit for the meteorological sector? Could the benefit be the additional income from delivering custom-made weather and climate information for the energy sector?

The objective of this chapter is to spell out the current and prospective linkages between the energy sector and meteorology service providers in order to further enhance the efficient supply and use of energy in an environment where demand is growing and is challenged by climate change related concerns. This is why it is important to understand options for service provision of meteorological information; while considering this issue in the next section we will include a brief historical background.

2 Current Forms of Meteorological Service Provision

Governments across many countries are seeking ways to improve process efficiency and to reduce the cost of public services. Strategies have ranged from the outsourcing of public services to the transfer of public sector functions to more independent government entities that operate at arm's length from their originating departments. Services associated with public safety have remained largely within the control of government departments. One exception has been the NMHSs, which despite having internationally-recognised responsibilities for the protection of lives and livelihoods, have become targets for quasi-commercialization in many countries (Rogers and Tsirkunov 2011).

2.1 Basic and Special Meteorological Services

On the basis of the major user groups and key product/service characteristics, the meteorological information provided by the NMHSs across many countries can be categorised into two broad groups: basic meteorological services and special meteorological services.

The basic services are those made freely available to the broader community generally through the mass media (including the world-wide web), in the public interest. Examples include public weather forecasts and severe weather warnings. In general, the basic meteorological services constitute what, in economic terms, are referred to as public goods. Use of these services by one user does not reduce their availability to others. Furthermore, once the service is available, it is available to all and it would be impossible or costly to exclude potential users from accessing the service and using it for their own benefit. Nor could it be efficient to charge for the basic services since the marginal cost of serving an additional user is close to zero. Basic services with public good characteristics currently account for between 70 and 90 % of the total volume of meteorological information provided by the NMHSs across countries.

Special meteorological services include the provision of value added services to meet the specific needs of particular users. Examples include weather services tailored to the needs of remote offshore oil and gas operators, and value added services to meet specific needs of users such as certain energy generators and distributors and energy regulators (e.g. Australian Energy Market Operator [AEMO]).

Special meteorological services, normally provided to a single user or a small collection of users combining as a group, are regarded as private goods. There are some value added specialised meteorological services (for example, specialised meteorological information provided to specific airlines in support of their particular operations) which have mixed goods properties of non-rival consumption and low cost of exclusion (see Zillman and Freebairn (2001) for a detailed discussion of the public/private/mixed good characteristics of meteorological services).

2.2 Alternative Service Delivery Models

There has been a trend in many countries to find ways to reduce the costs of providing meteorological services as a part of an overall strategy to reduce public spending. Common approaches have been to transfer NMHSs from government departments to more independent public sector agencies or to outsource the provision of meteorological services to the private sector (Rogers and Tsirkunov 2011). It could be argued that it is economical for the NMHS to provide the public good component of meteorological services, including the basic infrastructure with full government funding.

Despite the increasing pressure to reduce government expenditure in general, public sector NMHSs continue to provide meteorological services in virtually all countries. However, a number of countries (such as New Zealand and some European countries such as the United Kingdom and France) have decided to purchase these services through a purchaser-provider system. There are two separate but interrelated issues here. The first issue relates to the funding of the public good component of meteorological services, and here the government is usually the answer. The second issue relates to the actual supply of services. This may come from a public provider (the NMHS), a subsidised private provider or a supplier chosen through tender.

In looking at alternative service delivery types, it is convenient to assume that the overall services fall into three categories: basic infrastructure or basic systems; basic service; and special services. Several alternative types can be identified. In all different types it is assumed that the basic infrastructure and basic systems (including a large part of the research) are funded by the government (that is all taxpayers bear the infrastructure costs). However, depending on the degree of cost recovery and charging policy, it is possible to ensure that some of these costs are recouped from specific users.

Rogers and Tsirkunov (2011) have discussed four interrelated alternative service models for the management and delivery of hydro-meteorological services. In general, many NMHSs combine some of the elements of each adapted to their unique circumstances.

Publicly funded NMHSs (through budget appropriation)—These are NMHSs that provide 'public good' meteorological and hydrological services, including the basic infrastructure and general forecasts and warnings as a service funded through general taxation. The United States National Weather Service (NWS) is a good example here, which is budget funded to provide all of its products and services at public expense for the public good. This model has also helped to develop a relatively strong private sector that is not in direct competition with the publicly funded NMHS, thus maintaining a balance between the public sector's role in providing basic information for public safety and economic security, and sustaining specialized, value-added private-sector services (as in the US). Such an approach allows a distinction between public goods, private goods and mixed goods.

In some countries, the publicly funded NMHSs have been able to provide special services through a 'ring fenced' or structurally separated entity complying with competition policy and competitive neutrality requirements. The Australian Bureau of Meteorology adopted such an approach for some years although, in recent times, it has followed a more integrated approach with many parts of the Bureau providing both general public services and tailored forecast services to support operations and strategic planning in weather-sensitive industries such as mining (both onshore and offshore), energy production, construction, tourism, agriculture, and shipping using state-of-the-art numerical weather prediction models and a variety of observational data.

Given the strong public good characteristics of many of the meteorological services, there is a strong case for maintaining publicly funded NMHSs in general. It is important to recognise that in many developing countries the publicly funded NMHSs are the main source of meteorological services, despite declining funding for such services in these countries.

Government Business Enterprises—Several countries have corporatized some of their public sector service delivery agencies to reduce costs and improve service delivery and accountability. Meteorological services have also fallen into this category. Consequently, there are NMHSs operating in the form of a Government Business Enterprise/corporatized entity in some countries. These entities retain the responsibility of the 'publicly funded NMHS model' to provide a public service. However, they are required to compete for the additional funds needed to run the entire service, usually through short-term contracts to address the specific needs of a government department or state-owned enterprise. Potentially this could influence the effective delivery of some of the public good meteorological services if too much attention is paid to attracting additional funds though contract work.

The government purchases the basic services from the corporatized NMHS through a purchaser-provider framework. Special services are paid for by the clients and sold on the free market. The corporatized NMHS competes with the private sector firms in providing special services in a level playing field by complying with competition policy and competitive neutrality requirements. Many of the corporatized NMHSs are in developed countries.

Fully commercialised meteorological service providers—They are similar to the publicly funded entities except that all the special services are provided by the private sector and are paid for by the clients and sold on the free market. The NMHS does not provide any special services. Essentially, this form is a subset of the previous form.

These private sector service providers focus predominantly on customised services and not on meteorological services that have strong public good characteristics.

Public-Private Partnership (PPP) or Build-Operate-Transfer (BOT)—Under this model NMHSs partner with the commercial sector to create an entity, which can serve as a means to broker information and data exchange between the NMHSs and the private sector. It may also be a conduit for acquisition of observing stations to strengthen national networks while enhancing observations of highest value to the commercial sector.

PPP and BOT approach to service delivery is common in infrastructure services such as roads and railway networks. But their large-scale applicability to meteorological services is still at a very early stage of development.

3 Challenges Faced by the National Meteorological and Hydrological Services

Although there is a growing demand to reduce weather, climate and water related risks to the public, key economic sectors and national economies, many NMHSs face substantial institutional challenges. These include coping with limited and declining public sector investment, retaining qualified staff in adequate numbers, implementing and sustaining new forms of technologies essential to delivering the meteorological services that users demand, and competition from the private sector service providers.

In many developing country NMHSs the capacity to deliver the expected level of services is quite limited. The consequent information gaps are sometimes partially filled by other agencies, the private sector and academia. However, the approach is often disorganized and disaggregated. This leads to ill-informed decisions and strained relationships between the various groups, each trying to support the need for better weather and climate services using diverse approaches (Rogers and Tsirkunov 2011).

The dominant role of NMHSs as national weather, climate and hydrological service providers can lead them to take advantage of their monopoly on access to weather, climate and hydrological data to limit activities and opportunities for other groups to develop additional and innovative products and services. This is a major source of tension between the NMHSs and various other participants in the meteorology/hydrology sector, particularly in developing countries.

Another key challenge facing NMHSs, particularly in developing countries is the limited ongoing financial and human resources for the continuing sustainability of operations, maintenance and service delivery.

Historically, NMHSs have focussed on the production and dissemination of meteorological and hydrological products and services from a supply-driven perspective. However, there is an increasing need to focus on their services and delivery from a demand-driven perspective. This emphasis on demand-driven services is a departure from the traditional role of many NMHSs. Under a demand-driven service provision perspective there is an increasing need to understand how meteorological and related products and services are used and what can be done to increase the benefit to the user. In other words it is no longer sufficient to produce a good forecast of severe weather, for example, for the electricity sector, but the forecast must be used properly and the benefit in terms of improved safety and security of, say, the power lines must be realized (see also "In Search of the Best Possible Weather Forecast for the Energy Industry" by Mailier et al. (2014)). This is critical to developing an effective operating model for the meteorological and hydrological service providers across many regions.

It is important to recognise that implicit in the provision of highly demand-driven and impact-specific meteorological services is that many of these are services for specific economic sectors with a clear commercial value to that sector. In this context balancing the provision of public services and potential commercial services of the NMHSs is a challenge.

3.1 Relevance of Alternative Models of Meteorology Services to the Energy Sector

It is important to recognise that most weather- and climate-sensitive businesses such as the energy utilities do not want to become private weather service providers. Hence there is significant scope for increased opportunities for cooperation between the NMHSs and other meteorological service providers such as the private sector to find a cost-effective and mutually beneficial solution to provide more effective services to the energy sector.

In the past, NMHSs collected nearly all of the weather, climate and hydrological data and information and ran nearly all of the weather and climate models. At present, local agencies other than NMHSs, universities and private firms can deploy their own meteorological and hydrological instruments, and some run their own weather and climate models or models developed by others, and provide meteorological and hydrological services to users. These advances and innovations will continue into the future and will further influence the combined role of public and private sector meteorological service provision.

In the absence of government supported networks, the private sector and academia are creating their own private observing networks and using and sharing meteorological data. These separate networks may eventually provide more capacity to the national observing network than the NMHSs could provide on their own. Instead of pursuing an adversarial position, which is the case in some instances, it is important for both the government services and private sector meteorological service providers to work together, for the benefit of both groups and final consumers including the energy sector. A key motivation for the service providers to work together is to take advantage of the potential economies of size and scale associated with meteorological infrastructure and observation networks and the related R&D.

Given the long history and some similarities, the experience between aviation and meteorology may represent an exemplar from which energy and meteorology could draw lessons for a strengthened collaboration. In the majority of countries, one of the main tasks of NMHSs is the provision of services to the aviation industry. This is because meteorological information is critical to the safety, regularity and efficiency of aviation planning and operations. By their nature, air traffic services, aeronautical tele-communications and search and rescue, which heavily rely on meteorological information, also have a cross-border connotation. To formalize the provision for serving international civil aviation, the International Civil Aviation Organization (ICAO) was established in 1947 and funded by international aviation usually through the collection of fees for landing at airports in a State (landing fees) and fees charged for overflying a State's territory (en-route charges) (WMO 2007). Although probably not to the extent of aviation, there are many cases also in the energy industry in which cross borders issues emerge. Think of the gas pipelines connecting Russia and Ukraine with Western Europe, or the grid being built in the North Sea to transport large amount of electricity generated by say wind power.

4 Energy Services: Current and Future Trends

With the deregulation of electricity markets across many countries since the 1990s, it became crucial for the power generators and distributors to operate more efficiently in a competitive market environment. In that environment, consumers have a choice as to where they purchased their electricity and their selection was primarily based on price. Hence, electricity companies were motivated to operate more efficiently. If they had excess electricity it was more cost effective if they sold it to another region that needed it. The best electricity trades occurred when the weather situation was anticipated well in advance, using the best available forecasts, and electricity "futures" were purchased at the lowest possible price.

That same electricity could be sold to the weather -impacted region at a higher price when competing energy companies supplying that region realised they did not have the capacity to meet their customers' demand (Pirone 2007).

Furthermore, energy trading expanded beyond just electricity to natural gas and oil. The continuing volatility of these markets has attracted many participants in addition to the energy producers themselves. The growth and sustainability of energy trading across many countries is also influenced by the meteorology sector to generate the relevant weather and climate products and services that support its activities.

Some energy companies in several developed countries also have 'industrial meteorologists' on their staff involved in applied research, product and software development and management. These meteorologists directly cater for the needs of the parent companies' core business of energy generation and distribution (Dubus 2010; "Combining Meteorological and Electrical Engineering Expertise to Solve Energy Management Problems" by Pirovano et al. (2014), "Weather and climate and the power sector: Needs, recent developments and challenges" by Dubus (2014)). The role of the industrial meteorologists has become easier in recent years due to several factors (see Pirone 2007). First, the cost of accessing public sector generated meteorological data has fallen due to wide spread availability of such data via the internet and NMHS web servers. Second, software processing tools and computer equipment for generating weather and climate products required for specific businesses including the energy sector are no longer cost prohibitive. Third, there are new technologies being developed to deliver the value added meteorological information more efficiently and effectively to end users at low cost (e.g. "A Probabilistic View of Weather, Climate, and the Energy Industry" by Dutton et al. (2014)).

5 Energy and Meteorology Interaction

Energy is one of the key economic sectors dependent upon meteorological services for provision of critical information contributing to decision-making on all time scales. For example, use of weather and climate information in the energy sector varies from the shortest pertinent to generation and delivery of energy on a minute-to-minute basis, to the longest pertinent to decadal and longer-term planning for plant development and energy security. Economic and population growth is changing the requirements for energy. For example, electrical power generation, which is fundamental for economic security and sustainable development, is vulnerable to a range of weather and climate hazards including changing rainfall patterns affecting hydropower generation, availability of wind energy, solar and the impact of extreme heat and cold affecting network stability and demand (see also Johnston et al. 2012, in which some practical engineering and non-engineering climate risk management options are discussed).

NMHSs are being called upon to provide an ever-widening range of products and services with increasing quality to satisfy developing demands in numerous sectors including the energy sector. But, as has been discussed above, in many countries the funding available to NMHSs is failing to keep abreast with demands, and in many cases is falling, especially in the developing world.

As has been illustrated by different sub-sectors of the energy industry, innovative forms of meteorological services are needed at present and in the near future. The growing importance of energy supply and demand management in the presence of climate variability and change and projected longer term expansion in non-fossil fuel energy sources such as wind and solar requires greater emphasis on the delivery of innovative weather and climate services. Strategic cooperation between meteorology and energy sectors to achieve this goal varies from country to country.

However, there are several key common aspects in such strategic cooperation: they must be inclusive of both meteorology and energy sector stakeholders with expertise in their individual sectors and the linkages between the two groups; they should include both public and private meteorological services and their expertise in monitoring and prediction, and they should connect to the appropriate parts of government and regulatory agencies tasked with responsibilities for competitive provision of relevant meteorological services to be used in the energy sector. As highlighted by Rogers and Tsirkunov (2011), any far reaching success among service providers and users without extensive partnerships is unlikely.

6 Way Forward

The integration of social, economic and environmental information is central to sound decision-making. Timely, accurate and user friendly weather, climate, hydrological and other environmental information and forecasts and related services have many applications including those in the energy sector. However, the utility of these information and services is often poorly understood, resulting in low demand and lack of public investment in NMHSs. Unless demand-driven services can be created, new and additional socio-economic benefits, which could be provided by NMHSs, will always be viewed as a low priority for public sector spending amidst the needs and costs of other public goods and services (Rogers and Tsirkunov 2011).

The energy-meteorology interaction can be considered an end-to-end process in the sense that information tends to flow from the producers to the users.[2] Such a process has four areas where further improvements could be made, namely:

---

[2] The end-to-end process can actually be extended to improve coordination of all activities within the delivery and application chain in order to develop a more integrated decision-making package (Harrison et al. 2008).

1. By improving the existing meteorological services being used;
2. By developing innovative and new products and services to be used in a growing and dynamic energy market;
3. By further improving communication between the stakeholders; and
4. By improving the decision-making processes involving both the meteorology and energy sectors.

The end-to-end process of energy-meteorology interaction is not entirely the responsibility of the NMHSs alone. It requires an active partnership between the NMHSs and other producers of meteorological products and services and the energy industry. It is clear that a lot more effort is needed on the part of the meteorological service provider and the energy sector user to understand each other's capabilities and constraints in order to optimize the use of relevant meteorological information to improve the socio-economic performance of the energy sector. Such is the case for instance for the use of probabilistic forecasts, instead of the traditional deterministic ones.

The decision-support within the energy-meteorology interaction requires new and different capabilities than those found traditionally in NMHSs. Cross-sectoral training is needed to increase the capacity and capability of the meteorological service producers and the energy sector consumers of such information to work together (Troccoli et al. 2013). For example, organizations such as the International Research Institute (IRI) for Climate and Society are currently teaching courses aimed at employees of health and meteorological and related services. This type of training is relatively new in sectors such as energy, health and planning, but more common in agriculture, aviation and marine transportation. Similar teaching courses are required at the interface of meteorology and energy that will lead to better decision tools and more effective outcomes for the industry.

There are some NMHSs which effectively deploy several of their staff within their key customer organizations (e.g. road authorities, aviation industry, defence sector, agriculture sector and health sector). They generally develop a more effective producer and consumer relationship leading to greater innovation in the customer's sector and closer alignment of the NMHS with the expected outcomes of that provided service. There is scope for similar arrangements to deploy some of the NMHS staff in key energy sector user entities. It is also important to recognise that this is an area where there is considerable capacity within private meteorological service providers and a major area of tension between the public and private sector, which may be in direct competition.

Taking the aviation case as an example, one option for energy-meteorology interaction could be a tripartite collaboration between public, private and academic sector stakeholders in both the energy and meteorology services under the auspices of WMO and the International Energy Agency (IEA). The collaboration could begin by fostering advances in scientific understanding and technology relating to the energy–meteorology interface. This could allow more energy providers/users to interact with meteorology service providers in terms of what the latter is capable of providing to meet the demands of the former. The success of such collaborations will be influenced by the protocols relating to access and sharing of infrastructure and models, and usability and reusability of information and data.

A pathway for the energy–meteorology collaboration could be developed by undertaking several steps to formulate an appropriate operating model. These steps could include: assessing the existing operating arrangements of meteorology service providers in producing and delivering services to the energy utilities; reviewing governance and legal arrangements between meteorology service providers and energy utilities; surveying the energy sector user needs and identifying the gaps in services; estimating the costs of providing the required services to the energy sector; and choosing an appropriate energy–meteorology operating model out of the following four different types. The four different operating models could include:

1. Where existing business interactions between energy utilities and meteorology service providers are further improved via enhanced communication and dissemination of user-relevant information;
2. Where complementary roles of public and private meteorological service provision to the energy utilities are better defined and adopted;
3. Developing a special and commercial service provider within a public meteorology services provider to cater specifically for the energy utilities;
4. Formulating a formal public-private partnerships to deliver meteorological services to energy utilities.

Finally, whenever feasible and relevant, collaborations between energy and meteorology may be framed within existing international agreements, such as the Global Framework for Climate Services ("Weather and Climate Information Delivery within National and International Frameworks", see also WMO 2011) or the Desertec Industrial Initiative—Europe, the Middle East and North Africa (DII–EUMENA) (Zickfeld et al. 2012).

Acknowledgments Helpful comments from John Zillman and Ferenc Toth are gratefully appreciated.

References

Dubus L (2010) Practices, needs and impediments in the use of weather/climate information in the electricity sector. In: Troccoli A (ed) Management of weather and climate risk in the energy industry., NATO science seriesSpringer Academic Publishers, Dordrecht, The Netherlands, pp 175–188

Dubus L (2014) Weather and climate and the power sector: needs, recent developments and challenges. In: Troccoli A, Dubus L, Haupt SE (eds) Weather matters for energy. Springer, New York

Dutton et al. (2014) A probabilistic view of weather, climate and the energy industry. In: Troccoli A, Dubus L, Haupt SE (eds) Weather matters for energy. Springer, New York

Fischer M (2010) Modelling and forecasting energy demand: principles and difficulties. In: Troccoli A (ed) Management of weather and climate risk in the energy industry., NATO science seriesSpringer Academic Publishers, Dordrecht, The Netherlands, pp 207–226

George et al. (2014) Weather and climate impacts on Australia's national electricity market. In: Troccoli A, Dubus L, Haupt SE (eds) Weather matters for energy. Springer, New York

Harrison M, Troccoli A, Williams JB, Coughlan M (2008) Seasonal forecasts in decision-making. In: Troccoli A, Harrison M, Anderson DLT, Mason SJ (eds) Seasonal climate: forecasting and managing risk., NATO science seriesSpringer Academic Publishers, Dordrecht, The Netherlands, pp 13–42

Johnston PC, Gomez JF, Laplante B (2012) Climate risk and adaptation in the electric power sector. Asian development bank publication. Available at: http://www.iadb.org/intal/intalcdi/PE/2012/12152.pdf

Lior N (2012) Sustainable energy development: the present (2011) situation and possible paths to the future. Energy 43:174–191

Mailier et al. (2014) In Search of the Best Possible Weather Forecast for the Energy Industry. In: Troccoli A, Dubus L, Haupt SE (eds) Weather matters for energy. Springer, New York

Majithia S (2014) Improving resilience challenges and linkages of the energy industry in a changing climate. In: Troccoli A, Dubus L, Haupt SE (eds) Weather matters for energy. Springer, New York

Marquis M (2011) Weather, climate and the new energy economy', *BAMS*, November, ES38–ES39

Pirone MA (2007), The private sector in meteorology: the next 10 years, Paper presented at the WMO international symposium on public weather services: a key to service delivery, World Meteorological Organisation, Geneva, 3–5 December 2007. Available at: http://www.ametsoc.org/boardpges/cwce/docs/Economic-Study/2007-Pirone-WMO.pdf and also at http://www.wmo.int/pages/prog/amp/pwsp/documents/Symposium_Proceedings_Final.pdf

Pirovano et al. (2014) Combining Meteorological and Electrical Engineering Expertise to Solve Energy Management Problems. In: Troccoli A, Dubus L, Haupt SE (eds) Weather matters for energy. Springer, New York

PMSEIC Independent Working Group (2010), *Challenges at energy-water-carbon intersections*, report prepared for the Prime Minister's Science, Engineering and Innovation Council (PMSEIC), Canberra. Available at: http://www.innovation.gov.au/Science/PMSEIC/Documents/ChallengesatEnergyWaterCarbonIntersections.pdf

Rogers D and Tsirkunov V (2011), Managing and delivering national meteorological and hydro-meteorological services, WCIDS Report 2011, Global facility for disaster reduction and recovery

Ronalds et al. (2014) A new era for energy. In: Troccoli A, Dubus L, Haupt SE (eds) Weather matters for energy. Springer, New York

Schaeffer R, Szklo AS, de Lucena AFP, Borba BSMC, Nogueira LPP, Fleming FP, Troccoli A, Harrison M, Bouahya MS (2012) Energy sector vulnerability to climate change: a review. Energy 38:1–12

Shiel et al. (2014) Reducing the energy consumption of existing, residential buildings, for climate change and scarce resource scenarios in 2050. In: Troccoli A, Dubus L, Haupt SE (eds) Weather matters for energy. Springer, New York

Troccoli et al. (2013) Promoting new links between energy and meteorology. Bull Amer Meteorol Soc. New York (in press)

U.S. Energy Information Administration (2011) International energy outlook 2011, Washington. (http://www.eia.gov/forecasts/ieo/pdf/0484(2011).pdf)

WMO (2007) Guide to aeronautical meteorological services cost recovery: principles and guidance, world meteorological organization report No 904 (2nd edn) Geneva, Switzerland. Available at: www.wmo.int/pages/prog/amp/aemp/documents/904_en.pdf

WMO (2011) Climate knowledge for action: a global framework for climate services—empowering the most vulnerable. World meteorological organization report No 1065, Geneva, Switzerland. Available at: http://www.wmo.int/hlt-gfcs/downloads/HLT_book_full.pdf

Zickfeld F, Wieland A, Blohmke J, Sohm M, Yousef A (2012) Desert power 2050: perspectives on a sustainable power system for EUMENA. A Desertec Industrial Initiative (DII)–Europe, the Middle East and North Africa (EUMENA) publication. Available at: http://www.dii-eumena.com/dp2050/perspectives-on-a-sustainable-power-system-for-eumena.html

Zillman J, Freebairn JW (2001) Economic framework for the provision of meteorological services. WMO Bull 50(3):206–215

Index

A
Access, 177–179, 185, 192, 195, 196
Adaptive capacity, 180
Advanced algorithm, 410
Advanced lead-acid, 405
Aeronautical tele-communications, 505
Affordable, 177, 179, 196
Africa, 177
African countries
    countries, 187, 188, 192, 194–196
Aggregate, 181, 183
Agricultural crops, 178
Agriculture, 106
Agro-ecological, 192
AIDS, 179
Air-conditioning, 14
Air masses, 357
Air pollution, 179
Air traffic services, 505
Alcohol, 184, 185, 191
Algae, 195
Algal, 185
Anaerobic digestion, 185
Analog Kalman Filter, 302
Analog method, 387
Animal fats, 184
Animal feed, 186
AnKF, 302
Annual cycle, 76
AO-GCM, 474, 475, 490
Antarctic ice-sheet, 84
Application, 238, 242, 253
Applications of Meteorology Programme, 212
Arab Maghreb Union, 180
Arctic warming, 85
Arid, 180, 190–192
Artificial intelligence, 295, 302, 303, 305, 308, 315
Artificial Neural Networks

ANN, 269, 271, 272
Asia, 179, 186
Assimilation, 278, 279
ASTER Global Digital Elevation Model, 280
Atmosphere, 178, 185
Atmospheric informatics, 373, 374
Atmospheric aadiation measurement
    U.S. Department of Energy, 266
Australian Bureau of Meteorology, **230**, 502
Australian Energy Market Operator, 155, 157
Australian Wind Energy Forecasting System, 156, 166
Average
    arithmetic, 356
Aviation, 508
Aviation and meteorology, 505
Aviation industry, 505
AWEFS. *See* Australian Wind Energy Forecasting System

B
Backward and forward
    linkages, 191, 193
Balance portfolio, 313
Balancing, 260–262, 267, 268
Balancing authority, 260, 261, 297, 307
Baltic Sea, 284
Bankability, 457
Base load
    plants, or facilities, 267
Baseload facilities, 312
Basic service, 207, 209, 210
Beacon power, 405
Beef tallow, 184
Benefits, 178, 185, 186, 190, 192–194
Benefit stacking, 404
Bias, 327, 328, 331
Bias correction, 327, 328

Biodiesel, 184, 185, 188, 190
Biodiversity, 184, 186, 195
Bioenergy, 107, 177, 178, 178, 180, 183–188, 191–196
Bioenergy development
  development, 193, 195, 196
Bioenergy policy, 191, 193–195
Bioethanol, 185, 193
Biofuels, 79, 102, 105, 177, 184, 186, 189, 194, 195
Biogas, 105, 107, 185
Bio-hydrogen, 185
Biological process, 185
Biomass, 93, 96, 104, 106, 107, 179, 184
Biomethane, 105
Blending targets, 194
BouLac PBL, 287
Boundary conditions, 298, 300, 301
Brazil, 185, 193, 194
Breakeven price, 194
Briquettes, 184
Building energy management, 10
Business
  optimum results, 368
Business model, 370
  expected revenue and variance, 371
  framework for decisions, 375

C
California State Energy Storage Bill AB2514, 404
Camera
  fisheye, 268, 269
Capacity building, 195, **233**
Capacity factor, 158, 457, 461–464
Carbon, 178, 185, 186, 192
Carbon capture and storage (CCS), 10, 95, 99, 107, 252, 498
Carbon dioxide, 178, 185
Carbon penalty, 4
Carbon price, 157
Carrier to Noise (CNR) ratio, 288
Cassava, 184, 189, 190
Castor, 184, 190
Castor oil, 190
Catastrophic events, 6
CCI, 242
Cellulosic, 185, 191, 195
Central Africa, 180, 187, 188, 190, 191
Central station, 258–260, 272
Centralised energy, 106

Certification, 195
Challenges, 395
Chaos, 358
Charcoal, 183, 185
Chips, 184
Clean, 177, 179, 196
Clear sky model, 435
  clear sky index k^*, 437
  deterministic daily course, 444
Climate, 5, 202, 203, 212, 214, 216, 217
  1990 baseline, 469–471, 479, 480, 484
  adaptation, 116, 177, 195
  current, 319, 321–323, 329
  future projections, 319–322, 329
  reports, 116
  temperate, 469
Climate analyst, 289
Climate change, 65, 81, 133, 135, 136, 152, 153, 177, 178, 180, 181, 183, 187, 192
  2050 scenario A1FI. See Scenario, climate Change
  change, 195, 227, 319–321, 323, 326, 330, 331, 392
Climate Change Act 2008, 114
Climate change mitigation, 109, 277
Climate change projections, 204
Climate change risk assessment
  risk assessment, 114, 130
Climate change scenario
  heat stress, 483
Climate impacts, 86
Climate mitigation
  mitigation, 177
Climate model, 68, 78, 84, 87. See AO-GCM
  coupled, 320, 323, 324, 327, 330
  global, 319, 320
  INM-CM3.0, 475, 478, 488
  regional, 319, 320
  stretched-grid, 322, 324, 326
Climate predictions, 204
Climate projections, 71, 79, 392
Climate risk, 353
Climate risk assessment, 124
Climate risk management, 86
Climate science, 116, 127–129
  information gap, 127
Climate services, 242
Climate stress, 180
Climate variability, 11, 65, 507
Climatic conditions, 67
Climatological distribution, 390
Climatology, 224, 227

Index 515

Cloud, 257, 258, 263, 264, 266, 268–270, 272, 273
Cloud cover, 81
Cloud motion, 429, 431, 434–436, 438, 452
 advection, 434
 cloud information, 435
 extrapolation, 431, 435
 matching cloud structures, 435
 vectors, 269–271, 434–436
 stable cloud structures, 435
Cloud shadows
 passage of, 263
CMV. See CMVs
CMVs. See Cloud Motion Vectors
CO_2 fixation, 190
Coal, 177, 185, 222
 power stations, 4
Coal gasification industries, 185
Coal mines
 disruption, 7
Coal resources, 9
Coastal wind profiles, 283
Coconut, 184
Combination of different models, 440
 optimized weighting. See Statistical or learning approaches
Comfort approach, 470, 473
 NatHERS, 473, 473–75, 484–89
 SET, 475–479, 482, 485–489
Comité Maghrebin de l' Electricité, 180
Communication, 396, 508
Communication and dissemination
 of user-relevant information, 509
Community-based discussion, 10
Community perceptions, 180
Composite observing systems, 230
Compressed air, 405
Computational expense, 320, 323, 324, 326, 327
Concentrating Solar Power (CSP), 73, 258, 259, 262
Confidence level, 341, 342, 344, 346, 347
Connection node power forecasts, 304
Consensus forecast, 303, 304
Consumer demand, **224**
Consumption, 471
 and economic growth, 498
 and population growth, 506
 energy, 468
 resource, 489
Containers, 263
Contingency tables, 359, 366

Conventional bioethanol, 185
Conventional ethanol, 184
Conversion of natural habitats, 186
Cooking, 184, 193
Cooking oils, 185
Cooking stoves, 184
Cooling, 105, 106
Cooling degree days, 67, 80
Cooperation, 504
 between meteorology and energy sectors, 507
Copernicus, 243
CORINE, 281
Corn, 184, 185, 194, 195
Corporate Resilience Programme, 121
Correlation, 265
Cost, 103, 336, 337, 342–349
Cost of energy, 100
Cost of exclusion, 501
Cotton seed, 184
Country wind resource assessments, 282
Coupled Model Intercomparison Project 3 (CMIP3), 71
Cow dung
 dung, 177, 180
Crisis, 179, 180
Crop cultivations, 82
Crop cycle, 183
Crop growth, 79
Crop growth cycle, 180
Cross-disciplinary science, 3
Cross-sectoral training, 508
Crowding out, 178, 186
CSIRO Energy Transformed Flagship, 408
Curtailments, 262
Customer organizations, 508
Customised services, 503
Cyclones and anticyclones, 203

D

Dams, 181, 184, 191
Darfur, 182
Data assimilation, 298, 320
Data bankability, 260
Data exchange, 503
Dataset resolution, 320, 321, 326
Day-ahead market, 313
Decadal predictions, 392
Decadal variability, 456, 457
Decision-making, 507
Decision-making processes, 391, 508

Decision strategies, 368
Decision-support, 508
Decision support system, 297, 298
De-correlation factors
  cloud, 263
Deforestation, 183, 195
Degree day, 222, 226, 227, 357
  forecasts, 366
Delivery, 77
Demand, 222, 260, 261, 380
Demand and production forecasts, 382
Demand-driven, 504
Demand forecasting system, 164
Demand gradient, 381
Democratic Republic of Congo, 179
Demographic, 179, 222
Deregulation
  of electricity markets, 505
Design principle, 374
Desulfurization, 185
Deterministic forecasts, 335–337, 347–349, 383
Developed countries, 499, 503
Development, 183–185, 192–196
Development sectors, 180
Development strategies, 191
DFS. *See* Demand Forecasting System
DICast, 299, 303–305
Diesel, 179, 183, 185
Diffuse irradiance, 73
Direct beam, 73
Direct normal irradiance (DNI), 73, 263, 439
  concentrating solar power plants, 439
Disaster preparedness, 196
Dispatchable, 262
Dispersion factor, 266
Distillation, 184
Distribution, 177, 187, 191
  gamma, 357
  Gaussian, 356
  normal, 356
  probability, 356
  Weibull, 357
Distribution network operators, 117
DMOS, 304, 305
Downramps, 262
Down-scaling, 87
  Dynamical, 319, 320, 323, 325
  morphing approach, 475, 488, 489
  Multiple, 323, 327
  Statistical, 320, 326
  Statistical-dynamical, 320

Downscaling technique, 278
Drier conditions, 74
Drought, 86, 180, 184, 188, 190
DTU wind energy, 278
Dynamical Integrated Forecast, 299
Dynamical model output statistics, 304

E

EAAP. *See* Energy Adequacy Assessment Projection
Earth observation, 237–239, 243
East Penn Manufacturing Co., 406
Eastern Africa, 180
Eastern Australia, 77
Eastern Brazil, 77
Eastern Europe, 75
ECMWF, 387
ECMWF/ERA Interim, 279
Economic activity, 380
Economic and population growth, 506
Economic assessment, 390
Economic benefit, 387
Economic growth, 498
Economic security, 506
Ecosystem, 184, 186, 187, 190, 195, 196
Ecoult, 406
EDF, 380, 382–384, 387, 392
Effective Temperature (ET*). *See* New Effective Temperature (ET*)
Efficiency, 177, 179
Efficient, 178, 188, 190
Egypt, 184
El Niño, 326, 328
Electrical grid, 142, 143, 151, 152
Electrical storage, 399
Electricity, 3, 177–180, 183, 193
Electricity companies, 505
Electricity generation mix, 9
Electricity Grid Management, 224
Electricity network resilience and weather, 123
Electricity Network Strategy Report (ENSG), 128
Electricity system, 11, 104
Electrification rate, 178
Emissions, 5, 178, 183, 185, 186, 192
  greenhouse gas, 5
Emission scenarios, 79, 321, 330, 331
Empirical power conversion, 299, 304
Empirical power curve, 304
Empower, 178

Index

End users, 391
Energy, 237, 238, 242, 243, 245, 247, 249, 253, 254
  access to, 3
  global demand, 3
  goals, 8
Energy access, 178
Energy Act 2008, 2010 and 2011, 114
Energy Adequacy Assessment Projection, 168
Energy and meteorology, 497
Energy balances, 192
Energy, climate change, 113
  resilience, infrastructure, sustainability, natural hazards, 113
Energy consumption, 177
  and required energy, 473
  consumption, 177
  world, 498
Energy cost
  global, 354
  U.S., 354
Energy crops, 108
Energy demand, 4, 5, 79, 312, 319, 320, 329–331
Energy demand management, 499
Energy efficiency, 4, 109
Energy facility
  life cycle, 11
Energy generation
  variability, 10
Energy industry, 353, 355, 396
  challenges, 355
  mission, 354
Energy infrastructure, 3, 76
Energy infrastructure planning, 71
Energy market, 297, 312, 313, 380
Energy-meteorology collaboration, 509
Energy-meteorology interaction, 507, 508
Energy-meteorology linkages, 14
Energy mix, 4
Energy policies
  policies, 179
Energy price, 8
Energy production, 499
Energy-relevant variables, 74
Energy sector, 69, 391, 508
  vulnerability, 497
Energy security, 178
Energy services
  services, 177
Energy stock, 387
Energy storage, 13
Energy strategies
  strategies, 177

Energy suppliers, 499
Energy supply-demand, 6
Energy system, 5, 65, 82, 116, 123, 499
  transition, 4
  climate models, energy applications, 123
  efficiency and economics, 6
  vulnerability, 6
Energy traders, **227**
Energy trading, 506
Energy transformation, 3
Engagement, 87
Ensemble, 295, 299–303, 315, 358
Ensemble forecasting, 391
Ensemble predictions, 323, 330, 331
Ensemble prediction system, 384
Environment, 191, 195
Environmental benefits
  benefits, 193
Environmental factors
  factors, 194
Environmental impacts, 186
Enzymatic process, 185
EP1 (Energy Phase 1), EP2 (Energy Phase 2), 116
EPRI, 402
Equator, 184
Equatorial Pacific, 82
ERA Interim, 279
ERA Interim reanalysis, 285
ESA, 237, 239–244, 246
ESA GlobCover, 281
Ethanol, 102, 107, 184, 188, 193, 194
Ethiopia, 182, 184, 186–188
Ethnic tensions
  tensions, 182
ETR°138, 116, 126
EU, 178, 193, 194
European Centre for Medium-Range Weather Forecasts (ECMWF), 270, 387
European Wind Atlas, 277
Evapotranspiration, 191
Expert system, 295, 299, 307, 308, 310, 311, 315
Explicit physical modeling. *See* Numerical Weather Prediction, PV simulation
Exploration, 67
Exports, 183
Extreme events, 85, 223, 319, 331. *See* Extreme weather
  atmospheric moisture, 120
  coastal erosion, 120
  combined events, 120
  definition, 85
  heat waves, 120

Extreme events (cont.)
  ice storms, 120
  lightning, 120
  pluvial and fluvial flooding, 120
  river erosion, 120
  summer drought, 120
  tidal surges, 120
  vegetation changes, 120
  wind storms, 120
Extremes, 67, 391
Extreme temperatures. See Temperature
Extreme weather, 226
Extreme weather events, 142, 173

F
Farm, 181, 183
Feature extrapolation, 308
Federal Land Bank, 186
Feed-in Tariff (FiT), 259
Feedstock, 178, 185–188, 191–196
Fermentation, 184
Fertilizers, 190, 193
Field study, 263
FINO 1, 284, 285
FINO 2, 284, 285
Fleet, 266, 267
Flexibility, 260, 272
Flooding, 86
Fluctuating loads, 260, 272
FNL analysis, 289
Food, 178, 181, 183, 185, 188, 190–192, 194
Food crops, 194
Food insecurity
  insecurity, 181
Food prices, 178, 192
Footprints, 192
Forcings, 323, 330
Forecast, 227, 257, 260, 262, 267, 268, 268, 270, 271, 271, 272
  acting on, 368
  binary, 359
  computer probability, 358, 361
  degree days, 366
  ensembles, 358
  fraction correct, 359
  multi-model
    DEMETER, 358
    ENSEMBLES, 358
    National Multi-Model Ensemble, 358
    World Climate Service, 358
  multi-model ensembles, 358

  multi-scale ensemble, 365
  of winds and solar radiation, 499
  one- to four-week, 361
  probability, 357
  reliability, 361
  seasonal, 358
    ECMWF Seasonal Forecast System, 358
    NWS Climate Forecast System, 358
    World Climate Service multi-model, 358
  statistical, 358
  success ratio, 359
  two- to four-week, 358
Forecast accuracy, 301, 315, 338, 349, 437–439, 441, 452
  bias, 438, 440, 442
  climatological mean. See trivial reference model
  correlation coefficient, 437, 442, 447
  frequency distributions of the forecast errors, 447
  improvement scores, 443
  MAE, 442
  maximum possible forecast errors, 447
  persistence. See trivial reference model, 442
  RMSE, 442–445, 451
  scatterplot, 446
  skill score, 443
  trivial reference model, 442
Forecast error, 336
Forecast goodness, 336, 337, 348
Forecasting, 262, 267–269, 271–273
  solar variability, 257, 258, 267–271
Forecast quality, 336, 337, 396
Forecast uncertainty, 336, 347
Forecast value, 337, 342, 343, 345, 346, 349
Foreign exchange earnings, 183
Foreign investment, 191
Fossil fuels, 5
  fuels, 177, 178
Frequencies vs probabilities, 356
Frequency distribution, 72
FROGFOOT, 279
Furukawa, 407
Future Earth System, 374
Future network resilience, 124
  climate risk assessment, 124
  climate risk assessment, weather-related faults. See climate risk assessment

Index

G
Gas, 177, 178, 184, 186, 190, 192
  liquefied natural, 3
  natural, 9
  shale or coal seam, 4
Gasified, 185
GCM. *See* Climate model, global
GDP, 179, 182, 183
Generalised wind climate, 278, 279
Generation, 177
Generators, 183
Geo-engineering, 92
Geoscience Australia, **230**
Geothermal, 10, 99, 102, 104, 109, 183, 184
GHG, 178, 183, 185
GIS, 242
Global circulation, 203
Global consumption, 177
Global electricity supply, 259
Global Environmental Multiscale (GEM), 269, 271
Global Forecast System (GFS), 269, 270
Global framework for climate services, 214, **232**, 509
Global horizontal irradiance (GHI), 263
Global linear trends, 74
Global reanalysis data, 277
Global reanalysis datasets, 278
Global warming
  and buildings, 74
Global Wind Atlas, 276, 278, 282, 291
Global wind resource, 276
GMES, 240, 242, 243, 245, 247, 453, 254
Governance, 180, 186, 195
Government Business Enterprise, 502
Government policies, 9
Grand challenges of atmospheric science, 373
Grand Ethiopian Renaissance Dam, 180
Grand Inga Dam, 179
Graphical user interface, 306
Greenhouse, 178, 186, 190, 192
Green-house gas (GHG), 79, 92, 93, 95, 178, 183, 185
Greenhouse warming, 319, 329, 330
Grid, 10, 400
Gridded global dataset, 320
Grid-integration, 430, 447
  balancing power, 430
  day-ahead auction, 430
  day-ahead power market, 433
  demand side management, 430
  energy market, 430
  grid-integration in Germany, 430
  intraday trading, 430
  new supply structure, 429
  power plant scheduling, 429
  ramps. *See* Ground based sky images, 436
  transmission system operators, 430
Ground based sky imagers, 431
  red- to-blue ratio, 436
Ground based sky images, 429, 435
Ground nuts, 184

H
Hail on solar panels, 79
Hailstones events, 79
Hampton wind, 407
Hazardous meteorological phenomena, 202
Hazards in the energy sector, 127
Heat, 101, 104–106, 109
Heating degree days, 67, 80
Heating, 184, 193
Heat wave, 224, 381
Hedge, 368
  effect on revenue, 369
  in business model, 370
  reduction of variance, 369
Higher order statistics, 72
Highlands, 180
High latitudes, 81
High-Resolution Rapid Refresh (HRRR), 269, 271
Horn of Africa, 180, 188, 191
Host data, 324, 330
Host model, 320, 323, 326–328
House
  AccuRate simulation package, 470
  base case, 474, 479–87
  energy savings. *See* Retrofit actions, energy savings
  existing, 468, 469, 486, 487
  temperature control. *See* Retrofit actions, temperature control
House energy efficiencymeasures. *See* Retrofit actions
Household, 180, 181, 183–185, 190, 193, 194
Housing stock, 80
Høvsøre, 285, 288
Hub height, 73, 461
Human-induced climate change, 204, 214
Hurricanes, 7, 86
Hydro, 97, 104, 109
Hydroelectricity, 9

Hydrogen, 106
Hydrological cycle, 78
Hydrological functions, 190
Hydrological model, 389
Hydrological services, 204
Hydro-meteorological, 85
Hydro power, 77, 179–181, 183, 245, 246, 382, 387

I
Ice load, 149
Ice sheets, 77
Icing on power lines, 79
Impact, 178, 180, 187, 189, 196
Incentives, 193
In-country processing, 194
Independent system operator, 296
India, 278
Indoor, 179
Indoor air pollution, 179
Industry, 191, 193, 194
Information Gap Analysis, 127
Infrastructure, 179, 185, 191
Infrastructure network, 113
Infrastructure services, 503
Infrastructure vulnerability, 119
Initialisation, 320
In situ measurements, 277
Integration, 103, 104, 106, 180
Interaction
 energy and meteorology, 6
Interannual
 variability, 76
Interannual changes, 77
Interannual standard deviation, 76
Interannual variability, 76, 82, 86, 230
Intergovernmental Panel on Climate Change (IPCC), 71. *See* IPCC
Intermittent renewable energy, 402
International Civil Aviation Organization (ICAO), 505
International Conference Energy & Meteorology (ICEM), 258
International Energy Agency (IEA), 4, 177–179, 193, **231**, 257, 258, 260, 268, 508
Investments, 178–180, 184, 187, 194, 392
Investors, 226
IPCC, 469–471
Irradiance
 solar, 265, 266, 268, 269, 271
ISO, 296

J
Jatropha, 184, 187, 190

K
Kalahari, 187, 191
Kalealoa Airport
 Oahu, Hawaii, 265
Kenya, 184
Knowledge development
 development, 195
Knowledge gap
 gap, 195
Knowledge, 129
 information gap, climate science, 129
Kolmogorov, Andre, 355

L
Labor productivity, 181
LAM. *See* Regional climate simulations, limited area model
Land, 178, 186, 191, 192, 194, 195
Land grabs
 grabs, 178
Land use planning
 planning, 192
Land use, 320–324
Laplace, Pierre Simon, 355
Large-scale climate, 68
Large-scale filter, 326, 327, 329
Lateral boundary conditions, 325–327, 330
Learning by doing, 10
Leitmotif, 6
Liberia, 187
Life cycle, 11, 191, 192
Life cycle analysis (LCA), 99
Lighting, 184, 193
Lignocellulosic, 185, 196
Linear trend, 73
Linseed, 184
Liquid, 177, 184
Livestock mortality, 181
Load balancing, 296
Local climate
 local, 192
Local climate knowledge
 knowledge, 196
Logs, 184
Long term wind time series, 283
Long-term forecasts, 382
Lorenz, Edward N., 358

Index

Loss, 335–337, 347, 349
Low-level wind structure, 279
LPG, 191

M

Macroeconomic policies
    policies, 191
Madagascar, 188
Madrid Action Plan, 214
Maize, 182, 184, 187–190
Malaria, 179
Manufacturer's power curve, 304
Marginal cost, 501
Margins
    decreasing, 354
Market, 181, 192, 193, 195
Market design, 164
Market forces, 9
Market mitigation, 127
    Smart networks, 127
Mean distribution, 73
Mean sea level, 67, 70, 73, 77, 82
Mean value, 72
Measurements, 395
Mediterranean Sea, 77
Mellor-Yamada-Janjic scheme, 287
MERRA, 279
MESoR, **231**
Mesoscale, 87, 263, 268, 269, 271, 272
Mesoscale modelling, 276, 456
Météo-France, 394
MétéoGroup, 358
Meteorological data
    data, 196, 505
Meteorological events, 13
Meteorological factors, 66
Meteorological information, 202, 208, 211, 218, 497, 499, 506, 508
    provision, 500
Meteorological service, 201, 202, 204–209, 211, 218, 501, 502, 507, 508
    basic, 500
    special, 500, 501
Meteorological stations, 192
Meteorological variables, 65, 67
Meteorologists, 506
    industrial, 506
Meteorology, 65
Meteorology, climate and energy, 128, 129
    extreme events, risk assessment, hazards, 128
Methane, 85
Met-masts, 457

Met Office Hadley Centre, 121
Met offices
    offices, 192, 196
Microscale, 263
Microscale models, 279
Mid-merit, 312
    plants, or facilities,, 261, 267, 267
MINES ParisTech, 269
Mining, 239, 251
Mission success
    probability of, 355
Mitigation scenarios, 95
Model output statistics
    MOS, 269, 271, 272, 304
Model quality, 389
Model resolution, 327
    coarse, 319–321
    fine, 321–324, 329, 331
    variable, 327, 328
Modern energy
    energy, 177
Modern sector
    sector, 177
Moisture, 183, 187, 190, 191, 195
Monitoring, 87
Monocultures, 178, 186, 190
Monthly forecasts, 384–386
Morocco, 180, 184
Mozambique, 184, 187
Multidecadal variability, 226
Multidisciplinary analyses, **228**
Multi-linear regression, 302

N

Nakanishi and Niino level 2.5, 287
Nakanishi and Niino level 3, 287
NASA, 279
National electricity market
    NEM, 155, 156, 165
National Hydrological Service (NHS), 499
National Infrastructure Plan (NIP) 2011, 115
National Meteorological and Hydrological Services (NMHSs), 390, 497, 499, 501
    corporatized, 503
    publicly funded, 502
    publicly funded, 502
National Meteorological Service (NMS), 208, 209, 499
National, 186, 191–193, 196
Native plant diversity, 192
Natural gas, 3
Natural hazards, 113
Natural resource, 191

Navier Stokes equations, 300
NCAR/CFDDA, 279
NCEP, 387
NCEP/NCAR reanalysis, 278
NCEP/NCAR, 278
Negative trends, 75
NEMMCO, 400
NEPAD, 177, 179
Network, 382
New Effective Temperature (ET*), 476
    air speed cooling, 476, 478, 487
Newspaper weather services, 206
NGK, 405
Nitrate, 178
NOAA/CFSR, 279
Non-dispatchable, 262
Non-food crops, 194
Non-linear operations, 357
Non-tariff, 193, 195
Normalized mean absolute error, 306
North America, 75
North Sea, 77, 284
Northeastern China, 278
Nowcasting, 297–299, 307, 308
Nuclear power stations, 8
Nudging, 326–329
Numerical weather prediction (NWP), 68, **231**, 268, 272, 296–301, 303, 304, 307, 315, 384, 431
Numerical weather prediction models, 433
    COSMO-EU model, 440
    direct model output, 434
    ECMWF Integrated Forecast Syste, 440
    ensemble prediction systems, 444, 452.
        See Uncertainty
    global NWP model, 433
    grid resolution, 433
    mesoscale model, 433
    parameterizations, 434, 444, 452
    spatial and temporal discretization, 433
    Weather Research and Forecasting (WRF) model, 434
    weather services, 433
NWP forecasts
    NWP, 413
NWP models, 394

O

Observations, 320, 321, 329
Observed climate, 68
Observed sea-level, 84
Observing networks, 203, 211, 218
    private, 505

Obstacles, 277
Ocean energy, 99, 102, 109
Oceanic currents, 78
Oceans, 74
Offline optimising, 410
Offshore, 282, 283
Offshore wind resource, 464
Offshore windfarms, 457
Oil, 3
    global trade, 4
    peak, 4
Oil and gas, 237, 242, 247
Oil and gas platforms, 11
Oil palm, 184, 186, 187, 190, 191
Oil rigs, 77
Oilseeds, 184
OISST, 285
Okinawa Yanbaru, 405
On-shore wind, 93
Operating model, 509
Operational planning, 312
Opportunity
    sources, 355
Opportunity costs, 295, 314, 315
Optimization, 380, 383, 384, 387, 391, 396
Optimum energy mix, 184
Organisation for Economic Cooperation and Development (OECD), 5
Orography *See* Topography, 321
Outsourcing, 500
Overhead lines (OHL), **118**
    cables, 118
    cables, substations, transformers, 118

P

Partnerships, 396, 499
PBL schemes, 291
Peak load, 261
Peakers, 312
Peaking
    plants, or facilities, 267, 272
Pellets, 184
Penetration
    of PV, 258, 259, 261, 262
    of VRE, 257, 258, 260, 262, 267
Per capita, 179
Permafrost, 77
Permafrost extent, 67, 84
Persistence forecast, 298
Petroleum, 177, 193, 194
Petroleum exploration, 12
Photovoltaic (PV), 4, 73
Photovoltaic power, 393

Index

Photovoltaic power systems, 226
Physical parameterizations, 278
Physical processes, 74, 84
Pig lard, 185
Pitt review, 121
Planning, 192
Planting season, 183
Plants, 178, 185, 188, 191, 194
Pluvial and fluvial flooding, 120
Policy development, 192
Policy initiatives
   government, 259
Policy makers, 226
Political instability
   instability, 182
Pollutants, 178
Pollution, 179
Population, 497
   growth, 4
Population displacement
   displacement, 184
Post-processing, 278
Poultry fats, 185
Poverty, 181
Power, 179, 183, 184
Power conversion system, 304
Power curve, 461
Power demand, 381
Power demand profile, 10
Power outages
   outages, 179, 183
Power plants
   lifetime, 4
Power pools, 180
Power ramp, 307
Power sector, 179, 379, 380, 382, 391, 395
Power system, 155–157, 160, 168, 169, 171–173
Power system management, 387
Power system optimization, 395
Power yield, 76
Precipitation, 81, 357, 382, 388
Predictability, 262, 391
Prescient weather, 358
Price, 102
Primary energy, 95, 96
   energy 179
Probabilistic forecast, 301, 302, 383, 508
Probabilities
   transforming, 375
   vs frequencies, 356

Probability
   distribution, 356, 384
   forecasts, 357
   mathematical foundations, 355
   methods, 355
   of mission success, 355
Probability density function, 301
Probability forecast, 336, 337, 341
Process efficiency, 500
Processing, 177, 192–195
Producer gas, 185
Production, 177, 178, 183, 185–188, 190–195, 381
Projected climate changes, 86
Projects
   grid-tied, 258, 259, 262
Prospecting. *See* Solar resource assessment
Prosperity Solar Energy Storage Project, 406
Providers, 396
Public acceptance, 10
Public good, 207, 500, 502
Public investment
   in NMHSs, 507
Public meteorological services, **224**
Public-private partnerships, 509
Public safety, 500
Public safety and economic security, 502
Public sector funding, 497
Public sector investment, 503
Public spending, 501
Public services, 500
Public Weather Services Programme, 212
Pulsed wind lidar, 288
Pumped hydro, 405
Purchaser-provider system, 501
PV
   array, 258–260, 262–264, 266–269
PV capacity, 258
PV production, 258
PV simulation, 429, 431, 447, 448
   alternating current (AC), 448
   module temperature, 447, 448
   array plane, 447
   diffuse irradiance. *See* Direct normal irradiance, 448
   direct current (DC), 448
   distribution of radiance over the sky *See* Diffuse irradiance, 448
   inverter, 448
   maximum power point (MPP), 448
   module efficiency, 431

PV simulation (cont.)
  mounting type, 448
  nominal power, 431, 448–451
  PV simulation model, 431
  snow covered modules, 450
  temperature coefficients, 448
  tilt and orientation, 431
  tilted irradiance, 448
PV systems, 258, 260, 266

Q

Quantile regression, 302
Quasi-Normal Scale Elimination PBL, 287
QuikSCAT, 282

R

Radiation transfer, 434, 439
Radiation transfer modeling, 434
Radio weather services, 206
Rainfall, 180, 182, 183, 186–191, 223, 226
Rainfall distribution, 180
Rain-fed, 180, 187, 188, 190, 191
Rainforest, 186, 187, 190
Ramp, 395
  events, 261–266, 272
Ramp forecasting, 308
Ramp rates, 8
Rapeseed, 184
Rapid fluctuations in generation, 401
Rayleigh distribution, 457
RCM. *See* Climate model, regional
Real time four-dimensional data assimilation, 298
Real-time market, 313
Real-time observations, 227, 300
Real-time weather observations, 395
Reanalyses, 68, 87
Reanalysis, 278, 321
Recycled, 178
Reference meteorological year (RMY), 490
Refineries, 185, 192, 194
Regional climate simulations, 319, 321, 329, 331
  limited area model, 323, 325
  stretched-grid, 323, 327
  time-slice, 324
Regional prediction, 432
  cross-correlation of forecast errors, 444
  cumulative PV power generation, 432
  distributed stations, 444
  representative set, 432, 449, 450
  spatial averaging, 438, 439, 444, 446
  up-scaling, 432, 447, 449–451
Regression tree, 305
Regulation power, 402
Regulation services, 400
Reliability of seasonal forecasts, 361
REN-21
  Global Status Report, 258
Renewable energy, 92, 93, 134, 153, 155, 157, 174, 175, 237, 242, 245, 312, 379, 380, 382, 395, 396, 455, 456, 464
  capacity, 5
  energy, 180, 183–185, 193
Renewable energy generation, 399
Renewables, 392, 498. *See* Renewable energy
Required energy, 473, 474, 479–82, 484
  and energy consumption, 473, 487
Research and development, 393
Resilience, 113
Resilience challenges, 113
Resilience planning, 114
Resource assessment, 262, 464
Resource characterisation, 11
Resources, 97
Retrofit actions, 470, 473
  added thermal mass, 472, 473, 483–85, 488
  ceiling fans, 473, 484
  ceiling insulation, 472, 484–85
  conservatory, 473, 483, 485
  cost-effective, 469, 470, 484
  deciduous trees, 472, 482, 484, 485, 487, 488
  energy savings, 474, 480, 483–85
  innovative, 487, 488
  low-cost, 468, 484, 488
  Pergola, 472
  temperature control, 474, 485
  temporary double glazing, 472, 473, 484, 485, 487
  thermally-lined curtains, 472
  weather-stripping, 473, 484, 487
  window size reduction, 472, 484
Reunion Island, 269, 393
Reversal of the wind regime, 290
Rift valley, 184
Risk management, 14
  management, 196
Risks, 178, 180, 184, 185, 193, 384
  climate, 353
  sources, 355
  weather, 353
River flow, 8, 381, 387
River runoff, 78
RMSE, 271
ROC, 385

Index 525

Rooftop solar, 157, 168, 171, 172, 259
Rotating Shadowband Radiometer
    RSR, 263
RTFDDA, 298, 300, 308
RTG_SST, 285
Rural sector
    sector, 177, 178, 194

S

Sahara, 180, 187, 191
Sahel, 180, 187, 188
Saltwater intrusion, 78
Sandia Labs, 402
SAR, 240, 243
Satellite, 229, 238–241, 243, 249, 250, 253,
    254, 395, 429, 431, 433, 435, 437–440,
    442, 445, 446, 452
    cloud index. *See* Cloud motion, cloud
        information
    geostationary Meteosat second generation,
        435
    Heliosat method, 435
Satellite altimetry, 78
Satellite-derived winds, 283
Satellite images, 431, 435, 452
Satellite missions, 70
Satellite oceanography, 78
Scale of operations, 192
Scale parameter, 289
Scales
    of motion, 262, 263, 267, 268, 272
Scarce resources, 468, 471, 489
Scenario, 95, 99
    2050, 480–84, 489
    climate Change, 468–71, 478, 479–85, 489
    scarce resources, 469, 471, 478, 483, 484,
        489
Scores, 384
Sea breeze, 456, 458, 460–464
Sea ice cover, 67
Seasonal climate outlooks, 228
Seasonal outlooks. *See* Seasonal Climate
    Outlooks
Sea-surface temperatures, 327, 328
Second and third generation bioenergy, 178
Sector, 183, 185, 186, 190, 192, 193, 195
Semi-arid, 180, 191, 192
Sensors
    solar, 263, 265
Sequestration, 192
Service models, 502

Services
    to the energy sector, 504
Sesame seed, 184
Severe events, 66
Severe weather, 224
Severe weather warnings, 500
SGRCM. *See* Climate model, Stretched-grid
Shape parameter, 289, 290
Short-term forecasting, 297
Short-term forecasts, 383
Shuttle Radar Topography Mission, 280
Siberian region, 77
Skewness, 72
Small farmers, 178
Smallholder, 190
Smart networks, 116
Socioeconomic, 178, 192
Socio-economic benefits, 507
Socio-economic-environmental context, 497
Sodium sulphur, 405
Soil conditions, 79
Solar, 97, 102, 107, 109, 183, 183, 184, 228.
    See Renewable energy
Solar forecasting, **231**
Solar global irradiance, 73
Solar Heating and Cooling (SHC), 257, 258
Solar power plants, 82
Solar power prediction, 430
Solar power yield, 77
Solar PV, 93
Solar PV Technology Roadmap, 258
Solar radiation, 67, 81, 184
Solar resource
    information, 258–260, 268, 270
Solar resources. *See* Solar resource
Solar resource assessment, 229, 258
Solar renewable energy, 382
Solid, 177, 184
Solid fuels, 184
Solid wood
    wood, 177
Somalia, 188
Sorghum, 184, 188, 189
South Africa, 184, 187, 188, 278
South East Asia, 77
Southern Africa, 187
Soybean, 184, 187, 189
Spatial averaging
    smoothing filters, 438, 439
Special services, 207, 208, 210
Spinning reserves, 267, 400
Spin-up, 298, 307

SRTM Water Body Data (SWBD), 281
SST, 285
Stability, 180
Standard and resilience of infrastructure in the UK, 115
Standard deviation, 72, 81, 356
Standard Effective Temperature (SET), 470, 476, 482, 484, 485, 487
   air speed cooling, 478, 477, 486
Staple food, 185
Starchy, 184
Statistical and learning approaches
   predictors, 437, 438
Statistical forecasting, 296
Statistical learning, 302
Statistical methods, 392
Statistical or learning approaches, 431
   artificial intelligence methods, 437, 438
   Artificial Neural Networks (ANN), 437
   autocorrelation, 436
   autoregressive (AR) models, 437
   bias correction, 438, 440
   exogenous data, 437
   historic data, 436
   k Nearest Neighbors, 437
   Kalman-filtering, 438
   linear regression, 440
   model output statistics, 437
   pedictands, 436
   predictors, 436
   statistical post-processing, 431, 449
   support vector regression (SVR), 437
   times series models, 431, 436
   training phase, 436
Statistical significance, 74
Stern review, 114, 130
Storage, 399
Storm surges, 78
Sub-grid scale processes, 278
Sub-regional, 180, 191
Subsidies, 94, 192–195
Subtropics, 81
Sucrose synthesis, 190
Sudan, 186, 187
Sugar, 184, 188, 191
Sugarcane, 184–189, 191, 195
Sulfur dioxide, 178
Sunflower, 184
Supply-driven, 504
Surface roughness, 277
Sustainability, 113, 114, 177, 195
Sustainable, 178, 192, 194–196
Sustainable development, 506
Sweet sorghum, 185, 187, 188

Synoptic-scale, 263
Synthetic aperture radar, 282

T

Tall wind, 292
Target mitigation, 126
Targets
   renewable energy, 259
Tariff, 193, 195
Task , 36, 46, 257, 258, 268
Taum Sauk, 405
Teaching courses, 508
Technical potential, 92, 93, 99
Technological development, 4
Teleconnections, 328
Telephone services, 206
Temperature forecasts, 383
Temperatures, 67, 79, 180, 190, 380
Tercile, 388
Terrain elevation, 277
Terrestrial water storage, 470, 475, 478, 485
Terrestrial water storage, 84
Thermal contrast, 285
Thermal expansion, 77
Third World Climate Conference, 216
Thunderstorms, 86
Time series, 264, 272
Time-slice simulations. *See* Regional climate simulations
   time-slice, 323
Topography, 320–322, 324, 325
Total sky imagers, 268, 269, 272
Trading returns, 360
Traditional, 177, 178, 180, 183, 184, 188, 193
Traditional biomass, 177, 178, 180
Transition, 177, 178, 193
Transmission, 177
Transmission planning, 172
Transmission power line, 13
Transport, 105
Transportation, 67, 77
Tropical areas, 74
Tropical climate
   climate, 185, 191, 193
Tropical forest, 186–188, 190, 195
Tropical regions, 81
Tuberculosis, 179
Tunisia, 184
Turbine power, 306
TV weather services, 206
Twigs, 177
Typical meteorological year (TMY), 474. *See* RMY

Index

U
UK wind climate, 457
UKCP09, 115, 123, 130
UltraBattery, 408
Uncertainties, 4
Uncertainty, 82, 86, 296, 299, 301, 307, 312, 313, 353, 436, 443, 444, 452
   distribution function, 444
   prediction intervals, 443
   probability density functions, 443
   quantile regression, 444
UNECA, 178, 179, 184, 188, 191, 195
UNEP, 180, 182
United States, 185
University of California at San Diego (UCSD), 268
University of Oldenburg, 271
Urban, 178, 183
Urban heat island, 123
US Federal Energy Regulatory Commission (FERC), 404
User/provider collaboration, 396
Users, 396
Utility function, 371
Utility operating paradigm, 312, 314

V
Value added services, 501
Value of weather forecasts, 396
Variability, 82, 257, 260–263, 266, 269, 272, 312, 399
Variable, 104
Variable cost, 312
Variable energy sources, 13
Variable generation, 295, 297
Variable renewable energy, 257, 258, 260
Variable resolution models. *See* Climate model, stretched-grid
Variational doppler radar analysis system, 295, 299, 307, 315
VDRAS, 299, 307–310
Verification, 278, 385
Very short-term prediction, 394
Virtual earth system, 373
Virtual network, 266
VRE, 257, 258, 260–263, 267, 268, 272. *See* Variable renewable energy
VRE resources, 260, 267
VRE supply, 260, 262, 267
Vulnerabilities, 177, 497
Vulnerability, 178, 180, 183
Vulnerability metric, 124
Vulnerability to climate change, 178, 183, 497

Vulnerable, 506
Vulnerable to meteorological events, 65

W
Wake effect, 457
Warming, 180
WAsP, 277, 279
Waste, 179, 185, 193, 194
Water availability
   availability, 181
Water cycle, 387
Water tables, 181
Watershed, 382
Weather, 177, 178, 180, 183, 187, 191, 192, 196, 201–203, 206, 211, 212, 216
Weather and climate information, 499
Weather and climate services, 201, 208, 211, 218
Weather and climate, 201–204, 208, 210–212, 214–216, 218
   severe events, 6
Weather bureau
   bureau, 192
Weather derivatives, 360
Weather events, 65, 499
Weather forecasts, 12, 204, 500
Weather regime, 315
Weather-related faults, 124
Weather Research and Forecasting, 283, 300
Weather risk, 353
Weather windows, 12
Weibull distribution, 283, 287, 291, 457, 458
West Africa, 75, 187, 188
Wheat, 184, 189
WHO, 179
Wind, 67, 98, 99, 102, 107, 109, 183, 184, 223, 228, 382. *See* Renewable energy
Wind Atlas Analysis and Application Program, 289
Wind climate, 456, 457, 464
Wind energy, 77, 184, 413, 427, 456–458
Wind energy assessment, 275
Wind energy forecasting, 295, 296
Wind farm, 223, 243, 456, 460, 464
Wind farm smoothing, 407
Wind generation, 156–161, 164, 166, 169, 171
Wind integration, 159, 170
Wind penetration, 164, 171
Wind power, 295–299, 302–305, 307, 308, 315
   available, 357
Wind power density, 281

Wind power forecasting, 295, 303
Wind power forecasting system, 295, 297, 299–301, 303, 315
Wind resource, 460, 464
Wind resource assessment., 278
Wind resource atlas, 283
Wind roses, 457, 459
Wind speed, 73, 81, 82, 357
Wind turbines, 184, 395
WLS70, 288
WMO Strategy for Service Delivery, 214
WMO. *See* World Meteorological Organization
World Climate Programme, 212
World Climate Service, 358
World Energy Outlook, 4
World Meteorological Organization, 201, 204, 209, 210, 213, 216, 217
World Weather Watch, 211
WRF, 285, 291, 300, 301, 308
 model, 269, 271

Y
1:200 year flood event, 127
1:100 year fluvial and tidal flood eventExtreme events, 127. *See* 1:200year flood event
Yonsei University scheme, 287

Z
Zambia, 188
Zimbabwe, 188